Methods in Molecular Biology

For further volumes:
http://www.springer.com/series/7651

Bioinformatics

Volume I: Data, Sequence Analysis, and Evolution

Second Edition

Edited by

Jonathan M. Keith

Monash University
Melbourne, VIC, Australia

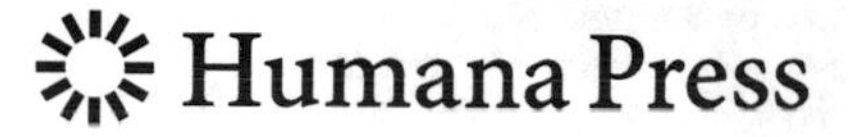

Editor
Jonathan M. Keith
Monash University
Melbourne, VIC, Australia

ISSN 1064-3745 ISSN 1940-6029 (electronic)
Methods in Molecular Biology
ISBN 978-1-4939-8252-3 ISBN 978-1-4939-6622-6 (eBook)
DOI 10.1007/978-1-4939-6622-6

Printed on acid-free paper

This Humana Press imprint is published by Springer Nature
The registered company is Springer Science+Business Media LLC
The registered company address is: 233 Spring Street, New York, NY 10013, U.S.A.

Preface

Bioinformatics sits at the intersection of four major scientific disciplines: biology, mathematics, statistics, and computer science. That's a very busy intersection, and many volumes would be required to provide a comprehensive review of the state-of-the-art methodologies used in bioinformatics today. That is not what this concise two-volume work of contributed chapters attempts to do; rather, it provides a broad sampling of some of the most useful and interesting current methods in this rapidly developing and expanding field.

As with other volumes in Methods in Molecular Biology, the focus is on providing practical guidance for implementing methods, using the kinds of tricks and tips that are rarely documented in textbooks or journal articles, but are nevertheless widely known and used by practitioners, and important for getting the most out of a method. The sharing of such expertise within the community of bioinformatics users and developers is an important part of the growth and maturation of the subject. These volumes are therefore aimed principally at graduate students, early career researchers, and others who are in the process of integrating new bioinformatics methods into their research.

Much has happened in bioinformatics since the first edition of this work appeared in 2008, yet much of the methodology and practical advice contained in that edition remains useful and current. This second edition therefore aims to complement, rather than supersede, the first. Some of the chapters are revised and expanded versions of chapters from the first edition, but most are entirely new, and all are intended to focus on more recent developments.

Volume 1 is comprised of three parts: Data and Databases; Sequence Analysis; and Phylogenetics and Evolution. The first part looks at bioinformatics methodologies of crucial importance in the generation of sequence and structural data, and its organization into conceptual categories and databases to facilitate further analyses. The Sequence Analysis part describes some of the fundamental methodologies for processing the sequences of biological molecules: techniques that are used in almost every pipeline of bioinformatics analysis, particularly in the preliminary stages of such pipelines. Phylogenetics and Evolution deals with methodologies that compare biological sequences for the purpose of understanding how they evolved. This is a fundamental and interesting endeavor in its own right but is also a crucial step towards understanding the functions of biological molecules and the nature of their interactions, since those functions and interactions are essentially products of their history.

Volume 2 is also comprised of three parts: Structure, Function, Pathways and Networks; Applications; and Computational Methods. The first of these parts looks at methodologies for understanding biological molecules as systems of interacting elements. This is a core task of bioinformatics and is the aspect of the field that attempts to bridge the vast gap between genotype and phenotype. The Applications part can only hope to cover a small number of the numerous applications of bioinformatics. It includes chapters on the analysis of genome-wide association data, computational diagnostics, and drug discovery. The final part describes four broadly applicable computational methods, the scope of which far exceeds that of bioinformatics, but which have nevertheless been crucial to this field. These are modeling and inference, clustering, parameterized algorithmics, and visualization.

Melbourne, VIC, Australia ***Jonathan M. Keith***

Contents

Contributors

FAISAL M. ABABNEH • *Department of Mathematics & Statistics, Al-Hussein Bin Talal University, Ma'an, Jordan*

SANNE ABELN • *Centre for Integrative Bioinformatics, Vrije Universiteit, Amsterdam, The Netherlands*

IMAD ABUGESSAISA • *Division of Genomic Technologies, RIKEN Center for Life Science Technologies, Yokohama, Kanagawa, Japan*

PUNTO BAWONO • *Centre for Integrative Bioinformatics, Vrije Universiteit, Amsterdam, The Netherlands*

ROBERT G. BEIKO • *Faculty of Computer Science, Dalhousie University, Halifax, Canada*

TIM BEIßBARTH • *Department of Medical Statistics, University Medical Center Göttingen, Göttingen, Germany*

BENJAMIN BUSBY • *National Center for Biotechnology Information, National Library of Medicine, National Institutes of Health, Bethesda, MD, USA*

CHEONG XIN CHAN • *Institute for Molecular Bioscience, The University of Queensland, Brisbane, QLD, Australia*

NATALIE L. DAWSON • *Institute of Structural and Molecular Biology, University College London, London, UK*

MAURITS DIJKSTRA • *Centre for Integrative Bioinformatics, Vrije Universiteit, Amsterdam, The Netherlands*

ANTON FEENSTRA • *Centre for Integrative Bioinformatics, Vrije Universiteit, Amsterdam, The Netherlands*

JAAP HERINGA • *Centre for Integrative Bioinformatics, Vrije Universiteit, Amsterdam, The Netherlands*

XIAOQIU HUANG • *Department of Computer Science, Iowa State University, Ames, IA, USA*

JAMES R.A. HUTCHINS • *Institute of Human Genetics, CNRS, Montpellier, France*

ANDREA ILARI • *CNR - Istituto di Biologia, Medicina molecolare e Nanobiotencologie c/o Dipartimento di Scienze Biochimiche "A. Rossi Fanelli", Università "Sapienza", Roma, Italy*

VIVEK JAYASWAL • *School of Biomedical Sciences, Queensland University of Technology, Brisbane, QLD, Australia*

LARS S. JERMIIN • *CSIRO Land & Water, Canberra, ACT, Australia*

TAKEYA KASUKAWA • *Division of Genomic Technologies, RIKEN Center for Life Science Technologies, Yokohama, Kanagawa, Japan*

HIDEYA KAWAJI • *Division of Genomic Technologies, RIKEN Center for Life Science Technologies, Yokohama, Kanagawa, Japan; RIKEN Preventive Medicine and Diagnosis Innovation Program, Wako, Saitama, Japan; Preventive Medicine and Applied Genomics Unit, RIKEN Advanced Center for Computing and Communication, Yokohama, Kanagawa, Japan*

JONATHAN M. KEITH • *School of Mathematical Sciences, Monash University, Clayton, VIC, Australia*

ARJUN KHOOSAL • *Computational Biology Group, Institute of Infectious Diseases and Molecular Medicine, University of Cape Town, Cape Town, South Africa*

ANDREAS KLOETGEN • *Department for Algorithmic Bioinformatics, Heinrich Heine University, Düsseldorf, Germany; Department of Pediatric Oncology, Hematology and Clinical Immunology, Heinrich Heine University, Düsseldorf, Germany*
FRANK KRAMER • *Department of Medical Statistics, University Medical Center Göttingen, Göttingen, Germany*
SAMARTH KULSHRESTHA • *Sri Venkateswara College, University of Delhi, New Delhi, India*
DEVI LAL • *Department of Zoology, Ramjas College, University of Delhi, Delhi, India*
RUSSELL L. MARSDEN • *Institute of Structural and Molecular Biology, University College London, London, UK*
DARREN P. MARTIN • *Computational Biology Group, Institute of Infectious Diseases and Molecular Medicine, University of Cape Town, Cape Town, South Africa*
LEONARDO DE OLIVEIRA MARTINS • *Department of Biochemistry, Genetics and Immunology, University of Vigo, Vigo, Spain; Department of Materials, Imperial College London, London, UK*
ALICE CAROLYN MCHARDY • *Department for Algorithmic Bioinformatics, Heinrich Heine University, Düsseldorf, Germany; Computational Biology of Infection Research, Helmholtz Centre for Infection Research, Braunschweig, Germany*
ILENE KARSCH MIZRACHI • *National Center for Biotechnology Information, National Library of Medicine, National Institutes of Health, Bethesda, MD, USA*
DAVID A. MORRISON • *Department of Organism Biology, Uppsala University, Uppsala, Sweden*
BREJNEV MUHIRE • *Computational Biology Group, Institute of Infectious Diseases and Molecular Medicine, University of Cape Town, Cape Town, South Africa*
BEN MURRELL • *Department of Integrative Structural and Computational Biology, The Scripps Research Institute, La Jolla, CA, USA*
CHRISTOPHER O'SULLIVAN • *National Center for Biotechnology Information, National Library of Medicine, National Institutes of Health, Bethesda, MD, USA*
CHRISTINE A. ORENGO • *Institute of Structural and Molecular Biology, University College London, London, UK*
WALTER PIROVANO • *Bioinformatics Department, BaseClear, Leiden, The Netherlands*
DAVID POSADA • *Department of Biochemistry, Genetics and Immunology, University of Vigo, Vigo, Spain*
AYUSH PURI • *Sri Venkateswara College, University of Delhi, New Delhi, India*
MARK A. RAGAN • *Institute for Molecular Bioscience, The University of Queensland, Brisbane, QLD, Australia*
JOHN ROBINSON • *School of Mathematics & Statistics, University of Sydney, Sydney, NSW, Australia*
CARMELINDA SAVINO • *CNR - Istituto di Biologia, Medicina molecolare e Nanobiotecnologie c/o Dipartimento di Scienze Biochimiche "A. Rossi Fanelli", Università "Sapienza", Roma, Italy*
IAN SILLITOE • *Institute of Structural and Molecular Biology, University College London, London, UK*
ANDERS GONÇALVES DA SILVA • *School of Biological Sciences, Monash University, Clayton, VIC, Australia*
EDWARD TASKER • *School of Mathematical Sciences, Monash University, Clayton, VIC, Australia*
MANSI VERMA • *Sri Venkateswara College, University of Delhi, New Delhi, India*
SIMON WHELAN • *Department of Ecology and Genetics, Uppsala University, Uppsala, Sweden*

Part I

Data and Databases

Chapter 1

Genome Sequencing

Mansi Verma, Samarth Kulshrestha, and Ayush Puri

Abstract

Genome sequencing is an important step toward correlating genotypes with phenotypic characters. Sequencing technologies are important in many fields in the life sciences, including functional genomics, transcriptomics, oncology, evolutionary biology, forensic sciences, and many more. The era of sequencing has been divided into three generations. First generation sequencing involved sequencing by synthesis (Sanger sequencing) and sequencing by cleavage (Maxam-Gilbert sequencing). Sanger sequencing led to the completion of various genome sequences (including human) and provided the foundation for development of other sequencing technologies. Since then, various techniques have been developed which can overcome some of the limitations of Sanger sequencing. These techniques are collectively known as "Next-generation sequencing" (NGS), and are further classified into second and third generation technologies. Although NGS methods have many advantages in terms of speed, cost, and parallelism, the accuracy and read length of Sanger sequencing is still superior and has confined the use of NGS mainly to resequencing genomes. Consequently, there is a continuing need to develop improved real time sequencing techniques. This chapter reviews some of the options currently available and provides a generic workflow for sequencing a genome.

Key words Sanger sequencing, Next-generation sequencing, Cyclic-array sequencing, Nanopore

1 Introduction

Sequencing technologies have developed rapidly in recent decades and have been compared with the engine of a car, allowing us to navigate the roadmap of various genomes [1], whether they be as primitive as viruses or as advanced as human beings. DNA sequencing has rapidly revolutionized molecular biology, medicine, genomics, and allied fields. First conceptualized by Frederick Sanger in 1977 [2], major advancements in sequencing technologies over the years have led to the development of new and improved DNA sequencing platforms. These technologies, along with various computational tools for analyzing and interpreting data, have helped researchers to better understand genomes of various economically important organisms. They have made sequencing a powerful yet feasible research tool that has evolved to the level at

Jonathan M. Keith (ed.), *Bioinformatics: Volume I: Data, Sequence Analysis, and Evolution*, Methods in Molecular Biology, vol. 1525, DOI 10.1007/978-1-4939-6622-6_1,

which it could easily be employed even in small labs with high efficacy, bypassing the need for large and cumbersome sequencing centers.

The progression of sequencing strategies can be divided into three generations [3], with Sanger's dideoxy method and Maxam-Gilbert's chain termination method constituting the first generation. Automation of the Sanger method led to it becoming the dominant methodology in the industry for more than 20 years. Major achievements of Sanger sequencing include the first ever sequenced microbial genome, *Haemophilus influenzae* [4] and completion of finished-grade human genome sequence [5]. Nevertheless, there have been dramatic transformations and innovations in genome sequencing in the last decade. Next-generation sequencing (NGS) technologies have been extensively utilized from macro- to micro-level applications, including de-novo whole-genome sequencing, genome re-sequencing, mRNA profiling, analyzing molecular evolution, detecting methylation patterns, solving criminal cases, exploring fossil records, and evaluating DNA binding proteins and chromatin structures [6–11]. Thus, sequencing has a wide range of applications that have resulted in an increased global demand for improved sequencing platforms.

In the following chapter, we first describe the general procedure for genome sequencing and parameters that guide the choice of sequencing strategy. We then describe each of the major sequencing technologies, advancement in their approaches, recent findings in the field, and current and emerging applications.

1.1 General Procedure for Genome Sequencing

The first step of any genome sequencing project is generation of an enormous amount of sequencing data (Fig. 1). The raw data generated by sequencing is processed to convert the chromatograms into the quality values (*see* **Note 1**). This procedure is known as **"base calling"** or fragment readout. The next step is optional as only those methods that involve library construction will utilize trimming of vector sequences. It is an important step as some parts of the vector are also sequenced by universal primers, along with the insert region. Further screening of good quality and bad quality regions is performed to ensure a more accurate sequence assembly. The processed data is then assembled by searching for fragment overlaps to form **"contigs."** However, because of repeat regions, the fragments may be wrongly aligned, leading to misassembly. Hence, assembly validation is done manually as well as by softwares to locate the correct position of reads. The contigs are then oriented one after the other with the help of mate-paired information to form ordered chains known as **"scaffolds"** [12]. This step is followed by the final and most crucial step of genome sequencing: **FINISHING**. It includes gap filling (also known as "minimum tiling path") and reassessing the assembly [13].

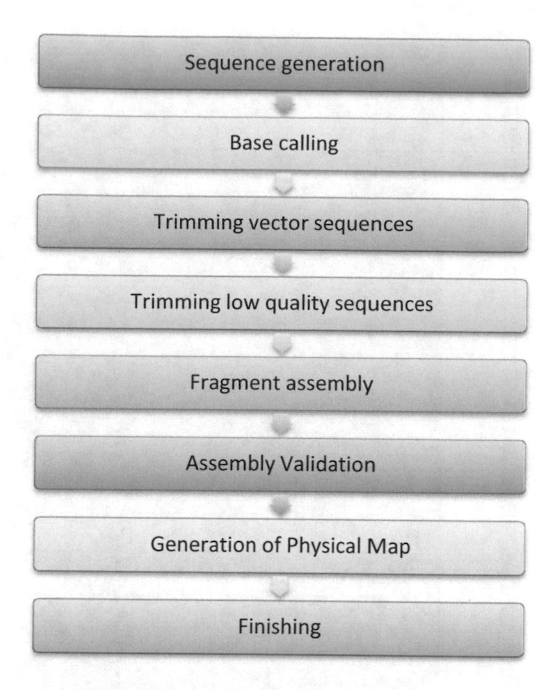

Fig. 1 Flow diagram to show the series of steps followed for sequencing any genome

1.2 Choosing a Sequencing Strategy

Given the rapid growth of sequencing technologies, each with unique advantages and shortcomings with regard to performance and cost, it is now becoming difficult to choose the best technology for a particular application. Some parameters that should be considered when selecting a sequencing method are the following [14, 15]:

1. *Read lengths and cost per read*: Read length is the number of contiguous bases sequenced in a given run. It remains one of the most important parameters for the selection of sequencing platforms. Increasing the read length facilitates assembly and reduces assembly error, because it provides unique context for repetitive sequences. Moreover, as the cost per read decreases, sequencing longer reads will become more economical (*see* Table 1).
2. *Raw accuracy*: This is an important quality control measure in massive DNA sequencing projects. Lower error rates result in better assembly, and better quality finished-sequence for the same coverage.
3. *Mate-paired reads*: Mate-paired reads are those reads that are separated by a known distance in the genome of origin [16]. Such reads are crucial for scaffolding de-novo genome assembly, for resolving repeat regions, and for many other downstream

Table 1
Details of first, second, and third generation sequencing technologies with respect to their cost per megabase, instrument cost, read length, and accuracy

Platform	Company	Cost per megabase (USD)	Cost per instrument (USD)	Read-length (bp)	Run time	Throughput	Raw accuracy
First generation							
Maxam-Gilbert	NA	–	–	–	2h	Low	–
Sanger	Applied Biosystems	2400	95,000	800	3h	Low	99.9999%
Second generation							
GS FLX	454 Life Sciences, Roche	~60.0	500,000	700	24 h	High	99.9%
SOLiD	Life Technologies	~0.13	495,000	35	8–14 days	Very high	99.94%
Genome Analyzer	Solexa, Illumina	~0.07	690,000	36	10 days	Very high	>98.5%
Polonator	Dover	~1.00	155,000	13	8–10 days	High	99.7%
HeliScope	Helicos Biosciences	~1.00	1,350,000	30	7 days	High	>99%
Third generation							
Ion Torrent	DNA Electronics Ltd.	1.00	80,000	200–400	3 h	Moderate	99.2%
CGA	BGI	~0.5–1.00	1200,000	10	6 h	Very high	99.99 %
Pacific Bio RS	Pacific biosciences	0.13–0.6	695,000	1400	0.5–2 h	Moderate	88.0%
Oxford Nanopore	Oxford technologies	Not yet calculated	750,000	Up to 4Tb	Upto 48h	Very high	99.99%

applications. The importance of this parameter is evident from its implementation in almost all NGS technologies.

4. *Preprocessing protocols*: Preprocessing protocols include preparation of the sample, in vitro library construction, loading of the sample, and maintenance of upscale mechanisms. These can also include eliminating low-quality sequences, removal of contaminant sequences, and piping unwanted features (*see* Table 2).
5. *Postprocessing protocols*: These protocols include creation of softwares and online tools for assembling immense quantities of short-read sequencing data, elimination of contig chimeras, and production of finished sequence.

2 Materials

DNA of the target organism can be isolated using commercially available kits. There are two different approaches that can be used to prepare DNA for any genome sequencing project. These are as follows:

(a) *Ordered-clone Approach*
This approach is also known as hierarchical shotgun sequencing and follows **"map first, sequence second"** progression [17]. In this approach, genomic DNA is fragmented and cloned into a suitable vector (usually BAC). Clones are analyzed by restriction digestion, Sequenced Tagged Sites (STSs) etc., to detect the unique DNA landmarks. Common landmarks between any two clones indicate whether they overlap (and hence are likely to form contigs) (Fig. 2). A series of overlaps are generated, thus producing a high-resolution BAC contig map. As the BAC clones are mapped onto the genome, it becomes easier to subject these clones to shotgun sequencing and assemble them with the help of the physical map. This technique is laborious, as it involves mapping and then sequencing, but still is a method of choice for large genomes to avoid confusion while assembling repeated regions.

(b) *Whole-genome shotgun*
In this approach, total genomic DNA of an organism is randomly sheared into short fragments of defined size (Fig. 2). These fragments are then cloned into a suitable vector whose sequence is known (*see* **Note 2**). Universal primers are designed from priming sites located at both ends of the vector and sequences are obtained from both ends of the insert region (mate-paired sequencing). Thousands of subclones are sequenced to get high coverage across the genome. This technique is faster than ordered clone approach, but requires exceedingly precise algorithms to assemble such enormous amount of data with high accuracy. The major drawback this

Table 2
Preprocessing protocols and uses of various sequencing techniques available till date

Platform	Sequencing reaction	Feature generation	Chemistry	Sequencing library	Pros	Cons	Uses
First generation							
Maxam-Gilbert	Cleavage	Chemicals	Specific base modifications	–	Sequencing from original fragment, no PCR, less susceptible to enzymatic mistakes	Use of hazardous chemicals, time-consuming, incomplete reactions	To sequence short DNA fragments, detect rare bases, chromatin structure and DNA methylation pattern
Sanger	Cleavage	Cloning/PCR	Enzymatic chain terminators (dideoxy-nucleotide triphosphates)		Long read lengths; high single-pass accuracy; good ability to call repeats	Low throughput; high cost of Sanger sample preparation, time consuming	De novo sequencing, target sequencing, shot-gun sequencing, methylation analysis. Gene expression analysis, mitochondrial sequencing, microbial sequencing
Second generation							
454	Pyro-sequencing	Emulsion PCR	Phospho-kinase fluorescent nucleotides	Linear adapters	Longer reads improved mapping in repetitive regions, faster run times	High reagent cost, rugged sample washing, problematic for homopolymeric repeats sequencing	Bacterial and insect genome de novo sequencing, resequencing of human genome 16S r RNA sequencing in metagenomics

eGenome Analyzer	Synthesis	Bridge PCR	Polymerase (reversible terminators)	Linear adapters	Generate billions of reads at a time, most widely used	Lower read length, time consuming, cumbersome process, high error rate	Whole-genome resequencing, gene regulation, epigenetic analysis, cytogenomics
SOLiD	Ligation	Emulsion PCR	Ligase (octamers with two-base encoding)	Linear adapters	Accuracy, Error correction via two-base encoding	Short read assembly, longer run time	Detection of rare and somatic mutations, epigenetic studies, whole genome exome capture
Polonator	Ligation	Emulsion PCR	Ligase (nonamers)	Polony libraries	Open source software, reagents and protocols, less expensive	Short read length	Personal genome project, target resequencing
Third generation							
Ion Torrent	Sequencing by synthesis	Ion-sensitive field effect transistor ion sensor	Proton/pH detection	NA	Rapid sequencing protocol, low operating cost, lower instrumentation cost	Difficult to enumerate homopolymer sequence, shorter read length	Microbial genome sequencing, amplicon sequencing, targeted sequencing, microbial transcriptome sequencing

(continued)

Table 2
(continued)

Platform	Sequencing reaction	Feature generation	Chemistry	Sequencing library	Pros	Cons	Uses
CGA platform	Ligation	Rolling circle amplification	Probe-anchor capture and ligation	Circular adapters	Highest throughput compared to third generation platforms, lowest reagent cost, Each sequencing step is independent, minimizing accumulation of errors	Short read lengths, laborious, not commercialized instrument available in the market	Research findings related to human diseases, study of genetic diseases
Pacific Bio	Synthesis, real-time	Single molecule	Real-time single polymerase	Bubble adapters	Sample preparation is fast, average read length is high, turnover rate is also high	Lower throughput and highest error rates compared to second generation chemistries	full-length transcriptome sequencing, identify structural and cell type variation, detect DNA base modifications, gene expression analysis, de novo sequencing

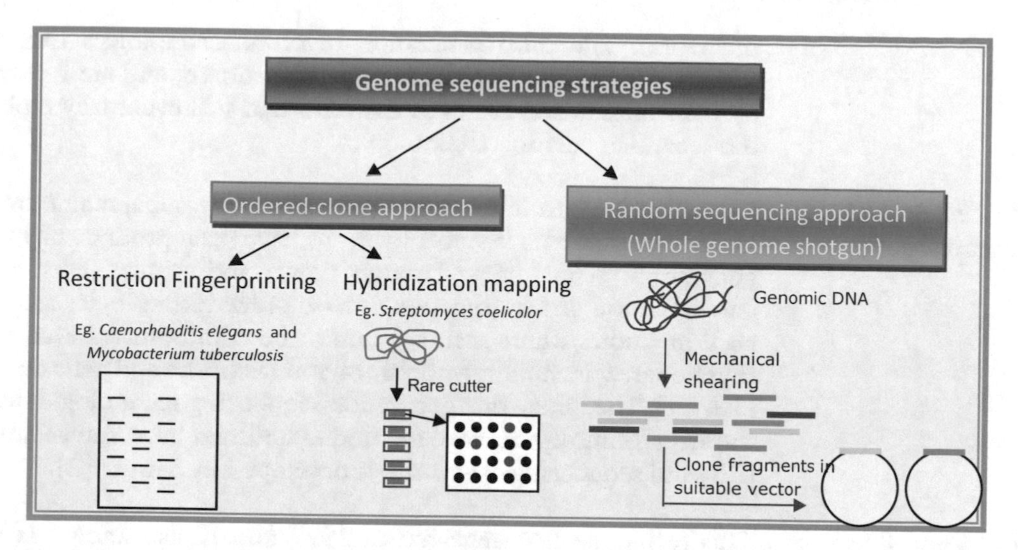

Fig. 2 Strategies for sequencing whole genome. Ordered clone approach is used for large insert size library for constructing physical map, whereas whole genome shotgun strategy is used for sequencing small insert size library for generating enormous amount of data

strategy faces is assembling repeat regions. This problem can be partially overcome by generating reads from both the ends of subclones whose insert size is known [17]. The softwares then place both the reads from the same subclone based on the expected physical distance, and hence resolve most problems of misassembly in repeat regions. This technique has been extensively used for the sequencing of bacterial genomes, including *Haemophilus influenzae*.

The above-mentioned strategies are described for Sanger sequencing. In next-generation sequencing techniques, no cloning is required as DNA is directly fragmented and ligated with adaptors or to the chip. But these advanced sequencing technologies also use the random shotgun approach.

3 Methods

Although most sequencing techniques are quite distinct, their procedures can be summarized into a general work flow that is fundamental in achieving low cost and high throughput sequencing. As mentioned above, sequencing techniques can be categorized into three generations. Each generation has been developed mainly to ameliorate the shortcomings of the previous generation. The first generation marked the beginning of the era of sequencing technologies and set the basis on which subsequent generations developed. The second generation technologies that were a huge improvement to their former counterparts are currently thriving in labs around the world. The third generation involves technologies that are currently being developed, have shown promise, and are foreseen by their makers and critics as the ones that will eventually replace the second generation technologies.

3.1 First Generation Technologies

Even after the introduction of novel techniques, a major amount of DNA sequence data is still credited to have been produced by first generation technologies. Though slower and costlier (even after parallelization and automation), these older technologies are still used in studies where accuracy cannot be compromised even to a slight degree, including synthetic oligonucleotides and gene targets [18, 19]. The initial, first generation sequencing technologies were the sequencing-by-cleavage method established by Maxam-Gilbert [20] and sequencing by synthesis developed by Sanger [2].

3.1.1 Maxam-Gilbert Method

This technique first appeared in 1977 and is also known as the "chemical-degradation" method [20]. The chemical reagents act on specific bases of existing DNA molecules and subsequent cleavage occurs (Fig. 3). In this technique, dsDNA is labeled with radiolabeled phosphorus at the 5′ end or 3′ end. The next step is to obtain ssDNA. This can be done by restriction digestion leading to sticky ends or denaturation at 90 °C in the presence of DMSO,

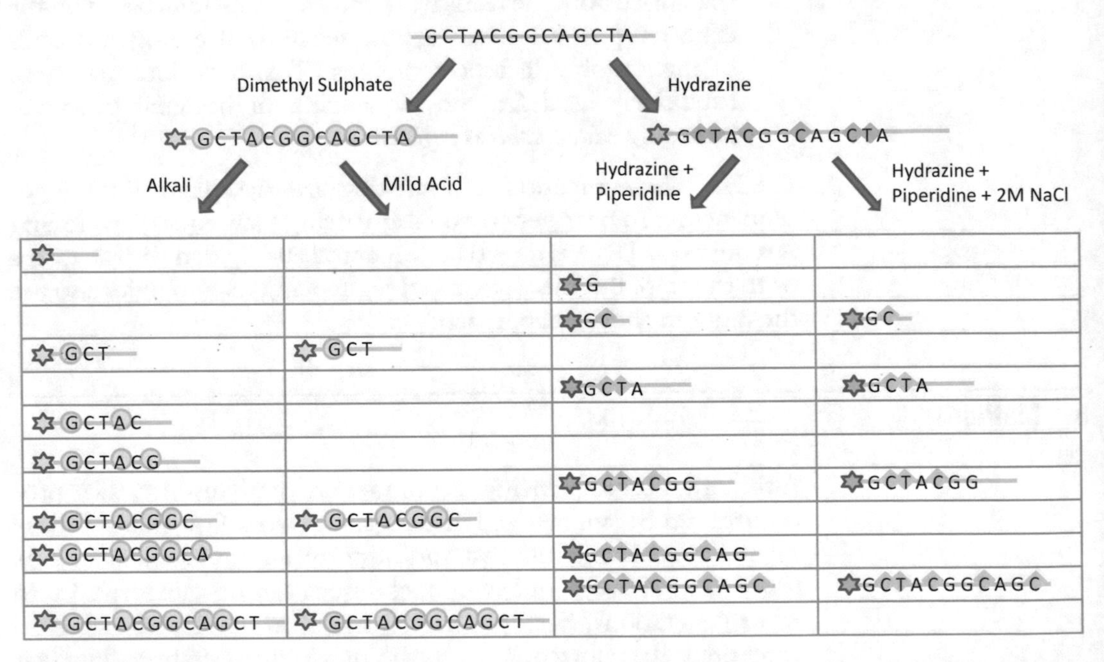

Fig. 3 Sequencing an oligonucleotide by Maxam-Gilbert method: This method largely depends upon treatment of oligonucleotides by different chemicals that cleave at specific sites. 5′ end of the oligonucleotide is radiolabeled. When cleaved with specific chemicals, only the radiolabeled part generates signals on an autoradiograph. Hence, by series of partial cleavages, nucleotide sequence can be determined

Table 3
Reagents used in Maxam-Gilbert method for sequencing

Base cleaved	Adenine/Guanine	Adenine	Cytosine/Thymine	Cytosine
Reagent required	Treatment with dimethyl sulphate followed by alkali treatment	Treatment with Dimethyl sulphate followed by mild acid treatment	Treatment with Hydrazine followed by a partial hydrazinolysis in 15–18 M aqueous hydrazine and 0.5 M piperidine	Treatment with hydrazine in the presence of 2 M NaCl

resulting in ssDNA. Only one strand is purified and divided into four samples. Each sample is treated with a different chemical reagent as explained in Table 3.

Hence, all four reaction products when run separately on 20 % polyacrylamide gel containing 7 M urea, reveal the point of breakage and the pattern of bands can be read directly to determine the sequence. However, the use of hazardous chemicals and incomplete reactions make this method unsuitable for large-scale DNA sequencing.

3.1.2 Sanger Sequencing

This technique is also known as chain termination sequencing or "dideoxy sequencing." Sanger sequencing has played a crucial role in understanding the genetic landscape of the human genome. It was developed by Frederick Sanger in 1975, but was commercialized in 1977 [21].

The technique is based on the principal usage of dideoxy ribonucleoside triphosphates that lack 3′ hydroxyl group (Fig. 4a). The technique comprises seven different components to perform the sequencing. These include ssDNA template to be sequenced, primers, Taq polymerase to amplify the template strand, buffer, deoxynucleotides (dNTPs), fluorescently labeled dideoxynucleotides (ddNTPs) and DMSO (to resolve the secondary structures if formed in the template strand) (*see* **Notes 3** and **4**). Because the incorporated ddNTP lacks a 3′OH group, the phosphodiester bond between the C3′OH of the latter sugar moiety and C5′ of the next dNTPs will not form, resulting in the termination of the chain at that point [2, 22].

When run on a polyacrylamide gel with the products of each reaction in four parallel lanes, the fragments of different lengths get separated (Fig. 4b and c). After its introduction, Sanger sequencing underwent further modifications for the automated sequencing protocol. One such advance was dye-terminator sequencing [23]. The main advantage of using this method lies in its greater accuracy and speed. In 1987, Applied Biosystems, Inc. (ABI) launched the first automated DNA sequencing machine, ABI model 370,

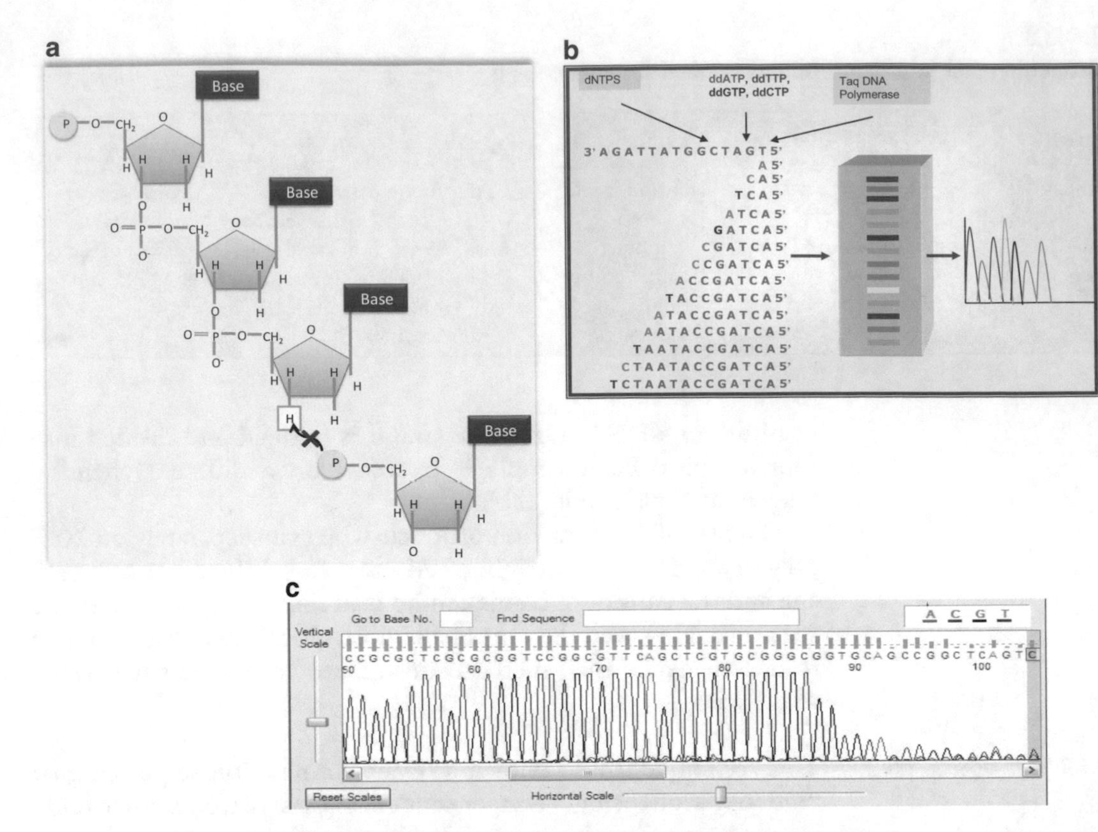

Fig. 4 Dideoxy chain termination method. (**a**) DNA synthesis by addition of dNTP (that has free 3′-OH) in the synthesizing strand. But once a ddNTP is incorporated, the strand synthesis stops as no 3′-OH is available to form a phosphodiester bond with the next dNTP. (**b**) formation of a series of chains by incorporation of ddNTPs which can by separated by capillary electrophoresis. (**c**) Electropherogram of a DNA sequence

developed by Leroy Hood and Mike Hunkapiller which could generate read lengths of up to 350 bp per lane. Also in 1995, ABI launched another model ABI PRISM 310 Genetic Analyzer to simplify the inconvenient and troublesome process of pouring gels, mounting on the instrument, and loading of samples. Swerdlow and Gesteland developed this machine, known as the capillary sequencer, using polyacrylamide gel-filled capillaries instead of using slab gels. Currently, available sequencers in the market encompass 4, 16, 48, 96, or 384 capillaries. With the increase in number of capillaries, the read length increased and with it the speed of sequencing [24].

3.2 The Birth of Next-Generation Sequencing (NGS) Techniques

The advent of next-generation sequencing techniques has made the once herculean task of sequencing much simpler and faster [25]. Gradually, the rapid and cost-effective next-generation sequencing technologies have emerged as the popular choice over their slow and cumbersome first generation counterparts. They have enabled bypassing the bottleneck caused by preparation of individual DNA templates, as required by the first generation. When these technologies are supplemented by tools and softwares from potent fields such as bioinformatics and computational biology, they result in a massive increase in the rate at which data is generated [26].

The techniques used in NGS can be grouped into several categories. These include:

1. *Cyclic-array sequencing*: In this technique, target sequences are physically separated on an array and are sequenced by repeatedly adding a set of enzymes in each cycle to detect base incorporation. The technique is based on sequencing-by-synthesis because as the base is incorporated, the sequence is detected either by using reversibly blocked nucleotide substrates or by supplying only one kind of nucleotide per cycle [27].
2. *Sequencing by hybridization*: In this technique, a large number of probes are fixed on an array. The fragmented target DNA (whose sequence is to be deciphered) is added to the array, where it binds to complementary probes. These probe sequences can be the DNA from a reference genome. This technique avoids all complicated steps of library construction and is more effective when a reference genome is available [28].
3. *Microelectrophoretic methods*: As the name suggests, a microelectrophoretic method is the miniaturization of the conventional Sanger' dideoxy method. The concept behind this technique is to retain the accuracy and read length of Sanger's method but to introduce parallelism so as to reduce time and cost [29].
4. *Real time sequencing of single molecules*: Several institutions and organizations are developing techniques to achieve ultra-fast DNA sequencing in which the sequence will be detected as soon as a nucleotide is incorporated. Such approaches include nanopore sequencing and real-time monitoring of DNA polymerase activity. The former involves sequencing nucleic acids by analyzing each base as it passes through a biological membrane or a synthetic pore known as a nanopore. The latter includes detection of nucleotide incorporation by FRET (fluorescence resonance energy transfer) or with zero mode waveguides [30].

As compared to conventional sequencing, advantages of NGS include the following:

1. Removal of cumbersome and time consuming techniques such as transformation of *E. coli* and colony picking [31].
2. Generation of hundreds of millions of sequencing reads in parallel [32].
3. Dramatic decrease in the effective reagent volume per reaction (ranging from picoliters to femtoliters) [18].

Collectively, these differences have enabled NGS technologies to produce a large amount of data at far lower costs and in much less time.

3.3 Second Generation Sequencing

Second generation sequencing primarily encompasses three technologies: Roche 454 pyrosequencing, Solexa technology by Illumina and sequencing by ligation by ABI/SOLiD. Using emulsion PCR and cyclic-array sequencing as the foundation, these platforms perform a crucial role in making whole genome sequencing projects less time consuming and more economical [33].

3.3.1 454 Pyrosequencing

The 454 (Roche) was the first of the NGS platforms introduced in 2005 [34]. The process of pyrosequencing, in short, involves generation of light from phosphates by employing an enzymatic cascade. This light is released while a polymerase replicates the template DNA and its detection is vital in accurately sequencing the DNA [35].

In this technique, the duplex DNA is first sheared into smaller fragments that are followed by ligation of adapters on both sides of the fragment that are complementary to the primer sequences. These adapters act as the site for primer binding and initiate the sequencing process. Each DNA fragment is bound to the emulsion microbeads such that the ratio of DNA:microbead is 1:1 (Fig. 5). This is followed by amplification of each strand using emulsion PCR and after a few cycles, many copies of these DNA molecules are obtained per bead [35].

Immobilized enzymes including DNA polymerase, ATP sulphurylase, luciferase, and apyrase are added to the wells in a fiber optic plate containing one amplified bead each. Fluidic assembly supplies the four dNTPs one by one to the wells of the fiber optic slide. These dNTPs are then incorporated, complementary to the template, with the help of DNA polymerase. This process releases PPi, on which ATP sulphurylase acts, converting it into ATP. In the presence of the generated ATP, luciferase converts luciferin to oxyluciferin and a light signal proportional to the amount of ATP is produced. Intensity of the signal produced is captured by the CCD camera. Once the signal is produced and captured, enzyme apyrase degrades the existing nucleotides and ATP and the next nucleotide is then added through fluidic assembly. The technique has recently been supplemented with "paired-end" sequencing where adapters are bound to both the ends of the fragmented DNA therefore, leading to sequencing from both ends [34, 35].

A major advantage of this technique is its long read length. Unlike other NGS technologies, 454 yields read length of 400 bases and can generate more than 1,000,000 individual reads per run. This feature is particularly suited for de novo assembly and metagenomics [36].

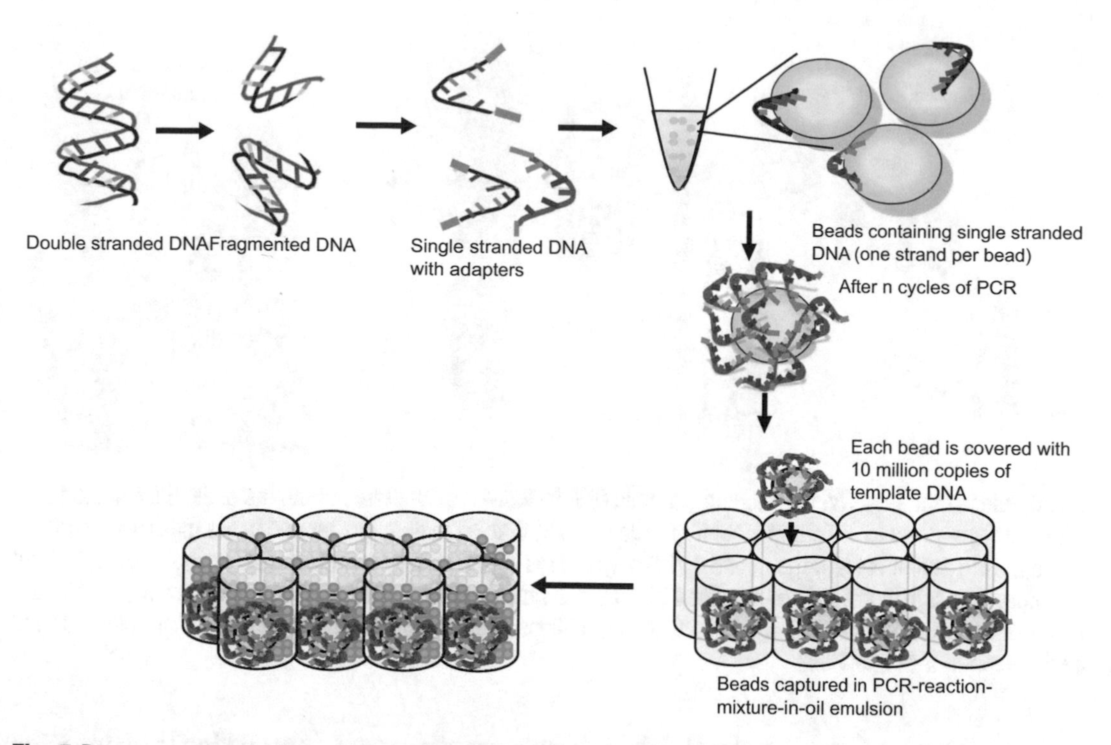

Fig. 5 Pyrosequencing. In this method, the double stranded DNA is fragmented into small pieces and then denatured to get single strands, which are ligated with adaptors at both the ends. Each DNA fragment gets bound to microbead and is amplified by emulsion PCR, generating many copies of a single molecule

3.3.2 The Illumina (Solexa) Genome Analyzer

The Solexa platform for sequencing DNA was first developed by British chemists Shankar Balasubramanian and David Klenerman and was later commercialized in 2006. It works using the principle of "sequencing-by-synthesis." Also known as cycle reversible ter mination (CRT) [37], this technique has been found to generate 50–60 million reads with an average read length of 40–50 bases in a single run. The genomic DNA is fragmented into small pieces and adapters are ligated at both the ends. The ligated DNA fragments are then loaded on the glass slide (flow cell) where one end of the ligated DNA fragment hybridizes to the complementary oligo (oligo 1) that is covalently attached to the surface. The opposite end of each of the bound ssDNAs is still free and hybridizes with a nearby complementary oligo (oligo 2) present on the slide to form a bridge. With the addition of DNA polymerase and dNTPs, the ssDNA are bridge-amplified as strand synthesis starts in the 5′–3′ direction from oligo 2. For the next PCR cycle, the template strand and the newly synthesized complementary strand are denatured to again start the amplification. As a result of repeated cycles of bridge amplification, millions of dense clusters of duplex DNA are generated in each channel of the flow cell (Fig. 6). Once such an ultra-high density sequencing flow cell is created, the slide is ready for sequencing [33, 37–39].

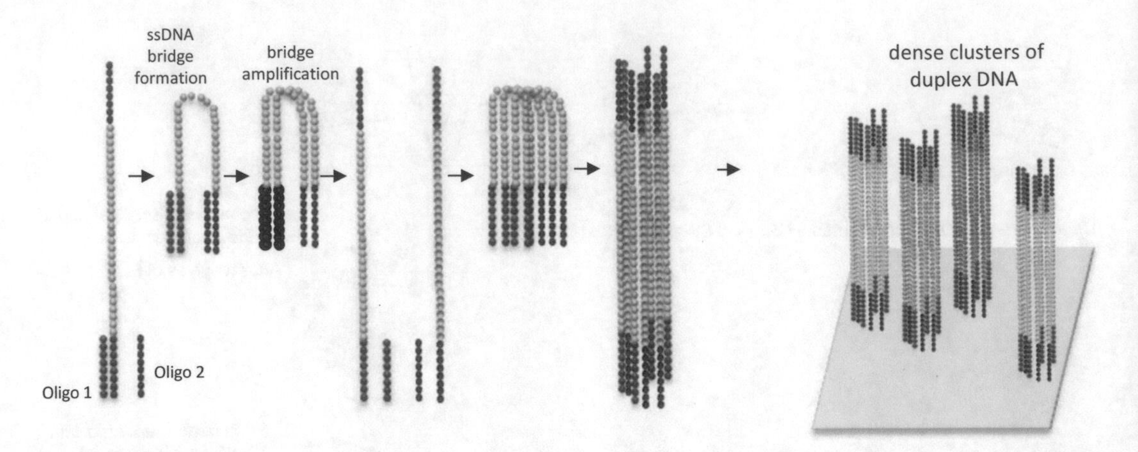

Fig. 6 Solexa Platform. The fragmented genomic DNA is ligated to adapters at both the ends. Each ligated DNA fragment hybridizes to the complementary oligo (oligo 1) that is covalently attached to the glass slide surface. The opposite end of each of the bound ssDNAs hybridizes with a nearby complementary oligo (oligo 2) to form a bridge. The ssDNA are bridge-amplified as strand synthesis starts in the 5′–3′ direction from oligo 2. As a result of repeated cycles of bridge amplification, millions of dense clusters of duplex DNA are generated in each channel of the flow cell

Amplified fragments are denatured, annealed with sequencing primers and synthesis starts with the incorporation of complementary dNTPs (each dNTP is tagged with different reversible flourophores). For the detection of the next base, tris (2-carboxyethyl) phosphine (TCEP) is added to remove the flourophore moiety from the incorporated dNTP, thus deblocking the 3′ OH end. As compared to Sanger sequencing as well as the other NGS systems, the Illumina system stands out in its ability to produce a larger quantity of data in less time and with less cost [32, 39].

3.3.3 Applied Biosystems SOLiD Sequencer

ABI SOLiD platform was described by J. Shendure and colleagues in 2005. This technology involves a series of hybridization-ligation steps. The di-base probes are actually octamers, containing two bases of known positions, three degenerate bases, and three more degenerate bases that are attached to the flourophore (and which are excised along with flourophore cleavage). Sixteen combinations of di-base probes are used, with each base represented by one of four fluorescent dyes (Fig. 7a) [40, 41].

The sheared genomic DNA is ligated with adaptors on both sides. The adaptor-ligated fragments get bound to microbeads and each molecule is clonally amplified onto beads in emulsion PCR (similar to pyrosequencing). After amplification, beads get covalently attached to the glass slide and are provided with universal primers, ligase, and di-base probes [42].

In the first round, universal primers (of length *n*) are added that hybridize with adaptors. Following this, one of the di-base probes (complementary to the sequence) binds to the template and is ligated with a universal primer with the help of ligase. As explained above, the di-base probe contains a fluorophore that is excited and the signal is captured. Each signal is a representative of one of the four alternatives. After excitation, the fluorophore gets cleaved along with the three degenerate nucleotides associated with it, exposing the 5′ phosphate of the bound probe. Now the next di-base probe is ready to bind to the template strand. Once this probe is bound, ligase joins it with the adjoining bound probe, emission is recorded and it is again cleaved off. This cycle is repeated seven times. In this way, signals for nucleotides 1–2, 6–7, 11–12, 16–17, 21–22, etc. are detected (leaving three unidentified nucleotides in between).

In the second round, one base from the universal primer is deducted ($n-1$). This time, the first base (0 position) of the synthesizing strand is known (as the primer is $n-1$). The above-mentioned steps are repeated. After seven cycles, the signals for nucleotides 0–1, 5–6, 10–11, 15–16 etc. are detected (Fig. 7b).

In the third round, the universal primer is reduced by two bases ($n-2$). Hence, after another seven cycles we get signals for -1–0, 4–5, 9–10, 14–15 nucleotide positions respectively.

As the "0" position is known, the first base can be readily predicted out of the four options for the same color. For example, if the "0" position is "A" and the signal for 0–1 positions is red, this means that the "1" position is "T" (Fig. 7a and b). Similarly, if the signal for 1–2 positions is again red, the next nucleotide is "A." By this method, each base is detected twice making this the most accurate of the NGS systems [39, 40, 42].

3.3.4 Polonator

This ligation based sequencer was developed by Dover in collaboration with Church Laboratory of Harvard Medical School. In addition to high performance, it has become popular as a cheap, affordable instrument for DNA sequencing. The system uses open source software and protocols that can be downloaded for free. Based on ligation detection sequencing (similar to SOLiD), the Polonator G.007 identifies the nitrogenous base by a single base probe, unlike SOLiD that uses a dual base probe. Nucleotides having single base probes known as nonanucleotides or nonamers are tagged with a fluorophore. The genomic DNA is sheared into many fragments that are used for the construction of a genomic paired tag shotgun library. Using emulsion PCR, clonal amplification of templates results in tens of thousands of such copies attached to the beads called polonies. This is followed by enrichment of beads having such amplified templates bound to them. After insertion of such beads in the flow cell, the Polonator is used to generate millions of short reads in parallel by using the cyclic sequencing array technology [3, 31].

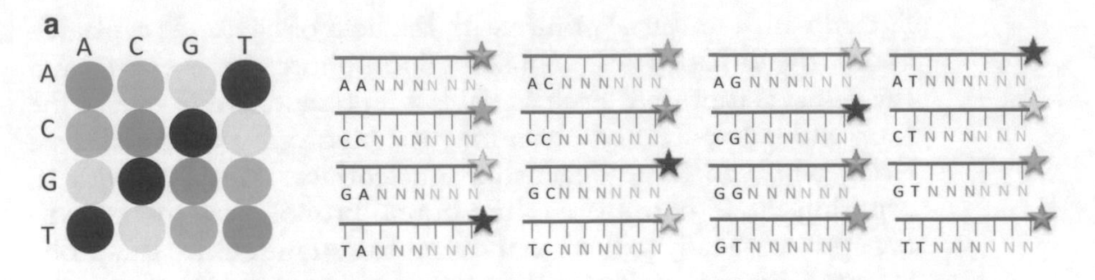

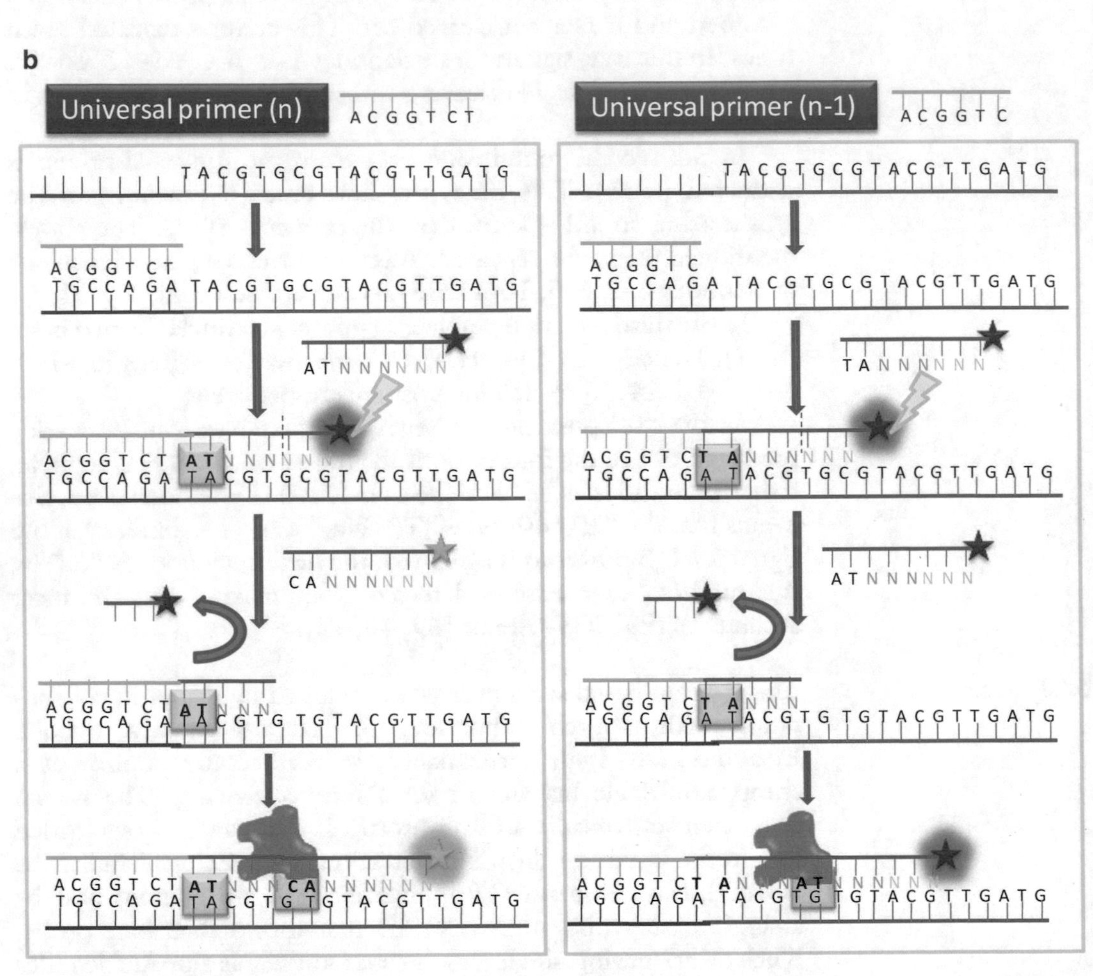

Fig. 7 (**a**) Diagrammatic representation of the 16 di-base probes used in ABI-SOLiD: The di-base probes consist of two known nitrogeneous bases, six degenerate bases, three of which are tagged with a fluorophore. Depending on the combination of known bases used, four different colors are emitted when these probes bind to the template DNA, signaling the correct sequence of the DNA strand. (**b**) First cycle is continued with a universal primer (*n*) which leads to the detection of nucleotide 1–2, 6–7, 11–12, 16–17, 21–22 etc. In the second cycle, the same universal primer with one base less (*n*−1) helps in detecting nucleotide position 0–1, 0–11, 15–16 etc.

3.3.5 Shortcomings of Second-Generation DNA Sequencing Technologies

Currently, a number of disadvantages plague second-generation DNA sequencing technologies and thus, counterbalance many of its advantages. The most crucial of these include shorter read-length and low accuracy of base calls as compared to the conventional sequencing strategies. The 454 technology is handicapped by its inability to resolve homopolymer-containing DNA segments, and the error rate is contributed by insertion-deletions (rather than substitutions). Pyrosequencing is favored because of its longer read length. However, this technique has a high cost per base as compared to other NGS techniques. Another drawback common to all second generation sequencing techniques is the use of emulsion PCR, which is cumbersome and technically challenging. Palindromes were also reported to have posed some challenges to the sequencing-by-ligation method used by the ABI SOLiD technology. The drawbacks of Illumina Solexa sequencing include high error rates, shorter read lengths, and difficulty in sequencing repeat regions [3, 43, 44].

But, as conventional sequencing gradually progressed over a period of time to reach its current level of performance, we have reason to believe that these technologies will improve over the years as scientists further analyze these limitations to reduce or eliminate them.

3.4 Third Generation Sequencing

Although the above-mentioned technologies have revolutionized genomics and have made the generation of massively parallel data possible, yet the drive to explore even more efficient techniques has given birth to third generation technologies. Considering the limitations and bias of PCR, the TGS technologies employ the use of single molecule sequencing (SMS). While it is difficult to define the boundaries between second and third generation techniques, the techniques listed here have been shown to enhance throughput and read lengths, to ease the sample preparations and to decrease the time and cost of sequencing [3].

3.4.1 The Ion Torrent Sequencing Technology

Ion torrent, a transitional technique between the second and third generations, is a high-throughput DNA sequencing technology introduced by Life technologies in 2010. The sequencing chemistry of Ion Torrent is based on the principle of covalent bond formation in a growing DNA strand, catalyzed by DNA polymerase. This results in release of pyrophosphate and a proton. The release of a proton decreases the pH of the environment which in turn is detected to determine the sequence [3, 31, 45].

The sequencer contains microwells on a semiconductor chip containing a clonally amplified single-stranded template DNA molecule that needs to be sequenced. The wells are then sequentially flooded with DNA polymerase and unmodified deoxynucleoside triphosphates (dNTPs). When flooded with a single species of dNTP, a biochemical reaction will occur if the nucleotide is

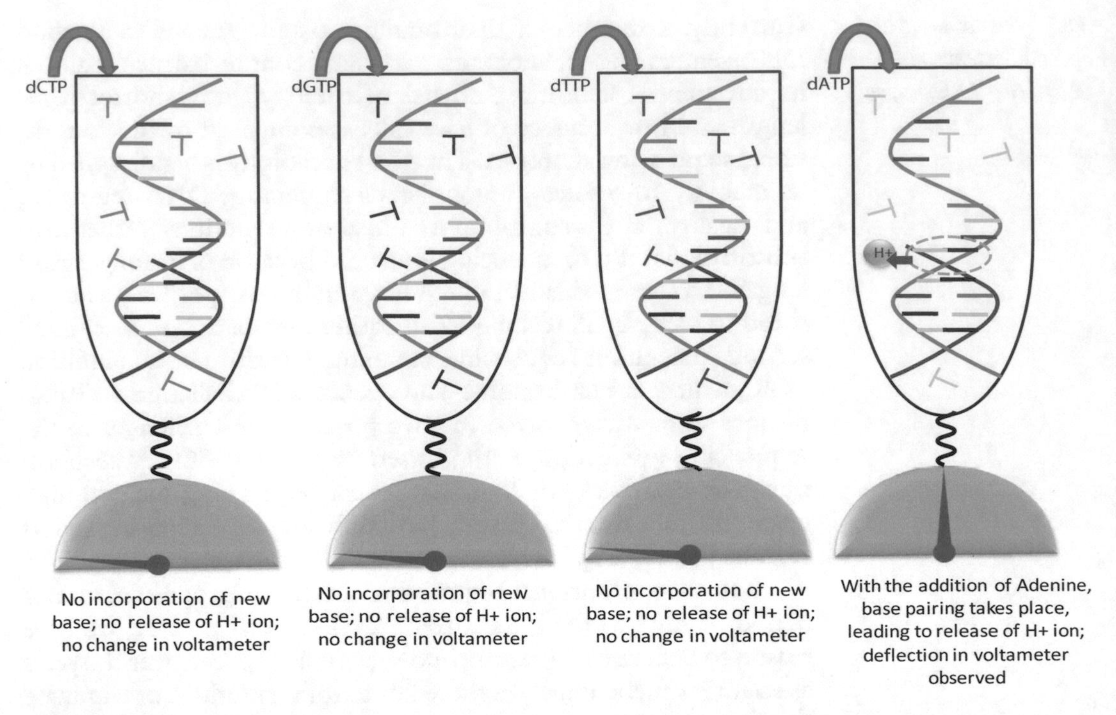

Fig. 8 Ion torrent sequencing technology. With the incorporation of a complementary base, a proton is released which leads to the change in pH of the environment. Thus, the sequence of incorporated base can be deciphered

incorporated in the growing chain with the liberation of hydrogen ions. Underneath each microwell, an Ion-Sensitive Field Effect Transistor (ISFET) is positioned to detect the pH change (due to release of a hydrogen ion) via potential difference and records each nucleotide incorporation event (Fig. 8) [46]. Before the next cycle begins, the unbound dNTP molecules are washed out. The next dNTP species is introduced and the cycle is repeated.

3.4.2 DNA Nanoball Sequencing

In 2010, Complete Genomics Inc published a combinatorial approach of probe-anchor hybridization and ligation (cPAL) that performs sequencing by unchained ligation. It is reputed to have the highest throughput among third generation sequencers. Its affect on the market was dramatic, and resulted in the reduction in sequencing cost from one million dollars in 2008, to $4400 in 2010. It relies on the use of rolling circle amplification of clonally amplified small target DNA sequences, referred to as “nanoballs” [3, 31, 42, 47].

The target DNA to be sequenced is fragmented into 500–600 base sequences. Four adaptors of known sequence are then inserted sequentially into each genomic fragment on either side of the flanks. This is done by repeated digestion with restriction enzymes followed by intramolecular ligation. These molecules are then amplified via rolling circle replication, resulting in the formation of DNA

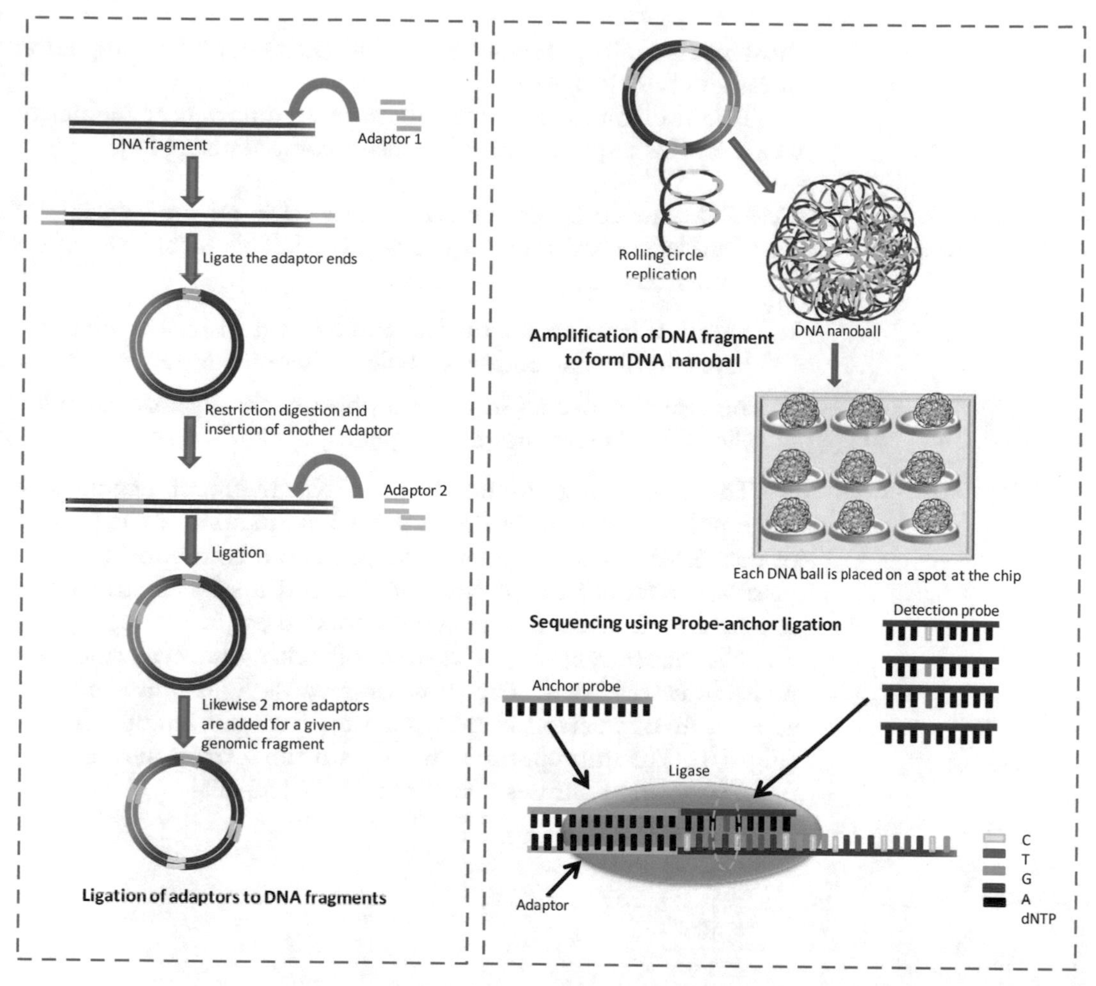

Fig. 9 DNA nanoball sequencing. The fragmented genomic DNA is ligated with four adaptors and the DNA nanoball is formed by rolling-circle replication. Each DNA nanoball is placed on a spot at the chip and sequencing is performed by probe-anchor ligation followed by the binding of detection probe

nanoballs (coils of repeated copies of original single-stranded DNA in solution) that are then randomly attached to a dense array such that each spot is attached to a single nanoball only (Fig. 9) [48].

Once the DNA nanoballs are attached to the spots, anchor probes (to determine adaptor sites) and detection probes (to decode the sequence of target DNA) are used for sequencing. The detection probe consists of nine bases (also known as nonamers) and four dyes, each connected to a known base in the nanomer probe at a specific location. An anchor probe hybridizes to one of the four adapters, followed by the binding of a detection probe adjacent to the bound anchor probe. The system resets once the signal is detected and the next cycle is followed by formation of a new anchor-probe complex. As a result, each base call is unchained, which improves the quality of the sequence because the detection of the previous base is independent of the completion of earlier cycles [48, 49].

This overcomes the error caused in reading repeat regions as well as avoids requirement of extensive computation [44].

3.4.3 Pacific Bioscience RS (SMRT Sequencing)

SMRT developed by Pacific Biosciences relies on the principle of single molecule real time sequencing [46]. This technique differs from other techniques in two ways:

1. Instead of labeling the nucleic acid bases themselves, the phosphate end of a nucleotide is labeled differently for all four bases.
2. The reaction occurs in a nano-photon visualization chamber called ZMW (zero mode waveguides).

The sequencing reaction for a DNA fragment begins with DNA polymerase that resides in the detection zone at the bottom of each ZMW. The chamber is designed to accommodate a polymerase, which is fixed at the bottom, and a ssDNA strand as a template. When a nucleotide is incorporated by the DNA polymerase, the fluorescent tag is cleaved off with the formation of a phosphodiester bond. The machine records light pulses emitted as a result of nucleotide incorporation into the target template (Fig. 10). The fluorophore released with the formation of a phosphodiester bond diffuses out of the ZMW [50–52].

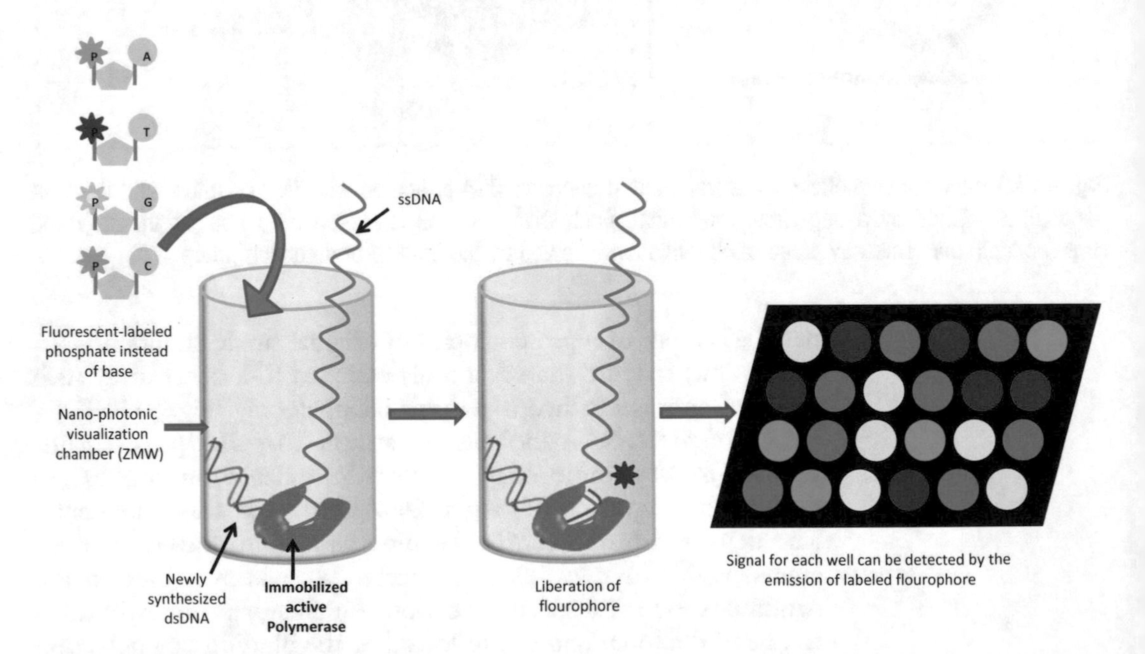

Fig. 10 SMRT sequencing. DNA polymerase is immobilized at the base of each nano-photonic visualization chamber (ZMW). Phosphate of each nucleotide species is tagged with a unique flourophore. As a new base is incorporated complementary to a ssDNA, the pyrophosphate is cleaved with the liberation of fluorophore and the light intensity is captured at each liberation

3.4.4 Nanopore Sequencing

The concept of nanopores and its use in sequencing technique emerged in the mid-1990s. With years of advancement and development, Oxford Nanopore technologies licensed the technology in 2008 [53].

Nanopores are channels of nanometer width, which can be of three types: (1) biological: pores formed by a pore-forming protein in a membrane, for example alpha-hemolysin; (2) solid-state: pores formed by synthetic material or derived chemically, for example silicon and grapheme; or (3) hybrid: pores formed by a biological agent such as pore-forming protein are encapsulated in a synthetic material [54].

Unlike all the above-mentioned sequencers, a Nanopore DNA sequencer does not require the labeling or detection of nucleotides. This technique is based on the principle of modulation of the ionic current generated when a DNA molecule passes through the Nanopore. This helps decipher various characteristics of the molecule such as its length, diameter, and conformation. Initially, an ionic current of known magnitude is allowed to flow through the nanopore. Since different nucleotides have different resistance, and therefore block current for a specific time period, measuring this time period one can determine the sequence of the molecule in question (Fig. 11). Further improvements in the technique may lead to the development of a rapid nanopore-based DNA sequencing technique [55–58].

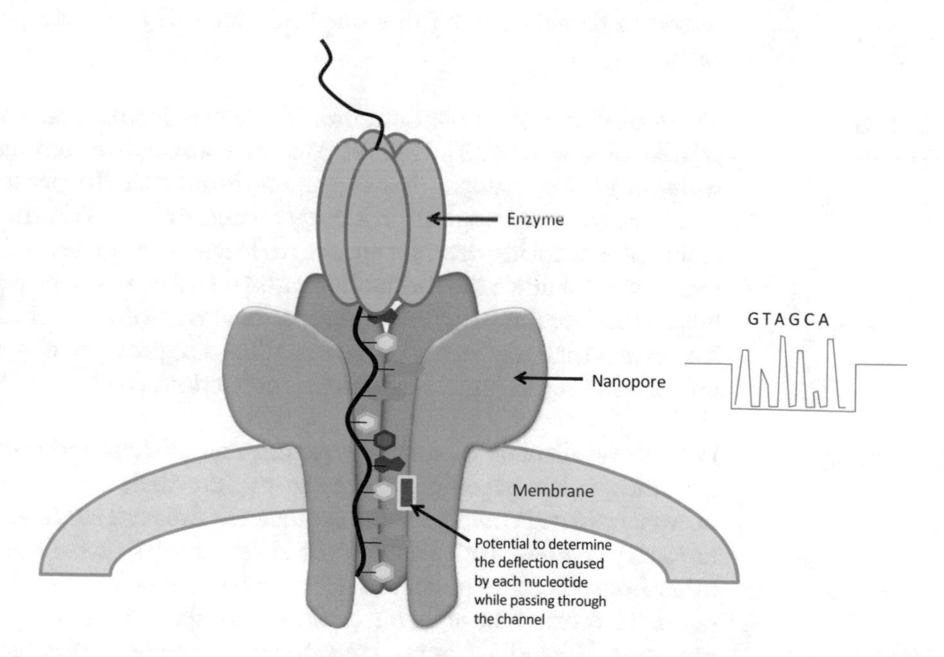

Fig. 11 Nanopore sequencing. A nanopore formed in the biological membrane. Each nucleotide of a single stranded DNA when passing through the nanopore leads to a characteristic change in the membrane potential, which can be readily recorded to decipher the sequence

4 Applications of Sequencing

Sequencing technologies have facilitated a large number of applications. An indispensable part of modern day research in many life sciences, they have enabled scientists working in the smallest of laboratories to answer any query posed by the DNA sequence. Sequencing has helped scientists extract genetic information from a range of biological entities with utmost clarity. Using the power of sequencing, it is possible to decode an entire genome from the minutest of DNA samples (for example, those collected from crime scenes), predict various attributes of a foetus before it is actually born, detect congenital diseases, identify functions of each part of the given DNA sequence, compare different genomes, study evolutionary patterns, and much more. These technologies have also been successfully extended to allied fields such as epigenomics, transcriptomics, and metagenomics [42]. The following section describes some of the applications of sequencing technology and considers some of its future directions.

4.1 Whole-Genome Sequencing

After the successful completion of the first genome, *Haemophilus influenzae*, in 1995 [4], all sequencing techniques have been applied to whole genome sequencing (WGS) of various organisms [18]. More than 20,000 complete genomes have been sequenced so far (including 11427 bacterial, 617 archeal, 5874 viral, and 2052 eukaryotic genomes). The rapid rise in the quantity of WGS data stored in databases is an obvious outcome of NGS technologies. [59].

4.2 Comparative Genomics

The availability of whole genomes has led to extensive studies on related organisms [33]. The impact of comparative genomics is widespread; for example, it has helped distinguish the pathogenic and nonpathogenic regions in closely related strains. With the availability of sequencing data, we are able to locate novel genes and other previously unknown functional elements [60]. The raw data available helps elucidate the structure and functional relationships of biomolecules and their regulatory processes. It has also provided a wealth of information regarding DNA-protein interactions [61].

4.3 Evolutionary Biology

With the availability of such a large number of draft and complete genomes, it has become possible to explore their detailed evolutionary history. Whole genome sequences advanced studies in evolutionary biology from single gene based studies (16S rRNA and other house-keeping genes) to genome based studies [62]. Various genomic features have been explored to shed light on genome evolution, including gene concatenation, gene order, genome based alignment-free phylogeny, and many others. Sequencing provides valuable information for studying molecular phylogenies to understand molecular clocks, to study the patterns of molecular evolution, to test hypotheses regarding the mechanisms that drive evolution, and to study the processes causing homology, analogy, and homoplasy [34].

4.4 Forensic Genomics

Forensic studies were previously conducted using RFLP, PCR, and other techniques. But reduction in the cost of sequencing has opened the way for more accurate methods to be used in forensics, for example in studies related to short tandem repeat (STR) typing [63], mitochondrial DNA analysis, and single nucleotide polymorphisms (SNPs) [64]. These studies are proving to be of immense value and providing additional evidences in criminal cases [65].

4.5 Drug Development

The availability of genome sequences has advanced drug discovery in two ways. First, computer aided drug discovery (CADD) has become easier with the availability of reference sequences, as these help in homology modeling [60]. Second, it aids in analyzing the interaction of a drug with the host genome (molecular simulations). Nowadays, we can not only identify pathogenic strains and virulence factors, we can also counter them using appropriate genetic and protein engineering [66, 67].

4.6 Personal Genomes

The completion of the human genome project in 2001 with ~99.99 % accuracy was a landmark that took more than a decade. But with the availability of NGS platforms, the reduced cost per read has paved the way for sequencing personal genomes [34]. Although the reference human genome still contains 0.01 % error, yet it provides the basis for resequencing individual genomes. Sequencing personal genomes facilitates: detection of single-nucleotide polymorphisms (SNPs); detection of insertions and deletions (InDels); large structural variations (SVs); new variations at the individual level; and variations in genotype/haplotype [68]. Personal genomics is used to study diseases, genetic adaptations, epigenetic inheritance, and quantitative genetics [69].

4.7 Cancer Research

The high-throughput upcoming sequencing technologies are leading us toward the discovery of novel diagnostic and therapeutic approaches in the treatment of cancer [67]. In cancer genomics, current technology empowers speedy identification of patient-specific rearrangements, tumor specific biomarkers, mutations responsible for cancer initiation, and micro-RNAs responsible for inhibition of translation in tumors cells [70]. Further, technology is able to provide "personalized biomarkers"—a set of distinct markers for an individual, used clinically to select optimal treatments for diseases and oncogenic factors [71].

4.8 Microbial Population Analysis

Low cost sequencing technologies have paved the way for simultaneous sequencing of viral, bacterial, and other small genomes in environmental samples, owing to its high throughput, depth of sequencing, and the small size of most microbial genomes [18]. It has given birth to a new field of study called, "metagenomics" [25]. It is now possible to detect unexpected disease-associated viruses and budding human viruses, including cancer-related ones [72]. Sequencing data significantly strengthen our perception of host pathogen communications: including via the discovery of new splice variants, mutations, regulatory elements, and epigenetic controls [73].

5 Future Aspects

The future of biology and genomics is highly likely to benefit from the advancing area of ultra-deep genome sequencing, which will in turn promote revolutionary changes in the medical field, for instance, in analyzing the causes, development and drugs for a disease. It has great potential to facilitate the understanding and analysis of mental and developmental disorders. The use of these technologies for sequencing whole genomes is likely to produce major clinical advancements in the coming few decades. Novel sequencing technologies will be crucial in understanding the role of beneficial microbes. The Human Microbiome Project (also called The Second Human Genome Project) was proposed to analyze all microbes inhabiting the human body, and is expected to lead to a better comprehension of human health and diseases.

A massive amount of comparative genome analysis is currently being performed with a goal to relate genotypes and phenotypes at the genomic level [74, 75]. These projects will not only provide insights into the structural chemistry of organisms, but will also open gateways for discovery of novel genes. This will increase our resources to design unique proteins and other molecules to help mankind.

6 Notes

1. Sequencing reads with poor quality always interfere with the accurate assembly. Hence, filtering of good quality data from whole sequencing data is required. For this purpose, the low-quality data is sieved by "base-calling." In this, the processed trace (chromatogram/pyrogram) is translated into a sequence of bases. Many softwares have integrated base-callers that generate confidence scores as an integral part of the chromatogram file. Phred is most commonly used base-caller for the analysis of larger genomes [76]. Phred defines the quality value q assigned to a base-call to be

Table 4
Probability score of phred value. The acceptable error rate for any genome project is 1/10,000 nucleotides (= phred quality score 40)

Phred quality score	Error rate	Accuracy of base call
10	1 in 10	90 %
20	1 in 100	99 %
30	1 in 1000	99.9 %
40	1 in 10,000	99.99 %
50	1 in 100,000	99.999 %

$$q = -10 \times \log_{10}(p)$$

where p is the estimated error probability for that base-call. Thus, a base-call having a probability of $1/1000$ of being incorrect is assigned a quality value of 30 (*see* Table 4).

Other frequently used base callers are Paracel's Trace Tuner (www.paracel.com) and ABI's KB (www.appliedbiosystems.com), but phred is more reliable than others. Once a phred quality file is generated, the low quality sequence is trimmed from both the ends.

2. *Shotgun cloning*: Genomic DNA can be fragmented either by restriction digestion (but this usually generates fragments of unequal lengths) or by using hydroshear apparatus (useful for generating fragments of equal size). Hydroshearing usually generates protruding ends; therefore end-repairing can be done using T4 DNA polymerase. Fragments can be ligated to the vector of your choice (pUC18/19 can be used for smaller fragments) in a vector:insert molar ratio of 3:1.

3. *Primer designing*: Primers can be designed manually or using primer designing softwares (Primer3, PrimerBLAST etc.). Certain parameters should be considered while designing primers.
   (a) Primers should be 15–28 bases in length;
   (b) Base composition should be 50–60 % (G + C);
   (c) Primers should end (3′) in a G or C, or CG or GC: this prevents "breathing" of ends and increases efficiency of priming;
   (d) Tms between 55–80°C are preferred (even 80°C is too high but many of the gaps were found to be GC rich);
   (e) 3′-ends of primers should not be complementary (i.e., base pair), otherwise primer dimers will be synthesized preferentially to any other product;
   (f) Primer self-complementarity (ability to form secondary structures such as hairpins) should be avoided;

(g) Runs of three or more Cs or Gs at the 3′-ends of primers may promote mispriming at G or C-rich sequences (because of stability of annealing), and should be avoided.

4. *PCR reaction for cycle sequencing*: Direct sequencing can be done on freshly isolated DNA. Routine cycle PCR chemistry can be as follows:

Components	Volume
DNA	50–150 ng/μl
Primer (specific)	1 μl (10 μM)
BigDye Terminator mix (v3.1)	0.5 μl
Buffer (5×)	2 μl
DMSO	0–0.3 μl
Sterile deionized water	To make up the final volume of 10 μl

The PCR can be set up for 25–30 cycles in which the denaturation, annealing, and extension conditions are as follows:

Denaturation:	95 °C—10 s
Annealing:	50 °C—5 s
Extension:	60 °C—4 min

Variation in these conditions can be introduced depending upon the length of template, G + C content, type of repeats etc.

References

1. Mardis EM (2011) A decade's perspective on DNA sequencing technology. Nature 470:198–203F
2. Sanger F, Nicklen S, Coulson AR (1977) DNA sequencing with chain-terminating inhibitors. Proc Natl Acad Sci U S A 74:5463–5467
3. Pareek CS, Smoczynski R, Tretyn A (2011) Sequencing technologies and genome sequencing. J Appl Genet 52:413–435
4. Fleischmann RD, Adams MD, White O, Clayton RA, Kirkness EF, Kerlavage AR, Bult CJ, Tomb JF, Dougherty BA, Merrick JM et al (1995) Whole-genome random sequencing and assembly of *Haemophilus influenzae* Rd. Science 269:496–512
5. Venter JC, Adams MD, Myers EW et al (2001) The sequence of the human genome. Science 291:1304–1351
6. Koboldt DC, Steinberg KM, Larson DE, Wilson RK, Mardis EM (2013) The next generation sequencing revolution and its impact on genomics. Cell 155:27–38
7. Bormann Chung CA, Boyd VL, McKernan KJ, Fu YT, Monighetti C, Peckham HE, Barker M (2010) Whole methylome analysis by ultra-deep sequencing using two-base encoding. PLoS One 5:1–8
8. Nowrousian M (2010) Next-generation sequencing techniques for eukaryotic microorganisms: sequencing-based solutions to biological problems. Eukaryot Cell 9:1300–1310
9. Koboldt DC, Larson DE, Chen K, Ding L, Wilson RK (2012) Massively parallel sequencing approaches for characterization of

structural variation. Methods Mol Biol 838:369–384

10. Brautigam A, Gowik U (2010) What can next generation sequencing do for you? Next-generation sequencing as a valuable tool in plant research. Plant Biol 12:831–841
11. Thudi M, Li Y, Jackson SA, May GD, Varshney RK (2012) Current state-of-art sequencing technologies for plant genomics research. Brief Funct Genomics 2:3–11
12. Pop M, Kosack D, Salzberg SL (2002) A hierarchical approach to building contig scaffolds. In: Second annual RECOMB satellite meeting on DNA sequencing and characterization. Stanford University
13. Shultz JL, Yesudas C, Yaegashi S, Afzal AJ, Kazi S, Lightfoot DA (2006) Three minimum tile paths from bacterial artificial chromosome libraries of soyabean (*Glycine max* cv Forrest): tools for structural and functional genomics. Plant Methods 2:9
14. Liu L, Li Y, Li S, Hu N, He Y, Pong R, Lin D, Lu L, Law M (2012) Comparison of next-generation sequencing systems. J Biomed Biotechnol 2012:1–11
15. Edwards A, Caskey T (1991) Closure strategies for random DNA sequencing. Methods 3:41–47
16. Chaisson MJ, Brinza D, Pevzner PA (2010) De novo fragment assembly with short mate-paired reads: does the read length matter? Genome Res 19:336–346
17. Green P (1997) Against a whole-genome shotgun. Genome Res 7:410–417
18. Stranneheim H, Lundeberg J (2012) Stepping stones in DNA sequencing. Biotechnol J 7:1063–1073
19. Hutchison CA (2007) DNA sequencing: bench to bedside and beyond. Nucleic Acids Res 35:6227–6237
20. Maxam MA, Gilbert W (1977) A new method for sequencing DNA. Proc Natl Acad Sci U S A 74(2):560–564
21. Zimmermann J, Voss H, Schwager C, Stegemann J, Ansorge W (1989) Automated Sanger dideoxy sequencing reaction protocol. FEBS Lett 223:432–436
22. Ansorge W, Voss H, Wirkner U, Schwager C, Stegemann J, Pepperkok R, Zimmermann J, Erfle H (1989) Automated Sanger DNA sequencing with one label in less than four lanes on gel. J Biochem Biophys Methods 20:47–52
23. Rosenthal A, Charnock-Jones DS (1992) New protocols for DNA sequencing with dye terminators. DNA Seq 3:61–64
24. Franca LTC, Carrilho E, Kist TBL (2002) A review of DNA sequencing techniques. Q Rev Biophys 35:169–200
25. Bubnoff AV (2008) Next-generation sequencing: the race is on. Cell 132:721–723
26. Metzker ML (2005) Emerging technologies in DNA sequencing. Genome Res 15:1767–1776
27. Shendure J, Ji H (2008) Next-generation DNA sequencing. Nat Biotechnol 26:1135–1145
28. Mardis EA (2013) Next-generation sequencing platforms. Annu Rev Anal Chem 6:287–303
29. Blazej RG, Kumaresan P, Mathies RA (2006) Micro fabricated bioprocessor for integrated nanoliter-scale Sanger DNA sequencing. Proc Natl Acad Sci U S A 103:7240–7245
30. Augustin MA, Ankenbauer W, Angerer B (2001) Progress towards single-molecule sequencing: enzymatic synthesis of nucleotide-specifically labeled DNA. J Biotechnol 86:289–301
31. Hui P (2014) Next- generation sequencing: chemistry, technology and application. Top Curr Chem 336:1–18
32. Hert DG, Fredlake CP, Annelise E (2008) Advantages and limitations of next-generation sequencing technologies: a comparison of electrophoresis and non-electrophoresis methods. Electrophoresis 29:4618–4626
33. Mardis EA (2008) Next-generation DNA sequencing methods. Annu Rev Genomics Hum Genet 9:387–402
34. Ansorge WJ (2009) Next-generation sequencing techniques. New Biotechnol 25:195–203
35. Ronaghi M, Uhlen M, Nyren P (1998) A sequencing method based on real-time pyrophosphate. Science 281:363–365
36. Keijser BJ, Zaura E, Huse SM, van der Vossen JM, Schuren FH, Montijn RC, ten Cate JM, Crielaard W (2008) Pyrosequencing analysis of the oral microflora of healthy adults. J Dent Res 87:1016–1020
37. Bentley DR, Balasubramanian S, Swerdlow HP, Smith GP, Milton J, Brown CG, Hall KP, Evers DJ, Barnes CL, Bignell HR et al (2008) Accurate whole human genome sequencing using reversible terminator chemistry. Nature 456:53–59
38. Turcatti G, Romieu A, Fedurco M, Tairi AP (2008) A new class of cleavable fluorescent nucleotides: synthesis and optimization as reversible terminators for DNA sequencing by synthesis. Nucleic Acids Res 36:1–13
39. Pettersson E, Lundeberg J, Ahmadian A (2009) Generations of sequencing technologies. Genomics 93:105–111

40. Niedringhaus TP, Milanova D, Kerby MB, Snyder MP, Barron AE (2011) Landscape of next-generation sequencing technologies. Anal Chem 83:4327–4341
41. Voelkerding KV, Dames SA, Durtschi JD (2009) Next-generation sequencing: from basic research to diagnostic. Clin Chem 55 (4):641–658
42. Metzker ML (2010) Sequencing technologies—next generation. Nat Rev Genet 11:31–46
43. Glenn TC (2011) Field guide to next-generation DNA sequencers. Mol Ecol Resour 11:759–769
44. Delsenya M, Han B, Hsing YI (2010) High throughput DNA sequencing: the new sequencing revolution. Plant Sci 179:407–422
45. Rothberg JM, Hinz W, Rearick TM, Schultz J, Mileski W, Davey M, Leamon JH, Johnson K, Milgrew MJ, Edwards M et al (2011) An integrated semiconductor device enabling non-optical genome sequencing. Nature 475:348–352
46. Eid J, Fehr A, Gray J, Luong K, Lyle J, Otto G, Peluso P, Rank D, Baybayan P, Bettman B et al (2010) Real-time DNA sequencing from single polymerase molecules. Science 323:133–138
47. Kaji N, Okamoto Y, Tokeshi M, Baba Y (2010) Nanopillar, nanoball, and nanofibers for highly efficient analysis of biomolecules. Chem Soc Rev 39:948–956
48. Drmanac R, Sparks AB, Callow MJ, Halpern AL, Burns NL, Kermani BG, Carnevali P, Nazarenko I, Nilsen GB, Yeung G et al (2010) Human genome sequencing using unchained base reads on self-assembling DNA nanoarrays. Science 327:78–81
49. Porreca GJ (2010) Genome sequencing on nanoballs. Nat Biotechnol 28:43–44
50. Korlach J, Marks PJ, Cicero RL, Gray JJ, Murphy DL, Roitman DB, Pham TT, Otto GA, Foquet M, Turner SW (2008) Selective aluminum passivation for targeted immobilization of single DNA polymerase molecules in zero-mode waveguide nanostructures. Proc Natl Acad Sci U S A 105:1176–1181
51. Korlach J, Bjornson KP, Chaudhuri BP, Cicero RL, Flusberg BA, Gray JJ, Holden D, Saxena R, Wegener J, Turner SW (2010) Real-time DNA sequencing from single polymerase molecules. Methods Enzymol 472:431–455
52. Schadt E, Turner S, Kasarskis A (2010) A window into third-generation sequencing. Hum Mol Genet 19(2):227–240
53. Venkatesan BM, Bashir R (2011) Nanopore sensors for nucleic acid analysis. Nat Nanotechnol 6:615–624
54. Clarke J, Wu HC, Jayasinghe L, Patel A, Reid S, Bayley H (2009) Continuous base identification for single-molecule nanopore DNA sequencing. Nat Nanotechnol 4:265–270
55. Stoddart D, Heron AJ, Mikhailova E, Maglia G, Bayley H (2009) Single-nucleotide discrimination in immobilized DNA oligonucleotides with a biological nanopore. Proc Natl Acad Sci U S A 106:7702–7707
56. Astier Y, Braha O, Bayley H (2006) Toward single molecule DNA sequencing: direct identification of ribonucleoside and deoxyribonucleoside 5′-monophosphates by using an engineered protein nanopore equipped with a molecular adapter. J Am Chem Soc 128:1705–1710
57. Maitra RD, Kim J, Dunbar WB (2012) Recent advances in nanopore sequencing. Electrophoresis 33:3418–3428
58. Haque F, Li J, Wu HC, Liang XJ, Guo P (2013) Solid state and biological nanopore for real time sensing of single chemical and sequencing of DNA. Nano Today 8:56–74
59. Lim JS, Choi BS, Lee JS, Shin C, Yang TJ, Rhee JS, Lee JS, Choi IY (2012) Survey of the applications of NGS to whole genome sequencing and expression profiling. Genomics Inform 10:1–8
60. Thompson JF, Milos PM (2011) The properties and applications of single-molecule DNA sequencing. Genome Biol 12:217
61. Zhou X, Ren L, Meng Q, Li Y, Yu Y, Yu J (2010) The next generation sequencing technology and application. Protein Cell 1:520–536
62. Buermans HPJ, Dunnen JTD (2014) Next generation sequencing technology: advances and applications. Biochim Biophys Acta 1842:1932–1941
63. Warshauer DH, Lin D, Hari K, Jain R, Davis C, Larue B, King JL, Budowle B (2013) STRait Razor: a length-based forensic STR allele-calling tool for use with second generation sequencing data. Forensic Sci Int Genet 7 (4):409–417
64. Kumar S, Banks TW, Cloutier S (2012) SNP discovery through next-generation sequencing and its applications. Int J Plant Genomics 2012:1–15
65. Berglund EC, Anna Kiialainen A, Syvänen AN (2011) Next generation sequencing technologies and applications for human genetic history and forensics. Investigative Genet 2:1–15
66. Ozsolak F (2012) Third generation sequencing techniques and applications to drug discovery. Expert Opin Drug Discov 7:231–243

67. Yadav NK, Shukla P, Omer A, Pareek S, Singh RK (2014) Next-generation sequencing: potential and application to drug discovery. Scientific World J 2014:1–7
68. Snyder M, Du J, Gerstein M (2010) Personal genome sequencing: current approaches and challenges. Genes Dev 23:423–431
69. Yngvadottir B, MacArthur DG, Jin H, Tyler-Smith C (2009) The promise and reality of personal genomics. Genome Biol 10:237.1–237.4
70. Grumbt B, Eck SH, Hinrichsen T, Hirv K (2013) Diagnostic applications of next generation sequencing in immunogenetics and molecular oncology. Transfus Med Hemother 40:196–206
71. Haimovich AD (2011) Methods, challenges and promise of next generation sequencing in cancer biology. Yale J Biol Med 84:439–446
72. Xuan J, Yu Y, Qing T, Guo L, Shi L (2013) Next-generation sequencing in the clinic: promises and challenges. Cancer Lett 340:284–295
73. Hall N (2007) Advanced sequencing technologies and their wider impact in microbiology. J Exp Biol 209:1518–1525
74. Dijk ELV, Auger H, Jaszczyszyn Y, Thermes C (2014) Ten years of next-generation sequencing technology. Trends Genet 30 (9):418–426
75. Morey M, Fernández-Marmiesse A, Castiñeiras D, Fraga JM, Couce ML, Cocho JA (2013) A glimpse into past, present, and future DNA sequencing. Mol Genet Metab 110:3–24
76. Ewing B, Green P (1998) Base-calling of automated sequencer traces using phred: II. Error probabilities. Genome Res 8(3):186–194

Chapter 2

Sequence Assembly

Xiaoqiu Huang

Abstract

We describe an efficient method for assembling short reads into long sequences. In this method, a hashing technique is used to compute overlaps between short reads, allowing base mismatches in the overlaps. Then an overlap graph is constructed, with each vertex representing a read and each edge representing an overlap. The overlap graph is explored by graph algorithms to find unique paths of reads representing contigs. The consensus sequence of each contig is constructed by computing alignments of multiple reads without gaps. This strategy has been implemented as a short read assembly program called PCAP.Solexa. We also describe how to use PCAP. Solexa in assembly of short reads.

Key words Short read assembly, Hashing, Contig construction

1 Introduction

Sequence assembly programs [1–4] were first developed for DNA sequences (reads) produced by the Sanger method or automated capillary sequencing machines, where the lengths of these capillary reads are between 500 and 1000 nucleotides (nt). Most of these *de novo* assembly programs are based on the overlap-layout-consensus strategy, in which overlaps between reads are computed by fast comparison techniques, the layouts of contigs are generated by using overlaps in a decreasing order of quality, and the consensus sequences of contigs are produced by fast multiple alignment methods [2–6]. Programs of this kind were used to generate good-quality genome assemblies in whole-genome random shotgun sequencing projects [7–11]. A major reason for the success of these programs was that capillary reads were long enough to make it possible to distinguish true overlaps from false overlaps.

With advances in second-generation (massively parallel) sequencing technologies, genome projects have produced millions or billions of short reads of 50–150 nt in length. The huge datasets make it too expensive in computation time and space to compute overlaps by dynamic programming algorithms in narrow matrix

Jonathan M. Keith (ed.), *Bioinformatics: Volume I: Data, Sequence Analysis, and Evolution,* Methods in Molecular Biology, vol. 1525, DOI 10.1007/978-1-4939-6622-6_2,

bands; the short read lengths make it difficult to distinguish true overlaps from false overlaps in generating contig layouts. To avoid these difficulties, assembly programs based on the detection of overlaps as exact matches of a fixed length have been developed [12–20]. Overlaps of this kind form a structure called a *de Bruijn* graph, in which every unique string of length k occurring in a read, called a k-mer, represents an edge from its $k - 1$-mer prefix as a vertex to its $k - 1$-mer suffix as a vertex [21]. Along with the graph structure, the frequency distribution of the strings in the reads is used to distinguish strings with errors from strings without. Strings with errors are either unused, or corrected and used in the assembly. Although the de Bruijn graph can be efficiently constructed and used, read sequences longer than k-mers are not directly used in the construction of contigs.

To use read sequences directly, we have developed an efficient method for computing overlaps between short reads. The method allows mismatches in overlaps, but not indels (insertions and deletions), which occur much less frequently than mismatches in short reads. Only short reads with indel errors are not used in assembly. In the overlap graph, each vertex represents a read and each edge represents an overlap. The overlap graph is explored by graph algorithms to find unique paths of reads representing contigs. The consensus sequence of each contig is constructed by computing alignments of multiple reads without gaps. This strategy has been implemented as a short read assembly program called PCAP.Solexa.

In the rest of this chapter, we first discuss how to compute overlaps between reads efficiently and how to construct contigs. Then we describe how to use PCAP.Solexa in assembly of short reads.

2 Materials

2.1 Hardware

PCAP.Solexa runs at the command line on a 64-bit Linux/Unix/MacOSX system. A system with 100 Gb of memory and 500 Gb of hard drive is required for the assembly of a fungal genome. Much smaller amounts of memory and hard drive (one fifth or less) are recommended for the assembly of a microbial genome, while five times more memory and hard drive are necessary for the assembly of an insect genome. PCAP.Solexa is not recommended for the assembly of a large plant or animal genome.

2.2 Software

PCAP.Solexa is available at http://seq.cs.iastate.edu. It is free to academic users. A licensing agreement is required for commercial users. See http://seq.cs.iastate.edu for information on how to obtain PCAP.Solexa.

2.3 Files

PCAP.Solexa takes as input any number of files of short reads in fastq format with each paired-end dataset provided as two separate fastq files of the same size, a file named fofn of all short read file names (one per line), a file named fofn.con of all pairs of paired-end file names (two names and insert size range and library names per line), and a file named fofn.lib of library names and mean and standard deviation of insert sizes (one name, one mean, and one standard deviation per line). The files for the example used in this unit are included in the PCAP.Solexa package.

3 Methods

3.1 Algorithm

First we describe a method for computing overlaps between reads. The method is based on a data structure called a superword array [22], which is named in a similar way to another data structure called suffix array [23]. A word of length w is a string of w characters, and a superword with v words is a string of $v * w$ characters, obtained by concatenating the v words in order. The v word positions of the superword from left to right are referred to as word positions 1 through v. For example, the word of the superword at word position v refers to the rightmost word of the superword. The positions in the word (with one-based numbering) are divided into two types called checked and unchecked. Two words of length w form a match if they have identical bases at each checked word position. For example, consider two words of length 15: ACCATACCATAGCAC and ACTATTCCATAACAC. Assume that only word positions 3, 6, and 12 are unchecked. Then the two words form a match, because they have identical bases at each of the other positions. The word length w is often set to 12 or a larger value such that the number of checked positions in the word is 12, which ensures that a lookup table for all strings of length 12 can fit into the main memory. Here the word is turned into a string of length 12 by removing bases at each unchecked position.

The value for the parameter v is selected such that the superword length $v * w$ (also called the minimum overlap length) is smaller than the length r of each read. For example, for reads of length 150, we can set w to 15, and v to 6, resulting in a superword length of 90. A read of length r has $r - v * w + 1$ superwords, starting at positions 1, 2, ..., $r - v * w + 1$. Two superwords form a match or are identical if they have identical regular bases at each checked word position. One superword is less (greater) than another superword if the string of bases at each checked position of the first superword, in lexicographic order, comes before (after) the string of bases at each checked position of the second superword. Each read is given a unique nonnegative integer, called a read index. Then each superword in each read can be given a unique nonnegative integer, called a superword index, computed from the start position of the superword in the read and the read index. There is also an inverse function that is efficiently used to produce, from a superword index, the start position of the superword in the read and the read index. The superword array for a set of reads is an array of all superword indexes that are sorted in the lexicographic order of the superwords.

Below is an example of eight superwords in a sorted order. Each superword is made up of four words of length 15, where each checked position of the word is indicated by the bit 1, and each unchecked position by the bit 0. The last position of each word in the superword is marked by a pound sign. The top three superwords are considered identical because they have identical bases at each checked position. The middle two superwords are also identical, and so are the bottom three superwords. Superwords in each block may have different bases or the undetermined base N at an unchecked position. The top block comes before the middle block because the top block has the base G at a checked position marked by an asterisk and the middle block has the base T at the same position. Likewise, the middle block comes before the bottom block, as determined by the bases at another checked position marked by an asterisk.

```
110110111110111110110111110111110110111110111110110111110111
              #              #              #              #
ATNAGCCCAGTTATCCTAGTCAGACTCAGGTTNCATCATTCNTCCGANCAGACTGACCAG
ATGAGGCCAGTCATCCTTGTGAGACTTAGGTTACATCATTCATCCGAACAAACTGANCAG
ATTAGNCCAGTAATCCTCGTAAGACTAAGGTTACANCATTCTTCCGACCANACTGATCAG
                  *
ATCAGTCCAGTNATCCTTTTTAGACTTAGGTTGCAACATTCGTCCGAGCATACTGACCAG
ATGAGACCAGTNATCCTATTGAGACTGAGGTTACATCATTCTTCCGACCAAACTGAACAG
                                        *
ATGAGACCAGTNATCCTNTTNAGACTCAGGTTCCAACATTGCTCCGANCATACTGAGCAG
ATNAGTCCAGTAATCCTTTTAAGACTTAGGTTGCAGCATTGGTCCGAGCANACTGACCAG
ATCAGNCCAGTNATCCTATTTAGACTNAGGTTGCATCATTGATCCGATCACACTGANCAG
```

The superword array is constructed in v rounds of sorting. First the array is initialized to the superword indexes in an increasing order of their values. Then for each word position p from v to 1 (from right to left in the superword), the array is sorted in the lexicographic order of words of every superword at word position p. The sorting is performed by using a lookup table along with a location array as buckets for all strings of length 12, where the location array is an integer array with its index running from 0 to the largest superword index. The lookup table is initially set to a negative value (denoting the empty bucket) at each index, the code of each string of length 12. The bucket sorting is done by placing the elements of the superword array from left to right into their buckets and then by copying the elements in the buckets in reverse order back to the superword array from right to left. Specifically, the current element, a superword index, of the superword array is placed into its bucket as follows. The current superword index is used to find its word at word position p. Then the word of length w is turned into a string of length 12 by using bitwise operations to remove bases at every unchecked position. Next the code of the resulting string is used as an index into the lookup table, the lookup table value at the index is saved in the location array at the superword index, and the lookup table at the index is set to the superword index. After all the elements of the superword array are entered into the buckets constructed with the lookup table and the location array, the elements in the buckets, in reverse order starting with the largest bucket, are copied back to the superword array from right to left.

After its construction, the superword array is partitioned into sections with each section composed of identical superwords. Then each section is processed to generate overlaps between reads as follows. Consider the current section with s superwords. Any two reads with superwords in the current section have a potential overlap. So there are a quadratic number $((s * (s - 1))/2)$ of potential overlaps in the current section. However, it is not necessary to compute all these overlaps; we just need to compute and save a linear number of overlaps (called a subset of overlaps) so that the rest of the overlaps can each be inferred from the subset of overlaps. This can be done by sorting the superwords in the section by the lengths of their prefixes in the reads and selecting adjacent pairs of superwords for computation of overlaps. If the two superwords are from two distinct reads, then the overlap between the two reads is computed. If the number of base mismatches in the overlap is bounded by a cutoff, the overlap is saved.

A large data set of reads is divided into subsets (numbered 0 through $j - 1$) of reads such that every read in the same subset has a superword whose code at word position p is equal to the subset number when the code is divided by the number j of subsets. Here the word position p is a constant between 1 and v. Note that it is possible that a read belongs to more than one subset. These j subsets can be assigned to j processors for computation of overlaps in parallel, where each processor computes overlaps between reads in its subset by constructing a superword array for the subset. A separate file of overlaps is produced for each subset. Once all overlap files are produced, they are screened to remove redundant overlaps, resulting in another set of non-redundant files of overlaps.

The files of overlaps are read one by one, with each overlap of percent identity above a cutoff saved in the main memory. An overlap graph is constructed with each vertex being a read and each edge being an overlap from one read to another. The overlap graph is examined to find long overlap paths of non-repetitive vertexes, where a vertex is repetitive if there are two long paths that end at the vertex but have no overlap between their long prefix paths. This is done in the framework of Dijkstra's algorithm. Each long overlap path of non-repetitive vertexes is reported as a contig of reads. The generation of the layout of each contig is performed on a single processor with a large amount of main memory. The output of this step is a number of files of contig layouts.

The files of contig layouts are processed in parallel with each file handled by a different processor. The reads in each contig are arranged to form a multiple alignment of reads. Then the multiple alignment is used to generate a consensus sequence for the contig. For each file of contig layouts, a file of contig consensus sequences in fasta format is generated along with a file of contig consensus base quality scores. In addition, an .ace file of contigs is produced for viewing and editing in Consed [24]. All the files of contig consensus sequences are merged into a single file of contig consensus sequences, and a single file of contig consensus base quality scores is generated similarly.

3.2 Using PCAP.Solexa

Download the PCAP.Solexa package for your computer system at http://seq.cs.iastate.edu.

Unpack the tar file for Linux and move to the `pcap.solexa` directory:

```
tar -xvzf pcap.solexa.linux.tgz
cd pcap.solexa
```

The `pcap.solexa` directory contains a number of executable code files, which are run in the proper order by a Perl script named `pcap.solexa.perl`. This Perl script needs to be modified in a text editor to let it know the location of the `pcap.solexa` directory and to set important parameter values as follows:

Find the path name of the current `pcap.solexa` directory with the command:

```
pwd
```

Enter the full path name in double quotes as the definition for the variable `$CodeDirPath` in the `pcap.solexa.perl` file. For example, if the path name is

```
/home/xqhuang/551/pcap.solexa,
```

then `$CodeDirPath` is defined with the following command (ending with a semicolon):

```
$CodeDirPath = "/home/xqhuang/551/pcap.solexa";
```

The other parameters in the `pcap.solexa.perl` file are set to their default values. See the comment sections for these parameters in the file for their effects on the assembly in order to select more appropriate values for your data set.

If necessary, make the `pcap.solexa.perl` file executable with the command:

```
chmod ugo+x pcap.solexa.perl
```

A small example of input files is provided in a subdirectory called `test`. It contains a pair of paired-end files named `s_4_1_sequence.fastq` and `s_4_2_sequence.fastq`, each with 7572 Illumina reads of 151 bp. The subdirectory also includes the corresponding `fofn`, `fofn.con`, and `fofn.lib` files. The `fofn` file contains the names of the two read files, one per line. The `fofn.con` file provides an insert size range and an insert library name as well as the two read file names on a line with six columns:

s_4_1_sequence.fastq s_4_2_sequence.fastq 100 700 tmp clone

Columns 3 and 4 give the lower and upper bounds (in bp) of the insert size range. Column 6 is the name of the insert library; column 5 is a placeholder. The `fofn.lib` file gives the mean (in column 2) and standard deviation (in column 3) of insert sizes for the insert library on a line with three columns:

clone 300 50

The PCAP.Solexa program is run on this example as follows:

Copy the `pcap.solexa.perl` file into the `test` subdirectory and move into it with the commands:

```
cp pcap.solexa.perl test
cd test
```

Now the subdirectory becomes the current directory. View the files in it with the command:

```
ls
```

Start the assembly job in the background on `fofn` (file of all read file names) with the command:

```
pcap.solexa.perl fofn &
```

If the job succeeds in producing an assembly (it took 15 s on this example), then the assembly results and statistics are in the following files:

(a) `contigs.bases`: contig base sequences in fasta format

(b) `contigs.quals`: contig quality sequences in fasta format

(c) `reads.placed`: the positions of reads used in the assembly

(d) `fofn.pcap.scaffold*.ace`: ace files of contigs for Consed

(e) `readpairs.contigs`: the lengths of contigs along with major unused read pairs (paired-end reads or mate pairs) between contigs

(f) `fofn.con.pcap.results`: the status of each read pair along with a summary

(g) `z.n50`: the N50 length statistics of contigs and scaffolds

(h) `fofn.pcap.contigs*.snp`: a list of potential SNPs along with their alignment columns.

These output files, except for the `fofn.pcap.scaffold*.ace` files, are described as follows: the `.ace` file format is explained in the documentation for the assembly viewer/editor Consed (http://www.phrap.org/). The `contigs.bases` file contains the sequences (in fasta format) of 4 contigs: Contig0.1, Contig1.1, Contig2.1, and Contig3.1. The name "Contig0.1" means that this contig is the first of scaffold 0. The scaffold index is 0-based; the contig index is 1-based. The `contigs.quals` file contains the Phred quality score sequences (in fasta format) of these contigs. The lengths of the contigs are given in the `readpairs.contigs` file, one contig per line. The first three columns on each line are an abbreviated contig name (e.g., C0.1 for Contig0.1), a bit representing the orientation of the contig (e.g., 1 for given orientation), and a number indicating the length of the contig. Note that columns in any output file are separated by spaces, with 1-based column indexes. For example, the first line of the file is

C0.1 1 23144

The N50 contig length for a set of contigs is defined as the largest number *L* such that the contigs of length at least *L* have a total length above half of the sum of the lengths of all contigs. The N50 contig number for a set of contigs is defined as the smallest number *N* such that the *N* longest contigs have a total length above half of the sum of the lengths of all contigs. The major N50 contig length and number are similarly defined by including only contigs of length at least 1 kb. The N50 contig statistics are reported in the `z.n50` file. The `z.n50` file produced on the example contains the following lines with the N50 contig statistics (ctg stands for contig):

Ctg N50 length: 23134, Ctg N50 number: 1

Major ctg N50 length: 23134, Major ctg N50 number: 1

The information on whether each pair of paired-end reads occurs in a scaffold or contig in proper orientation and with their insert size in the given range is reported in the `fofn.con.pcap.results` file, one pair per line. Each line has at least six columns. Columns 1 and 2 give the indexes of the two reads in the pair; in the example, the read index in the `s_4_1_sequence.fastq` file ranges from 0 to 7571, and that in the `s_4_2_sequence.fastq` file ranges from 7572 to 15,143. Columns 3 and 4 give the lower and upper bounds for the insert size range. Column 5 is a placeholder. Column 6 is a keyword or number reporting the status of the pair: the keyword "singlet" means that at least one of the reads in the pair was not placed in the assembly; "redundant" means that the pair was ignored because it was similar to another pair in the orientation and location of their reads in the assembly; "short" means that one of the reads was placed in a short scaffold or near the end of a scaffold. A number reported in column 6 means that the two reads were placed in a scaffold or contig in opposite orientations with the number indicating the distance between the reads. If the distance is within the insert size range, then a description like "satisfied in a contig" or "satisfied in a scaffold" is reported in column 7. Counts of read pairs in each category are reported at the end of this file.

The `fofn.pcap.contigs*.snp` files are used to report candidate single nucleotide polymorphisms (SNPs). The number of such files is the value V for the parameter $NoCutJobs in the `pcap.solexa.perl` script, with the files indexed by a number N between 0 and $V - 1$. The `fofn.pcap.contigsN.snp` file is used to report SNPs in scaffolds N, $N + V$, $N + 2 \times V$, etc. Each SNP is reported in a section of lines beginning with an SP line containing the key "SP" in column 1. Column 3 of the SP line gives the position of the SNP in the contig; column 4 is the name of the contig. Column 2 is the read coverage depth of the SNP, with each read specified on a BS line in the rest of this section. Thus, column 2 gives the number of BS lines in the rest of this section. The BS lines represent the column of the multiple read alignment at the position of the SNP in the contig. Each read in the alignment is shown on a BS line with the key "BS" in column 1, the read base at the SNP position in column 2, the base quality score in column 3, the read length in column 4, and the read name in column 5.

The `reads.placed` file is used to report the position of each read placed in the assembly, one read per line. The file is organized by scaffolds (also called supercontigs). For each read in a contig of a scaffold, column 2 of its line is the name of the read; column 4 is the length of the read; column 5 is a bit indicating the orientation (e.g., 0 for given orientation) of the read in the contig; columns 6 and 7 give the names of the contig and scaffold; columns 8 and 9 give the positions of the read in the contig and scaffold, respectively. The other columns are placeholders.

3.3 Troubleshooting

If the variable $CodeDirPath in the `pcap.solexa.perl` script is not defined as the full path name of the `pcap.solexa` directory, then the following error message is shown on the screen upon running `pcap.solexa.perl` on `fofn`:

The definition ("/home/xqhuang/551/pcap.solexa") for the variable $CodeDirPath in this Perl script is incorrect.

Enter the path name (of the pcap.solexa directory) in double quotes as the definition for the variable by editing this script.

If the variables $NoPcapJobs and $NoCutJobs in the `pcap.solexa.perl` script are set to a value larger than the number of scaffolds, then failures in opening scaffold output files will occur because the number of scaffold output files cannot be larger than the number of scaffolds. For example, when these variables are set to 5, running the script on the example (with 4 scaffolds) produces the following error messages:

Cannot open fofn.pcap.scaffold4.

Cannot open fofn.pcap.contigs4.

Cannot open fofn.pcap.contigs4.

Cannot open contigs.bases.

A large number of intermediate and output files are produced by PCAP.Solexa. We suggest that all output files from a previous assembly attempt be removed with the `pcap.solexa.clean` script before making another assembly attempt. Otherwise, some old output files may be used in the current assembly attempt. This script is run without arguments:

```
pcap.solexa.clean
```

Finally, if the input files contain errors (e.g., the number of reads in one of a pair of paired-end files is different from the number of reads in the other file), then PCAP.Solexa may fail with a strange error message. Thus, it is helpful to check to see if paired-end files are of the same size.

Acknowledgements

The author thanks Shiaw-Pyng Yang for invaluable feedback on using PCAP.Solexa on various datasets of short reads.

References

1. Dear S, Staden R (1991) A sequence assembly and editing program for efficient management of large projects. Nucleic Acids Res 19:3907–3911
2. Huang X (1992) A contig assembly program based on sensitive detection of fragment overlaps. Genomics 14:18–25
3. Kececioglu JD, Myers EW (1995) Combinatorial algorithms for DNA sequence assembly. Algorithmica 13:7–51
4. Green P (1995) http://www.phrap.org
5. Huang X, Madan A (1999) CAP3: a DNA sequence assembly program. Genome Res 9:868–877
6. Sutton GG, White O, Adams MD et al (1995) TIGR assembler: a new tool for assembling large shotgun sequencing projects. Genome Sci Tech 1:9–19
7. Myers EW, Sutton GG, Delcher AL et al (2000) A whole-genome assembly of *Drosophila*. Science 287:2196–2204
8. Aparicio S, Chapman J, Stupka E et al (2002) Whole-genome shotgun assembly and analysis of the genome of *Fugu rubripes*. Science 297:1301–1310
9. Mullikin JC, Ning Z (2003) The Phusion assembler. Genome Res 13:81–90

10. Jaffe DB, Butler J, Gnerre S et al (2003) Whole-genome sequence assembly for mammalian genomes: ARACHNE 2. Genome Res 13:91–96
11. Huang X, Wang J, Aluru S et al (2003) PCAP: a whole-genome assembly program. Genome Res 13:2164–2170
12. Pevzner PA, Tang H, Waterman MS (2001) An Eulerian path approach to DNA fragment assembly. Proc Natl Acad Sci USA 98:9748–9753
13. Chaisson M, Pevzner PA (2008) Short read fragment assembly of bacterial genomes. Genome Res 18:324–330
14. Butler J, MacCallum I, Kleber M et al (2008) ALLPATHS: De novo assembly of whole-genome shotgun microreads. Genome Res 18:810–820
15. Zerbino DR, Birney E (2008) Velvet: Algorithms for de novo short read assembly using de Bruijn graphs. Genome Res 18:821–829
16. Simpson JT, Wong K, Jackman SD et al (2009) ABySS: a parallel assembler for short read sequence data. Genome Res 19:1117–1123
17. Li R, Zhu H, Ruan J et al (2010) De novo assembly of human genomes with massively parallel short read sequencing. Genome Res 20:265–272
18. Boisvert S, Laviolette F, Corbeil J (2010) Ray: simultaneous assembly of reads from a mix of high-throughput sequencing technologies. J Comput Biol 17:1519–1533
19. Liu Y, Schmidt B, Maskell DL (2011) Parallelized short read assembly of large genomes using de Bruijn graphs. BMC Bioinform 12:354
20. Bankevich A, Nurk S, Antipov D et al (2012) SPAdes: a new genome assembly algorithm and its applications to single-cell sequencing. J Comput Biol 19:455–477
21. Compeau PEC, Pevzner PA, Tesler G (2011) How to apply de Bruijn graphs to genome assembly. Nat Biotechnol 29:987–991
22. Huang X, Yang S-P, Chinwalla A et al (2006) Application of a superword array in genome assembly. Nucleic Acids Res 34:201–205
23. Gusfield D (1997) Algorithms on strings, trees, and sequences: computer science and computational biology. Cambridge University Press, New York
24. Gordon D, Abajian C, Green P (1998) Consed: a graphical tool for sequence finishing. Genome Res 8:195–202

Chapter 3

A Practical Approach to Protein Crystallography

Andrea Ilari and Carmelinda Savino

Abstract

Macromolecular crystallography is a powerful tool for structural biology. The resolution of a protein crystal structure is becoming much easier than in the past, thanks to developments in computing, automation of crystallization techniques and high-flux synchrotron sources to collect diffraction datasets. The aim of this chapter is to provide practical procedures to determine a protein crystal structure, illustrating the new techniques, experimental methods, and software that have made protein crystallography a tool accessible to a larger scientific community.

It is impossible to give more than a taste of what the X-ray crystallographic technique entails in one brief chapter and there are different ways to solve a protein structure. Since the number of structures available in the Protein Data Bank (PDB) is becoming ever larger (the protein data bank now contains more than 100,000 entries) and therefore the probability to find a good model to solve the structure is ever increasing, we focus our attention on the Molecular Replacement method. Indeed, whenever applicable, this method allows the resolution of macromolecular structures starting from a single data set and a search model downloaded from the PDB, with the aid only of computer work.

Key words X-ray crystallography, Protein crystallization, Molecular replacement, Coordinates refinement, Model building

1 Introduction

1.1 Protein Crystallization

The first requirement for protein structure determination by X-ray crystallography is to obtain protein crystals diffracting at high resolution. Protein crystallization is mainly a "trial and error" procedure in which the protein is slowly precipitated from its solution. As a general rule, the purer the protein, the better the chances to grow crystals. Growth of protein crystals starts from a super-saturated solution of the macromolecule, and evolves toward a thermodynamically stable state in which the protein is partitioned between a solid phase and the solution. The time required before the equilibrium is reached has a great influence on the final result, which can go from an amorphous or microcrystalline precipitate to large single crystals. The super-saturation conditions can be obtained by the addition of precipitating agents (salts, organic solvents, and polyethylene glycol

Jonathan M. Keith (ed.), *Bioinformatics: Volume I: Data, Sequence Analysis, and Evolution*, Methods in Molecular Biology, vol. 1525, DOI 10.1007/978-1-4939-6622-6_3, © Springer Science+Business Media New York 2017

polymers) and/or by modifying some of the internal parameters of the solution, such as pH, temperature, and protein concentration. Since proteins are labile molecules, extreme conditions of precipitation, pH, and temperature should be avoided. Protein crystallization involves three main steps [1]:

1. Determination of protein degree of purity. If the protein is not at least 90–95 % pure, further purification will have to be carried out to achieve crystallization. A pure protein might have multiple oligomeric forms, which have a direct effect on its crystallization. Thus, it could be important to evaluate the homogeneity of the sample and oligomeric state of the protein before performing crystallization trials by size exclusion chromatography or Dynamic Light Scattering experiments.
2. The proteins are dissolved in a suitable solvent from which it must be precipitated in a crystalline form. The solvent is usually a water buffer solution at low concentration.
3. The solution is brought to super-saturation. In this step, small aggregates are formed, which are the nuclei for crystal growth. Once nuclei have been formed, actual crystal growth begins.

1.2 Crystal Preparation and Data Collection

1.2.1 Protein Crystals

A crystal is a periodic arrangement of molecules in three-dimensional space. Molecules precipitating from a solution tend to reach the lowest free energy state. This is often accomplished by packing in a regular way. Regular packing involves the repetition in the three space dimensions of the unit cell, defined by three vectors a, b, c and three angles α, β, γ between them. The unit cell contains a number of asymmetric units that coincide with our macromolecule or with more copies of it, related by symmetry operations such as rotations with or without translations. There are 230 different ways to combine the symmetry operations in a crystal, leading to 230 space groups, a list of which can be found in the *International Table of Crystallography*, Vol A [2]. Nevertheless, only 65 space groups are allowed in protein crystals, because the application of mirror planes and inversion points would change the configuration of amino acids from L to D, and D-amino acids are never found in natural proteins.

Macromolecule crystals are loosely packed and contain large solvent-filled holes and channels, which normally occupy 40–60 % of the crystal volume. For this reason, protein crystals are very fragile and have to be handled with care. In order to maintain its water content unaltered, protein crystals should always be kept in their mother liquor or in the saturated vapor of their mother liquor [3, 4]. During data collection (*see below*) X-rays may cause crystal damage due to the formation of free radicals. The best way to avoid damage is crystal freezing. In cryo-crystallography protein crystals are soaked in a solution called "cryo-protectant" so that, when frozen, vitrified water, rather then crystalline ice, is formed. In these conditions, crystals exposed to X-rays undergo negligible radiation damage. Cryo-crystallography usually allows a complete data set to be collected from a single crystal and results in generally higher quality and higher resolution diffraction data, while providing more accurate structural information. Normally, all measurements, both in house and using synchrotron radiation, are performed at 100 K.

1.2.2 X-ray Diffraction

X-ray scattering or diffraction is a phenomenon involving both interference and coherent scattering. Two mathematical descriptions of the interference effect were put forward by Max von Laue and W. L. Bragg [5, 6]. The simplest description, known as Bragg's law, is presented here. According to Bragg, X-ray diffraction can be viewed as a process similar to reflection by planes of atoms in the crystal. Incident X-rays are scattered by crystal planes, identified by the Miller indices *hkl* (*see* **Note 1**) with an angle of reflection θ. Constructive interference only occurs when the path-length difference between rays diffracting from parallel crystal planes is an integral number of wavelengths. When the crystal planes are separated by a distance *d*, the path length difference is $2d \cdot \sin\theta$. Thus, for constructive interference to occur the following relation must hold true: $n\lambda = 2d \cdot \sin\theta$. As a consequence of Bragg's law, to "see" the individual atoms in a structure, the radiation wavelength must be similar to interatomic distances (typically 0.15 nm or 1.5 Å).

1.2.3 X-ray Sources

X-rays are produced in the laboratory by accelerating a beam of electrons emitted by a cathode into an anode, the metal of which dictates what the wavelength of the resulting X-ray will be. Monochromatization is carried out either by using a thin metal foil, which absorbs much of the unwanted radiation or by using the intense low-order diffraction from a graphite crystal. To obtain a brighter source, the anode can be made to revolve (rotating anode generator) and is water-cooled to prevent it from melting. An alternative source of X-rays is obtained when a magnet bends a beam of electrons. This is the principle behind the synchrotron radiation sources that are capable of producing X-ray beams about a thousand times more intense than a rotating anode generator. A consequence of this high-intensity radiation source is that data collection time has been drastically reduced. A further advantage is that the X-ray spectrum is continuous from around 0.05–0.3 nm (*see* **Note 2**).

1.2.4 X-ray Detector

In an X-ray diffraction experiment, a diffraction pattern is observed that could be regarded as a three-dimensional lattice, reciprocal to the actual crystal lattice (*see* **Note 3** and Fig. 1). For a crystal structure determination the intensities of all diffracted reflections must be measured. To do so, all corresponding reciprocal lattice points must be brought to diffracting conditions by rotating the lattice (that is, by rotating the crystal) until the required reciprocal

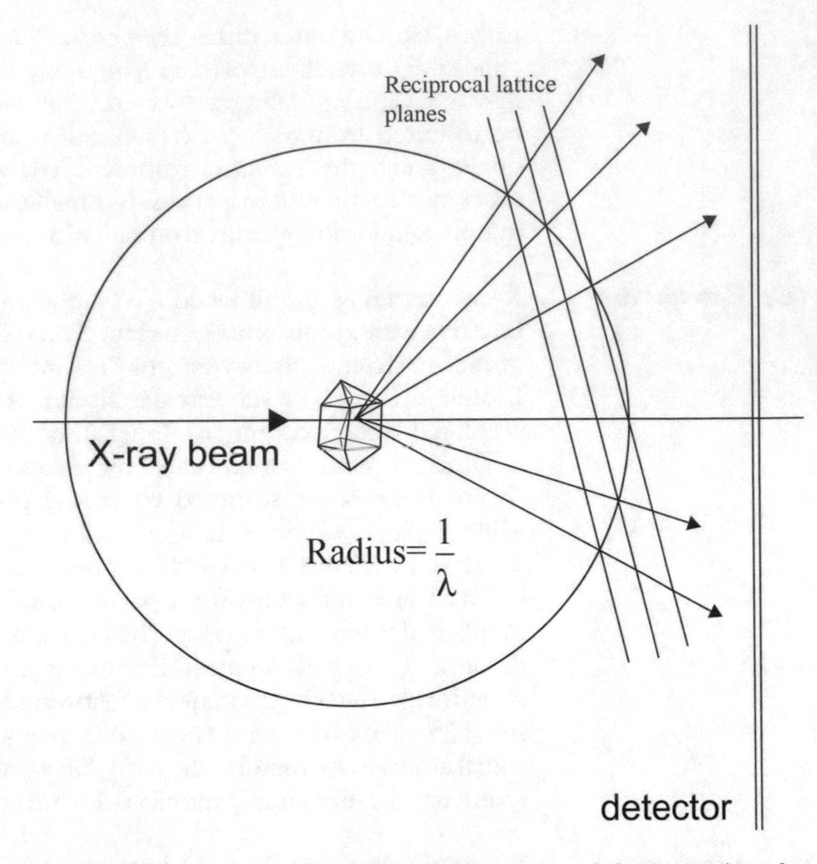

Fig. 1 Ewald sphere. A diagrammatic representation of the generation of an X-ray diffraction pattern (*see* **Note 3**)

lattice points are on a sphere with radius $1/\lambda$. It follows that an X-ray diffraction instrument consists of two parts:

1. a mechanical part for rotating the crystal;
2. a detecting device to measure the position and the intensity of the diffracted reflections.

For a protein structure determination the number of diffracted beams to be recorded is extremely high (of the order of 10^4–10^6) and requires highly efficient hardware. The first efficient and fast devices developed for data collection in protein crystallography are image plate, and CCD camera (*see* **Notes 4** and **5**). These instruments are much more sensitive than a X-ray film, reducing considerably the time for exposure and data processing and solving the time-consuming data collection problem.

The most recent generation of X-ray detectors are PILATUS detectors, developed at the Paul Scherrer Institut (PSI) for the Swiss Light Source (SLS) (*see* **Note 6**). PILATUS detectors display several advantages compared to current CCD and imaging plate detectors: no readout noise, superior signal-to-noise ratio, short read-out time (5 ms), good dynamic range (*see* **Note 7**), high detective quantum efficiency, and the possibility of suppressing fluorescence by an energy threshold. The short readout together with the fast framing time allows taking diffraction data in continuous mode without opening and closing the shutter for each frame.

1.2.5 Data Measurement and Data Processing

Successful data integration depends on the choice of the experimental parameters during data collection. It is therefore crucial that the diffraction experiment is correctly designed and executed. The essence of the data collection strategy is to collect every unique reflection at least once. The most important issues that have to be considered are:

1. The crystal must be single.
2. In order to have a good signal-to-noise ratio, it is recommended to measure crystal diffraction at the detector edge.
3. The exposure time has to be chosen carefully: it has to be long enough to allow collection of high resolution data (*see* **Note 8**), but not so long as to cause overload reflections at low resolution and radiation damage.
4. The rotation angle per image should be optimized: too large an angle will result in spatial overlap of spots, too small an angle will give too many partial spots (*see* **Note 9**).
5. High data multiplicity will improve the overall quality of the data by reducing random errors and facilitating outlier identification.

Data analysis, performed with modern data reduction programs, is normally performed in three stages:

1. (Auto)indexing of one or more images. The program deduces the lattice type, the crystal unit cell parameters, and crystal orientation parameters.
2. Indexing of all images. The program compares the diffraction measurements to the spots predicted on the basis of the autoindexing parameters, assigns the *hkl* indices to each spot.
3. Integration of the peaks. The program calculates the diffraction intensities for each spot in all the collected images.
4. Scaling. The program scales and merges together the reflections of all the collected images (identified by the indices *hkl*).

1.3 Structure Determination

The goal of X-ray crystallography is to obtain the distribution of the electron density which is related to the atomic positions in the unit cell, starting from the diffraction data. The electronic density function has the following expression:

$$\rho(x, y, z) = \frac{1}{V}\Sigma_{hkl}F_{hkl}e^{2\pi i(hx+ky+lz)} \tag{1}$$

where $\mathbf{F}_{hkl}$ are the structure factors, V is the cell volume, and h, k, l are the Miller indices. **F** is a complex number and can be represented as a vector with a module and a phase. It is possible to easily calculate the amplitude of **F** directly from the X-ray scattering measurements but the information on the phase value would be lost. Different experimental techniques can be used to solve the "phase problem," allowing the building of the protein three-dimensional structure: Multiple Isomorphous Replacement (MIR), Multiple Anomalous Diffraction (MAD), and Molecular Replacement (MR). The last one can be performed by computational calculations using only the native data set.

1.3.1 Molecular Replacement

The Molecular Replacement method consists in fitting a "probe structure" into the experimental unit cell. The probe structure is an initial approximate atomic model, from which estimates of the phases can be computed. Such a model can be the structure of a protein evolutionarily related to the unknown one, or even of the same protein from a different crystal form, if available. It is well known that the level of resemblance of two protein structures correlates well with the level of sequence identity [7]. If the starting model has at least 40 % sequence identity with the protein of which the structure is to be determined, the structures are expected to be very similar and molecular replacement will have a high probability of being successful. The chance of success of this procedure progressively decreases with a decrease in structural similarity between the two proteins. It is not uncommon that Molecular Replacement based on one search probe only fails. If this happens, the use of alternative search probes is recommended. Alternatively, a single pdb file containing superimposed structures of homologous proteins can be used. Lastly, if conventional molecular replacement is unsuccessful, models provided by protein structure prediction methods can be used as probes in place of the structure of homologous proteins.

The Molecular Replacement method is applicable to a large fraction of new structures since the Protein Data Bank (http://www.rscb.org) [8] is becoming ever larger and therefore the probability of finding a good model is ever increasing.

Molecular Replacement involves the determination of the orientation and position of the known structure with respect to the crystallographic axes of the unknown structure; therefore, the problem has to be solved in six dimensions. If we call **X** the set of vectors representing the position of the atoms in the probe and **X**′ the transformed set, the transformation can be described as:

$$\mathbf{X}' = [R]\,\mathbf{X} + T \tag{2}$$

where R represents the rotation matrix and T the translation vector. In the traditional Molecular Replacement method, Patterson functions (*see* **Note 10**), calculated for the model and for the experimental data, are compared. The Patterson function has the advantage that it can be calculated without phase information. The maps calculated through the two Patterson functions can be superimposed with a good agreement only when the model is correctly oriented and placed in the right position in the unit cell. The calculation of the six variables, defining the orientation and the position of the model, is a computationally expensive problem that requires an enormous amount of calculations. However, the Patterson function properties allow the problem to be divided into two smaller problems: the determination of (1) the rotation matrix and (2) the translation vector. This is possible because the Patterson map is a vector map, with peaks corresponding to the positions of vectors between atoms in the unit cell. The Patterson map vectors can be divided into two categories: intramolecular vectors (self-vectors) and intermolecular vectors (cross-vectors). Self-vectors (from one atom in the molecule to another atom in the same molecule) depend only on the orientation of the molecule, and not on its position in the cell, therefore they can be exploited in the rotation function. Cross-vectors depend both on the orientation of the molecule and on its position in the cell; therefore, once the orientation is known, these vectors can be exploited in the translation function.

1.3.2 Rotation Function

As mentioned above, the rotation function is based on the observation that the self-vectors depend only on the orientation of the molecule and not on its position in the unit cell. Thus, the rotation matrix can be found by rotating and superimposing the model Patterson (calculated as the self-convolution function of the electron density, *see* **Note 10**) on the observed Patterson (calculated from the experimental intensity). Mathematically, the rotation function can be expressed as a sum of the product of the two Patterson functions at each point:

$$F(R) = \int_r P_{\text{cryst}}(u) P_{\text{self}}(Ru)\, du \qquad (3)$$

where P_{cryst} and P_{self} are the experimental and the calculated Patterson functions respectively, R is the rotation matrix, and r is the integration radius. In the integration, the volume around the origin where the Patterson map has a large peak is omitted. The radius of integration has a value of the same order of magnitude as the molecule dimensions because the self-vectors are more concentrated near the origin. The programs most frequently used to solve X-ray structures by Molecular Replacement implement the fast rotation function developed by Tony Crowther, who realized that the rotation function can be computed more quickly using the Fast Fourier Transform, expressing the Patterson maps as spherical harmonics [9].

1.3.3 Translation Function

Once the orientation matrix of the molecule in the experimental cell is found, the next step is the determination of the translation vector. This operation is equivalent to finding the absolute position of the molecule. When the molecule (assuming it is correctly rotated in the cell) is translated, all the intermolecular vectors change. Therefore, the Patterson functions' cross-vectors, calculated using the observed data and the model, superimpose with good agreement only when the molecules in the crystal are in the correct position. The translation function can be described as:

$$T(t) = \int_{v} P_{\text{cryst}}(u) P_{\text{cross}}(ut) du \tag{4}$$

where P_{cryst} is the experimental Patterson function, whereas P_{cross} is the Patterson function calculated from the probe oriented in the experimental crystal, t is the translation vector, and u is the intermolecular vector between two symmetry-related molecules.

1.4 Structure Refinement

Once the phase has been determined (for example with the molecular replacement method) an electron density map can be calculated and interpreted in terms of the polypeptide chain. If the major part of the model backbone can be fitted successfully in the electronic density map, the structure refinement phase can begin. Refinement is performed by adjusting the model in order to find a closer agreement between the calculated and the observed structure factors. The adjustment of the model consists in changing the three positional parameters (x, y, z) and the isotropic temperature factors B (*see* **Note 11**) for all the atoms in the structure except the hydrogen atoms. Refinement techniques in protein X-ray crystallography are based on the least squares minimization and depend greatly on the ratio of the number of independent observations to variable parameters. Since the protein crystals diffract very weakly, the errors in the data are often very high and more than five intensity measurements for each parameter are necessary to refine protein structures. Generally, the problem is poorly overdetermined (the ratio is around 2) or sometimes under-determined (the ratio below 1). Different methods are available to solve this problem. One of the most commonly used is the Stereochemically Restrained Least Squares Refinement, which increases the number of the observations by adding stereo-chemical restraints [10]. The function to minimize consists in a crystallographic term and several stereochemical terms:

$$\begin{aligned} Q = & \sum w_{hkl}\{|F_{\text{obs}}| - |F_{\text{cal}}|\}^2 + \sum w_{\text{D}}(d_{\text{ideal}} - d_{\text{model}})^2 \\ & + \sum w_{\text{T}}(X_{\text{ideal}} - X_{\text{model}})^2 + \sum w_{\text{P}}(P_{\text{ideal}} - P_{\text{model}})^2 \\ & + \sum w_{\text{NB}}(E_{\text{min}} - E_{\text{model}})^2 + \sum w_{\text{C}}(V_{\text{ideal}} - V_{\text{model}})^2 \end{aligned} \tag{5}$$

where w terms indicate weighting parameters: "w_{hkl}" is the usual X-ray restraint, "w_D" restrains the distance (d) between atoms (thus defining bond length, bond angles, and dihedral angles), "w_T" restrains torsion angles (X), "w_P" imposes the planarity of the aromatic rings (P), "w_{NB}" introduces restraints for nonbonded and Van der Waals contacts (E), and finally "w_C" restrains the configuration to the correct enantiomer (V). The crystallographic term is calculated from the difference between the experimental structure factor amplitudes F_{obs} and the structure factor amplitudes calculated from the model F_{calc}. The stereochemical terms are calculated as the difference between the values calculated from the model and the corresponding ideal values. The ideal values for the geometrical parameters are those measured for small molecules and peptides. The refinement program minimizes the overall function by calculating the shifts in coordinates that will give its minimum value by the least squares fitting method. The classical least square method can produce over-fitting artifacts by moving faster toward agreement with structure factor amplitudes than toward correctness of the phases, because its shift directions assume the current model phases to be error-free constants. The refinement programs, developed more recently, use the Maximum Likelihood method, which allows a representation of the uncertainty in the phases so that the latter can be used with more caution (*see* **Note 12**).

Another popular refinement method is known as "simulated annealing" in which an energy function that combines the X-ray term with a potential energy function comprising terms for bond stretching, bond angle bending, torsion potentials, and van der Waals interactions, is minimized [11].

The parameter used for estimating the correctness of a model in the refinement process is the crystallographic R factor (R_{cryst}), which is usually the sum of the absolute differences between observed and calculated structure factor amplitudes divided by the sum of the observed structure factor amplitudes:

$$R_{cryst} = \frac{\Sigma|F_{obs} - F_{calc}|}{\Sigma|F_{obs}|} \quad (6)$$

Using R_{cryst} as a guide in the refinement process could be dangerous because it often leads to over-fitting the model. For this reason, it is recommended to also use the so-called R_{free} parameter, which is similar to R_{cryst} except for the fact that it is calculated from a fraction of the collected data that has been randomly chosen to be excluded from refinement and maps calculation. In this way, the R_{free} calculation is independent of the refinement process and "phase bias" is not introduced. During the refinement process both R factors should decrease reaching a value in the 10–20 % range.

1.4.1 Model Building

A key stage in the crystallographic investigation of an unknown structure is the creation of an atomic model. In macromolecular crystallography, the resolution of experimentally phased maps is rarely high enough so that the atoms are visible. However, the development of modern data collection techniques (cryo-crystallography, synchrotron sources) has resulted in remarkable improvement in the map quality, which, in turn, has made atomic model building easier. Two types of maps are used to build the model: the "$2F_o - F_c$" map and the "$F_o - F_c$" map (where F_o indicates the observed structure factor amplitudes and F_c the calculated ones). The first one is used to build the protein model backbone and is obtained by substituting the term $|2F_o - F_c|\exp(-i\varphi_{calc})$ (where φ_{calc} is the phase calculated from the model) to the structure factor term in the equation of the electronic density (Eq. 1). The "$F_o - F_c$" map helps the biocrystallographer to build difficult parts of the model and to find the correct conformation for the side chains. Moreover, it is used to add solvent and ligand molecules to the model. The "$F_o - F_c$" map is obtained by substituting the term $|F_o - F_c|\exp(-i\varphi_{calc})$ into the structure factor term in the equation of the electronic density.

2 Materials

2.1 Crystallization and Crystal Preparation

1. Hampton Research, Jena Bioscience, Molecular Dimensions, Qiagen or other crystallization kits.
2. Protein more than 95 % pure, at a concentration between 5 and 15 mg/ml.
3. VDX plates to set up crystallization trials by hanging drop method.
4. Siliconized glass cover slides and vacuum grease.
5. Magnetic crystal caps, and mounted cryoloops.
6. Cryo tong and crystal wand.
7. Dewars to conserve and transport crystals at nitrogen liquid temperature.

2.2 Data Measurement and Data Processing

1. Goniometer head.
2. Program to process the X-ray data: HKL2000 suite, not freely available (HKL Research Inc: http://www.hkl-xray.com); the program package XDS; the program package MOSFLM.

2.3 Molecular Replacement

1. One of the following programs: AMoRe, MolRep, CNS, Phaser.

2.4 Refinement and Model Building

1. Data manipulation : program of the Collaborative computational Project n°4 interactive (CCP4i) suite and /or of PHENIX package.
2. Model-building : Coot, O to build the model.
3. Refinement: Refmac5 (CCP4i package), CNS, PHENIX.

3 Methods

3.1 Crystallization and Crystal Preparation

3.1.1 Crystallization Procedure

Precise rules to obtain suitable single-protein crystals have not been defined yet. For this reason, protein crystallization is mostly a trial and error procedure. However, there are some rules that should be followed in order to increase the success probability in protein crystallization:

1. Check protein sample purity, which has to be around 90–95 %;
2. Slowly increase the precipitating agent concentration (PEGs, salts, or organic solvents) in order to favor protein aggregation;
3. Change pH and/or temperature.

It is usually necessary to carry out a large number of experiments to determine the best crystallization conditions while using a minimum amount of protein per experiment (crystallization trials). In these trials the aliquots of purified protein are mixed with an equal amount of mother solution containing precipitating agents, buffers, and other additives. The individual chemical conditions in which a particular protein aggregates to form crystals are used as a starting point for further crystallization experiments. The goal is optimizing the formation of single protein crystals of sufficient size and quality suitable for diffraction data collection. The protein concentration should be about 10 mg/ml, therefore, 1 mg of purified protein is sufficient to perform about 100 crystallization experiments. Crystallization can be carried out using different techniques, the most common of which are: liquid-liquid diffusion methods, crystallization under dialysis, and vapor diffusion technique. The latter is described in detail since it is easy to set up and allows the biocrystallographer to utilize a minimum protein amount. The vapor diffusion technique can be performed in two ways: the “hanging drop” and the “sitting drop” methods.

1. In the “hanging drop” method, drops are prepared on a siliconized microscope glass cover slip by mixing 1–5 μl of protein solution with the same volume of precipitant solution. The slip is placed upside-down over a depression in a tray; the depression is partly filled (about 1 ml) with the required precipitant solution (reservoir solution). The chamber is sealed by applying grease to the circumference of depression before the cover slip is put into place (Fig. 2a).

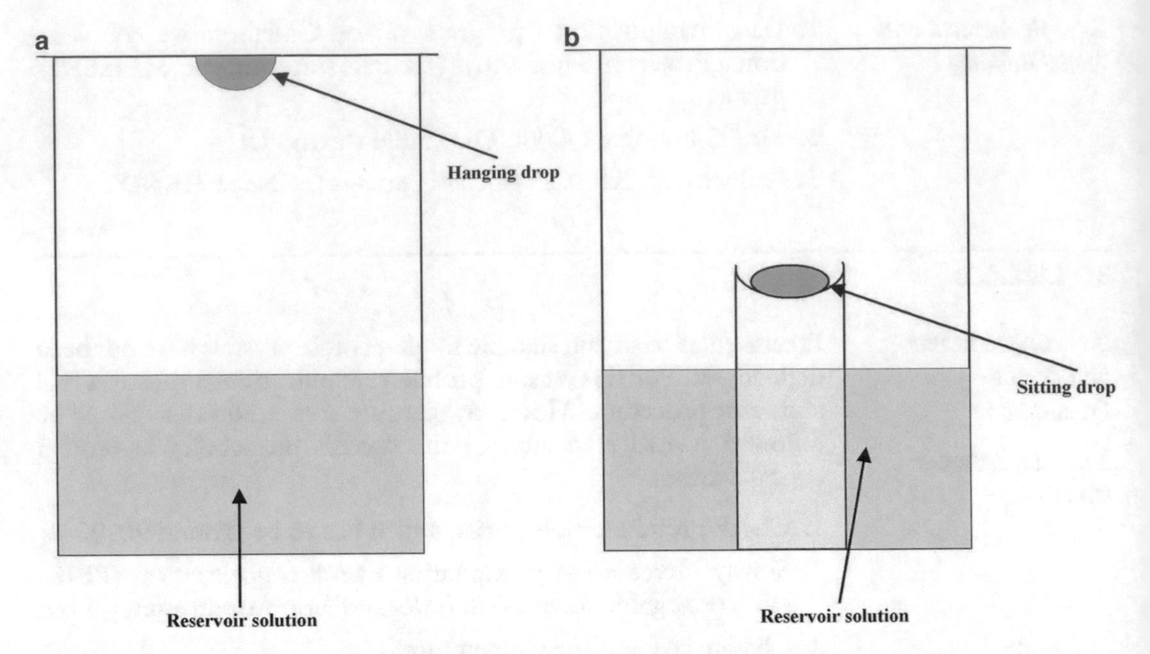

Fig. 2 (**a**) "Hanging drop" crystallization method. A drop of protein solution is suspended from a glass cover slip above a reservoir solution, containing the precipitant agent. The glass slip is siliconized to prevent spreading of the drop. (**b**) "Sitting drop" crystallization method. A drop of protein solution is placed in a plastic support above the reservoir solution

2. The "sitting drop" method is preferable when the protein solution has a low surface tension and the equilibration rate between drop solution and reservoir solution needs to be slowed down. A schematic diagram of a sitting drop vessel is shown in Fig. 2b.

The parameters that can be varied include: nature and concentration of the precipitating agent; buffers to explore the entire pH range; additional salts, detergents, and ligands that affect the solubility of the protein; and others.

3.1.2 High-Throughput Crystallization

Usually to obtain protein crystals several crystallization conditions should be explored. High Throughput (HT) crystallization has been developed to set up many crystallization trials in a short time, using small amounts of protein [12]. A fully integrated HT automated crystallization system consists of a crystallization robot able to screen many crystallization conditions and to dispense nano-drops. These robots are often associated with an automated system to visualize the crystallization plates [13].

The advantages of HT crystallization are tied to the automation that ensures speed and reliability, the miniaturization that allows using small quantities of pure protein, and the parallelization that integrates many trials in a short time. Success of this approach depends on the ability to standardize experimental setup and execution, and is therefore best suited for well-established workflows.

There are several consortia and facilities, open to the scientific community, offering fully automated protein crystallization services and some of these are close to synchrotron radiation sources.

3.1.3 Crystal Cryo-Protection

The most widely used cryo-mounting method consists of the suspension of the crystal in a film of an "antifreeze" solution, held by surface tension across a small diameter loop of fiber, and followed by rapid insertion into a gaseous nitrogen stream. The cryo-protected solution is obtained by adding cryo protectant agents such as glycerol, ethylene glycol, MPD (2-Methyl-2,4-pentandiol), or low molecular weight PEG (polyethylene glycol) to the precipitant solution. The crystal is immersed in this solution for a few seconds prior to being flash-frozen. This method places little mechanical stress on the crystal, so it is excellent for fragile samples. Loops are made from very fine (~10 μm diameter) fibers of nylon. As some crystals degrade in growth and harvest solutions, liquid nitrogen storage is an excellent way to stabilize crystals for long periods [14]. This system is particularly useful when preparing samples for data collection at synchrotron radiation sources, in that, by minimizing the time required by sample preparation, it allows using the limited time available at these facilities to collect data.

3.2 Data Measurement

Once the crystal is placed in the fiber loop, the latter must be attached to a goniometer head. This device has two perpendicular arcs that allow rotation of the crystal along two perpendicular axes. Additionally, its upper part can be moved along two perpendicular sledges for further adjustment and centering of the crystal. The goniometer head must be screwed onto a detector, making sure that the crystal is in the X-ray beam.

The data collection parameters used in diffraction experiments have a strong impact on the data quality and for this reason should be carefully chosen.

In agreement with Bragg's law, the crystal-to-detector distance should be as low as possible to obtain the maximum resolution together with a good separation between diffraction spots. Generally, a distance of 150 mm allows collection of high quality data sets with a good resolution (i.e., lower than 2.0 Å) for protein crystals with unit cell dimensions around 60–80 Å. Long unit cell dimensions (a, b, and/or c longer than 150 Å), large mosaicity (more than 1.0°) (*see* **Note 13**), and large oscillation range (more than 1.0°) are all factors affecting spot separations and causing potential reflection overlaps.

The availability in many beamlines of hybrid pixel detectors operating in single-photon counting mode (PILATUS), displaying different characteristics compared with CCD detectors, imposes on users different data collection strategies. In particular, recent studies have shown that, if the single photon counting pixel detectors are used, fine slicing ($\Delta\varphi = 0.1$–$0.2°$) is recommended for data collection. In particular, the best results are obtained by using a rotation angle comparable to half the mosaicity [15]. This strategy improves the quality and the resolution of the data.

Data collection is best performed interactively, with immediate data processing to get a fast feedback during data collection. This strategy avoids gross inefficiencies in the setup of the experiment, for example incomplete data sets and/or reflection overlaps and/or large percentages of overloaded reflections.

3.3 Data Processing

The basic principles involved in integrating diffraction data from macromolecules are common to many data integration programs currently in use. Here, we describe the data processing performed by the HKL2000 suite [16] and XDS [17]. The currently used data processing methods exploit automated subroutines for indexing the X-ray crystal data collection, which means assigning the correct *hkl* index to each spot on a diffraction image and for integrating images, which means measuring spot intensities (Fig. 3).

3.3.1 HKL2000

1. Peak search. The first automatic step is the peak search, which chooses the most intense spots to be used by the autoindexing subroutine. Peaks are measured in a single oscillation image, which for protein crystals requires 0.1–1.0 oscillation degrees.
2. Autoindexing of one image. If autoindexing succeeds a good match between the observed diffraction pattern and predictions is obtained. The autoindexing permits the identification of the space group and the determination of cell parameters (*see* **Note 14** and Table 1). Other parameters also have to be refined. The most important are the crystal and detector orientation parameters, the center of the direct beam, and the crystal-to-detector distance.
3. Autoindexing of all the images. The autoindexing procedure, together with refinement, is repeated for all diffraction images.
4. Integration of the peaks. In the integration step all spot intensities are measured.

Data are processed using a component program of the HKL2000 suite called Denzo. The scaling and merging of indexed data, as well as the global refinement of crystal parameters, is performed with the program Scalepack that is another HKL2000 suite component. The values of unit-cell parameters refined from a single image may be quite imprecise. Therefore, a postrefinement procedure is implemented in the program to allow for separate refinements of the orientation of each image while using the same unit cell for the whole data set. The quality of X-ray data is firstly assessed by statistical parameters reported in the scale.log file. The first important parameter is the I/σ value (I: intensity of the signal,

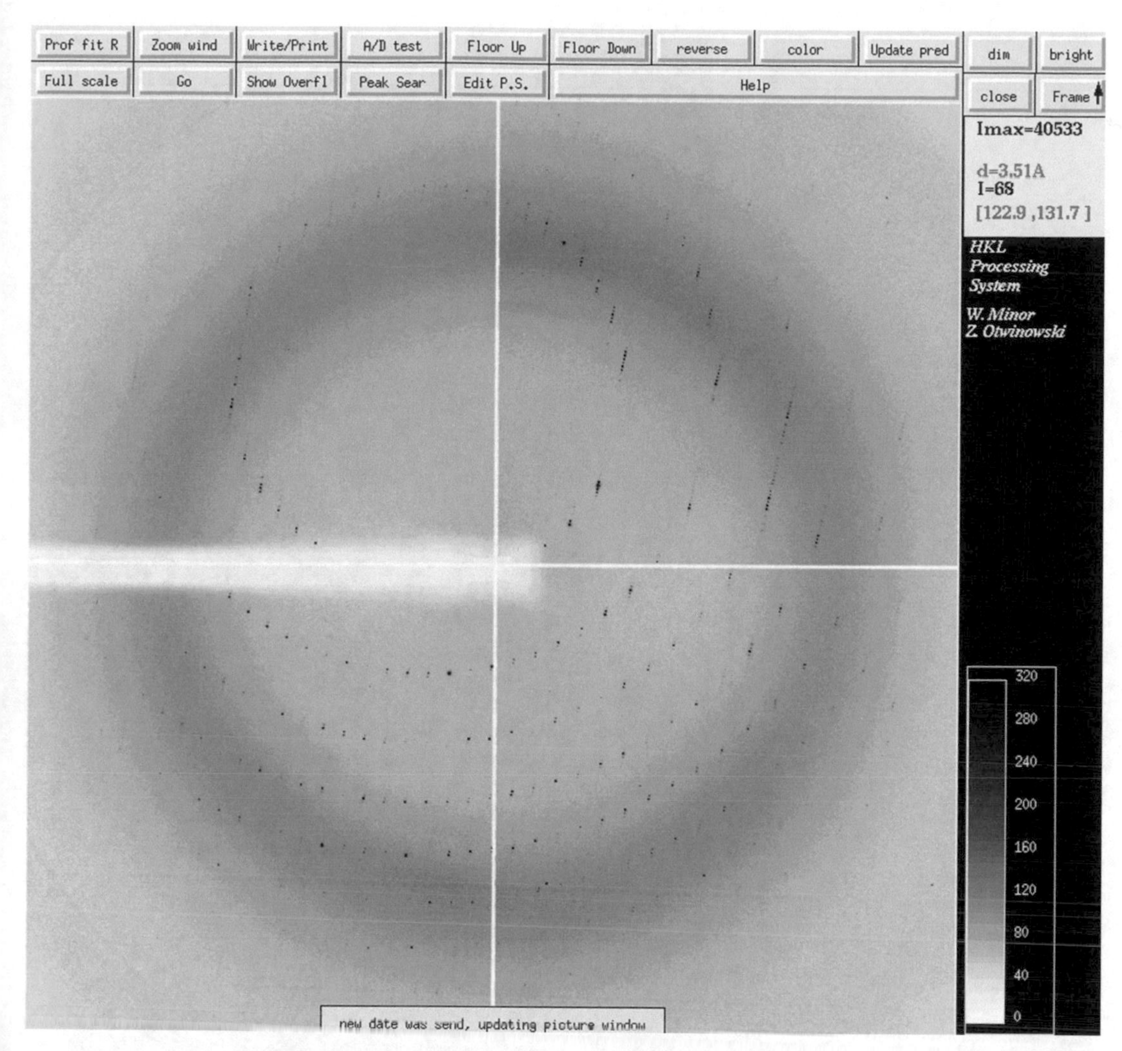

Fig. 3 Diffraction oscillation image visualized with the program Xdisp (HKL2000 suite) of the whole human sorcin collected at the ESRF synchrotron radiation source (Grenoble, FR). The spot distances from the image center are proportional to the resolution, so the spots at the image edge are the highest resolution spots

σ the standard deviation), that is the signal-to-noise ratio, which is also used to estimate the maximum resolution. The second parameter is χ^2, which is closely related to I/σ (*see* **Note 15**). The program tries to bring χ^2 close to 1.0 by manipulating the error model. Another important parameter is R_{sym}, which is a disagreement index between symmetry-related reflections and of which the average value should be below 10 % (*see* **Note 16**). The output of the data processing procedure is a file with suffix .hkl, containing all the measured intensities with their relative σ values and the corresponding *hkl* indices. Using the program Truncate implemented in the CCP4i suite [18], it is possible to calculate: the structure factor amplitudes from the intensities by the French and

Table 1
Output of the Denzo autoindexing routine. The lattice and unit cell distortion table, and the crystal orientation parameters are shown. The present results are obtained for a human sorcin (soluble Resistance related calcium binding protein) crystal

Lattice	Metric tensor distortion index	Best cell (symmetrized) Best cell (without symmetry restrains)					
Primitive cubic	60.04 %	64.84 148.55	64.96 148.55	315.85 148.55	90.09 90.00	89.92 90.00	60.43 90.00
I centered cubic	74.97 %	65.32 236.68	322.35 236.68	322.36 236.68	20.04 90.00	84.33 90.00	84.38 90.00
F centered cubic	79.51 %	335.11 326.68	322.36 326.68	322.56 326.68	23.24 90.00	157.19 90.00	157.36 90.00
Primitive rhombohedral	2.75 %	322.35 320.18 64.90	322.34 320.18 64.90	315.85 320.18 953.96	11.69 11.62 90.00	11.60 11.62 90.00	11.57 11.62 120.00
Primitive hexagonal	*0.22 %*	65.32 65.08	64.84 65.08	315.85 315.85	89.92 90.00	90.17 90.00	120.12 120.00
Primitive tetragonal	13.37 %	64.84 64.90	64.96 64.90	315.85 315.85	90.09 90.00	89.92 90.00	60.43 90.00
I centered tetragonal	13.68 %	64.84 64.90	64.96 64.90	634.88 634.88	87.11 90.00	92.88 90.00	60.43 90.00

Primitive orthorhombic	13.37 %	64.84 64.84	64.96 64.96	315.85 315.85	90.09 90.00	89.92 90.00	60.43 90.00
C centered orthorhombic	0.09 %	65.32 65.32	112.17 112.17	315.85 315.85	90.01 90.00	89.83 90.00	90.13 90.00
I centered orthorhombic	13.68 %	64.84 64.84	64.96 64.96	634.88 634.88	87.11 90.00	92.88 90.00	60.43 90.00
F centered orthorhombic	2.37 %	65.32 65.32	112.17 112.17	634.88 634.88	89.99 90.00	95.73 90.00	90.13 90.00
Primitive monoclinic	0.07 %	64.84 64.84	315.85 315.85	64.96 64.96	90.09 90.00	119.57 119.57	90.08 90.00
C centered monoclinic	0.05 %	65.32 65.32	112.17 112.17	315.85 315.85	89.99 90.00	90.17 90.17	90.13 90.00
Primitive triclinic	0.00 %	64.84	64.96	315.85	90.09	90.08	119.57
Autoindex unit cell	65.24	65.24	315.85	90.00	90.00	120.00	
Crystal rotx, roty, rotz	−8.400	55.089	70.885				
Autoindex Xbeam, Ybeam	94.28	94.90					

Wilson method [19]; the Wilson plot to estimate an overall B factor (*see* **Note 9**); an absolute scale factor; intensity statistics to evaluate the correctness of the data reduction procedure. The truncated output data are stored in a file that usually has the extension .mtz.

3.3.2 XDS

XDS is one of the few program packages which allows us to process datasets collected using the single-photon-counting detectors (PILATUS 2M and 6M).

Data processing with XDS requires an input file XDS.INP provided with the XDS PACKAGE. It consists of eight program steps, which in the input file are indicated as JOBS:

1. XYCORR that calculates the spatial corrections at each detector pixel;
2. INIT calculates the detector gain;
3. COLSPOT identifies the strong reflections to be used for indexing;
4. IDXREF identifies the space group of the crystal and perform the indexing;
5. DEFPIX identifies the detector surface area used to measure intensities;
6. XPLAN is a routine that supports the planning of data collection;
7. INTEGRATE integrates the reflection of the whole dataset;
8. CORRECT scales and merges the symmetry-related reflections.

After completing each individual step, the program writes a .LP file containing the results obtained running the program steps.

As for all the data processing programs, the indexing of the reflections should be carefully managed.

First, the standard deviation of spot position and of spindle position should be checked in the IDXREF.LP. The first parameter should be in the order of 1 pixel whereas the second one, which depends on both the crystal mosaicity and the data collection $\Delta\varphi$, is usually between 0.1° and 0.5°. If the spindle position is greater than 1°, it means that the indexing procedure has not worked properly.

Then the space group and crystal cell parameters should be checked. The correct space group is that one with the lowest QUALITY OF FIT value and the highest symmetry. If the correct lattice is determined, the reflections of the whole data set can be integrated and after that the reflections with the same symmetry can be scaled and merged. The CORRECT step, the last one, produces a file CORRECT.LP containing all the statistics and a file XDS_ASCII.HKL containing the integrated and scaled reflections.

The final table of the CORRECT.LP file reports the number of measured reflections and of the unique reflections, the completeness of dataset together with three other important parameters that must be checked to evaluate the correctness of the scaling and merging procedures: the $I/\sigma(I)$ (I: intensity of the signal, $\sigma(I)$ the standard deviation) value that should be greater than 1.5; the R_{merge} value that should be as low as possible and the CC(1/2) (*see* **Note 17**). The data can be additionally scaled with XSCALE, which allows us to cut the data set at the desired resolution and produces a file XSCALE.LP containing all the parameters present in the CORRECT.LP file (Table 2) and a .ahkl file containing all the scaled and merged reflections. This file can be converted (using the XDSCONV program) into a .hkl file readable by the CCP4 program f2mtz, which finally transforms it in a .mtz file.

3.4 Molecular Replacement

1. Search model. The first operation is searching the databases for probe structure similar to the structure to be solved. Since we do not know the structural identity of our protein with homologous proteins, we use sequence identity as a guide. Proteins showing a high degree of sequence similarity with our "query" protein can be identified in protein sequence databases using sequence comparison methods such as BLAST [20]. The protein of known three-dimensional structure showing the highest sequence identity with our query protein is generally used as the search model.
2. Files preparation. The Pdb file of the search probe has to be downloaded from the Protein Data Bank (http://www.rcsb.org). The file has to be manipulated before Molecular Replacement is performed. The water molecules as well as the ligand molecules have to be removed from the file. The structure can be transformed into a polyalanine search probe to avoid model bias during the Molecular Replacement procedure (*see* **Note 18**). The other file needed to perform the molecular replacement is the file with extension .mtz resulting from earlier data processing (*see* Subheading 3.3), containing information about crystal space group, cell dimensions, molecules per unit cell, and a list of the collected experimental reflections.
3. Molecular Replacement. The Molecular Replacement procedure consists in Rotation and Translation searches to put the probe structure in the correct position in the experimental cell. This operation can be done using different programs, belonging to the CCP4 suite [18] : AMoRe [21], Phaser [22] and Molrep [23]. In this chapter, we describe briefly the use of MolRep, which is automated and user-friendly.
4. The program performs rotation searches followed by translation searches. The only input files to upload are the .mtz and the .pdb files. The values of two parameters have to be chosen: the integration radius and the resolution range to be used for

Table 2
Example of an XSCALE output table. On the table, different parameters, allowing the evaluation of the dataset quality, are reported, namely completeness of the dataset, R factor (= R_{merge}), $I/\sigma(I)$ and CC1/2. As shown in the table, the parameters are calculated separately for the twenty different resolution shells. The data shells with CC1/2 marked by an asterisk display a correlation significant at the 0.1 %, and can be used for structure determination and model refinement

Resolution	Number of reflections			Completeness	R-Factor					
limit	Observed	Unique	Possible	of data	Observed	Expected	Compared	*I*/Sigma	*R*-meas	CC(1/2)
7.38	1718	457	469	97.4 %	1.4 %	1.7 %	1716	74.14	1.7 %	100.0*
5.22	3219	851	852	99.9 %	1.7 %	1.9 %	3214	65.40	2.0 %	99.9*
4.26	3629	1091	1100	99.2 %	1.8 %	2.1 %	3604	56.06	2.2 %	99.9*
3.69	4124	1283	1295	99.1 %	1.9 %	2.2 %	4063	49.51	2.3 %	99.9*
3.30	4931	1442	1450	99.4 %	2.2 %	2.3 %	4920	47.24	2.7 %	99.9*
3.01	5748	1627	1632	99.7 %	2.5 %	2.6 %	5735	41.15	3.0 %	99.9*
2.79	6292	1744	1750	99.7 %	2.6 %	2.7 %	6289	40.77	3.0 %	99.9*
2.61	6740	1904	1907	99.8 %	3.0 %	3.0 %	6724	35.62	3.5 %	99.9*
2.46	6559	1998	2007	99.6 %	3.4 %	3.5 %	6506	29.33	4.1 %	99.8*
2.33	6775	2104	2127	98.9 %	3.9 %	3.9 %	6692	26.29	4.6 %	99.8*
2.22	7334	2220	2237	99.2 %	4.6 %	4.6 %	7280	23.84	5.5 %	99.7*

2.13	8112	2359	2372	99.5 %	5.6 %	5.5 %	8075	21.17	6.6 %	99.6*
2.05	8492	2426	2433	99.7 %	7.5 %	7.4 %	8457	16.79	8.8 %	99.5*
1.97	8976	2542	2549	99.7 %	9.3 %	9.3 %	8953	14.38	10.9 %	99.2*
1.91	9292	2621	2627	99.8 %	12.8 %	12.8 %	9274	10.97	15.1 %	98.7*
1.84	9714	2731	2735	99.9 %	18.5 %	18.6 %	9701	8.19	21.8 %	97.7*
1.79	9011	2782	2799	99.4 %	23.9 %	23.8 %	8923	6.25	28.7 %	95.5*
1.74	9586	2862	2876	99.5 %	33.4 %	33.0 %	9513	4.82	39.9 %	93.2*
1.69	9272	2945	2964	99.4 %	39.0 %	39.0 %	9133	3.84	46.9 %	91.1*
1.65	10,092	3051	3063	99.6 %	55.9 %	56.4 %	9994	3.00	66.6 %	83.3*
total	139,616	41,040	41,244	99.5 %	3.3 %	3.4 %	138,766	21.57	3.9 %	99.9*

Patterson calculation. In the rotation function, only intramolecular vectors need to be considered. Since all vectors in a Patterson function start at the unit cell axes origin, the vectors closest to the origin will in general be intramolecular. By judiciously choosing a maximum Patterson radius, we can improve the chances of finding a strong rotation hit. Usually, a value of the same order of magnitude as the search probe dimensions is chosen. Regarding the second parameter, high resolution reflections (above 3.5 Å) will differ substantially because they are related to the residue conformations. On the other hand, low resolution reflections (below 10 Å) are influenced by crystal packing and solvent arrangement. Thus, the resolution range that should be used is usually within 3.5–10 Å.

5. Output files. The output files to check are the file with extension .log that lists all the operations performed by the program and the coordinates file representing the MR solution, which is the model rotated and translated in the real cell, in a pdb format. As shown in Table 3, after the rotational and translational searches are performed, the program lists all the possible solutions (*see* **Note 19**) followed by the rotation angles and the translation shifts necessary to position the model in the real cell, the crystallographic R factors, and finally the correlation coefficients (*see* **Note 20**). The first line of Table 3 represents a clear solution for an MR problem. In fact, the crystallographic R factor is below 0.5, and the correlation coefficient is very high (75.9 %).

Table 3
Output of the MolRep program after rotation and translation searches. The present results (data not published) have been obtained for the protein Dps (Dna binding proteins) from *Listeria monocitogenes* [32] using as search model the Dps from *Listeria innocua* (Pdb code 1QHG)

		Alpha	Beta	Gamma	Xfrac	Yfrac	Zfrac	TF/sig	R-fac	Corr
Sol_TF_7	*1*	*32.27*	*84.87*	*78.84*	*0.824*	*0.498*	*0.091*	*65.34*	*0.333*	*0.759*
Sol_TF_7	2	32.27	84.87	78.84	0.324	0.041	0.092	25.76	0.482	0.478
Sol_TF_7	3	32.27	84.87	78.84	0.324	0.454	0.091	24.18	0.477	0.481
Sol_TF_7	4	32.27	84.87	78.84	0.324	0.498	0.016	23.57	0.483	0.467
Sol_TF_7	5	32.27	84.87	78.84	0.422	0.498	0.091	23.37	0.479	0.478
Sol_TF_7	6	32.27	84.87	78.84	0.324	0.498	0.325	23.12	0.482	0.471
Sol_TF_7	7	32.27	84.87	78.84	0.238	0.498	0.092	23.01	0.481	0.473
Sol_TF_7	8	32.27	84.87	78.84	0.324	0.498	0.372	22.99	0.479	0.475
Sol_TF_7	9	32.27	84.87	78.84	0.324	0.498	0.400	22.97	0.480	0.473
Sol_TF_7	10	32.27	84.87	78.84	0.324	0.000	0.196	22.93	0.490	0.456

Moreover, there is a jump between the first possible solution (first line) and the second possible solution (second line).

3.5 Structure Refinement

Several programs can be used to perform structure refinement. The most common are: CNS written by Brünger [24], which uses conventional least square refinement as well as simulated annealing to refine the structure; REFMAC5 (CCP4i suite) written by Murshudov [25] that uses maximum likelihood refinement; and PHENIX that allows multistep complex refinement protocols in which most of the available refinement strategies can be combined with each other and applied to any selected part of the model [26]. Although CNS and PHENIX have been used with success, in this chapter we illustrate the use of REFMAC5 implemented in CCP4i because it provides a graphic interface to compile the input files; this feature is particularly helpful for beginners.

1. Rigid body refinement. Firstly, the initial positions of the molecules in the unit cell and the crystal cell provided by MR procedures have to be refined. For this purpose, Rigid Body refinement should be performed. This method assigns a rigid geometry to parts of the structure and the parameters of these constrained parts are refined rather than individual atomic parameters. The input files to be uploaded are the MR solution and the .mtz file containing the experimental reflections. The resolution at which to run rigid body refinement has to be specified (in general the rigid body refinement should start at the lowest resolution range) and the rigid entity should be defined (this can be an entire protein, a protein subunit, or a protein domain). To define the rigid entities in REFMAC5, simply select the chain and protein regions that are to be fixed.
2. Output files. The output files are: (a) the .log file that contains a list of all the operations performed, statistics (Table 4) about the geometrical parameters after each refinement cycle, crystallographic R factor and R free factor values, and finally the figure of merit (*see* **Note 21**); (b) the .pdb file containing the refined coordinates of the model; (c) the .mtz file containing the observed structure factors (F_{obs}), the structure factor amplitudes calculated from the model (F_{calc}), and the phase angles calculated from the model.
3. Coordinates and B factors refinement. The program REFMAC 5 refines the x, y, z, and B parameters using the maximum likelihood method. As for the Rigid Body refinement, the input files are the .mtz file containing the F_{obs} and the .pdb file containing the coordinates of the model. It is also necessary to restrain the stereochemical parameters using the maximum likelihood method. It is possible to choose a numerical value for the relative weighting terms or, more easily, to choose a single value for the so-called weight matrix that allows the program to

Table 4
Summary of ten cycles of DpsTe (*see* Note 25) coordinate refinement using REFMAC5. The Rfact, R_{free}, figures of merits (FOM) and root mean square deviation values of some stereo-chemical parameters are shown

Ncyc	Rfact	R_{free}	FOM	LLG	rmsBOND	rmsANGLE	rmsCHIRAL
0	0.213	0.213	0.862	1,165,259.2	0.004	0.734	0.055
1	0.196	0.210	0.865	1,151,022.5	0.010	1.022	0.074
2	0.191	0.209	0.867	1,146,576.9	0.011	1.106	0.080
3	0.188	0.209	0.868	1,144,297.8	0.011	1.144	0.083
4	0.187	0.209	0.869	1,142,920.2	0.011	1.166	0.085
5	0.186	0.209	0.870	1,142,088.8	0.011	1.178	0.086
6	0.185	0.209	0.870	1,141,496.4	0.011	1.186	0.087
7	0.184	0.209	0.870	1,141,031.5	0.011	1.190	0.088
8	0.184	0.209	0.871	1,140,743.6	0.011	1.192	0.088
9	0.184	0.209	0.871	1,140,461.8	0.011	1.195	0.088
10	0.183	0.209	0.871	1,140,311.0	0.011	1.196	0.088

restrain all the stereochemical parameters together. The value of the "weight matrix" should be between 0.5, indicating loose stereochemical restraints, and 0, indicating strong stereochemical restraints, which keep geometrical parameters of the macromolecules near the ideal values. In REFMAC5 NCS (*see* **Note 22**) restraints can also be used for refinement.

3.6 Model Building

After the Molecular Replacement and the first cycles of coordinates refinement, only a partial model has been obtained. In this model, the side chains are absent, and often parts of the model do not match the electronic density map. Therefore, the building of the first structural elements is followed by refinement cycles that should lead to an improvement on the statistics (that is, the R factor has to decrease and the figure of merit has to increase). The most common program used for model building is COOT [27], which permits the direct calculation of electronic density maps. Two maps are necessary to build a model: the $2F_o - F_c$ map contoured at 1σ which is used to trace the model and the $F_o - F_c$ map contoured at 3σ, which is necessary to observe the differences between the model and the experimental data.

1. Starting point. Firstly, a match between protein sequence and the $2F_o - F_c$ density map should be found. If the phases are good, this operation should not be too difficult. The electron density map should be clear (especially if it has been calculated

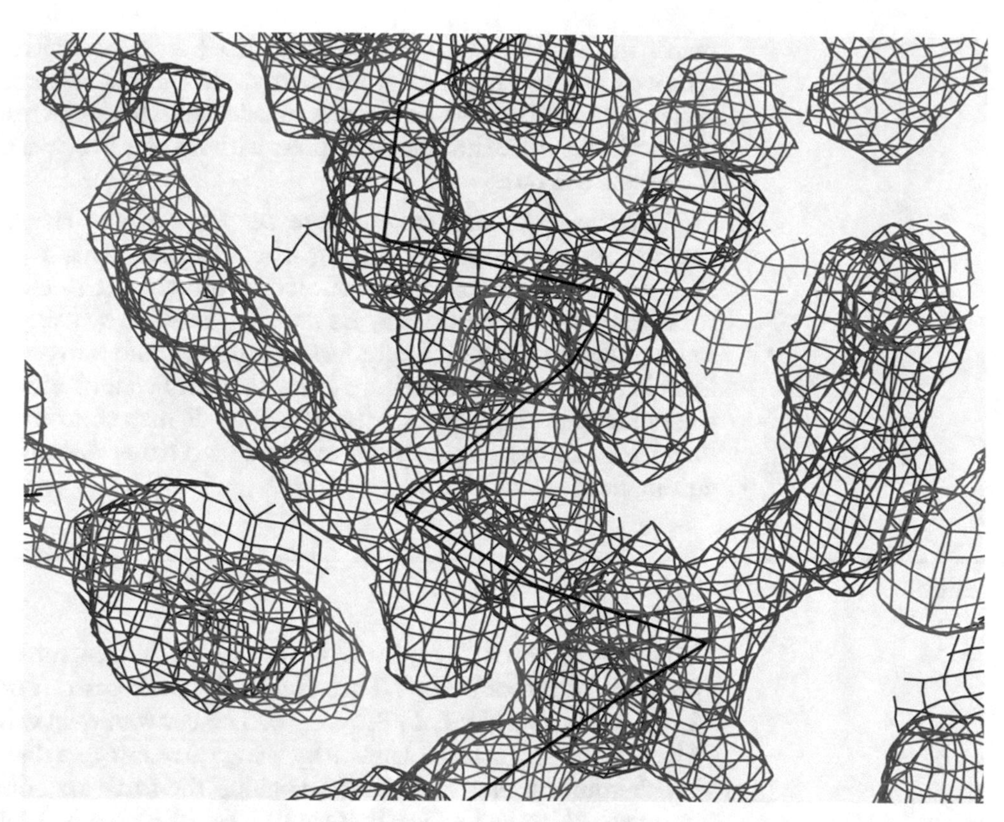

Fig. 4 Initial Electronic density map of Dps from *Thermosynechococcus elongatus* (*see* **Note 25**) calculated after Molecular Replacement. Cα trace of the model is superimposed on the map. The electronic density of a Trp residue and a Tyr residue are easily recognizable in the map

from high-resolution data) and should allow the identification of the amino acids (*see* **Note 23** and Fig. 4).

2. Initial building. Once the first residues have been identified and fitted into the electron density map, model building can be performed by fitting the whole protein sequence residue by residue in the map.
3. Building of the unfitted structure elements. If the initial model does not contain all the protein residues, it is possible to build the main chain of the protein region missing from the model "ab initio." As an example, it is possible to add amino acids, which are automatically bound to the rest of the protein after or before a selected residue in the correct position, according to the electron density map. The main chain of the missing protein region can also be constructed using the program database, i.e., using modular elements of secondary structure as α-helices and/ or β-sheets of different lengths.
4. Omit map. If a part of the structure does not match the map, this means that it is built incorrectly. Thus it is possible to use, as a major strategy for overcoming phase bias, the so-called omit

maps. In practice, the model region that has to be refitted is removed and the maps are recalculated after a few refinement cycles. This method allows the phases calculated from the rest of the model to phase the area of interest with no bias from parts of the model left out.

5. Optimization. At this stage, large sections of the structure should be approximately fitting the electron density map (Fig. 5). The next step is the choice of the correct side-chain rotamers. This operation may be done by hand or by using real space refinement tools. Finally, water molecules and ions and/or ligands bound to the protein have to be identified and added to the model. For this purpose only the $F_o - F_c$ map contoured at 3σ is used. The water molecules can be added either manually or automatically (*see* **Note 24**).

4 Notes

1. The intercepts of the planes with the cell edges must be fractions of the cell edge. Therefore, cell intercepts can be at $1/0$ ($= \infty$), $1/1$, $1/2$, $1/3$, ... $1/n$. The conventional way of identifying these sets of planes is by using three integers that are the denominators of the intercepts along the three axes of the unit cell, *hkl*, called Miller indices. If a set of planes had intercepts at $1/2$, $1/3$, $1/1$ then the planes would be referred to as the (2 3 1) set of planes.
2. Another advantage of synchrotron radiation is its tunability, which allows the user to select radiation wavelengths higher or lower than 1.5418 Å (copper radiation). Collection of data at wavelengths below 1.5418 Å results in a lower signal-to-noise ratio.
3. A crystal can be regarded as a three-dimensional grid and one can imagine that this will produce a three-dimensional X-ray diffraction pattern. As with electron microscope grids, the pattern is reciprocal to the crystal lattice. The planes that intersect the sphere in Fig. 1 are layers in a three-dimensional lattice, called reciprocal lattice because the distances are related reciprocally to the unit cell dimensions. Each reciprocal lattice point corresponds to one diffracted reflection. The reciprocal lattice is an imaginary but extremely convenient concept to determine the direction of the diffracted beams. If the crystal rotates, the reciprocal lattice rotates with it. In an X-ray diffraction experiment the direction of the diffracted beams depends on two factors: the unit-cell distances in the crystal, from which the unit-cell distances in the reciprocal lattice are derived, and the X-ray wavelength. As indicated in Fig. 1, diffraction conditions are determined not only by the reciprocal lattice but also

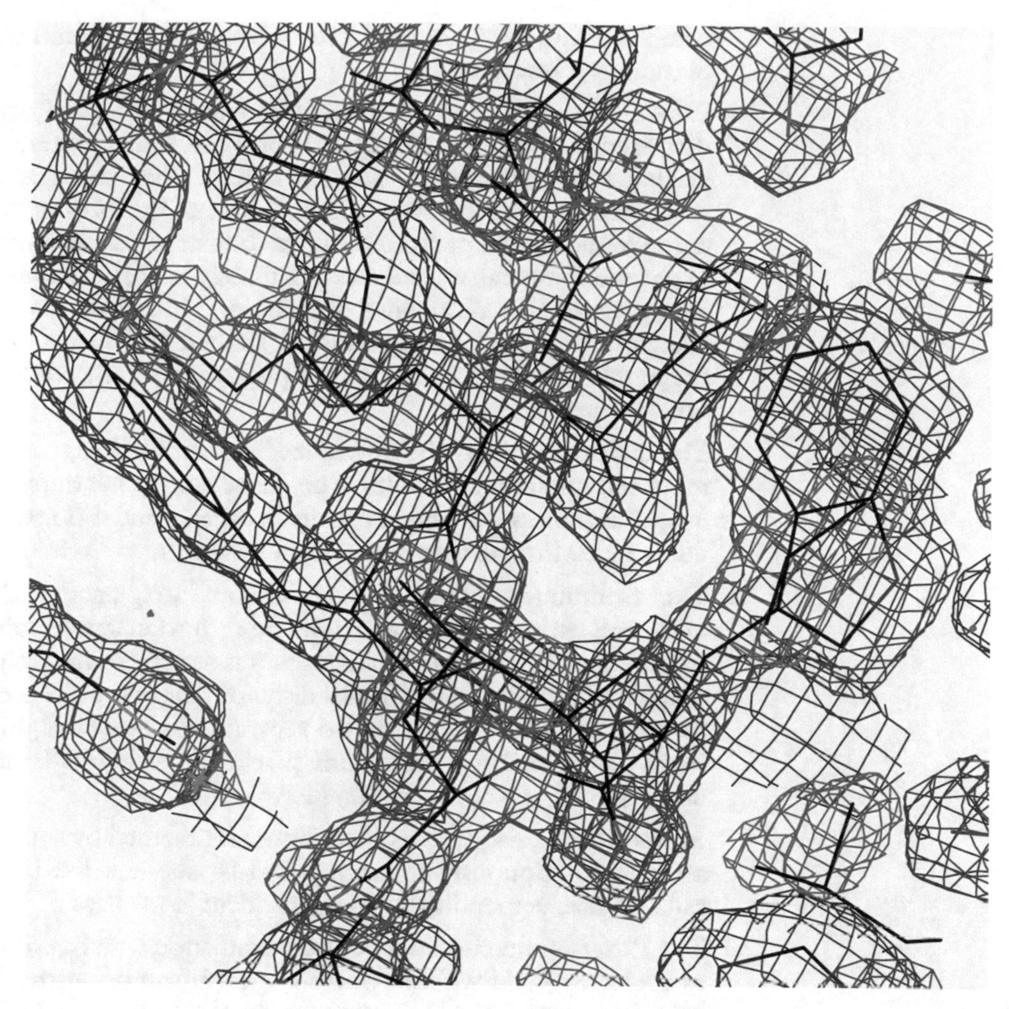

Fig. 5 Electronic density map contoured at 1.0 σ of Dps from *Thermosynechococcus elongatus* (*see* **Note 25**) calculated after many REFMAC5 refinement cycles. The final structure (*thick lines*) solved at 1.8 Å resolution is superimposed on the map

by the radius of the sphere of the reflection or "Ewald sphere," of which the radius is $1/\lambda$.

4. The imaging plate detector is formed by photosensitive plate, made of BaFBr:Eu. When hit by a radiation, the plate produces a latent image that can be excited by a laser operating at 633 nm, which generates 390 nm radiation corresponding to the fluorescence transition of Europium. This radiation is collected in the photomultiplier and converted to an electric signal.
5. The CCD camera (Charged Coupled Device) is another kind of area detector. The detector surface is constituted by voltage-sensitive elements (pixels). They have a high dynamic range,

combined with excellent spatial resolution, low noise, and high maximum count rate.

6. PILATUS detectors are two-dimensional hybrid pixel array detectors, formed by millions of hybrid pixels that operate in single-photon counting mode [28]. Each pixel comprises a preamplifier, a comparator, and a counter. The preamplifier enforces the charge generated in the counter by the incoming X-ray; the comparator produces a digital signal if the incoming charge exceeds a predefined threshold. Thus, this kind of detectors allows a complete digital storage and read-out of the number of detected X-rays per pixel without any read-out noise or dark current.
7. The dynamic range is the difference between the image maximum and minimum densities. The image density measures the image brightness and ranges usually between 0 and 4 (logarithmic scale). More density means less brightness.
8. The resolution is defined as the minimum interplanar spacing of the real lattice for the corresponding reciprocal lattice points (reflections) that are being measured. It is directly related to the optical definition, the minimum distance that two objects can be apart and still be seen as two separate objects. Thus, high resolution means low minimum spacing. Resolution is normally quoted in Ångstroms (Å) ($1\ \text{Å} = 10^{-10}$ m).
9. An oscillation image (also called frame) is obtained by rotating a crystal continuously through 0.1–1.0° about a fixed axis, called ϕ axis, perpendicular to the incident X-ray beam.
10. The Patterson function is a Fourier summation with intensities as coefficients and without phase angles. It can be written as: $P(u, v, w) = \Sigma|F(hkl)|^2 \cos 2\pi(hu + kv + lw)$. Further, it can be demonstrated that the Patterson function can be alternatively written as the self-convolution of the electronic density: $P(u, v, w) = \int_r \rho(r)\rho(r + u)dr$.
11. Macromolecules in crystals are not static. Atoms vibrate around an equilibrium position and, as a consequence, the intensities of the diffracted beams are weakened. This phenomenon is expressed by the temperature factor $\mathrm{B} = 8\pi^2 \times u^2$ where "u" is the mean square displacement of atoms around the atomic positions.
12. The maximum likelihood method involves determining the parameters of an assumed model that maximize the likelihood of the data. Thus, the most appropriate value for each variable (for example bond distances, angles, etc.) is that which maximizes the probability of observing the measured values.
13. Protein crystals are affected by lattice defects. Therefore, they are formed by different mosaic blocks with slightly different

orientations. As an ideal single crystal has a mosaicity equal to 0°, a good quality protein crystal should have a low mosaicity (between 0.1° and 0.5°).

14. Table 1 shows the output of the program Denzo after auto-indexing. In this table, all the 14 possible Bravais lattices are listed from the highest symmetry (primitive cubic) to the lowest (primitive triclinic) and allow the identification of the crystal lattice. After the lattice name, the table displays a percentage value that represents the amount of distortion that unit-cell parameters would suffer in order to fit the lattice. Next to this percentage, the "distorted-to-fit" unit-cell parameters are listed. Below these values, the undistorted unit-cell parameters are shown for comparison. The goal of the autoindexing procedure is to find the highest symmetry lattice that fits the data with minimal distortion. In the example shown in Table 1, the crystal lattice is primitive hexagonal, since 0.22 % is an acceptable amount of distortion, especially given that the unit-cell parameters were refined from a single frame. The crystal lattice should be confirmed by the overall Denzo data reduction and Scalepack scaling procedure.

15. χ^2 is a parameter related to the ratio between intensity and its standard deviation σ for all measurements and its value should be around 1. The χ^2 is mathematically represented by the following equation:

$$\chi^2 = \frac{\Sigma_{ij}\left(\left|I_{ij}(hkl) - \langle I_i(hkl)\rangle\right|\right)^2}{\sigma i^2 \frac{N}{N-1}}$$

where hkl are the Miller indices and N indicates the number of observations.

16. R_{sym} (or R_{merge}) is the parameter used to compare the intensity (I) of symmetry-related reflections for n independent observations:

$$R_{sym} = \frac{\sum_{hkl}\sum_{i}\left|Ii(hkl) - \overline{I(hkl)}\right|}{\sum_{hkl}\sum_{i} Ii(hkl)}.$$

The index i indicates the experimental observations of a given reflection. $\overline{I(hkl)}$ is the average intensity for symmetry-related observations.

17. CC(1/2) is the Pearson's correlation coefficient between two random half datasets, X and Y, and is equal to the covariance (a measure of how much two random variables change together) divided by the product between the two standard deviations. According to Karplus [29], this parameter is a quantity useful for assessing data quality and for defining high resolution cutoff in crystallography:

$$\rho_{X,Y} = \frac{cov(X,Y)}{\sigma_X \sigma_Y}$$

18. To avoid model bias often the model is transformed into a poly-Ala search probe. Only the coordinates of the polypeptide backbone and of Cβ atoms are conserved, whereas the side chains atoms are deleted.

19. The MolRep solutions represent the highest superposition peaks between the experimental Patterson function and the Patterson function calculated from the search probe, rotated and translated in the real cell.

20. Correlation coefficient value (CC_f) lies between 0 and 1 and measures the agreement between the structure factors calculated from the rotated and translated model and the observed structure factors. The correlation coefficient is calculated by REFMAC5 using the following formula:

$$CC_f = \frac{\left[\sum_{hkl} (|F_{obs}||F_{calc}|) - (\langle|F_{obs}|\rangle\langle|F_{calc}|\rangle)\right]}{\left[\sum_{hkl}\left(F_{obs}{}^2 - \langle F_{obs}\rangle^2\right)\left(\sum_{hkl}\left(F_{calc}{}^2 - \langle F_{calc}\rangle^2\right)\right)\right]^2}$$

21. Figure of merit. The "figure of merit" m is:

$$m = \frac{|F(hkl)\text{best}|}{|F(hkl)|}$$

where:

$$F(hkl)\text{best} = \frac{\sum_{\alpha} P(\alpha) F_{hkl}(\alpha)}{\sum_{\alpha} P(\alpha)},$$

$P(\alpha)$ is the probability distribution for the phase angle α and F_{hkl}(best) represents the best value for the structure factors. The m value is between 0 and 1 and is a measure of the agreement between the structure factors calculated on the basis of the model and the observed structure factors. If the model is correct the figure of merit approaches 1.

22. Noncrystallographic symmetry (NCS) occurs when the asymmetric unit is formed by two or more identical subunits. The presence of this additional symmetry could help to improve the initial phases and obtain interpretable maps for model building using the so-called density modification techniques [30].

23. Usually, the sequence region that contains the largest number of aromatic residues is chosen to start the model building. The aromatic residues (especially tryptophan) contain a high

number of electrons and display an electronic density shape that is easily identifiable (Fig. 4).

24. All the programs for model building are provided with functions that identify the maxima in the $F_o - F_c$ map above a given threshold (usually 3σ is used) and place the water molecules at the maxima peaks.

25. DpsTe. DpsTe is a member of the Dps family of proteins (DNA binding proteins from starved cells). DpsTe has been isolated and purified from the cyanobacterium *Thermosynechococcus elongatus*. The structure has been solved by Molecular Replacement at 1.81 Å resolution and has been deposited in the Protein Data bank with the accession number 2C41 [31].

References

1. Ducruix A, Giegè R (1992) In: Rickwood D, Hames BD (eds) Crystallization of nucleic acids and proteins. Oxford University Press, New York, pp 7–10
2. Hahn T (ed) (2002) International table of crystallography. Kluwer Academic Publishers, Dodrecht
3. McPherson AJ (1990) Current approach to macromolecular crystallization. Eur J Biochem 189:1–23
4. Matthews BW (1968) Solvent content of protein crystals. J Mol Biol 33:491–497
5. Friedrich W, Knipping P, Laue M (1981) In: Glusker JP (ed) Structural crystallography in chemistry and biology. Hutchinson & Ross, Stroudsburg, PA, pp 23–39
6. Bragg WL, Bragg WH (1913) The structure of crystals as indicated by their diffraction of X-ray. Proc R Soc Lond 89:248–277
7. Chothia C, Lesk AM (1986) The relation between the divergence of sequence and structure in proteins. EMBO J 5:823–826
8. Berman HM, Westbrook J, Feng Z, Gilliland G, Bhat TN, Weissig H, Shindyalov IN, Bourne PE (2000) The Protein Data Bank. Nucleic Acids Res 28:235–242
9. Crowther RA (1972) In: Rossmann MG (ed) The molecular replacement method. Gordon & Breach, New York, pp 173–178
10. Hendrickson WA (1985) Stereochemically restrained refinement of macromolecular structures. Methods Enzymol 115:252–270
11. Brunger AT, Adams PD, Rice LM (1999) Annealing in crystallography: a powerful optimization tool. Prog Biophys Mol Biol 72:135–155
12. Skarina T, Xu X, Evdokimova E, Savchenko A (2014) High-throughput crystallization screening. Methods Mol Biol 1140:159–168
13. Hui R, Edwards A (2003) High-throughput protein crystallization. J Struct Biol 142:154–161
14. Rodgers DW, Rodgers DW (1994) Cryocrystallography. Structure 2:1135–1140
15. Mueller M, Wang M, Schulze Briese C (2012) Optimal fine φ-slicing for single-photon-counting pixel detector. Acta Crystallogr D Biol Crystallogr D68:42–56
16. Otwinoski Z, Minor W (1997) Processing of X-ray diffraction data collected in oscillation mode. Methods Enzymol 276:307–326
17. Kabsch W (2010) XDS. Acta Crystallogr D Biol Crystallogr D66:125–132
18. CCP4 (Collaborative Computational Project, number 4) (1994) The CCP4 suite: programs for protein crystallography. Acta Crystallogr D50:760–763
19. French GS, Wilson KS (1978) On the treatment of negative intensity observations. Acta Crystallogr A34:517–525
20. Altshul SF, Koonin EV (1998) Iterated profile searches with PSI-BLAST—a tool for discovery in protein databases. TIBS 23:444–447
21. Navaza G (1994) AMORE: an automated package for molecular replacement. Acta Crystallogr A50:157–163
22. McCoy AJ, Grosse-Kunstleve RW, Adams PD, Winn MD, Storoni LC, Read RJ (2007) Phaser crystallographic software. J Appl Crystallogr 40:658–674
23. Vagin A, Teplyakov A (1997) MOLREP: an automated program for molecular replacement. J Appl Crystallogr 30:1022–1025
24. Brünger AT, Adams PD, Clore GM, DeLano WL, Gros P, Grosse-Kunstleve RW, Jiang JS, Kuszewski J, Nilges M, Pannu NS, Read RJ, Rice LM, Simonson T, Warren GL (1998) Crystallography & NMR system: a new

software suite for macromolecular structure determination. Acta Crystallogr D54:905–921

25. Murshudov GN, Vagin AA, Dodson EJ (1997) Refinement of macromolecular structures by the maximum-likelihood method. Acta Crystallogr D53:240–255
26. Adams PD, Afonine PV, Bunkóczi G, Chen VB, Davis IW, Echols N, Headd JJ, Hung L-W, Kapral GJ, Grosse-Kunstleve RW, McCoy AJ, Moriarty NW, Oeffner R, Read RJ, Richardson DC, Richardson JS, Terwilliger TC, Zwart PH (2009) *PHENIX*: a comprehensive Python-based system for macromolecular structure solution. Acta Crystallogr D66:213–221
27. Emsley P, Cowtan K (2004) Coot: model-building tools for molecular graphics. Acta Crystallogr D60:2126–2132
28. Kraft P, Bergamaschi A, Broennimann C, Dinapoli R, Eikenberry EF, Henrich B, Johnson I, Mozzanica A, Schleputz CM, Willmott PR, Schmitt B (2009) Performance of single-photon-counting PILATUS detector modules. J Synchrotron Radiat 16:368–375
29. Karplus PA, Diederichs K (2012) Linking crystallographic model and data quality. Science 336:1030–1033
30. Rossmann MG, Blow DM (1962) The detection of subunits within the crystallographic asymmetric unit. Acta Crystallogr 15:24–31
31. Franceschini S, Ceci P, Alaleona F, Chiancone E, Ilari A (2006) Antioxidant Dps protein from the thermophilic cyanobacterium Thermosynechococcus elongatus. FEBS J 273:4913–4928
32. Bellapadrona G, Chiaraluce R, Consalvi V, Ilari A, Stefanini S, Chiancone E (2007) The mutations Lys 114 → Gln and Asp 126 → Asn disrupt an intersubunit salt bridge and convert *Listeria innocua* Dps into its natural mutant Listeria monocytogenes Dps. Effects on protein stability at Low pH. Proteins 66 (4):975–983

Chapter 4

Managing Sequence Data

Christopher O'Sullivan, Benjamin Busby, and Ilene Karsch Mizrachi

Abstract

Nucleotide and protein sequences are the foundation for all bioinformatics tools and resources. Researchers can analyze these sequences to discover genes or predict the function of their products. The INSDC (International Nucleotide Sequence Database—DDBJ/ENA/GenBank + SRA) is an international, centralized primary sequence resource that is freely available on the Internet. This database contains all publicly available nucleotide and derived protein sequences. This chapter discusses the structure and history of the nucleotide sequence database resources built at NCBI, provides information on how to submit sequences to the databases, and explains how to access the sequence data.

Key words Sequence database, GenBank, SRA, INSDC, RefSeq, Next generation sequencing

1 Structure and History of Sequence Databases at NCBI

The National Center for Biotechnology Information (NCBI) is responsible for building and maintaining the sequence databases Sequence Read Archive (SRA), GenBank, and RefSeq. GenBank and SRA are primary archival resources that collect data from researchers as part of the publication process. RefSeq is a secondary database built from data submitted to the primary archives but with added curation. GenBank and SRA are part of the International Nucleotide Sequence Database Collaboration (INSDC) a centralized public sequence resource that encompasses raw sequence reads, assembled sequences and derived annotations.

1.1 INSDC

The INSDC [1] is a partnership between GenBank/SRA at NCBI in the USA [2] https://www.ncbi.nlm.nih.gov/, the European Nucleotide Archive (ENA) at EMBL-EBI in Europe [3] http://www.ebi.ac.uk/ena, and DNA Databank of Japan (DDBJ) in Japan [4] http://www.ddbj.nig.ac.jp/. For more than 25 years, the three partners have maintained this active, successful collaboration for building the nucleotide sequence databases. As new sequencing technologies have emerged, additional archives have been

Jonathan M. Keith (ed.), *Bioinformatics: Volume I: Data, Sequence Analysis, and Evolution*, Methods in Molecular Biology, vol. 1525, DOI 10.1007/978-1-4939-6622-6_4,

incorporated to store raw and aligned sequence read data. Representatives from the three databases meet annually to discuss technical and biological issues affecting the databases. Ensuring that sequence data from scientists worldwide is freely available to all is the primary mission of this group. As part of the publication process, scientists are required to deposit sequence data in a public repository; the INSDC encourages publishers of scientific journals to enforce this policy to ensure that sequence data associated with a paper is freely available from an international resource for research and discovery. For example, in 2011, analysis of genomic sequence data in INSDC led to the identification of the enterohemorrhagic *Escherichia coli* that caused numerous deaths in Germany [5].

The INSDC website, http://www.insdc.org, contains links to the member sites, data release policy, the Feature Table Document (which outlines features and syntax for sequence records), lists of controlled vocabularies and procedures that are important for the collaborators, data submitters and database users.

Though each of the three INSDC partners has its own set of tools for submission and retrieval, data is exchanged regularly so that all content is made available at each of the member sites.

The following table (Table 1) from the INSDC website contains links for retrieval of data at the three partner sites (*see* **Note 1**).

In this chapter, we discuss NCBI's contribution to INSDC, including information about submission processing and data usage. Similar tools and processes can be found at the other partner sites.

1.2 SRA/GEO

The NCBI SRA stores raw sequencing data and alignment information from high-throughput sequencing platforms, including Illumina, Applied Biosystems SOLiD, Complete Genomics, Pacific Biosciences SMRT, Nanopore, Ion Torrent, and Roche 454. The

Table 1
Links for data retrieval from DDBJ, EMBL-EBI, and NCBI

Data type	DDBJ	EMBL-EBI	NCBI
Next generation reads	Sequence Read Archive	European Nucleotide Archive (ENA)	Sequence Read Archive
Capillary reads	Trace Archive	European Nucleotide Archive (ENA)	Trace Archive
Annotated sequences	DDBJ	European Nucleotide Archive (ENA)	GenBank
Samples	BioSample	European Nucleotide Archive (ENA)	BioSample
Studies	BioProject	European Nucleotide Archive (ENA)	BioProject

data stored in the NCBI SRA are suitable for the reanalysis of data that supports publications as well as data that supports assembly, annotation, variation, and expression data submissions to other NCBI archives. As of October 2016, SRA contains more than 9 PetaBases (9×10^{15} bases) from nearly 49,000 different source organisms and metagenomes. Approximately half of the total is controlled-access human clinical sequence data supporting dbGaP studies. SRA growth is explosive, doubling approximately every 12 months. Statistics are updated daily and available from the SRA home page (https://trace.ncbi.nlm.nih.gov/Traces/sra/). SRA stores descriptive metadata and sequence data separately using distinct accession series: the SRA Experiment and Run. The SRA Experiment record is used for search and display. It contains details describing sequence library preparation, molecular and bioinformatics workflows and sequencing instruments. The SRA Run contains sequence, quality, alignment, and statistics from a specific library preparation for a single biological sample. Every SRA Experiment references a BioSample and a BioProject.

GEO [6], The Gene Expression Omnibus, is a public functional genomics data repository supporting MIAME-compliant data submissions. An integral part of NCBI's primary data archives, GEO stores profiles of gene expression, coding and noncoding RNA, Chromatin Immunoprecipitation, genome methylation, genome variation, and SNP arrays derived from microarrays and next-generation sequencing. Raw sequence data submitted in support of GEO profiles is stored in SRA. The profiles and underlying sequence data are linked via BioSample and BioProject references.

1.3 Bioproject

A BioProject is a collection of biological data related to a single initiative, such as a grant, manuscript, consortia project or other collection. A BioProject record provides users a single place to find links to the diverse data types generated for that project (Fig. 1). For instance, a multiisolate genome sequencing project for a foodborne pathogen *Salmonella enterica* subsp. enterica serovar Enteritidis contains links to the SRA reads, the genome assemblies, the BioSamples, and the publications.

1.4 BioSample

The BioSample database contains descriptive information about biological source materials from which data stored in INSDC sequence archives are derived. BioSample records describe cell lines, whole organisms, and environmental isolates using structured and consistent attribute names and values. The information is important to provide context to derived data to facilitate analysis and discovery. It also acts to aggregate and integrate disparate data sets submitted to any number of resources.

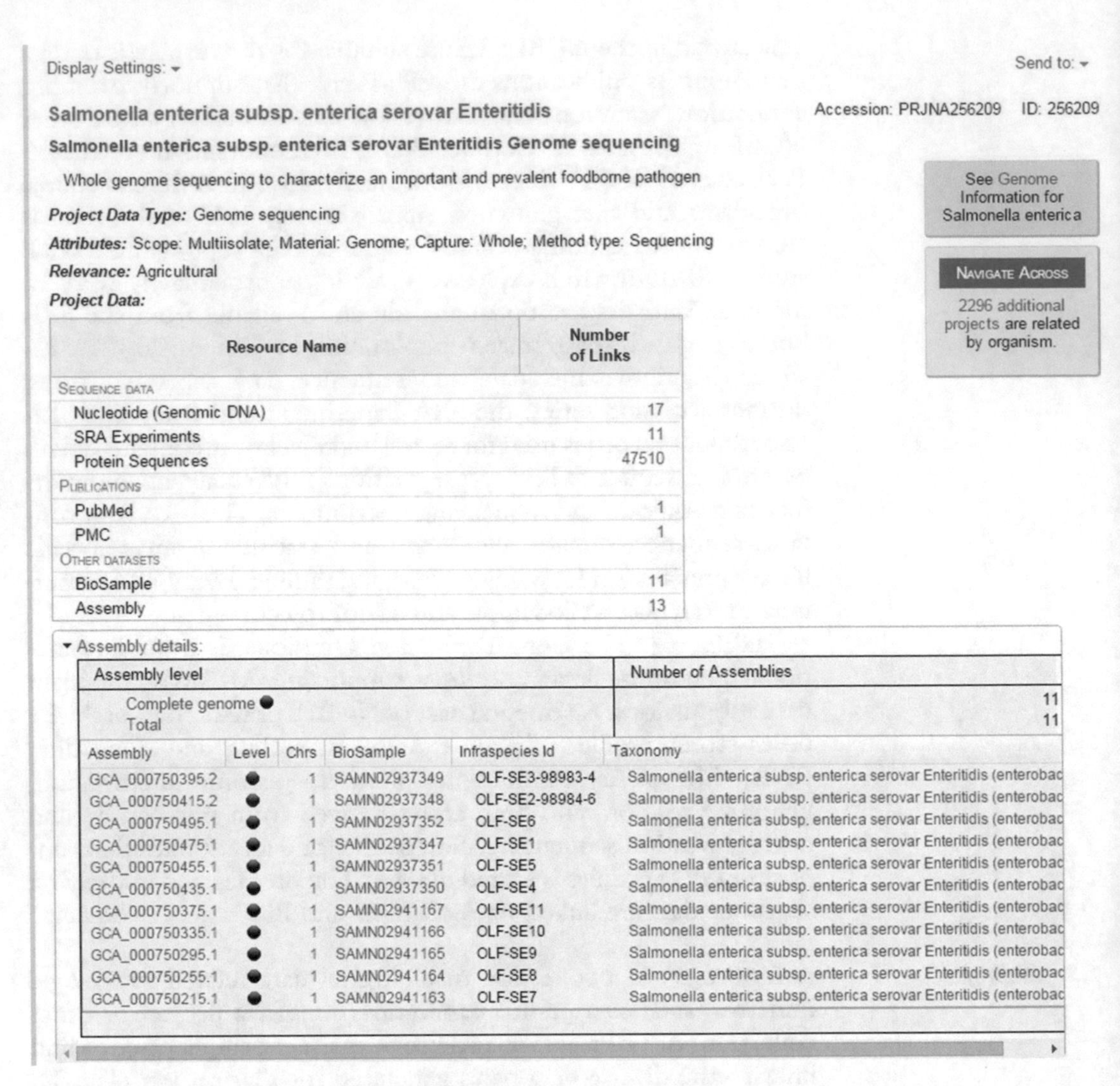

Fig. 1 BioProject report from Entrez. This report provides information about an initiative for sequencing *Salmonella enterica*, a common foodborne pathogen. The report includes links to the GenBank assemblies, the SRA reads, the BioSamples, and the publications. There are also links to other related projects

1.5 GenBank

GenBank [2] is a comprehensive public database of nucleotide sequences and supporting bibliographic and biological annotation. GenBank is maintained and distributed by the NCBI, a division of the National Library of Medicine (NLM) at the US National Institutes of Health (NIH) in Bethesda, MD. NCBI builds GenBank from several sources including the submission of sequence data from individual researchers and from the bulk submission of whole genome shotgun (WGS), transcriptome shotgun assembly (TSA) and other high-throughput data from sequencing centers. GenBank doubles in size every 18 months. In October 2016, Release 216 contained over 220 billion bases in 197 million sequences. WGS and TSA projects contribute another terabase of sequence data. GenBank contains sequences from over 373,000 named species. Full releases of GenBank are made every 2 months beginning in the middle of February each year. Between full releases, daily updates are provided on the NCBI FTP site. Detailed statistics for the current release can be found in the GenBank release notes ftp:/ftp.ncbi.nlm.nih.gov/genbank/README.genbank.

1.6 Genomes

The first complete microbial genome was submitted to GenBank in 1995 [7]. Since then, more than 151,000 cellular genomes have been released into the public archive. These genomes are derived from organisms from all branches of the tree of life. Microbial genomes, those from bacteria, archaea and fungi, are relatively small in size compared with their eukaryotic counterparts, ranging from hundreds of thousands to millions of bases. Nonetheless, these genomes contain thousands of genes, coding regions, and structural RNAs. Many of the genes in microbial genomes have been identified only by similarity to other genomes and their gene products are often classified as hypothetical proteins. Each gene within a genome is assigned a locus_tag: a unique identifier for a particular gene in a particular genome. Since the function of many of the genes is unknown, locus_tags have become surrogate gene identifiers. Submitters of prokaryote genomes are encouraged to use the NCBI Prokaryotic Genome Annotation Pipeline [8] to annotate genome submissions. This service provides standardized annotation for genomes, which can easily be compared across genomes. This pipeline is also used to annotate prokaryotic genomes for RefSeq.

The first genome sequences were built by sequencing overlapping clones and then assembling the genome by overlap. Many bacterial genomes and the first human genome were sequenced in this manner. In 2001, a new approach for sequencing complete genomes was introduced: Whole Genome Shotgun (WGS) data are generated by breaking the genome into random fragments for sequencing and then computationally assembling them to form contigs which can be assembled into larger structures called scaffolds. All of the contig sequences from a single genome assembly, along with the instructions for building scaffolds, are submitted together as a single WGS project. WGS has become such a dominant sequencing technique that more nucleotides of WGS sequence have been added to INSDC than from all of the other divisions since the inception of the database.

For each project, a master record is created that contains information that is common among all the records of the sequencing projects, such as the biological source, submitter information, and publication information. Each master record includes links to the range of accession numbers for the individual contigs in the assembly and links to the range of accessions for the scaffolds from the project. The NCBI Assembly resource (https://www.ncbi.nlm.nih.gov/assembly/) stores statistics about each genome assembly. Each genome assembly is also linked to a BioProject and BioSample.

In response to the anticipated rapid growth in the submission of highly redundant prokaryotic genome sequences from clinical samples and other large sequencing projects, a new model for prokaryotic protein sequences has been employed at NCBI in which a single nonredundant protein record is used for all identical protein sequences annotated on different genome records. The accession numbers for these nonredundant proteins begins with WP. In Entrez Protein, there is a link to an Identical Protein report (Fig. 2) that displays the database source, chromosomal localization, protein accession, protein name, organism, and other taxonomic information for this protein. At the time of this writing only RefSeq bacterial genomes are being annotated with the nonredundant proteins but in the future, these will be used for GenBank annotation, as well. However, the identical protein report encompasses both GenBank and RefSeq genomes.

1.7 Metagenomes and Environmental Sample Sequencing

The microbial biodiversity of an environment can be discerned by studying the genome and transcriptome sequences of the microorganisms living in that environment. This field of study is known as metagenomics. Sequencing of 16S ribosomal RNA is a popular way to detect bacterial and archaeal species present in a community.

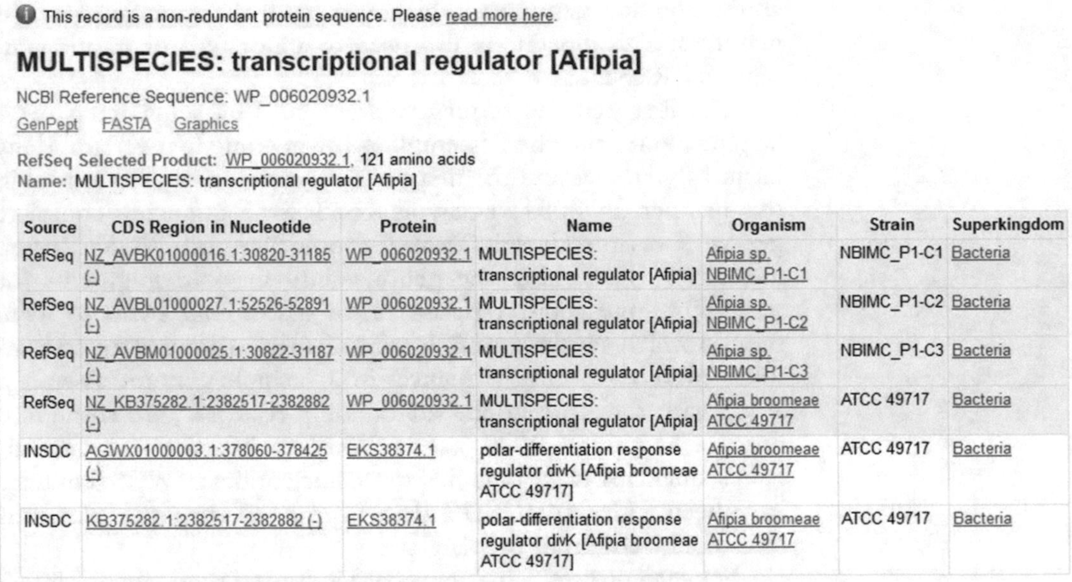

This record is a non-redundant protein sequence. Please read more here.

MULTISPECIES: transcriptional regulator [Afipia]

NCBI Reference Sequence: WP_006020932.1

GenPept FASTA Graphics

RefSeq Selected Product: WP_006020932.1, 121 amino acids
Name: MULTISPECIES: transcriptional regulator [Afipia]

Source	CDS Region in Nucleotide	Protein	Name	Organism	Strain	Superkingdom
RefSeq	NZ_AVBK01000016.1:30820-31185 (-)	WP_006020932.1	MULTISPECIES: transcriptional regulator [Afipia]	Afipia sp. NBIMC_P1-C1	NBIMC_P1-C1	Bacteria
RefSeq	NZ_AVBL01000027.1:52526-52891 (-)	WP_006020932.1	MULTISPECIES: transcriptional regulator [Afipia]	Afipia sp. NBIMC_P1-C2	NBIMC_P1-C2	Bacteria
RefSeq	NZ_AVBM01000025.1:30822-31187 (-)	WP_006020932.1	MULTISPECIES: transcriptional regulator [Afipia]	Afipia sp. NBIMC_P1-C3	NBIMC_P1-C3	Bacteria
RefSeq	NZ_KB375282.1:2382517-2382882 (-)	WP_006020932.1	MULTISPECIES: transcriptional regulator [Afipia]	Afipia broomeae ATCC 49717	ATCC 49717	Bacteria
INSDC	AGWX01000003.1:378060-378425 (-)	EKS38374.1	polar-differentiation response regulator divK [Afipia broomeae ATCC 49717]	Afipia broomeae ATCC 49717	ATCC 49717	Bacteria
INSDC	KB375282.1:2382517-2382882 (-)	EKS38374.1	polar-differentiation response regulator divK [Afipia broomeae ATCC 49717]	Afipia broomeae ATCC 49717	ATCC 49717	Bacteria

Fig. 2 The identical protein report that can be accessed from Entrez Protein (https:/www.ncbi.nlm.nih.gov/protein) record. It is a tabular display of all of the protein sequences in GenBank with the identical sequence. It includes links to coding regions, the genome records, the protein product name as it is cited in the genome record and the source organisms that have this protein

Alternatively, one can sequence the whole metagenome or whole transcriptome of the environmental sample and assemble genomes and transcripts from the sample to understand the diversity in the environment. Clustered and assembled 16S rRNA, in addition to assembled metagenome and metatranscriptome sequences, can be submitted to GenBank. It is preferable that the raw reads are submitted to SRA as well. All of the datasets from a single environment, such as mouse gut or seawater, should cite a single BioProject so that the entirety of the project will be detectable to users simultaneously from the NCBI BioProject Database.

RNA sequence analysis or transcriptome analysis is an important mechanism for studying gene expression of a particular organism or tissue type. The transcriptome provides insight into the particular genes that are expressed in specific tissues, disease states or under varied environmental conditions. The transcriptome can also be used to map genes and understand their structure in the genome. Computationally assembled transcript sequences should be deposited into TSA and the underlying sequence reads in SRA so that researchers can use this data to make important scientific discoveries. The structure of TSA records is similar to WGS with a similar accessioning scheme, a master record and sequence overlap contigs.

1.8 The GenBank Sequence Record

A GenBank sequence record is most familiarly viewed as a flat file where the data and associated metadata is structured for human readability. The format specifications are as follows.

1.8.1 Definition, Accession and Organism

The top of a GenBank flat file is shown in Fig. 3.

The first token of the LOCUS field is the locus name. At present, this locus name is the same as the accession number, but in the past, more descriptive names were used. For instance, HUMHBB is the locus name for the human beta-globin gene in the record with accession number U01317. This practice was abandoned in the mid-1990s when the number of sequences deposited had increased to a point where it was too cumbersome to generate these manually. Following the locus name is the sequence length, the molecule type, the topology (linear or circular), the division

```
LOCUS       KM527068                 963 bp    DNA     linear   BCT 13-DEC-2014
DEFINITION  Salinispora tropica strain CNS197 Cas1 protein I-E (cas1) gene,
            complete cds.
ACCESSION   KM527068
VERSION     KM527068.1
KEYWORDS    .
SOURCE      Salinispora tropica
  ORGANISM  Salinispora tropica
            Bacteria; Actinobacteria; Micromonosporales; Micromonosporaceae;
            Salinispora.
```

Fig. 3 The top of a GenBank flat file

code and the date of last modification. Records are assigned to divisions based on the source taxonomy or the sequencing strategy that was used (*see* **Note 2** for list of GenBank divisions). The DEFINITION line gives a brief description of the sequence including information about the source organism, gene(s) and molecule information. The ACCESSION is the database-assigned identifier, which has one of the following formats: two letters and six digits (e.g., KM123456) or one letter and five digits for INSDC records; four letters and eight digits for WGS and TSA records (e.g., ABCD01012345); and two letters, an underscore, and six to eight digits for RefSeq records (e.g., NM_123456). The VERSION line contains the accession number and sequence version.

WGS and TSA projects are assigned a four-letter project ID which serves as the accession prefix for that project. This is followed by a two-digit assembly version, and a six to eight-digit contig id. For example, the *Neurospora crassa* genome is stored in project accession AABX, the first version of the genome assembly is AABX01000000 and AABX01000111 is the 111th contig.

Historically, the KEYWORD field in the GenBank record was used as a summary of the information present in the record. It was a free text field and may have contained gene name, protein name, tissue localization, etc. Information placed on this line was more appropriately placed elsewhere in the sequence record. GenBank strongly discourages the use of keywords to describe attributes of the sequence. Instead, GenBank uses a controlled vocabulary on the KEYWORD line to describe different submission types or divisions. The list of INSDC sanctioned keywords is available at http://www.insdc.org/documents/methodological-keywords.

The SOURCE and ORGANISM lines contain the taxonomic name and the taxonomic lineage, respectively, for that organism.

1.8.2 Reference Section

The next section of the GenBank flat file contains the bibliographic and submitter information (Fig. 4):

The REFERENCE section contains published and unpublished references. Many published references include a link to a PubMed ID number that allows users to view the abstract of the cited paper in PubMed. The last REFERENCE cited in a record reports the names of submitters of the sequence data and the location where the work was done.

The COMMENT field may have submitter provided comments about the sequence or a table that contains structured metadata for the record. This example has sequencing and assembly methodology but there are other structured comments with isolation source or other phenotypic information. In addition, if the sequence has been updated, then the COMMENT will have a link to the previous version.

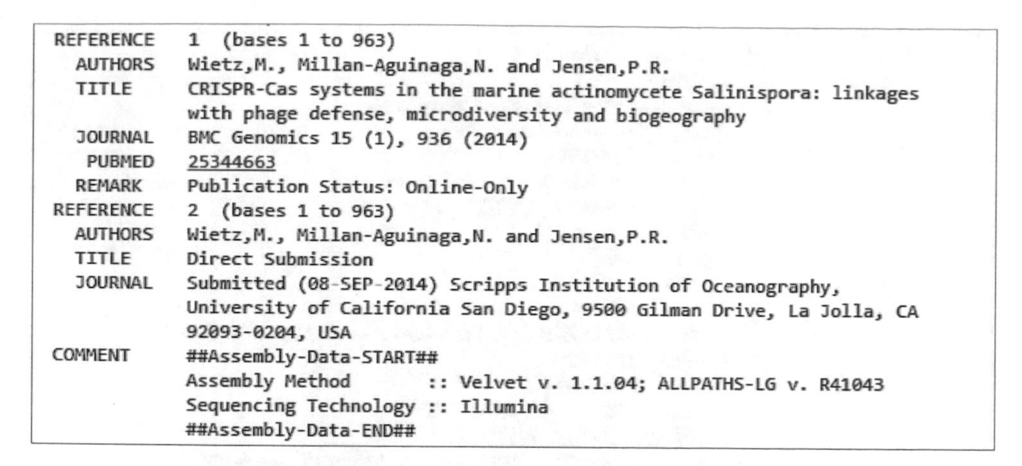

```
REFERENCE   1  (bases 1 to 963)
  AUTHORS   Wietz,M., Millan-Aguinaga,N. and Jensen,P.R.
  TITLE     CRISPR-Cas systems in the marine actinomycete Salinispora: linkages
            with phage defense, microdiversity and biogeography
  JOURNAL   BMC Genomics 15 (1), 936 (2014)
   PUBMED   25344663
  REMARK    Publication Status: Online-Only
REFERENCE   2  (bases 1 to 963)
  AUTHORS   Wietz,M., Millan-Aguinaga,N. and Jensen,P.R.
  TITLE     Direct Submission
  JOURNAL   Submitted (08-SEP-2014) Scripps Institution of Oceanography,
            University of California San Diego, 9500 Gilman Drive, La Jolla, CA
            92093-0204, USA
COMMENT     ##Assembly-Data-START##
            Assembly Method       :: Velvet v. 1.1.04; ALLPATHS-LG v. R41043
            Sequencing Technology :: Illumina
            ##Assembly-Data-END##
```

Fig. 4 Reference and comment sections of a GenBank flat file

1.8.3 Features and Sequence

The FEATURES section contains a source feature, which has additional information about the source of the sequence and the organism from which the DNA was isolated. There are approximately 50 different standard qualifiers that can be used to describe the source. Some examples are /strain, /chromosome, and /host. Following the source feature are annotations that describe the sequence, such as gene, CDS (coding region), mRNA, rRNA, variation, and others. Like the source feature, other features can be further described with feature-specific qualifiers. For example, a mandatory qualifier for the CDS feature is a /translation which contains the protein sequence. Following the Feature section is the nucleotide sequence itself. The specification for the Feature Table can be found at http://www.insdc.org/documents/feature-table (*see* **Note 3**). An example is included as Fig. 5.

1.9 Updates and Maintenance of the Database

An INSDC record can be updated by the submitter any time new information is acquired. Updates can include: adding new sequence, correcting existing sequence, adding a publication, or adding new annotation. Information about acceptable update formats can be found at https://www.ncbi.nlm.nih.gov/genbank/update. The new, updated record replaces the older one in the database and in the retrieval and analysis tools. However, since GenBank is archival, a copy of the older record is maintained in the database. The sequence in the GenBank record is versioned. For example, KM527068.1 is the first version of the sequence in GenBank record KM527068. When the sequence is modified or updated, the accession version gets incremented. So the accession in the sample record will become KM527068.2 after a sequence update. The base accession number does not change as this is a stable identifier for this record. A COMMENT is added to the updated GenBank flat file that indicates when the sequence is

```
FEATURES             Location/Qualifiers
     source          1..963
                     /organism="Salinispora tropica"
                     /mol_type="genomic DNA"
                     /strain="CNS197"
                     /isolation_source="marine sediment"
                     /db_xref="taxon:168695"
     gene            1..963
                     /gene="cas1"
     CDS             1..963
                     /gene="cas1"
                     /note="CRISPR/Cas system-associated protein Cas1"
                     /codon_start=1
                     /transl_table=11
                     /product="Cas1 protein I-E"
                     /protein_id="AIZ06592.1"
                     /translation="MSTSAQRRLAAPTLAMLPRVADSLSFLYADIVRIVQDDTGVLAQ
                     VDTTKGTERVYLPTAALSCLLLGPGTSITHHALSTLARHGTTVVCVGSGVVRCYAGIT
                     PTSLTTNWLEKQARCWADDNTRLQVAVRMYEHRFGEAVPEGTTLAQLRGMEGQRMKVL
                     YRLLAQKYRTGKFRRNYDPSKWDTQDPVNLALSAASACLYGVVHAVVLALGCSPALGF
                     VHSGTQHAFVYDIADLYKAKVTVPLAFAMSTSAQPERDVRRKLCDDFRLLKLMPTIVT
                     DIQRLLDPDSTPKQRQPVAEVTALWDPEMGAMPSGVNYSSDPWD"
ORIGIN
        1 atgagcacca gcgcccagcg gcgactcgcc gcaccgaccc tggccatgct gccccgcgtg
       61 gcggactcgc tcagcttcct ctacgccgac atcgttcgga tcgtccaaga cgacaccgga
      121 gtcctcgccc aggtcgacac aaccaagggg accgaacgcg tctacctacc caccgccgcc
      181 ctgagttgcc ttctcctcgg acccggcacc tcgatcaccc accacgccct gtccaccctc
      241 gcccgccacg gcaccaccgt cgtctgcgtc ggctccggtg tcgtccgctg ttacgccggc
      301 atcaccccca cctccctgac caccaactgg ctggaaaagc aggcccgctg ctgggccgac
      361 gacaacaccc gcctacaggt agcagtacgg atgtatgagc atcgcttcgg cgaagccgtg
      421 cccgaaggca ccacgctggc ccagcttcgt ggcatggaag gccagcgcat gaaagtgctc
      481 taccgcctgc tggcccagaa atatcgaacc ggcaaattcc gccgcaacta tgaccccagc
```

Fig. 5 Features section of a GenBank flat file

updated and provides a link to the older version of the sequence. Annotation changes do not increment the accession version. However, these changes can be detected using the Sequence Revision History, which can be accessed from the Display Setting menu while viewing a sequence record in Entrez (Fig. 6).

1.10 Pitfalls of an Archival Primary Sequence Database

1.10.1 Bad Annotation and Propagation

Because INSDC is an archival primary sequence database, submitters "own" their annotation and are primarily responsible for ensuring that it is correct. Database staff review records for accuracy and inform submitters of any problems. If entries with poor annotation appear in the database, they may be used by other scientists to validate their own data and possibly to prepare a submission to the database which can to lead to the propagation of bad data to subsequent entries in the database. During processing, the GenBank annotation staff checks the sequences and annotation for biological validity. For instance, does the conceptual translation of a coding region match the amino acid sequence provided by the submitter? Does the sequence cluster with the taxonomically related sequences? Does the submitter's description of the

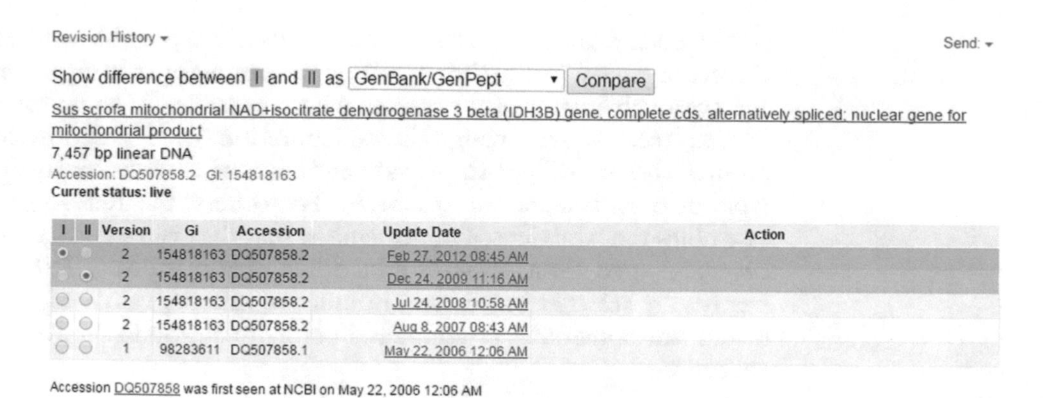

Fig. 6 Sequence Revision History page allows users to retrieve older versions of a sequence record prior to it being updated. Sequence changes are indicated by incrementing the version number. One can view the modifications that were made during an update for two versions of a sequence record by choosing a version in columns I and II and then clicking the Show button

sequence agree with the results of a BLAST similarity search? If problems are detected, the submitter is contacted to correct their submission. Since 2013, if a submitter is unable or unwilling to correct their entry, the sequences are flagged as UNVERIFIED with a comment indicating that the "GenBank staff is unable to verify source organism and sequence and/or annotation provided by the submitter." These unverified sequences are excluded from the BLAST databases. While this practice attempts to minimize problematic sequences in GenBank, there is still legacy data and submissions processed through automated pipelines that may not be flagged unverified even though they may have problems.

1.10.2 Sequence Contamination

The GenBank staff actively removes vector and linker contamination from sequence submissions when it is discovered. GenBank submissions are screened using a specialized BLAST database—UniVec (https://www.ncbi.nlm.nih.gov/tools/vecscreen/)—to detect vector contamination. Assembled genomes are also screened for contamination by sequence from other organisms, for instance, detection of stretches of human DNA in a bacterial assembly.

1.11 RefSeq

The Reference Sequence (RefSeq) database at NCBI (https://www.ncbi.nlm.nih.gov/RefSeq/) provides a comprehensive, integrated, nonredundant well-annotated set of reference sequences including genomic DNA, transcript (RNA), and protein sequences. RefSeq records are derived from INSDC records and can be a synthesis of information from a number of sources. The relationship between the primary data in GenBank and RefSeq is analogous to the relationship between a research paper and a review article. Each sequence is annotated as accurately as possible with the correct organism name, the correct gene symbol for that organism, and reasonable names for proteins where possible. In some cases, RefSeq records are created in collaboration with authoritative groups who are willing to provide annotations or links to phenotypic or organism-specific resources. For others, the RefSeq staff assembles the best view of the organism that they can put together based on data from INSDC and other public sources. INSDC records are selected based on a number of criteria, validated, corrected, and sometimes re-annotated before inclusion in the RefSeq collection.

2 Submission of Sequence Data to NCBI Archives

Submission of sequence data to INSDC is required by most journals as a condition of publication. A unified portal for the submission of all sequence related data and metadata is being developed. The Submission Portal (https://submit.ncbi.nlm.nih.gov) offers both Web wizards to guide a user through the process of submitting and a programmatic interface for the more advanced user. At the time of publication, wizards are available for the submission of SRA, genomes, transcriptomes, ribosomal RNA sequences and their associated BioProject and BioSample. Submitters create BioSample and BioProject records for SRA and GenBank submissions at the beginning of the process. While additional wizards are developed for other sequence submission types, existing submission tools, like BankIt, and Sequin for GenBank are still supported.

2.1 SRA Submissions

SRA processes multiple TeraBytes of sequence data every day. Like GenBank, SRA submissions come from a variety of sources including small labs, core facilities, and genome sequencing centers. Low to mid volume submissions are initiated via NCBI sra submission portal (https://submit.ncbi.nlm.nih.gov/subs/sra/) where submitters enter descriptions of the samples and libraries that they intend to upload. Data files may be uploaded from the browser but are typically delivered separately using ftp or Aspera FASP protocol (https://downloads.asperasoft.com/connect2/). High volume automated submission pipelines use dedicated upload accounts to deliver data files and bulk metadata submission via programmatically generated xml files.

NCBI works to help labs that do significant amounts of sequencing, covered under the Genomic Data Submission policy, comply with that policy. Submission to SRA, GEO, or dbGaP (https://www.ncbi.nlm.nih.gov/gap) or GenBank qualify as acceptable submissions. Resource specific help desks assist individuals with technical issues regarding those submissions (*see* **Note 4**).

SRA accepts a variety of file formats (*see* **Note 4**). Following best formatting practices will help prevent delays or errors during data loading. Multiple sequencer runs from a library should be split into distinct SRA Runs or submitted as distinct Read Groups in bam format in order to retain batch information. It is best to submit fastq files with the original header formatting. Modification or replacement of the systematic identifiers generated by the instrument may lead to errors or delays in Submission processing. Bam files containing reads aligned to a reference genome are the preferred submission format. Please ensure that submitted bam files have robust header information, including Program (@PG) and Read group (@RG) fields. In addition, alignments to high quality (chromosome level) genomic reference assemblies are strongly recommended. Methods for pre-submission validation of data files can be found in SRA File Format Guide and should ensure successful loading to SRA. Additional detail and the most current information on SRA submissions can be found on SRA documentation and home pages (*see* **Note 4**).

2.2 GenBank Submissions

GenBank processes thousands of new sequence submissions per month from scientists worldwide. GenBank submissions come from a variety of submitters, from small laboratories doing research on a particular gene or genes to genome centers doing high-throughput sequencing. The "small scale" submissions may be a single sequence or sets of related sequences, and with annotation. Submissions include mRNA sequences with coding regions, fragments of genomic DNA with a single gene or multiple genes, a viral or organelle complete genome or ribosomal RNA gene clusters. If part of the nucleotide sequence encodes a protein, a coding sequence (CDS) feature and resulting conceptual translation are annotated in the record. Each nucleotide and protein sequence is assigned an accession number that serves as a permanent identifier for the sequence records.

Submitters can specify that their sets of sequences that span the same gene or region of the genome are biologically related by classifying them as environmental sample, population, or phylogenetic sets. Each sequence within a set is assigned its own accession number and can be viewed independently in Entrez. However, each set is also included in Entrez PopSet (https://www.ncbi.nlm.nih.gov/popset/), allowing scientists to view the relationship among the set's sequences through an alignment.

BankIt (https://www.ncbi.nlm.nih.gov/WebSub/?tool=genbank) has a series of wizards to guide a submitter through the submission process to ensure that the appropriate data, like sequence and annotations, metadata, sample isolation information and experimental details, are submitted. Users upload files containing sequences and source information and annotations in tabular format. These web-based tools have quality assurance and validation checks that report the any problems back to the submitter for resolution.

NCBI maintains two submission tools that a user can download to their own computer to prepare their submission, Sequin (https://www.ncbi.nlm.nih.gov/Sequin/) and tbl2asn (https://www.ncbi.nlm.nih.gov/genbank/tbl2asn2). Sequin is a stand-alone application that can be used for the annotation and analysis of nucleotide sequences. It also has a series of wizards to guide users through the submission process. tbl2asn is a command-line executable that combines input sequences and tables to generate files appropriate for submission to GenBank (*see* **Note 5**). The input files necessary for submission include: nucleotide and amino acid sequences in FASTA format (*see* **Note 6**), tables for source organism information, tables for source information and feature annotation (*see* **Note 7**). For submitting multiple, related sequences (e.g., those in a phylogenetic or population study), these tools accept the output of many popular multiple sequence-alignment packages, including FASTA + GAP, PHYLIP, and NEXUS. Submitters can use the alignments to propagate annotation from a single record to the other records in the set. Prior to submission to the database, the submitter is encouraged to validate their submission and correct any errors that may exist.

Direct submissions to GenBank are analyzed by the annotation staff. The first level of review occurs before accession numbers are assigned to ensure that the submissions meet the minimal criteria. Sequences should be of a minimal length, sequenced by, or on behalf of, the group submitting the sequence, and they must represent molecules which exist in nature (not a consensus sequence or a mix of genomic or mRNA sequence).

The GenBank annotation staff checks all submissions for:

(a) Is the sequence of the appropriate length and derived from a single molecule type (not a mix of genomic DNA and mRNA)?

(b) Biological validity: Does the conceptual translation of a coding region match the amino acid sequence provided by the submitter? Is the source organism name present in NCBI's taxonomy database? Does the sequence cluster with those of closely related organisms? Does the submitter's description of the sequence agree with the results of a BLAST similarity search against other sequences?

(c) Is the sequence free of vector contamination?

(d) Is the sequence or accession published? If so, can a PubMed ID be added to the record so that the sequence and publication can be linked in Entrez?

(e) Formatting and spelling.

If there are problems with the sequence or annotation, the annotator works with the submitter by email to correct the problems.

Completed entries are sent to the submitter for a final review before their release into the public database. Submitters may request that their sequence remain confidential until a specified future release date. Records will be held until that date or when the accession number or the sequence is published, whichever is first.

As the volume of submissions is increasing, the GenBank staff is no longer able to manually review every submission that is received. Classes of sequence data, such as 16S ribosomal RNA from bacteria and archaea are processed and released automatically after undergoing a series of rigorous validation checks.

Microbial genomes are subjected to checks for completeness, the quality of the assembly and the quality of the annotation.

3 Finding Sequence Data in SRA and GenBank

3.1 Direct Entrez SRA Query

Entrez (https://www.ncbi.nlm.nih.gov/sra/) is the primary NCBI Search and Retrieval system, and one of the primary points of access to SRA data. Since Entrez SRA indexing includes descriptive metadata contained in submitted SRA experiments, you can query Entrez using terms found in SRA experiment metadata. Metadata contained in an SRA experiment includes a description of sequencing libraries using controlled vocabularies for library concepts such as strategy, source material, and capture techniques as well as sequencing platform and instrument models (*see* **Note 8**).

3.2 Simple Entrez Query

To perform a direct Entrez query using SRA metadata search terms, go to the top of the NCBI home page (https://www.ncbi.nlm.nih.gov/) and select "SRA" from the drop-down list of available databases and enter a query (for example, "salmonella") in the search box. The SRA display page of results for your query will include records for each matching study, with links to read/run data, project descriptions, etc.

3.3 Advanced Entrez Query

Go to the top of the NCBI home page (https://www.ncbi.nlm.nih.gov/) and select "SRA" from the drop-down list of available databases, then click the "Advanced" link located just below the search bar at the top of the page. The SRA Advanced Search Builder page (https://www.ncbi.nlm.nih.gov/sra/advanced) will appear and on this page you can construct a complex SRA query by selecting multiple search terms from a large number of fields and qualifiers such as accession number, author, organism, text word, publication date, and properties (paired-end, RNA, DNA, etc.). See the Advanced Search Builder video tutorial (https://www.youtube.

com/watch?v=dncRQ1cobdc&feature=relmfu) for information about how to use existing values in fields and combine them to achieve a desired result.

3.4 SRA Home Page Query

You can also search SRA through the coordinated use of the "Browse" and "Search" tabs on the SRA home page (https://www.ncbi.nlm.nih.gov/Traces/sra/). The SRA Web interface allows the user to:

(a) Access any data type stored in SRA independently of any other data type (e.g., accessing read and quality data without the intensity data).

(b) Access reads and quality scores in parallel.

(c) Access related data from other NCBI resources that are integrated with SRA.

(d) Retrieve data based on ancillary information and/or sequence comparisons.

(e) Retrieve alignments in "vertical slices" (showing underlying layered data) by reference sequence location.

(f) Review the descriptions of studies and experiments (metadata) independently of experimental data.

3.5 The SRA Run Selector

The SRA Run Selector (https://www.ncbi.nlm.nih.gov/Traces/study/) can be used to view, filter and sort a metadata table from Entrez search results, or a list of accessions (e.g., SRA BioSample, BioProject, or dbGaP accession) pasted into the field at the top of the page. The Run Selector will dynamically generate a metadata table from library preparation and sample attributes.

The SRA Run Selector displays those attributes common to all selected Runs at the top of the page and displays variable values as columns. The columns are sortable and values can be used to filter the table content using the "Facets" box. Once you have a filtered set of data that you wish to work with, there are three options for saving the results:

1. Click "permalink" from the top of the page and copy the URL for later use or embedding.
2. Click the "RunInfo Table" button to download a tab-delimited text file of all or only selected metadata.
3. Click the "Accession List" button to download a list of Run accessions that can be used with the SRA toolkit to analyze and download the sequence and alignment data.

3.5.1 Putting It All Together with Bioproject

Bioproject can aggregate several data types (e.g., a genome assembly and a transcriptome) by study (e.g., NIH grant). You can access BioProject records by browsing, querying, or downloading in Entrez, or by following a link from another NCBI database.

3.6 BioProject Browsing

To browse through BioProject content, go to the "By project attributes" (https://www.ncbi.nlm.nih.gov/bioproject/browse/) hyperlink from the BioProject home page (https://www.ncbi.nlm.nih.gov/bioproject/). You can browse by major organism groups, project type (umbrella projects vs. primary submissions), or project data type. The table includes links to the NCBI Taxonomy database (https://www.ncbi.nlm.nih.gov/taxonomy/), where additional information about the organism may be available, and to the BioProject record.

3.7 BioProject Query

You can perform a search in BioProject like you would in any other Entrez database, namely by searching for an organism name, text word, or BioProject accession (PRJNA31257), or by using the Advanced Search page to build a query restricted by multiple fields. Search results can be filtered by Project Type, Project Attributes, Organism, or Metagenome Groups, or by the presence or absence of associated data in one of the data archives.

Table 2 contains some representative searches:

3.8 BioProject Linking

You can also find BioProject records by following links from archival databases when the data cites a BioProject accession. You can find links to BioProject in several databases including SRA, Assembly, BioSample, dbVar, Gene, Genome, GEO, and Nucleotide (which includes GenBank and RefSeq nucleotide sequences). Large consortia also LinkOut (https://www.ncbi.nlm.nih.gov/projects/linkout/) from BioProject to their resources.

Table 2
Some example BioProject searches

Find BioProjects by ...	Search text example(s)
A species name	Escherichia coli [organism]
Project data type	"metagenome" [Project Data Type]
Project data type and Taxonomic Class	"transcriptome" [Project Data Type] AND Insecta [organism]
Publication	"19643200" [PMID]
Submitter organization, consortium, or center	JGI [Submitter Organization]
Sample scope and material used	"scope environment" [Properties] AND "material transcriptome" [Properties]
A BioProject database Identifier	PRJNA33823 or PRJNA33823 [bioproject] or 33823 [uid] or 33823 [bioproject]

3.9 GenBank

GenBank and Refseq nucleotide sequences records can be retrieved from Entrez Nucleotide (https://www.ncbi.nlm.nih.gov/nuccore/). EST and GSS records are searched and retrieved from https://www.ncbi.nlm.nih.gov/est/ and https://www.ncbi.nlm.nih.gov/gss/ or by choosing the appropriate database form the pull-down menu at the top of most NCBI Web pages. Entrez Protein (https://www.ncbi.nlm.nih.gov/protein/) is a collection of protein sequences from a variety of sources, including translations from annotated coding regions in INSDC and RefSeq, UniProt and PDB. The Entrez retrieval system has a network of links that join entries from each of the databases. For example, a GenBank record in Nucleotide can have links to the Taxonomy, PubMed, PubMed Central, Protein, PopSet, BioProject, Gene, and Genome databases. Within a GenBank flat file hyperlinks to Taxonomy, BioProject, BioSample, and PubMed databases are displayed if the links are present. Links to external databases can be made by LinkOut or by cross-references (db_xrefs) within the entry. By taking advantage of these links, users can make important scientific discoveries. These links are critical to discovering the relationship between a single piece of data and the information available in other databases.

In Entrez, sequence data can be viewed in a number of different formats. The default and most readable format is the GenBank flat file view. The graphical view, which eliminates most of the text and displays just sequence and biological features, is another display option. Other displays of the data, for instance XML, ASN.1, or FASTA formats, are intended to be more computer-readable.

3.10 Genomes

Genome assembly sequences can be accessed in Entrez from a number of entry points including BioProject, Nucleotide, Genome and Assembly. However, sometimes it is not straightforward to understand which sequences contribute to a complete genome assembly from some of these resources. The Assembly database (https://www.ncbi.nlm.nih.gov/assembly) provides users with detailed information and statistics for each genome assembly including links to download the sequences and annotation. One can search for organism, and even strain, then use the links to our FTP site (for either GenBank or RefSeq) provided in the upper right hand corner (Fig. 7). The FTP site contains nucleotide and protein sequences in FASTA format, in addition to annotation data in .gff and GenBank flatfile format.

4 Downloading the Sequence Data in SRA and GenBank

4.1 The SRA Toolkit

The SRA Toolkit is a collection of tools and libraries for using the SRA archive file format. SRA utilities have the ability to locate and download data on-demand from NCBI servers, removing the need for a separate download step, and most importantly, downloading

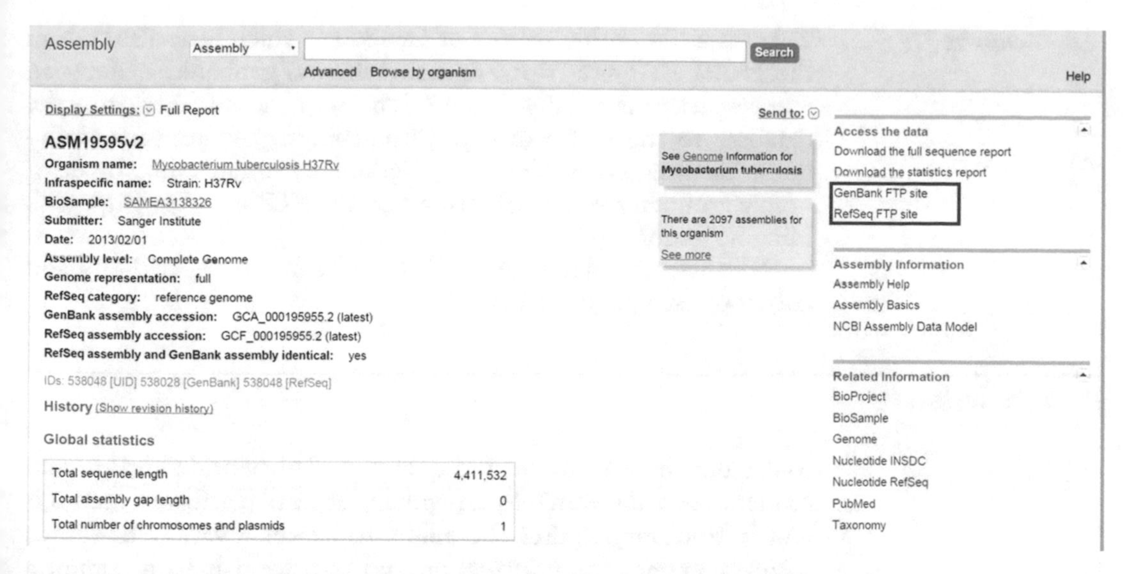

Fig. 7 The NCBI Assembly resource can be searched for taxonomic and sequence information pertaining to whole genomes and scaffolds submitted to NCBI. Sequences can be downloaded from the GenBank and RefSeq FTP sites that are accessible from the NCBI Assembly pages

only required data. This feature can reduce the bandwidth, storage, and the time taken to perform tasks that use less than 100 % of the data contained in a run. Utilities are provided for delivering data in commonly used text formats such as fastq and sam. Additional information on using, configuring, and building the toolkit is maintained on the NCBI github repository (*see* **Note 9**).

4.2 Programmatic Interaction with the SRA Toolkit

We have developed a new, domain-specific API for accessing reads, alignments and pileups produced from Next Generation Sequencing called NGS (*see* **Note 9**). The API itself is independent from any particular back-end implementation, and supports use of multiple back-ends simultaneously. It also provides a library for building new back-end "engines." The engine for accessing SRA data is contained within the sister repository ncbi-vdb.

The API is currently expressed in C++, Java, and Python languages. The design makes it possible to maintain a high degree of similarity between the code in one language and code in another—especially between C++ and Java.

4.3 BioProject Download

In addition to the Entrez Web interface and the BioProject browse page, you can download the entire BioProject database and the database .xsd schema from the FTP site: ftp://ftp.ncbi.nlm.nih.gov/bioproject/, or use Entrez Programming Utilities (E-utilities) to programmatically access public BioProject records.

4.4 GenBank FTP

There is a bimonthly release of GenBank, which is available from the NCBI FTP site (ftp://ftp.ncbi.nih.gov/genbank/). Between releases, there is a daily dump of the sequences loaded into the database to the FTP site (ftp://ftp.ncbi.nih.gov/genbank/daily-nc/). Genome assemblies are retrievable by organism or by taxonomic group in a variety of formats on the FTP site (ftp://ftp.ncbi.nih.gov/genomes/).

The assembly database has FTP URLs in the upper right hand corner of each page (*see* Fig. 7).

5 Conclusion

Managing the ever increasing quantity of nucleotide and protein sequence data generated by an evolving array of platforms, methods and institutions requires the ability to accept a variety of source formats, extract the information and transform it to a common archival standard for display and computational access. To achieve this, NCBI collects and validates sequence data and provides easily accessible public websites and application interfaces. Sequence data records are enhanced with links to other NCBI resources such as taxonomy and PubMed. INSDC defines standard elements for the sequence data records to ensure that the information can be submitted to and retrieved from any of the collaborating archives, regardless of the tools used to collect or display the data. The integrity of the sequence data and annotation is confirmed by validation steps both at the submission and the processing stages. GenBank provides methods for updating, correcting, or adding additional information to existing sequence records, before and after they become publicly available. Finally, sequence data is not useful unless it is easily available to researchers in their labs and at their desks so NCBI provides these public users with multiple tools to search, discover, retrieve, and analyze sequence data and educational resources to help users understand it (*see* **Note 10**).

6 Notes

1. The table on the INSDC homepage contains links to the resources at the partner websites.
   (a) DDBJ Sequence Read Archive: http://trace.ddbj.nig.ac.jp/dra/
   (b) DDBJ Capillary Reads: http://trace.ddbj.nig.ac.jp/dta/
   (c) DDBJ Annotated Sequences: http://www.ddbj.nig.ac.jp/
   (d) DDBJ Samples: http://trace.ddbj.nig.ac.jp/biosample/
   (e) DDBJ Studies: http://trace.ddbj.nig.ac.jp/bioproject/

(f) ENA Sequence Read Archive, Annotated Sequences, Samples and Studies: http://www.ebi.ac.uk/ena/submit/data-formats

(g) NCBI Sequence Read Archive: https://www.ncbi.nlm.nih.gov/sra/

(h) NCBI Capillary Reads: https://www.ncbi.nlm.nih.gov/Traces/

(i) NCBI Annotated Sequences: https://www.ncbi.nlm.nih.gov/genbank/

(j) NCBI Samples: https://www.ncbi.nlm.nih.gov/biosample/

(k) NCBI Studies: https://www.ncbi.nlm.nih.gov/bioproject/

2. GenBank records are grouped into 19 divisions; either by taxonomic groupings or by a specific technological approach, such as WGS or TSA. Sequences in the technique-based divisions often have a specific keyword in the record from a controlled list http://www.insdc.org/documents/methodological-keywords. Entrez can be specifically queried for sequence records in these divisions. For example, if one wanted to retrieve actin sequences for all non-primate mammalian species, one could search for

   Actin AND "gbdiv MAM" [prop]

   The GenBank divisions are listed in Table 3.

3. The DDBJ/EMBL/GenBank Feature Table: Definition, which can be found at http://insdc.org/feature_table.html, lists all allowable features and qualifiers for a DDBJ/EMBL/GenBank record. This document gives information about the format and conventions, as well as examples, for the usage of features in the sequence record. Value formats for qualifiers are indicated in this document. Qualifiers may be

   (a) free text

   (b) controlled vocabulary or enumerated values

   (c) sequence

   Other syntax related to the flat file is described in this document. The document also contains reference lists for the following controlled vocabularies:

   (a) Nucleotide base codes (IUPAC)

   (b) Modified base abbreviations

   (c) Amino acid abbreviations

   (d) Modified and unusual Amino Acids

   (e) Genetic Code Tables

4. Help emails and documentation, as well as fact sheets, are available from the following links:

Table 3
Traditional taxonomic GenBank divisions

Code	Description
BCT	Bacterial sequences
PRI	Primate sequences
MAM	Other mammalian sequences
VRT	Other vertebrate sequences
INV	Invertebrate sequences
PLN	Plant, fungal, and algal sequences
VRL	Viral sequences
PHG	Bacteriophage sequences
SYN	Synthetic and chimeric sequences
UNA	Unannotated sequences, including some WGS sequences obtained via environmental sampling methods
Nontraditional GenBank divisions	
PAT	Patent sequences
EST	EST division sequences, or expressed sequence tags, are short single pass reads of transcribed sequence
STS	STS division sequences include anonymous STSs based on genomic sequence as well as gene-based STSs derived from the 3′ ends of genes. STS records usually include primer sequences, annotations and PCR reaction conditions
GSS	GSS records are predominantly single reads from bacterial artificial chromosomes ("BAC-ends") used in a variety of clone-based genome sequencing projects
ENV	The ENV division of GenBank, for non-WGS sequences obtained via environmental sampling methods in which the source organism is unknown
HTG	The HTG division of GenBank contains unfinished large-scale genomic records that are in transition to a finished state. These records are designated as Phase 0–3 depending on the quality of the data. Upon reaching Phase 3, the finished state, HTG records are moved into the appropriate taxonomic division of GenBank
HTC	The HTC division of GenBank accommodates high-throughput cDNA sequences. HTCs are of draft quality but may contain 5′ UTRs and 3′ UTRs, partial coding regions, and introns
CON	Large records that are assembled from smaller records, such as eukaryotic chromosomal sequences or WGS scaffolds, are represented in the GenBank "CON" division. CON records contain sets of assembly instructions to allow the transparent display and download of the full record using tools such as NCBI's Entrez
TSA	Transcriptome shotgun data are transcript sequences assembled from sequences deposited in the NCBI Trace Archive, the Sequence Read Archive (SRA), and the EST division of GenBank

SRA Submission Quick Start Guide: https://www.ncbi.nlm.nih.gov/books/NBK47529/

SRA File Format Guide: https://www.ncbi.nlm.nih.gov/books/NBK242622/

GEO Submission Guide: https://www.ncbi.nlm.nih.gov/geo/info/submission.html

GenBank Submissions Handbook: https://www.ncbi.nlm.nih.gov/books/NBK51157/

SRA homepage: https://trace.ncbi.nlm.nih.gov/Traces/sra/

dbGaP homepage: https://www.ncbi.nlm.nih.gov/gap/

GEO homepage: http://www.ncbi.nlm.nih.gov/geo/

GenBank homepage: https://www.ncbi.nlm.nih.gov/genbank/

Questions about data archives and submissions can be sent to: submit-help@ncbi.nlm.nih.gov; gb-admin@ncbi.nlm.nih.gov

General NCBI questions can be sent to the NCBI helpdesk: info@ncbi.nlm.nih.gov

5. GenBank Submission tools are available on NCBI's FTP site (Sequin: ftp://ftp.ncbi.nih.gov/sequin/, tbl2asn: ftp://ftp.ncbi.nih.gov/toolbox/ncbi_tools/converters/by_program/tbl2asn/). Additional information about these programs can be found on the NCBI website (sequin: https://www.ncbi.nlm.nih.gov/Sequin/ tbl2asn: http://www.ncbi.nlm.nih.gov/genbank/tbl2asn2). Both of these submission utilities contain validation software that will check for problems associated with your submission. The validator can be accessed in Sequin from the Search->Validate menu item or in tbl2asn using –v in the command line.
6. When preparing a submission by Sequin or tbl2asn, information about the sequence can be incorporated into the Definition Line of the FASTA formatted sequence. FASTA format is simply the raw sequence preceded by a definition line. The definition line begins with a > sign and is followed immediately by the sequence identifier and a title. Information can be embedded into the title which Sequin and tbl2asn use to construct a submission. Specifically, you can enter organism and strain or clone information in the nucleotide definition line and gene and protein information in the protein definition line using name-value pairs surrounded by square brackets. Example:

   >myID [organism=Drosophila melanogaster] [strain=Oregon R] [clone=abc1].

7. For both submission and updates, the submitter should prepare tabular files with source metadata and feature information in the following formats. These files can easily be incorporated into the GenBank records using any of the submission tools.

   Source information (i.e., strain, cultivar, country, specimen_voucher) is to be provided in a two-column tab-delimited table, for example:

Sequence id.	Strain	Country
AYxxxxxx	82	Spain
AYxxxxxy	ABC	France

   Nucleotide sequence should be submitted in FASTA format:
   >AYxxxxxx

   cggtaataatggaccttggaccccggcaaagcggagagac

   >AYxxxxxy

   ggaccttggaccccggcaaagcggagagaccggtaataat

   Feature annotation should be submitted as a tab-delimited five-column feature table.

   Column 1: Start location of feature

   Column 2: Stop location of feature

   Column 3: Feature key

   Column 4: Qualifier key

   Example:

>Feature Sc_16				
1	7000	REFERENCE		
			PubMed	8849441
<1	1050	gene		
			gene	ATH1
<1	1009	CDS		
			product	acid trehalase
			product	Ath1p
			codon_start	2
<1	1050	mRNA		

   In the future, GFF3 format will be supported as well.

8. Table 4.

Table 4
SRA experimental enumeration values and definitions

Strategy	Sequencing strategy used in the experiment
WGA	Random sequencing of the whole genome following non-PCR amplification
WGS	Random sequencing of the whole genome
WXS	Random sequencing of exonic regions selected from the genome
RNA-Seq	Random sequencing of whole transcriptome
miRNA-Seq	Random sequencing of small miRNAs
WCS	Random sequencing of a whole chromosome or other replicon isolated from a genome
CLONE	Genomic clone based (hierarchical) sequencing
POOLCLONE	Shotgun of pooled clones (usually BACs and Fosmids)
AMPLICON	Sequencing of overlapping or distinct PCR or RT-PCR products
CLONEEND	Clone end (5′, 3′, or both) sequencing
FINISHING	Sequencing intended to finish (close) gaps in existing coverage
ChIP-Seq	Direct sequencing of chromatin immunoprecipitates
MNase-Seq	Direct sequencing following MNase digestion
DNase-Hypersensitivity	Sequencing of hypersensitive sites, or segments of open chromatin that are more readily cleaved by DNaseI
Bisulfite-Seq	Sequencing following treatment of DNA with bisulfite to convert cytosine residues to uracil depending on methylation status
Tn-Seq	Sequencing from transposon insertion sites
MRE-Seq	Methylation-sensitive restriction enzyme sequencing strategy
MeDIP-Seq	Methylated DNA immunoprecipitation sequencing strategy
MBD-Seq	Direct sequencing of methylated fractions sequencing strategy
OTHER	Library strategy not listed (please include additional info in the "design description")
Source	**Type of genetic source material sequenced**
GENOMIC	Genomic DNA (includes PCR products from genomic DNA)
TRANSCRIPTOMIC	Transcription products or non genomic DNA (EST, cDNA, RT-PCR, screened libraries)
METAGENOMIC	Mixed material from metagenome
METATRANSCRIPTOMIC	Transcription products from community targets
SYNTHETIC	Synthetic DNA
VIRAL RNA	Viral RNA

(continued)

Table 4
(continued)

Strategy	Sequencing strategy used in the experiment
OTHER	Other, unspecified, or unknown library source material (please include additional info in the "design description")
Selection	**Method of selection or enrichment used in the Experiment**
RANDOM	Random shearing or other "shotgun" method
PCR	Source material was selected by designed primers
RANDOM PCR	Source material was selected by randomly generated primers
RT-PCR	Source material was selected by reverse transcription PCR
HMPR	Hypo-methylated partial restriction digest
MDA	Multiple displacement amplification
MSLL	Methylation spanning linking library
cDNA	Complementary DNA
ChIP	Chromatin immunoprecipitation
MNase	Micrococcal nuclease (MNase) digestion
DNAse	Deoxyribonuclease (MNase) digestion
Hybrid Selection	Selection by hybridization in array or solution
Reduced Representation	Reproducible genomic subsets, often generated by restriction fragment size selection, containing a manageable number of loci to facilitate re-sampling
Restriction Digest	DNA fractionation using restriction enzymes
5-methylcytidine antibody	Selection of methylated DNA fragments using an antibody raised against 5-methylcytosine or 5-methylcytidine (m5C)
MBD2 protein methyl-CpG binding domain	Enrichment by methyl-CpG binding domain
CAGE	Cap-analysis gene expression
RACE	Rapid amplification of cDNA ends
Padlock probes capture method	Circularized oligonucleotide probes
other	Other library enrichment, screening, or selection process (please include additional info in the "design description")

9. SRA software tools documentation and downloads

   NCBI distributes a variety of software tools from our github repository:

   https://github.com/ncbi

   Common SRA utilities are in the sra-tools repo:

   https://github.com/ncbi/sra-tools/wiki

   APIs for software development and examples are in the ngs repo:

   https://github.com/ncbi/ngs/wiki

   https://github.com/ncbi/ngs-tools

10. NCBI presents educational resources in several formats. First, we provide video tutorials as live webinars—both "full length" (30–60 min) and in "NCBI minute" (5–10 min format). We make these products approximately monthly and weekly. Announcements of upcoming webinars can be accessed at https://www.ncbi.nlm.nih.gov/home/coursesandwebinars.shtml. The recorded webinars are made public 2-3 weeks after the conclusion of the webinar and are available on our YouTube channel https://www.youtube.com/user/NCBINLM. Short, standalone videos are also available on our YouTube channel. Readers are encouraged to subscribe to get updated information about our video resources. Additionally, we have fact sheets about each resource that are available at http://www.ncbi.nlm.nih.gov/home/documentation.shtml and more extensive documentation about each resource available in the NCBI Handbook at https://www.ncbi.nlm.nih.gov/books/NBK143764/.

Acknowledgement

This research was supported by the Intramural Research Program of the NIH, NLM, NCBI.

References

1. Karsch-Mizrachi I, Nakamura Y, Cochrane G (2012) The International Nucleotide Sequence Database Collaboration. Nucleic Acids Res 40 (Database issue):D33–D37
2. Benson DA, Clark K, Karsch-Mizrachi I, Lipman DJ, Ostell J, Sayers EW (2015) GenBank. Nucleic Acids Res 43(Database issue):D30–D35
3. Silvester N, Alako B, Amid C, Cerdeno-Tarraga A, Cleland I, Gibson R et al (2015) Content discovery and retrieval services at the European Nucleotide Archive. Nucleic Acids Res 43(Database issue):D23–D29
4. Kodama Y, Mashima J, Kosuge T, Katayama T, Fujisawa T, Kaminuma E et al (2015) The DDBJ Japanese Genotype-phenotype Archive for genetic and phenotypic human data. Nucleic Acids Res 43(Database issue):D18–D22
5. Mellmann A, Harmsen D, Cummings CA, Zentz EB, Leopold SR, Rico A et al (2011) Prospective genomic characterization of the

German enterohemorrhagic Escherichia coli O104:H4 outbreak by rapid next generation sequencing technology. PLoS One 6(7): e22751

6. Barrett T, Wilhite SE, Ledoux P, Evangelista C, Kim IF, Tomashevsky M et al (2013) NCBI GEO: archive for functional genomics data sets—update. Nucleic Acids Res 41(Database issue):D991–D995
7. Fleischmann RD, Adams MD, White O, Clayton RA, Kirkness EF, Kerlavage AR et al (1995) Whole-genome random sequencing and assembly of Haemophilus influenzae Rd. Science 269 (5223):496–512
8. Tatusova T, DiCuccio M, Badretdin A, Chetvernin V, Ciufo S, Li W (2013) Prokaryotic genome annotation pipeline. In: The NCBI Handbook, 2nd edn. [Internet], Bethesda, MD. Available from: http://www.ncbi.nlm.nih.gov/books/NBK174280/

Chapter 5

Genome Annotation

Imad Abugessaisa, Takeya Kasukawa, and Hideya Kawaji

Abstract

The dynamic structure and functions of genomes are being revealed simultaneously with the progress of technologies and researches in genomics. Evidence indicating genome regional characteristics (genome annotations in a broad sense) provide the basis for further analyses. Target listing and screening can be effectively performed in silico using such data. Here, we describe steps to obtain publicly available genome annotations or to construct new annotations based on your own analyses, as well as an overview of the types of available genome annotations and corresponding resources.

Key words Genome annotation, Gene functions, RNA-Seq, Epigenetic marks, Genome browser

1 Introduction

The completion of the full genome sequence of numerous eukaryotic and prokaryotic organisms has influenced the way basic research in molecular biology is conducted. At the same time advances in the fields of high-throughput genomic technologies, bioinformatics, and computing science have paved the road for biologists to approach a comprehensive catalog of functional elements in the genome and discover unrecognized patterns in the genomic context. The process of identifying genomic elements and their functions is referred to as genome annotation. Although the term "genome annotation" was used mainly for structures of genes (mRNAs) on genomes in a narrow sense, it has been used for any genomic elements in a broad sense recently. Accumulated genome annotations have demonstrated: the variety of functional elements found in human genomes [1], biochemical functions of genomic segments [2], and unexpected complexity of transcripts (a variety of mRNA isoforms and non-protein coding RNAs) [3]. Genome annotation is not only essential for interpretation of the genome sequences but also an active research area in the field of genomics.

Jonathan M. Keith (ed.), *Bioinformatics: Volume I: Data, Sequence Analysis, and Evolution*, Methods in Molecular Biology, vol. 1525, DOI 10.1007/978-1-4939-6622-6_5, © Springer Science+Business Media New York 2017

Full utilization of annotations is essential not only for comprehensive "-omics" and systems biology research but also in other research areas like cell biology, developmental biology, and translational research [4]. It will improve the process to select molecular targets (key genes, drug targets, biomarkers, etc.) and to approach underlying systems (pathways and pathogenesis). In "-omics" researches, in-depth examination of correlation among a variety of genome annotations is required to reveal nonobvious relationships. Here, we explain how to use various genome annotations and provide steps to make your own annotations. An overview of existing genome annotations and technical frameworks supporting the process of genome annotation are described in Subheading 2, and the practical steps and tools to handle genome annotations are discussed in Subheading 3.

2 Materials

2.1 Genome Annotation, An Overview

2.1.1 Transcription and Its Regulation

1. An important step in the process of generating a functional molecule from a particular gene is transcription, a process of RNA synthesis based on DNA as template. The complete structure of the resulting RNA (called a transcript), in terms of exons and introns (intervening sequences that will not be part of its matured form after splicing), is a baseline to understand how genetically inherited information are encoded. It can be derived from sequence alignments of full-length cDNAs with the genome, and partial structures can be derived from sequencing of expressed sequence tags (ESTs) or tag-based approaches [5–8]. Next generation sequencing (NGS) technologies allow us to further identify RNA partial structures with quantification on a genome-wide scale. For example, RNA-Seq [9] is a method to sequence randomly fragmented RNA, and the results can be used to profile exons and splice junctions, and to estimate complete RNA structures. CAGE (Cap Analysis Gene Expression) is a method to sequence only 5′-ends of capped RNAs, and these results can be used to identify transcription start sites (TSSs) at single base pair resolution and active enhancers based on bidirectional transcription [10, 11]. A variety of protocols (including those above) based on NGS have been developed recently to profile RNAs, and they produce genome annotations on the captured part of the RNAs (such as exons, splicing junctions, and TSSs).
2. The right structure of each transcript has to be present in the right amount at the right time within a cell. Thus gene expression level (abundance of transcribed RNAs) is an alternative aspect of a gene's behavior, providing a clue to its functions [12]. Methods for measuring the abundance of transcripts

include northern blots, quantitative real-time polymerase chain reaction (qRT-PCR), microarrays, and NGS-based methods (such as RNA-Seq and CAGE above). The quantified levels of RNAs are associated with genome annotation of transcript structure (the complete or partial structure of RNAs) and samples indicating biological states.

3. Transcription, the process of RNA polymerization on DNA, is regulated by proteins called transcription factors, which bind to DNA and interact with RNA polymerase. Identifying the genomic locations of transcription factor binding sites (TFBSs) is essential to identify cis-acting elements and to understand transcriptional regulation. Chromatin immunoprecipitation (ChIP) assay is widely used to identify these locations, in which an antibody targeting the protein of interest is used to enrich its associated DNAs in the target samples. Collected DNAs can be measured by microarrays (ChIP-chip) [13] and sequencing (ChIP-seq) [14]. The data quality of ChIP depends on the antibodies used in the experiment, the quantification method and subsequent data processing [15]. The binding sites are identified at dozens-base resolution in general, and more accurate resolution can be achieved by using exonuclease (ChIP-exo) [16]. Position-weight matrixes (PWMs) representing binding patterns of transcription factors can be complementarily used to scan potential TFBSs in the genome at base-pair resolution. All of these analyses produce genomic locations of (potential) TFBSs, while their resolution and reliability varies.

2.1.2 Epigenetic Modifications and Chromatin Structures

1. Eukaryotic genomic DNA is coupled with histones to form nucleosomes and higher level chromatin structures. The process of transcription regulated by transcription factors is affected by a variety of epigenetic modifications such as DNA methylation [17] and histone modifications [18]. In particular, the effects of acetylation and methylation at specific residues of histone H3 and H4 have been extensively studied [19]. Enrichment-based methods, such as those using Methyl-CpG-binding domain (MBD-seq) [20, 21] and immunoprecipitation (ChIP-seq) [15, 16] can be effectively used to identify the locations with targeted epigenetic modifications as genome annotations. A higher level of annotation for biological interpretation, such as active promoters and transcription elongation, can be generated by integration of multiple epigenetic markers [22].

2. DNA methylation occurring at CpG dinucleotides can be determined at one base pair resolution by bisulfite treatment of DNA. This process converts cytosine residues to uracil except for 5-methylcytosine. Sequencing of DNAs with and without bisulfite treatment enables identification of DNA methylation at single base-pair resolution, and high-throughput profiling by using

NGS provides genome-wide annotation of DNA methylation at fine resolution [23].

3. Nucleosome structures can be investigated using nucleases. Micrococcal nuclease (MNase) digests linker regions between two nucleosome cores, and sequencing of the resulting DNA identifies genomic DNA segments wrapping a histone, mono-nucleosomes, and complementary linker regions [24]. DNase I cleaves DNAs preferentially at nucleosome-free regions, and sequencing to identify DNase I hypersensitive sites (DHSs) has been widely used for identifying active regulatory elements in the genome [19].
4. Genomic DNAs are compacted in the nucleus, where individual DNA segments can interact with each other in the three-dimensional space even though they are distant in genomic coordinates. These DNA segment interactions can be probed with a variation of chromosomal conformation capture [25] and chromatin interaction analysis by paired-end tag sequencing (ChIA-PET) [26] followed by paired-end sequencing. The resulting data provide links (spatial proximity) between DNA segments.

2.1.3 Evolutionary Conservation and Variation

1. Genome sequences reflect the history of life, and are shaped by negative selection against disadvantageous genomic variations (alleles), positive selection of favored ones, and genetic drift on neutral ones. Conserved genomic segments among species are likely important and actually many of the protein coding sequences (CDSs) are conserved. Interestingly, promoter regions of noncoding transcripts (ncRNA) are conserved as well as promoters of protein-coding transcripts, while ncRNA exons tend to be less conserved [27]. Ultra-conserved regions, where an almost complete identity is observed between orthologous regions of human, rat, and mouse [28], are remarkable occurrences in terms of evolution. DNA sequence alignments, conserved genomic segments, and scores indicating the levels of conservation at a single base-pair resolution [29, 30] are relevant components of genome annotation.
2. Genetic variations among cells, individuals, and populations may provide explanations for variations in phenotypes. A large number of studies have published statistical associations between specific allele and diseases [31] and a variety of somatic mutations in cancers [32]. ClinVar [33] aims to collect genomic variation of clinical importance, not only for cancers but many other diseases. Besides coordinates and allelic frequencies of genetic variations [34–37], their associations to specific phenotypes may provide clues to understand the contribution of each genomic segment to individual phenotypes at the system level.

2.1.4 Collaboration Efforts, Genome Annotation Databases and Browser

1. The challenges and opportunities for annotating genome sequence have been recognized, and several parallel initiatives and collaborative projects have been established to pool resources for genome annotations from different perspectives. For example, HapMap [35, 36] and the 1000 genome [34] projects aim at constructing a complete catalog of common genetic variation in healthy people. Complementarily, Encyclopedia of DNA Elements (ENCODE) [38], Roadmap Epigenomics [39], and The Functional Annotation of the Mammalian Genome (FANTOM) [10] focus on functional elements across the genome, in terms of transcription factor binding, epigenetic modification, full-length structure of mRNAs and transcriptional initiation activities. In contrast, the International Cancer Genome Consortium (ICGC [40]), including The Cancer Genome Atlas (TCGA [41]) focuses on cancers specifically to obtain genomic descriptions of importance to clinics and societies.

2.2 Technical Frameworks to Use Genome Annotations

2.2.1 Data File Formats

One of the common ways to store genome annotations is to use tab-delimited text format, since it allows us to inspect the contents manually and to handle the data files in any computers. Use of predefined formats permits the end-user to use his/her own annotations with existing tools, hence be able to explore, query, or visualize the data. Several formats based on tab-delimited file have been proposed, including General Feature Format (GFF, *see* **Note 1**) [42], Gene Transfer Format (GTF), Browser Extensible Data (BED, *see* **Note 2**), Wiggle Track Format (Wig) [43], Variant Call Format (VCF [44]), and Sequence Alignment/Map format (SAM). Most of the file format specifications are easy to understand, parse, and to process using lightweight scripting languages such as Perl, Python, or Ruby, and several tools dedicated to handling genomic intervals are freely available, such as BedTools [45] and SAMtools [46]. Here we introduce BED, as an example of a genome annotation file format. Individual values are simply tab-delimited as in Table 1, which describes seven regions as genome annotations:

At minimum, a BED file has three mandatory columns and nine optional ones. The mandatory columns are:

1. The name of the chromosome in the format of chr#, e.g., chr1, chr3, chr5, and chrM.
2. The starting position in the chromosome, i.e., the starting position of the genomic region.
3. The ending position in the chromosome, i.e., the ending position of the genomic region.

To describe genome annotations, the following optional columns can be used:

Table 1
CAGE peaks stored in BED file format

chr1	3158609	3158612	chr1:3158609..3158612,-	53	-
chr1	3195689	3195690	chr1:3195689..3195690,-	16	-
chr1	3195719	3195730	chr1:3195719..3195730,-	57	-
chr1	3195904	3195916	chr1:3195904..3195916,-	29	-
chr1	3203939	3203941	chr1:3203939..3203941,-	33	-
chr1	3256891	3256893	chr1:3256891..3256893,-	69	-
chr1	3273270	3273273	chr1:3273270..3273273,-	30	-

4. The name of the track is a label to identify the name of the row in the BED file, usually displayed at the left side of the genome browser.
5. The score value: the range of this field is 0–1000. The content of this score determine the way the genomic feature will be displayed in the genome browser.
6. Strand value, containing either '+' or '–'.

2.2.2 Genome Browsers

Visualization of genomics data is an essential step in data analysis to explore genomic information, inspect quality of user-defined data, elucidate the meaning of data and generate hypotheses. A genome browser is a tool to visualize genomic sequences with annotations in a schematic view, where users are typically able to perform multiple operations across the genomic coordinates (e.g., zoom in, zoom out, search, and download) via an interactive graphical user interface. A variety of genome browsers have been developed so far and they can be classified into two groups:

1. Web-based genome browsers, such as the UCSC genome browser [43], Ensembl genome browser [47], Generic Genome Browser (Gbrowse [48]), and ZENBU [49] (*see* Figs. 1, 2, and 3), in which services are hosted in a remote computer and end-users access them with a web browser over the Internet. Typically a variety of genome annotations are preloaded and configured. End-users can display and browse annotations and upload their own data to compare with them.
2. Desktop genome browsers, such as IGV [50] and the NCBI genome workbench (http://www.ncbi.nlm.nih.gov/tools/gbench/) in which the software has to be installed in a local computer. Typically only a minimum set of genome annotations are pre-configured at the beginning, and end-users can browse the data without real time connection to the Internet.

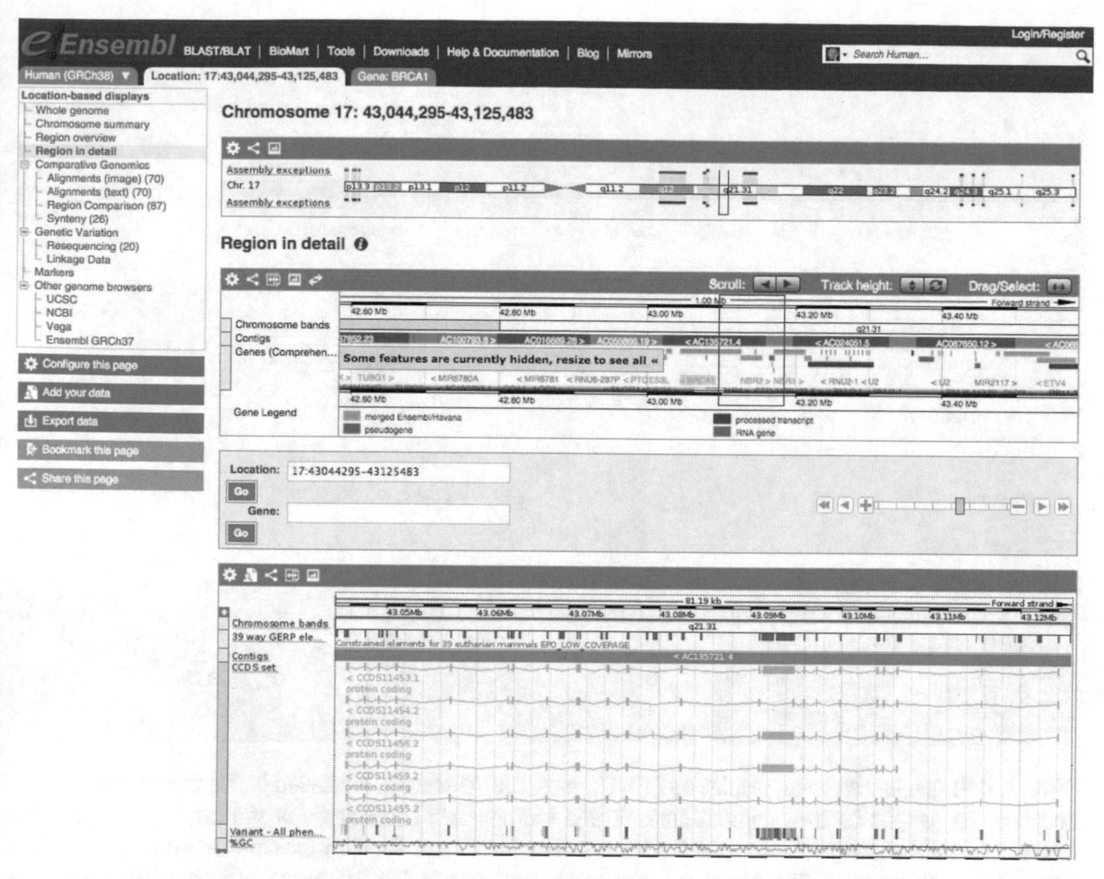

Fig. 1 Ensembl genome browser, displaying BRCA1 locus. BRCA1 locus is displayed in three panes: the top pane indicates the genetic band of chromosome 17, the middle pane indicates neighboring genes, and the bottom one displays more details, such as mRNA structure, at the locus. A keyword search box is located under the middle pane as well as in the top menu

2.2.3 Further Use of Genome Annotations

Genome browsers are the most commonly used tools to inspect genome annotations, and tab-delimited format is the simplest way to store them as described above. We introduce several complementary alternative tools and formats that are used for specific purposes.

1. End-users may wish to obtain a fraction of the entire genome annotations for their own analysis. Several systems support batch queries with graphical user interfaces, including BioMart [51] and the UCSC table browser [43].

2. Genomic analyses often require handling large sizes of genome annotations, in particular when using NGS-based approaches. Several compressed and binary formats have been proposed for efficient retrieval of individual records by using genomic coordinates, such as BAM—a binary version of SAM [46], BCF—a binary version of VCF [44], and Big Binary Indexed (BBI) files

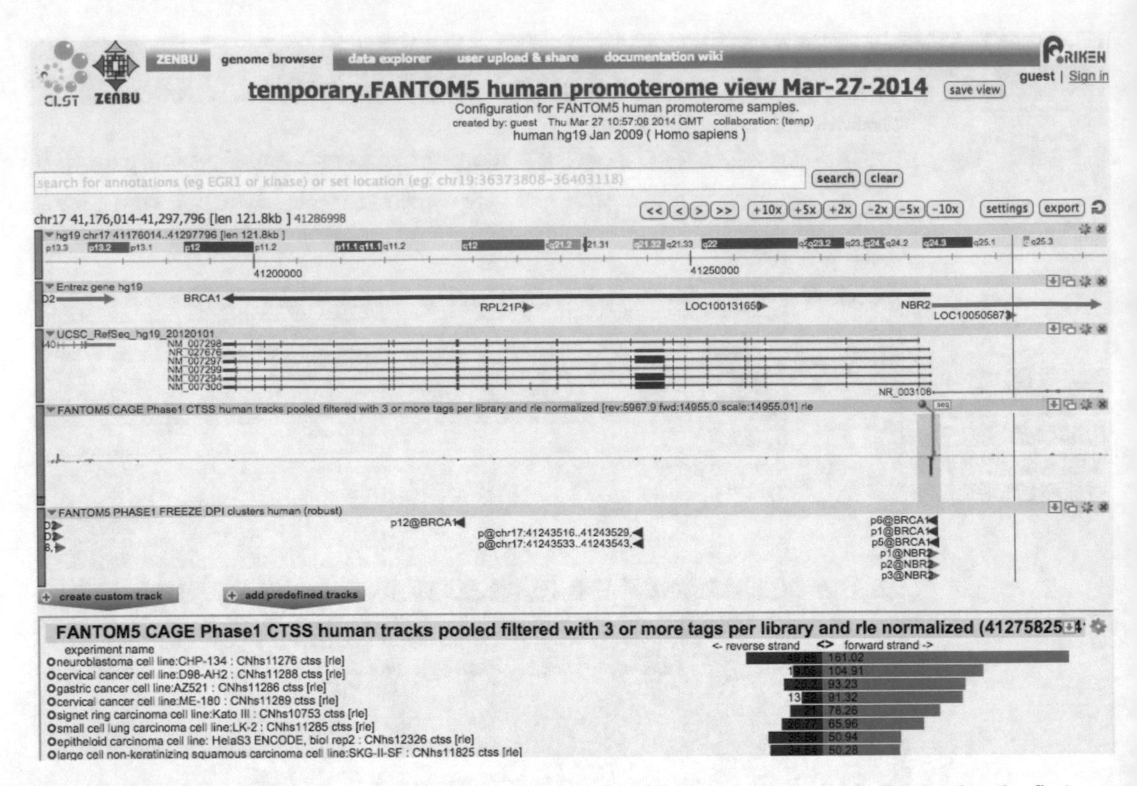

Fig. 2 ZENBU genome browser, displaying BRCA1 locus. BRCA1 locus is displayed in five tracks: the first one indicates the genetic band of chromosome 17 and the genomic coordinates of the displayed region, the second and third show gene and mRNA structure. The fourth and the fifth indicate transcription initiation activities and their peaks. The bar graph below the tracks indicates the levels of transcription initiation activities per sample. A keyword search box is located above the tracks

(such as bigBed and bigWig)—binary versions of BED and WIG files [43, 52],

3. Genomic analyses often generate a large number of genome annotations, which have to be inspected visually at some stage. The scheme of track data hub [52] allows data producers to generate a custom configuration to visualize their genome annotations based on the compressed binary formats above using the UCSC Genome browsers and the Ensembl genome browsers.

3 Methods

In this section we illustrate the steps to use genome annotations based on two scenarios: (I) finding elements of interest from genome annotations available publicly, and (II) generation of genome annotations, followed by their visualization.

3.1 Scenario I: Search and Obtain Published Genome Annotations

Assume that you find a tumor suppressor gene BRCA1 in an article or a list of your analysis results, and you are interested in its RNA structure (exon/intron), genetic variations, genome conservation and epigenetic content around the gene. Genome browsers, such as Ensembl Genome Browser including ZENBU, can be used for this purpose and we here use the UCSC Genome Browser as an example. Conceptually, the following steps are common in these genome browsers. For further reading see manuals for their details.

1. Access a genome browser with your favorite web browser, and open a window corresponding to the human genome assembly. In the case of the UCSC genome browser, open the URL http://genome.ucsc.edu/, followed by selection of "Genome Browser" in the left menu. Note that multiple genome assemblies have been published for human as a consequence of successive efforts to improve the reference genome sequences. You have to select an appropriate assembly version (usually the latest one) from those available.
2. Type the gene name "BRCA1" and press the submit button (the browser support text and sequence search). The resulting page shows a list of annotations including "BRCA1".
3. Click any of the genes in the above list. This will take you back to the genome browser displaying the BRCA1 gene. Figure 3 shows some of the annotations that may be displayed, including several isoforms of BRCA1, SNPs reported in the region, variants reported in cancer, genome conservation across the species, and epigenetic context such as transcription factor binding regions and transcription initiation sites. Several interfaces are provided for exploring neighboring regions, such as zoom in/out, move to right/left, and display/hide for each of the genome annotations.
4. Click "tools" in the top menu and select "Table Browser" to access the UCSC table browser. It enables the user to obtain genome annotations, whether in the displayed region or the whole genome, in a tabular form (*see* **Note 3**).

3.2 Scenario II: Make and Browse Your Own Annotations

Assume that you have produced your own ChIP-seq data by using anti-BRCA1 antibody, and you hope to generate your own genome annotations and inspect them visually. As an example we use a set of data generated by the ENCODE project [38], consisting of raw reads, their alignments with the reference genome, and identified binding sites of the protein in HeLa cells (*see* Table 2 for URLs of the example data). BRCA1 has been shown to regulate itself (autoregulation) negatively in several cell lines [53], thus we expect the presence of ChIP-seq signals at the locus.

1. Install a local genome browser, IGV, by following the instructions in http://www.broadinstitute.org/igv/ (this can be skipped if you have already installed it).

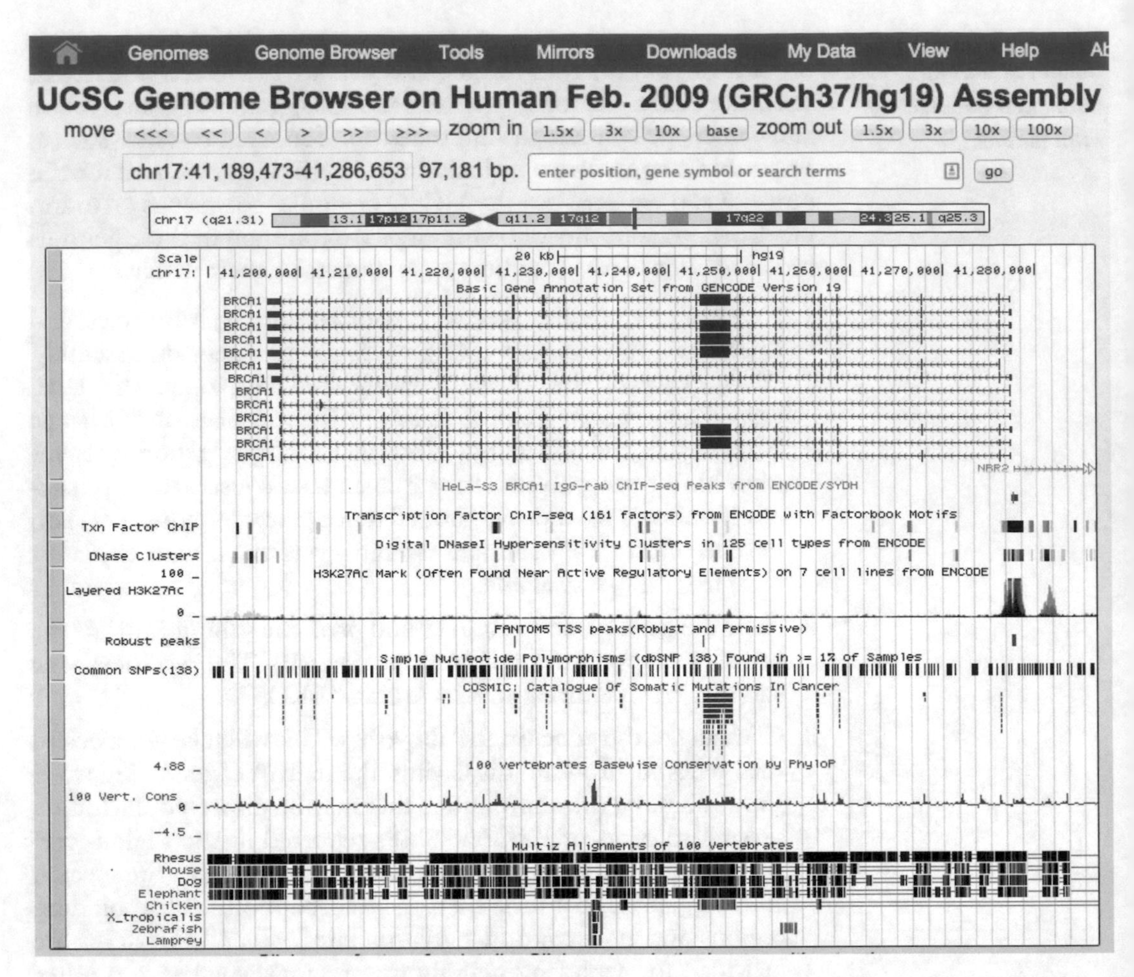

Fig. 3 The UCSC genome browser, displaying BRCA1 locus. BRCA1 locus is displayed with BRCA1 ChIP-seq results, somatic mutations in cancers, and other annotations. A keyword search box is placed above the genetic band. The view of the track is adjustable by right-clicking (click while pressing the control key on Mac) of the tracks or using the detailed interface below the tracks. The immediate view from the search results is very dense and gives an overview of the whole chromosome. Users can zoom in and click on one element to get detailed information about a particular gene, mRNA, etc.

2. Align the raw reads with the reference genome using an appropriate alignment program (the algorithms are reviewed in Ref. [54]), and produce alignment results in BAM format with index [46]. Note that the aligned results are available for this case as indicated in Table 2.
3. Start the local genome browser IGV, followed by loading the reference genome (hg19) and specifying the prepared BAM file. Note that the BAM file is located at a remote server in this case, but it can be at a local disk. Numerous reads are aligned at the promoter of BRCA1 (Fig. 4).

Table 2
Resulting files of ChIP-seq on BRCA1 protein

Data type	URL
Raw reads (sequencing results)	http://hgdownload.cse.ucsc.edu/goldenPath/hg19/encodeDCC/wgEncodeSydhTfbs/wgEncodeSydhTfbsHelas3Brca1a300IggrabRawDataRep1.fastq.gz
alignments with the genome (mapping results)	http://hgdownload.cse.ucsc.edu/goldenPath/hg19/encodeDCC/wgEncodeSydhTfbs/wgEncodeSydhTfbsHelas3Brca1a300IggrabAlnRep1.bam
index of the alignment file	http://hgdownload.cse.ucsc.edu/goldenPath/hg19/encodeDCC/wgEncodeSydhTfbs/wgEncodeSydhTfbsHelas3Brca1a300IggrabAlnRep1.bam.bai
binding sites (called peaks)	http://hgdownload.cse.ucsc.edu/goldenPath/hg19/encodeDCC/wgEncodeSydhTfbs/wgEncodeSydhTfbsHelas3Brca1a300IggrabPk.narrowPeak.gz

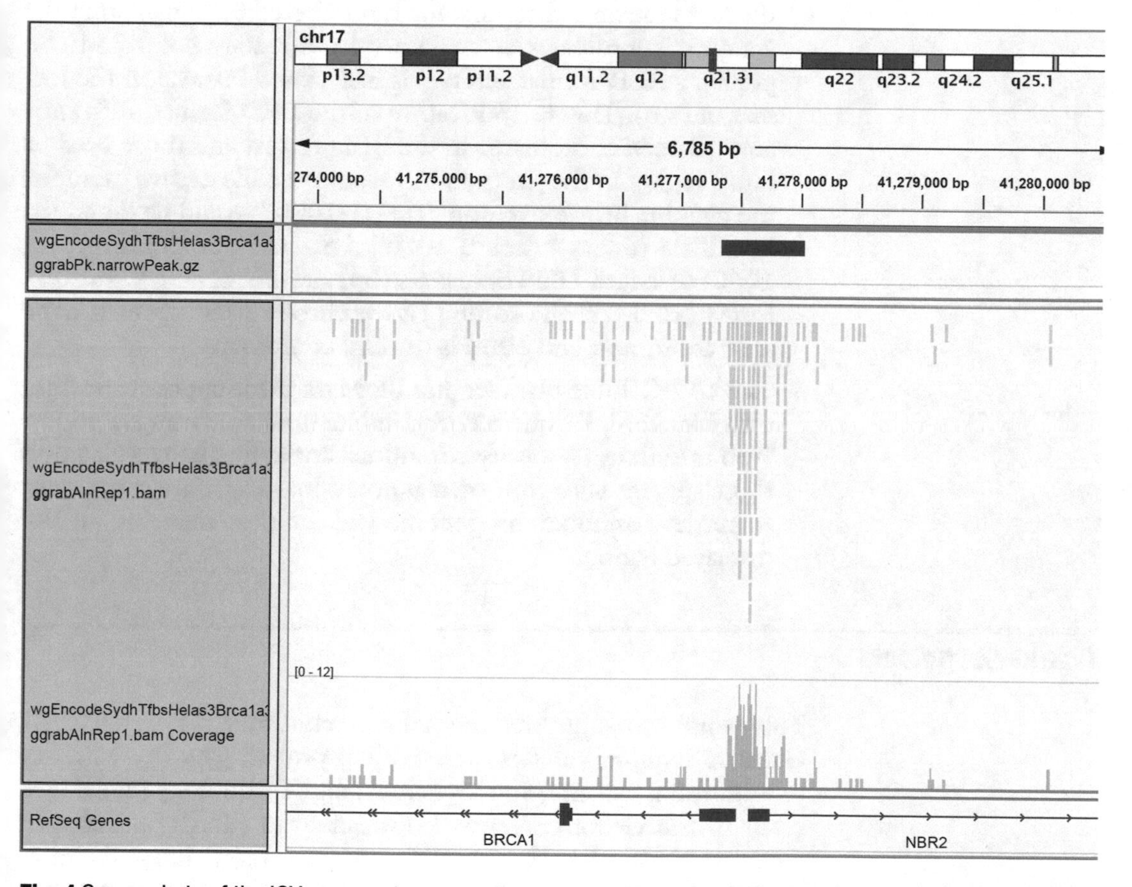

Fig. 4 Screenshots of the IGV genome browser, displaying BRCA1 locus. BRCA1 locus is displayed with BRCA1 ChIP-seq results. The top pane indicates genetic bands, the second indicates an identified binding site by the ChIP-seq, the third indicates individual reads and their frequency, and the fourth indicates the gene location

4. Identify BRCA1 binding sites by performing peak calls (several methods are reviewed in [15, 55], and produce the results in a BED file (a precomputed result is available as in Table 2). The file can be opened in IGV (Fig. 4).

4 Notes

1. Major variations of GFF are GFF version 1, 2 (http://www.sanger.ac.uk/Software/formats/GFF/), and 3 (http://song.sourceforge.net/gff3.shtml). The differences are only in the ninth column: version 1 requires just one string rather than a pair of name and value, and version 3 requires value name to be concatenated with its value by "=" (equal). Some keywords are reserved for specific use in GTF (Gene Transfer Format (http://genes.cs.wustl.edu/GTF2.html), a format of GFF version 2 with some special attribute names.
2. The genomic coordinates in a single BED row are 0-based and the first base on a chromosome is numbered 0. Considering the genomic coordinates as an interval (with start point and end point), a BED format interval is half-closed interval in the format of [a,b). This special feature of the BED format differentiates between coordinates in the BED record and those used to find a region in the genome browser. As an illustrative example, the genome browser region "chr1:1-1000" would be described in a BED row as "chr1 0 1000", i.e., half-closed interval [0, 1000) of length 1000 base pairs (bp) starting at base 0. In BED format, chr1 (chromosome 1) is the chromosome name, 0 is the start coordinate and 1000 is the end coordinate.
3. The UCSC Table Browser and BioMart [51] support exporting data with complex queries, or combinations of some conditions. Web interfaces to specify conditions are available by setting up those systems with your own annotations. These are convenient to retrieve annotations genome-wide, rather than just in the displayed region.

Acknowledgments

This work was supported by a Research Grant for the RIKEN Genome Exploration Research Project provided by the Ministry of Education, Culture, Sports, Science and Technology (MEXT), a grant to the Genome Network Project from MEXT, a Research Grant from MEXT to the RIKEN Center for Life Science Technologies, a Research Grant to RIKEN Preventive Medicine and a Diagnosis Innovation Program from MEXT to Yoshihide Hayashizaki.

References

1. Genomes Project Consortium, Abecasis GR, Auton A, Brooks LD, DePristo MA, Durbin RM et al (2012) An integrated map of genetic variation from 1,092 human genomes. Nature 491(7422):56–65
2. Li W, Manktelow E, von Kirchbach JC, Gog JR, Desselberger U, Lever AM (2010) Genomic analysis of codon, sequence and structural conservation with selective biochemical-structure mapping reveals highly conserved and dynamic structures in rotavirus RNAs with potential cis-acting functions. Nucleic Acids Res 38(21):7718–7735
3. Kageyama Y, Kondo T, Hashimoto Y (2011) Coding vs non-coding: translatability of short ORFs found in putative non-coding transcripts. Biochimie 93(11):1981–1986
4. Abugessaisa I, Saevarsdottir S, Tsipras G, Lindblad S, Sandin C, Nikamo P et al (2014) Accelerating translational research by clinically driven development of an informatics platform—a case study. PLoS One 9(9):e104382
5. Harbers M, Carninci P (2005) Tag-based approaches for transcriptome research and genome annotation. Nat Methods 2 (7):495–502
6. Cock PJ, Fields CJ, Goto N, Heuer ML, Rice PM (2010) The Sanger FASTQ file format for sequences with quality scores, and the Solexa/Illumina FASTQ variants. Nucleic Acids Res 38 (6):1767–1771
7. Kodzius R, Kojima M, Nishiyori H, Nakamura M, Fukuda S, Tagami M et al (2006) CAGE: cap analysis of gene expression. Nat Methods 3 (3):211–222
8. Shiraki T, Kondo S, Katayama S, Waki K, Kasukawa T, Kawaji H et al (2003) Cap analysis gene expression for high-throughput analysis of transcriptional starting point and identification of promoter usage. Proc Natl Acad Sci U S A 100(26):15776–15781
9. Wang Z, Gerstein M, Snyder M (2009) RNA-Seq: a revolutionary tool for transcriptomics. Nat Rev Genet 10(1):57–63
10. Forrest AR, Kawaji H, Rehli M et al (2014) A promoter-level mammalian expression atlas. Nature 507(7493):462–470
11. Andersson R, Gebhard C, Miguel-Escalada I, Hoof I, Bornholdt J, Boyd M et al (2014) An atlas of active enhancers across human cell types and tissues. Nature 507(7493):455–461
12. Lockhart DJ, Winzeler EA (2000) Genomics, gene expression and DNA arrays. Nature 405 (6788):827–836
13. Cawley S, Bekiranov S, Ng HH, Kapranov P, Sekinger EA, Kampa D et al (2004) Unbiased mapping of transcription factor binding sites along human chromosomes 21 and 22 points to widespread regulation of noncoding RNAs. Cell 116(4):499–509
14. Mikkelsen TS, Ku M, Jaffe DB, Issac B, Lieberman E, Giannoukos G et al (2007) Genome-wide maps of chromatin state in pluripotent and lineage-committed cells. Nature 448 (7153):553–560
15. Landt SG, Marinov GK, Kundaje A, Kheradpour P, Pauli F, Batzoglou S et al (2012) ChIP-seq guidelines and practices of the ENCODE and modENCODE consortia. Genome Res 22 (9):1813–1831
16. Rhee HS, Pugh BF (2011) Comprehensive genome-wide protein-DNA interactions detected at single-nucleotide resolution. Cell 147(6):1408–1419
17. Ndlovu MN, Denis H, Fuks F (2011) Exposing the DNA methylome iceberg. Trends Biochem Sci 36(7):381–387
18. Bannister AJ, Kouzarides T (2011) Regulation of chromatin by histone modifications. Cell Res 21(3):381–395
19. Huebert DJ, Bernstein BE (2005) Genomic views of chromatin. Curr Opin Genet Dev 15 (5):476–481
20. Lan X, Adams C, Landers M, Dudas M, Krissinger D, Marnellos G et al (2011) High resolution detection and analysis of CpG dinucleotides methylation using MBD-Seq technology. PLoS One 6(7):e22226
21. Aberg KA, McClay JL, Nerella S, Xie LY, Clark SL, Hudson AD et al (2012) MBD-seq as a cost-effective approach for methylome-wide association studies: demonstration in 1500 case–control samples. Epigenomics 4 (6):605–621
22. Hoffman MM, Ernst J, Wilder SP, Kundaje A, Harris RS, Libbrecht M et al (2013) Integrative annotation of chromatin elements from ENCODE data. Nucleic Acids Res 41 (2):827–841
23. Li Y, Tollefsbol TO (2011) DNA methylation detection: bisulfite genomic sequencing analysis. Methods Mol Biol 791:11–21
24. Portela A, Liz J, Nogales V, Setien F, Villanueva A, Esteller M (2013) DNA methylation

determines nucleosome occupancy in the 5′-CpG islands of tumor suppressor genes. Oncogene 32(47):5421–5428
25. Lieberman-Aiden E, van Berkum NL, Williams L, Imakaev M, Ragoczy T, Telling A et al (2009) Comprehensive mapping of long-range interactions reveals folding principles of the human genome. Science 326 (5950):289–293
26. Paulsen J, Rodland EA, Holden L, Holden M, Hovig E (2014) A statistical model of ChIA-PET data for accurate detection of chromatin 3D interactions. Nucleic Acids Res 42(18): e143
27. Carninci P, Kasukawa T, Katayama S, Gough J, Frith MC, Maeda N et al (2005) The transcriptional landscape of the mammalian genome. Science 309(5740):1559–1563
28. Bejerano G, Pheasant M, Makunin I, Stephen S, Kent WJ, Mattick JS et al (2004) Ultraconserved elements in the human genome. Science 304(5675):1321–1325
29. Kent WJ, Baertsch R, Hinrichs A, Miller W, Haussler D (2003) Evolution's cauldron: duplication, deletion, and rearrangement in the mouse and human genomes. Proc Natl Acad Sci U S A 100(20):11484–11489
30. Pollard KS, Hubisz MJ, Rosenbloom KR, Siepel A (2010) Detection of nonneutral substitution rates on mammalian phylogenies. Genome Res 20(1):110–121
31. Marigorta UM, Gibson G (2014) A simulation study of gene-by-environment interactions in GWAS implies ample hidden effects. Front Genet 5:225
32. Forbes SA, Bindal N, Bamford S, Cole C, Kok CY, Beare D et al (2011) COSMIC: mining complete cancer genomes in the Catalogue of Somatic Mutations in Cancer. Nucleic Acids Res 39(Database issue):D945–D950
33. Landrum MJ, Lee JM, Riley GR, Jang W, Rubinstein WS, Church DM et al (2014) ClinVar: public archive of relationships among sequence variation and human phenotype. Nucleic Acids Res 42(Database issue): D980–D985
34. Kuehn BM (2008) 1000 Genomes Project promises closer look at variation in human genome. JAMA 300(23):2715
35. International HapMap Consortium (2005) A haplotype map of the human genome. Nature 437(7063):1299–1320
36. International HapMap Consortium, Altshuler DM, Gibbs RA, Peltonen L, Altshuler DM, Gibbs RA et al (2010) Integrating common and rare genetic variation in diverse human populations. Nature 467(7311):52–58
37. Sherry ST, Ward MH, Kholodov M, Baker J, Phan L, Smigielski EM et al (2001) dbSNP: the NCBI database of genetic variation. Nucleic Acids Res 29(1):308–311
38. ENCODE Project Consortium (2012) An integrated encyclopedia of DNA elements in the human genome. Nature 489(7414):57–74
39. Bernstein BE, Stamatoyannopoulos JA, Costello JF, Ren B, Milosavljevic A, Meissner A et al (2010) The NIH Roadmap Epigenomics Mapping Consortium. Nat Biotechnol 28 (10):1045–1048
40. Zhang J, Baran J, Cros A, Guberman JM, Haider S, Hsu J et al (2011) International Cancer Genome Consortium Data Portal—a one-stop shop for cancer genomics data. Database 2011: bar026
41. Cancer Genome Atlas Research Network, Weinstein JN, Collisson EA, Mills GB, Shaw KR, Ozenberger BA et al (2013) The Cancer Genome Atlas Pan-Cancer analysis project. Nat Genet 45(10):1113–1120
42. Rastogi A, Gupta D (2014) GFF-Ex: a genome feature extraction package. BMC Res Notes 7:315
43. Kuhn RM, Haussler D, Kent WJ (2013) The UCSC genome browser and associated tools. Brief Bioinform 14(2):144–161
44. Danecek P, Auton A, Abecasis G, Albers CA, Banks E, DePristo MA et al (2011) The variant call format and VCFtools. Bioinformatics 27 (15):2156–2158
45. Quinlan AR, Hall IM (2010) BEDTools: a flexible suite of utilities for comparing genomic features. Bioinformatics 26(6):841–842
46. Li H, Handsaker B, Wysoker A, Fennell T, Ruan J, Homer N et al (2009) The Sequence Alignment/Map format and SAMtools. Bioinformatics 25(16):2078–2079
47. Stalker J, Gibbins B, Meidl P, Smith J, Spooner W, Hotz HR et al (2004) The Ensembl Web site: mechanics of a genome browser. Genome Res 14(5):951–955
48. Donlin MJ (2009) Using the Generic Genome Browser (GBrowse). Current protocols in bioinformatics/editoral board, Andreas D. Baxevanis [et al.] Chapter 9:Unit 9
49. Severin J, Lizio M, Harshbarger J, Kawaji H, Daub CO, Hayashizaki Y et al (2014) Interactive visualization and analysis of large-scale sequencing datasets using ZENBU. Nat Biotechnol 32(3):217–219
50. Thorvaldsdottir H, Robinson JT, Mesirov JP (2013) Integrative Genomics Viewer (IGV): high-performance genomics data visualization and exploration. Brief Bioinform 14 (2):178–192

51. Kasprzyk A (2011) BioMart: driving a paradigm change in biological data management. Database 2011:bar049
52. Raney BJ, Dreszer TR, Barber GP, Clawson H, Fujita PA, Wang T et al (2014) Track data hubs enable visualization of user-defined genome-wide annotations on the UCSC Genome Browser. Bioinformatics 30(7):1003–1005
53. De Siervi A, De Luca P, Byun JS, Di LJ, Fufa T, Haggerty CM et al (2010) Transcriptional autoregulation by BRCA1. Cancer Res 70 (2):532–542
54. Li H, Homer N (2010) A survey of sequence alignment algorithms for next-generation sequencing. Brief Bioinform 11 (5):473–483
55. Bailey T, Krajewski P, Ladunga I, Lefebvre C, Li Q, Liu T et al (2013) Practical guidelines for the comprehensive analysis of ChIP-seq data. PLoS Comput Biol 9(11):e1003326

Chapter 6

Working with Ontologies

Frank Kramer and Tim Beißbarth

Abstract

Ontologies are powerful and popular tools to encode data in a structured format and manage knowledge. A large variety of existing ontologies offer users access to biomedical knowledge. This chapter contains a short theoretical background of ontologies and introduces two notable examples: The Gene Ontology and the ontology for Biological Pathways Exchange. For both ontologies a short overview and working bioinformatic applications, i.e., Gene Ontology enrichment analyses and pathway data visualization, are provided.

Key words Data management, Knowledge management, Ontologies, BioPAX, rBiopaxParser, Gene ontology, topGO, GOstat

1 Introduction

Vast amounts of biomedical knowledge have been accumulated over the past decades. In order to store and work with this complex data, it has to be archived in an accessible and well-documented way. The advent of ontologies in computer science has offered a flexible, extensible way for modeling specific domains of knowledge [1, 2]. Nowadays, ontologies are used in numerous settings to manage biological and medical knowledge.

This chapter first introduces the theoretical background of ontologies, provides insights into encoding data via ontologies, and then illustrates a number of common applications and uses for ontologies in Bioinformatics. Subheading 2 describes collections and search engines for acquiring relevant ontologies and details the encoding and editing of ontologies. Subheading 3 provides two step-by-step examples for working with ontologies in Bioinformatics: On the one hand statistical analyses using the Gene Ontology [3] and on the other hand accessing and using pathways encoded using the BioPAX ontology [4].

Jonathan M. Keith (ed.), *Bioinformatics: Volume I: Data, Sequence Analysis, and Evolution*, Methods in Molecular Biology, vol. 1525, DOI 10.1007/978-1-4939-6622-6_6, © Springer Science+Business Media New York 2017

1.1 Theoretical Background

The term "Ontology" originates from philosophy, where it denotes the studies of existence and reality, known as a branch of metaphysics, founded on the work of the philosopher Aristotle [5]. In computer science an ontology can be defined as follows:

> "A specification of a representational vocabulary for a shared domain of discourse – definitions of classes, relations, functions and other objects – is called an ontology [6]."

An ontology is always based on a conceptualization, i.e., an abstract, simplified view of the domain which is to be modeled. An ontology is a specific implementation of this conceptualization; it defines existing classes of objects, as well as the relationships between them [1].

The main goals for developing an ontology are to formalize the structure of domain-specific information, to separate knowledge about the data structure and the data itself, and to enable the reuse and sharing of the structure and knowledge [7]. Furthermore, it is possible to model description logics, which enables automated reasoning and inference based on the knowledge base and logical operations [8].

Every ontology is made up of a number of core components: *classes* define types of objects or things, whereas *properties* define the respective attributes and features of these classes. Restrictions on these properties allow the modeling of assertions and predetermined values. Classes can be instantiated for specific objects and are called instances. Properties of objects can either reference objects or consist of numeric or textual facts, for example a *Surname* property [7]. Furthermore, rules in an if-then form and axioms (assertions) can be used to infer statements about a domain of knowledge.

In practice ontologies are often used to add a layer of abstraction when the underlying reality is very complex and the available knowledge can be detailed in very different granularity. An example of this would be a full-length research paper detailing the processes involved in drawing the conclusion that Gene A activates Gene B, compared to the general statement "Gene A activates Gene B." On a very high abstraction level these statements would be identical; however, this conclusion cannot be drawn by comparing the free text format of a research paper and the short statement [9].

1.2 Applications

First and foremost, ontologies are used for knowledge and data management by modeling the knowledge of a specific domain. Ontologies offer a generic and comprehensive framework for creating and documenting a specific data model and additionally ease knowledge exchange between users. A common application in bioinformatics is the use of statistical tests for associating experimental results to existing ontologies, i.e., differentially expressed genes, with functional annotations for genes [10, 11].

Furthermore, a number of metrics, e.g., semantic similarity, have been developed to associate results and ontology knowledge [12–14] or perform hypotheses-generation [15]. Another popular subject in bioinformatic research is the use of formal logic for causal inference [16–19].

2 Materials

A large number of ontologies have been suggested, defined, and published in the last decade. Examples of notable developments in the biomedical community are the ontologies Chemical Entities of Biological Interest (ChEBI) [20], the Gene Ontology (GO) [3], as well as the ontology for Biological Pathways Exchange (BioPAX) [4]. ChEBI is a dictionary of small chemical molecules and molecular entities commonly used in metabolic processes, as well as pharmaceuticals, laboratory reagents, and subatomic particles. However, more complex macromolecules like proteins are generally excluded. The idea behind ChEBI is to provide an extensive, cross-referencing dictionary of basic biochemical entities, their machine-readable structural information, their biological role (e.g., antibiotic or hormone), and their applications (e.g., pesticide or drug) [20]. Descriptions and exemplary uses for GO and BioPAX are given in Subheadings 3.1 and 3.2.

2.1 Finding and Accessing Ontologies

Users interested in integrating knowledge from ontologies into their work need access to this data. Websites provide lists and search engines that can help in finding and browsing relevant ontologies (*see* **Note 1**). The ChEBI and GO ontologies are part of the Open Biomedical Ontologies Foundry (OBO, www.obofoundry.org) [21], a collaboration to standardize the way biomedical ontologies are developed and to allow cross-ontology referencing between members of the OBO Foundry. The OBO Foundry currently contains ten ontologies and lists dozens of candidate ontologies. Several web sites are available which list and categorize biomedical ontologies, e.g., BioPortal (bioportal.bioontology.org) and the EMBL-EBI Ontology Lookup Service (http://www.ebi.ac.uk/ontology-lookup/) listing currently almost 400 and 100 ontologies, respectively [22–24].

2.2 Encoding Ontologies

Several tools and software solutions exist to help users define new and edit existing ontologies, and to encode and modify data according to an existing ontology-definition.

Commonly the encoding of ontologies is based on three specifications: The definition of classes and properties that make up an ontology can be defined via the Web Ontology Language (OWL), a World Wide Web Consortium (W3C) standard [25]. These OWL definitions can be encoded in an XML/RDF file format [26] based on the extensible markup language (XML) [27] and the Resource Description Framework (RDF) [28]. In short, XML is a markup language, which encodes data using tags (‘< >‘) for annotation, and RDF defines the so-called triples in the form of subject-predicate-object expressions to specify statements.

OWL comes in three flavors: OWL Full, OWL DL, and OWL Lite. In a nutshell, OWL Lite covers basic data modeling needs including cardinality restraints. OWL DL extends OWL Lite by introducing description logics and enabling inference, while retaining completeness and decidability of computations. OWL Full includes the complete OWL syntax and even allows extending and augmenting standing OWL vocabulary. In most cases OWL Lite is sufficient for users in order to encode and manage knowledge, while OWL DL is required when logical inference is desirable.

2.3 Editing Ontologies

Protégé is a very popular general software tool that allows definition of and data entry for ontologies and is widely used in industrial and academic settings alike [29] (*see* **Note 2**). The software is open-source and available for all platforms. Furthermore, a browser version is available for facilitating collaboration, online editing and data entry of ontologies [30]. The web application can be downloaded and hosted at a local web server and is also available directly at the University of Stanford (webprotege.stanford.edu). Various ontology-specific tools are available for some ontologies, enabling manipulation [31, 32] and visualization [33, 34] of specific ontologies.

3 Methods

Two of the most common use-cases for working with ontologies in Bioinformatics are presented here, the Gene Ontology and the Ontology for Biological Pathways Exchange.

3.1 Gene Ontology

The Gene Ontology (GO) emerged from a cooperation of three model organism databases: FlyBase, Mouse Genome Informatics (MGI), and the Saccharomyces Genome Database (SGD). A major goal of GO arose from the discovery that there are large amounts of DNA sequences that are identical between species, as well as functional conservation within these genes [3]. The desire for a common site of annotation for genes is a consequence of this finding. The idea of GO is to model the knowledge about genes and gene products across species and to provide access to this information. GO consists of three independent ontologies, each modeling a different domain: biological process, molecular function, and cellular component [3]. Aiming for a generalizing model, the cellular component ontology models the parts and pieces of eukaryotic cells

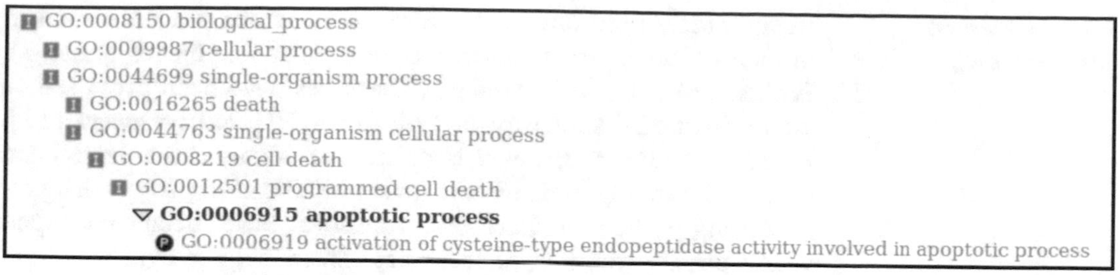

```
GO:0008150 biological_process
  GO:0009987 cellular process
  GO:0044699 single-organism process
    GO:0016265 death
    GO:0044763 single-organism cellular process
      GO:0008219 cell death
        GO:0012501 programmed cell death
          ▽ GO:0006915 apoptotic process
            GO:0006919 activation of cysteine-type endopeptidase activity involved in apoptotic process
```

Fig. 1 Tree-view of the annotation of the apoptosis GO term (GO:0006915). "I" denotes "is-a" relations while "P" denotes "part-of" relations

and their microenvironments. The biological process ontology contains all processes and events that take place within cells and organisms. Finally, the molecular function ontology describes the functional activities of proteins within a cell. GO is constructed in a manner that the ontologies can be understood as a directed acyclic graph. Each node in this graph represents one GO term, its name, annotations and references to other databases or GO domains. In this graph every GO term is connected via edges to its parents and children, representing the ancestry between these GO terms.

This hierarchical modeling enables GO to provide an open controlled vocabulary where the user is able to retrieve knowledge about a certain item, as well as more generalized or detailed knowledge about the GO term (Fig. 1). GO is not static, but continuously developed and curated as the biological knowledge increases [32, 35].

3.1.1 Accessing the Gene Ontology

The GO data can be accessed via a multitude of different tools. The official website (geneontology.org) offers functionality to explore, search and visualize GO terms. Additionally, the whole GO ontology can be downloaded as bulk (ftp://ftp.geneontology.org/pub/go/godatabase/archive/). GO is further available within Cytoscape [33, 36] and for the R Project for Statistical Computing [37] via the GO.db and RamiGO packages [38]. The following code will install the GO data within R and access the one GO term:

```
source("http://bioconductor.org/biocLite.R")
biocLite("GO.db")
library(GO.db)
GOID(xx[[1]])
> [1] "GO:0000001"
Term(xx[[1]])
> [1] "mitochondrion inheritance"
Defintion(xx[[1]])
> [1] "The distribution of mitochondria, [...]"
```

3.1.2 Working with the Gene Ontology

Being widely used and hierarchical in structure, GO has sparked numerous new approaches in bioinformatics (*see* **Notes 3** and **4**). For example, semantic similarity measures have been proposed to assess functional similarity of genes [13, 39] and pathways [12]. Based on these measures a large number of methods have been proposed, ranging from disease gene identification [40] to drug repurposing [41]. Furthermore, methods have been developed which aim at extending or reconstructing GO [42–44].

However, by far the most prominent application of the gene ontology is enrichment analysis for a list of differentially expressed genes [11]. The general idea behind this is to interpret the lists of genes resulting from high-throughput experiments by using statistical methods to find significantly over or underrepresented GO terms within the list of genes. Gene ontology annotations can be cut at different levels, allowing the definition of gene sets for each GO term, leading to a hierarchy of gene sets. The subsequent statistical test for whether a certain biological function (or gene set) is associated with the experimental results is referred to as gene set enrichment analysis [11]. This analysis allows testing for gene sets (or functional groups) that are represented significantly more frequently in a list of differentially expressed genes than in a comparable list of genes that would be selected randomly from all genes. A wide array of different testing strategies and approaches has been proposed [45–48], most notably those available via websites GOstat [10] (gostat.wehi.edu.au) and DAVID [49] (david.abcc.ncifcrf.gov), within Cytoscape using the plugins clueGO [33] or BiNGO [50] and within R using the topGO package.

The following code will install the topGO package and perform an enrichment analysis of example data within R:

```
source("http://bioconductor.org/biocLite.R")
biocLite("topGO")
library(topGO)
data(GOdata)              # load example data set
Godata                    # display example data set
res = runTest(GOdata, algorithm = "classic", statis-
    tic = "fisher")
head(sort(score(res))) # display top scoring GO terms
> GO:0006091 GO:0022900 GO:0009267 [...]
0.0002510612 0.0002510612 0.0004261651 [...]
```

3.2 Biological Pathway Exchange Ontology

Different approaches to standardizing the encoding pathways have been published, for example the Biological Pathway Exchange [4] (BioPAX), the Systems Biology Markup Language (SBML) [51], and the Human Proteome Organizations Proteomics Standards Initiative's Molecular Interaction format (PSI-MI) [52]. Overviews of the capabilities of these standards and assessments were published by Strömbäck and Lambrix [53], by Cary et al. [54] and by

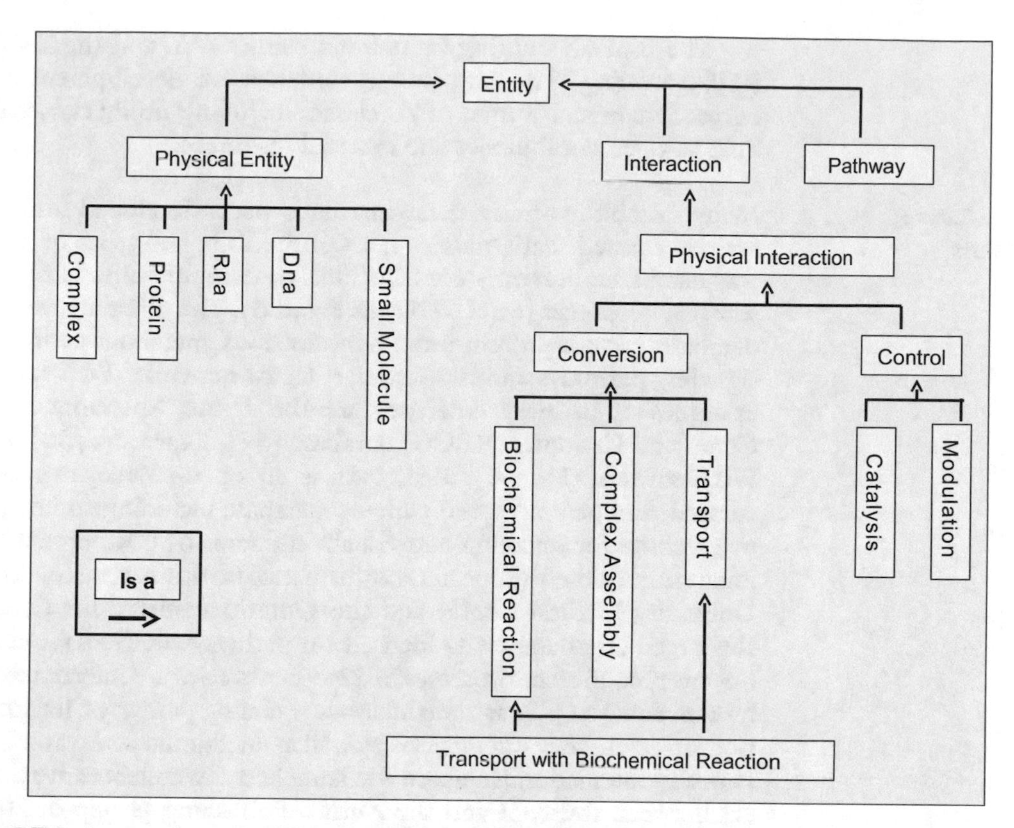

Fig. 2 This diagram shows the central classes and their inheritance relationships [4]. Reproduced according to the BioPAX specification (see www.biopax.org)

Kramer [55]. The following paragraph gives an introduction to the BioPAX ontology and its applications.

A strong advantage of encoding knowledge using an ontology is the fixed modeling space, which eases the exchange and portability of knowledge by ensuring compatibility. BioPAX aims at easing the sharing of pathway knowledge by offering a standardized knowledge model for the pathway domain. Research groups and database providers can use this common model to make their information easily accessible and sharable by users. This ontology defines entities (such as a protein), their properties (such as the name and sequence of a protein), and their relationships to other entities, by using predefined vocabulary. The main classes of BioPAX are physical entities, interactions, and pathways (Fig. 2). Physical entities are defined as all physically existing objects, for example proteins, small molecules, as well as RNA and DNA fragments. The interaction class and its subclasses define all biological processes and events within pathways, e.g., complex assembly, cell transport and regulatory events. Depending on the interaction, its participants are physical entities, interactions, and whole pathways. The pathway class models pathways, which are made up of a number of interaction instances.

The BioPAX ontology is defined using OWL and the XML/RDF encoding. The ontology is under active development and currently contains a total of 70 classes including utility classes for links to open vocabularies and external resources.

3.2.1 Pathway Databases

Many notable pathway databases have been developed and are actively curated. Pathguide.org, a website listing all types of pathway databases, currently contains links to over 500 different pathway data resources [56] (*see* **Notes 5** and **6**). The different types of databases include protein-protein interactions, metabolic pathways, signaling pathways and transcription factor networks. Well known examples of pathway databases are the Kyoto Encyclopedia of Genes and Genomes (KEGG) database [57], Reactome [58], and WikiPathways [59, 60]. Reactome is an open-source, manually curated, and peer-reviewed pathway database including an interactive website for querying and visualizing data [61]. Reactome is a joint effort of the European Bioinformatics Institute, the New York University Medical Center and the Ontario Institute for Cancer Research. The database is focused on pathways in *Homo sapiens*; however, equivalent processes in 22 other species are inferred from human data [61]. Reactome includes signaling pathways, information on regulatory interactions as well as metabolic pathways. The Pathway Interaction Database was launched as a collaborative project between the NCI and the Nature Publishing Group in 2006 [62]. It uses *Homo sapiens* as a model system and offers well annotated and curated signaling pathways. Its data includes molecules annotated using UniProt identifiers and posttranslational modifications [63]. WikiPathways is a community approach to pathway editing [59, 60]. It allows everyone to join and share new pathways or curate existing ones. Pathway Commons is a meta-database aiming at providing a single point of access to publicly available pathway knowledge [64]. It is a collection of pathway databases covering many aspects and common model organisms trying to ease access to a large number of different sources.

These four databases are all freely available for download as BioPAX-export.

3.2.2 Accessing Pathway Knowledge via rBiopaxParser

A number of software tools can be used to access BioPAX-encoded date, e.g., general tools like Protégé [29] or software written specifically for BioPAX, for example Paxtools [31] or rBiopaxParser [34]. The R package rBiopaxParser [34] is specifically implemented to make pathway data that is encoded using the BioPAX ontology accessible within the R Project for Statistical Computing [37]. The software package has been published as open source and has been released as part of Bioconductor (www.bioconductor.org) [65].

The readBiopax function reads in a BioPAX .owl file and generates the internal data format used within this package. As this function has to traverse the whole XML-tree of a database export, it is computationally intensive and may have a long run-time depending on the size of the BioPAX files. The rBiopaxParser can parse BioPAX encoded data into R from the local file system using the XML library and also includes a function to directly download BioPAX-encoded databases. The following R code installs rBiopaxParser from Bioconductor, downloads the Biocarta database from the NCI servers, parses it into R and displays a summary of the data:

```
source("http://bioconductor.org/biocLite.R")
biocLite("rBiopaxParser")
library(rBiopaxParser)
file = downloadBiopaxData("NCI","biocarta")
biopax = readBiopax(file)
print(biopax)
```

A number of convenience functions are available, which aid the user in selecting specific parts or instances of the BioPAX model. Generally, these functions require the parsed BioPAX object as parameter and other parameters that differ from function to function. Examples for these functions are listPathways, listPathwayComponents, and selectInstances. Additionally, functions are available that allow users to merge and modify BioPAX databases and write out changed BioPAX ontologies to the file system. Furthermore, the function pathway2RegulatoryGraph transforms the BioPAX-encoded knowledge of a signaling pathway and compiles it into an interaction graph, which can be used as prior knowledge input for methods for network reconstruction. These graphs rely solely on the available BioPAX information about activations and inhibitions, by classes of or inheriting from, class control. Involved molecules, represented by nodes, are connected via edges depending on the encoded knowledge. The following code selects the WNT signaling pathway encoded in the parsed database and displays its interaction graph via the R plotting interface (Fig. 3):

```
pw_list = listInstances(biopax, class="pathway")
id = "pid_p_100002_wntpathway"
pw_graph = pathway2RegulatoryGraph(biopax, id,
splitComplexMolecules=T)
plotRegulatoryGraph(pw_graph)
```

These generated graph objects provide a starting point for integrating pathway knowledge in all kinds of bioinformatic analyses within R [66] (*see* **Note 7**). Additionally, different plotting options, for example using RCytoscape [67] or the iGraph [68] package are possible.

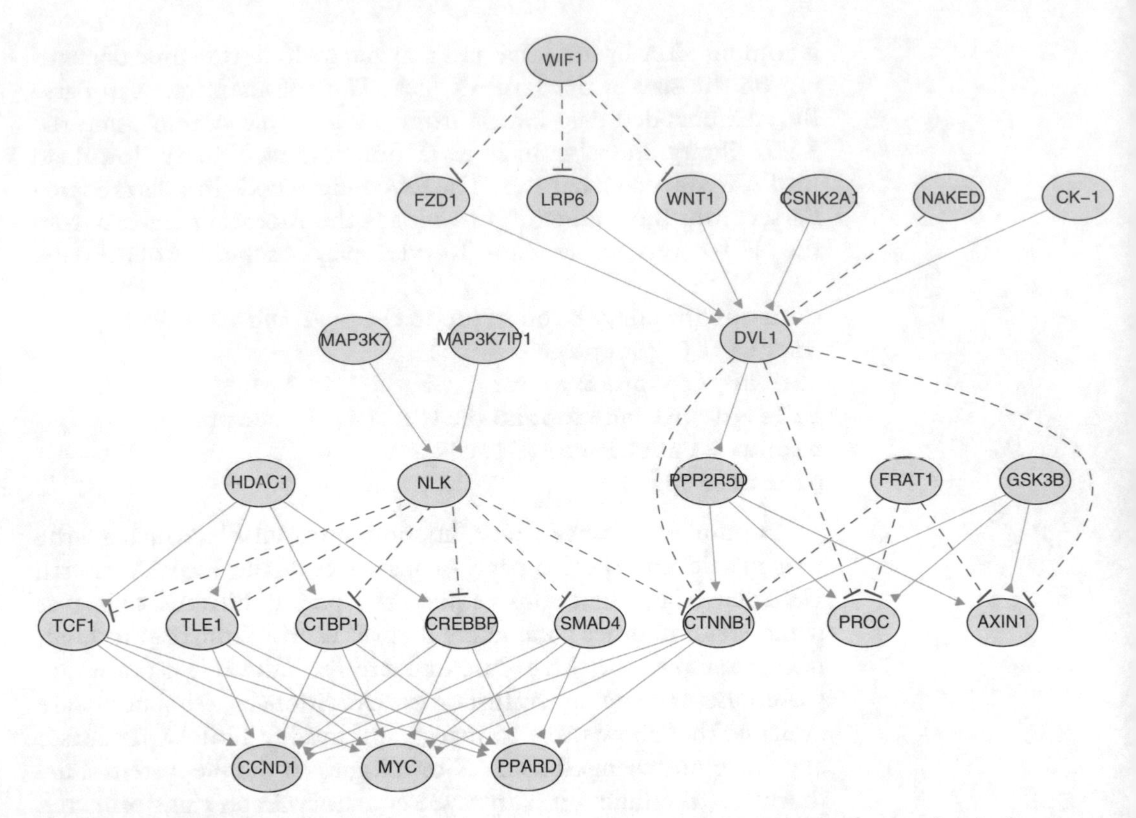

Fig. 3 Generated R plot of the WNT signaling pathway. *Green edges* denote activations and *red edges* denote inhibitions

4 Notes

Naturally, this chapter cannot introduce every ontology available in biomedical research and aims at introducing the theory and offering use cases for popular ontologies as a guideline for fellow researchers. It can be seen that ontologies are powerful and popular tools to encode knowledge. Furthermore, ontologies were the basis for notable developments in knowledge encoding and management and have additionally lead to interesting new approaches in bioinformatic analyses. Some additional notes will help the user to get the most out of ontologies:

1. Discussing and communicating newly developed ontologies with your peers might offer new views on the existing structures and dependencies. Furthermore, publishing and linking a new ontology in journal articles and ontology collections alike will help fellow researchers and create visibility and appraisal for your own work.
2. Tools like Protégé (and even more so its web-based interface) provide a point-and-click approach to ontology creation and make it easy to create a new ontology from scratch. However, it is advisable (a) to read through a few basics of ontology

development (e.g., Noy et al. [7]) and (b) to research via search engines whether an already existing ontology can be used or extended for your own work.

3. The Gene Ontology can be browsed and used for enrichment analysis online at geneontology.org. This is a good place to get started; however, it is usually not feasible to use in (semi-) automated pipelines found in bioinformatic service facilities, where often R-scripts and/or Cytoscape visualizations are commonly run.
4. As shown in "Working with the Gene Ontology", GO allows a wide range of analyses. When in doubt, consult articles published in your field of research in order to understand the conclusions and results that can stem from GO analyses.
5. The advice given in **Note 4** is also pertinent to pathway enrichment analyses: research which methods others are using and how they are applied. Pathway enrichment analyses have gained in popularity compared to GO analyses. Three commonly used databases are KEGG [57], Reactome [58], and PID [62].
6. The focus and granularity of knowledge in pathway databases differs extremely. For example Reactome pathways are focused on cellular mechanisms and processes while KEGG pathways are also disease-specific. Furthermore, Reactome has a highly hierarchical structure; its thousands of pathways are nested into a total of 23 top-tier pathways. Each of these pathways includes dozens of sub-pathways, including sub-sub-pathways and so on.
7. Visualizing BioPAX-encoded pathway data is usually easiest with Cytoscape, while automated analyses are often developed within R [3].

Acknowledgements

This work was funded by the German ministry of education and research (BMBF) grants FKZ01ZX1508 and FKZ031L0024A.

References

1. Gruber TR (1995) Toward principles for the design of ontologies used for knowledge sharing? Int J Hum Comput Stud 43:907–928
2. Berners-Lee T, Hendler J, Lassila O et al (2001) The semantic web. Sci Am 284:28–37
3. Ashburner M, Ball CA, Blake JA et al (2000) Gene ontology: tool for the unification of biology. Nat Genet 25:25–29
4. Demir E, Cary MP, Paley S et al (2010) The BioPAX community standard for pathway data sharing. Nat Biotechnol 28:935–942
5. Burkhardt H, Smith B (1991) Handbook of metaphysics and ontology. Philosophia Verlag, Muenchen
6. Gruber TR (1993) A translation approach to portable ontology specifications. Knowl Acquis 5(2):199–220
7. Noy NF, McGuinness DL et al (2001) Ontology development 101: a guide to creating your first ontology. Stanford knowledge systems laboratory technical report KSL-01-05 and Stanford medical informatics technical report SMI-2001-0880

8. Hitzler P, Krotzsch M, Rudolph S (2011) Foundations of semantic web technologies. CRC Press, Boca Raton, FL
9. du Plessis L, Škunca N, Dessimoz C (2011) The what, where, how and why of gene ontology—a primer for bioinformaticians. Brief Bioinform 12:723–735
10. Beißbarth T, Speed TP (2004) GOstat: find statistically overrepresented Gene Ontologies within a group of genes. Bioinformatics 20:1464–1465
11. Beißbarth T (2006) Interpreting experimental results using gene ontologies. In: Kimmel A, Oliver B (eds) Methods Enzymol. Academic, Waltham, pp 340–352
12. Guo X, Liu R, Shriver CD et al (2006) Assessing semantic similarity measures for the characterization of human regulatory pathways. Bioinformatics 22:967–973
13. Fröhlich H, Speer N, Poustka A, Beißbarth T (2007) GOSim—an R-package for computation of information theoretic GO similarities between terms and gene products. BMC Bioinformatics 8:166
14. Cheng L, Li J, Ju P et al (2014) SemFunSim: a new method for measuring disease similarity by integrating semantic and gene functional association. PLoS One 9:e99415
15. Hoehndorf R, Hancock JM, Hardy NW et al (2014) Analyzing gene expression data in mice with the Neuro Behavior Ontology. Mamm Genome Off J Int Mamm Genome Soc 25:32–40
16. Xu Q, Shi Y, Lu Q et al (2008) GORouter: an RDF model for providing semantic query and inference services for Gene Ontology and its associations. BMC Bioinformatics 9(Suppl 1): S6
17. Chi Y-L, Chen T-Y, Tsai W-T (2015) A chronic disease dietary consultation system using OWL-based ontologies and semantic rules. J Biomed Inform 53:208–219
18. Nadkarni PM, Marenco LA (2010) Implementing description-logic rules for SNOMED-CT attributes through a table-driven approach. J Am Med Inform Assoc 17:182–184
19. Rector AL, Brandt S (2008) Why do it the hard way? The case for an expressive description logic for SNOMED. J Am Med Inform Assoc 15:744–751
20. Degtyarenko K, de Matos P, Ennis M et al (2008) ChEBI: a database and ontology for chemical entities of biological interest. Nucleic Acids Res 36:D344–D350
21. Smith B, Ashburner M, Rosse C et al (2007) The OBO Foundry: coordinated evolution of ontologies to support biomedical data integration. Nat Biotechnol 25:1251–1255
22. Noy NF, Shah NH, Whetzel PL et al (2009) BioPortal: ontologies and integrated data resources at the click of a mouse. Nucleic Acids Res 37:W170–W173
23. Rubin DL, Shah NH, Noy NF (2008) Biomedical ontologies: a functional perspective. Brief Bioinform 9:75–90
24. Côté R, Reisinger F, Martens L et al (2010) The Ontology Lookup Service: bigger and better. Nucleic Acids Res 38:W155–W160
25. McGuinness DL, Van Harmelen F et al (2004) OWL web ontology language overview. W3C Recomm 10
26. Beckett D, McBride B (2004) RDF/XML syntax specification (revised). W3C Recomm 10
27. Bray T, Paoli J, Sperberg-McQueen CM et al (1997) Extensible markup language (XML). World Wide Web J 2:27–66
28. Klyne G, Carroll JJ, McBride B (2004) Resource description framework (RDF): concepts and abstract syntax. W3C Recomm 10
29. Gennari JH, Musen MA, Fergerson RW et al (2003) The evolution of Protégé: an environment for knowledge-based systems development. Int J Hum Comput Stud 58:89–123
30. Horridge M, Tudorache T, Nuylas C et al (2014) WebProtégé: a collaborative Web-based platform for editing biomedical ontologies. Bioinformatics 30:2384–2385
31. Demir E, Babur Ö, Rodchenkov I et al (2013) Using biological pathway data with Paxtools. PLoS Comput Biol 9:e1003194
32. The Geno Ontology Consortium (2014) Gene Ontology Consortium: going forward. Nucleic Acids Res 43(Database issue):D1049–D1056
33. Bindea G, Mlecnik B, Hackl H et al (2009) ClueGO: a Cytoscape plug-in to decipher functionally grouped gene ontology and pathway annotation networks. Bioinformatics 25:1091–1093
34. Kramer F, Bayerlová M, Klemm F et al (2013) rBiopaxParser—an R package to parse, modify and visualize BioPAX data. Bioinformatics 29:520–522
35. The Geno Ontology Consortium (2008) The Gene Ontology project in 2008. Nucleic Acids Res 36:D440–D444
36. Shannon P, Markiel A, Ozier O et al (2003) Cytoscape: a software environment for integrated models of biomolecular interaction networks. Genome Res 13:2498–2504
37. R Core Team (2013) R: a language and environment for statistical computing, Vienna, Austria
38. Schröder MS, Gusenleitner D, Quackenbush J et al (2013) RamiGO: an R/Bioconductor package providing an AmiGO Visualize interface. Bioinformatics 29:666–668

39. Pesquita C, Faria D, Bastos H et al (2008) Metrics for GO based protein semantic similarity: a systematic evaluation. BMC Bioinformatics 9:S4
40. Jiang R, Gan M, He P (2011) Constructing a gene semantic similarity network for the inference of disease genes. BMC Syst Biol 5:S2
41. Andronis C, Sharma A, Virvilis V et al (2011) Literature mining, ontologies and information visualization for drug repurposing. Brief Bioinform 12:357–368
42. Kramer M, Dutkowski J, Yu M et al (2014) Inferring gene ontologies from pairwise similarity data. Bioinformatics 30:i34–i42
43. Dutkowski J, Ono K, Kramer M et al (2014) NeXO Web: the NeXO ontology database and visualization platform. Nucleic Acids Res 42: D1269–D1274
44. Dutkowski J, Kramer M, Surma MA et al (2013) A gene ontology inferred from molecular networks. Nat Biotechnol 31:38–45
45. Zheng Q, Wang X-J (2008) GOEAST: a web-based software toolkit for Gene Ontology enrichment analysis. Nucleic Acids Res 36: W358–W363
46. Huang DW, Sherman BT, Lempicki RA (2009) Bioinformatics enrichment tools: paths toward the comprehensive functional analysis of large gene lists. Nucleic Acids Res 37:1–13
47. Bauer S, Grossmann S, Vingron M, Robinson PN (2008) Ontologizer 2.0—a multifunctional tool for GO term enrichment analysis and data exploration. Bioinformatics 24:1650–1651
48. Eden E, Navon R, Steinfeld I et al (2009) GOrilla: a tool for discovery and visualization of enriched GO terms in ranked gene lists. BMC Bioinformatics 10:48
49. Huang DW, Sherman BT, Tan Q et al (2007) DAVID Bioinformatics Resources: expanded annotation database and novel algorithms to better extract biology from large gene lists. Nucleic Acids Res 35:W169–W175
50. Maere S, Heymans K, Kuiper M (2005) BiNGO: a Cytoscape plugin to assess overrepresentation of Gene Ontology categories in Biological Networks. Bioinformatics 21:3448–3449
51. Hucka M, Finney A, Sauro HM et al (2003) The systems biology markup language (SBML): a medium for representation and exchange of biochemical network models. Bioinformatics 19:524–531
52. Hermjakob H, Montecchi-Palazzi L, Bader G et al (2004) The HUPO PSI's Molecular Interaction format—a community standard for the representation of protein interaction data. Nat Biotechnol 22:177–183
53. Strömbäck L, Lambrix P (2005) Representations of molecular pathways: an evaluation of SBML, PSI MI and BioPAX. Bioinformatics 21:4401–4407
54. Cary MP, Bader GD, Sander C (2005) Pathway information for systems biology. FEBS Lett 579:1815–1820
55. Kramer F (2014) Integration of pathway data as prior knowledge into methods for network reconstruction. Georg-August-Universitat Göttingen, Göttingen
56. Bader GD, Cary MP, Sander C (2006) Pathguide: a pathway resource list. Nucleic Acids Res 34:D504–D506
57. Ogata H, Goto S, Sato K et al (1999) KEGG: Kyoto encyclopedia of genes and genomes. Nucleic Acids Res 27:29–34
58. Joshi-Tope G, Gillespie M, Vastrik I et al (2005) Reactome: a knowledgebase of biological pathways. Nucleic Acids Res 33:D428–D432
59. Kelder T, van Iersel MP, Hanspers K et al (2011) WikiPathways: building research communities on biological pathways. Nucleic Acids Res 40:D1301–D1307
60. Pico AR, Kelder T, van Iersel MP et al (2008) WikiPathways: pathway editing for the people. PLoS Biol 6:e184
61. Vastrik I, D'Eustachio P, Schmidt E et al (2007) Reactome: a knowledge base of biologic pathways and processes. Genome Biol 8: R39
62. Schaefer CF, Anthony K, Krupa S et al (2009) PID: the pathway interaction database. Nucleic Acids Res 37:D674–D679
63. Bauer-Mehren A, Furlong LI, Sanz F (2009) Pathway databases and tools for their exploitation: benefits, current limitations and challenges. Mol Syst Biol 5:290
64. Cerami EG, Gross BE, Demir E et al (2011) Pathway Commons, a web resource for biological pathway data. Nucleic Acids Res 39:D685–D690
65. Gentleman RC, Carey VJ, Bates DM et al (2004) Bioconductor: open software development for computational biology and bioinformatics. Genome Biol 5:R80
66. Kramer F, Bayerlová M, Beißbarth T (2014) R-based software for the integration of pathway data into bioinformatic algorithms. Biology 3:85–100
67. Shannon PT, Grimes M, Kutlu B et al (2013) RCytoscape: tools for exploratory network analysis. BMC Bioinformatics 14:217
68. Csardi G, Nepusz T (2006) The igraph software package for complex network research. Int J Complex Syst 1695

Chapter 7

The Classification of Protein Domains

Natalie Dawson, Ian Sillitoe, Russell L. Marsden, and Christine A. Orengo

Abstract

The significant expansion in protein sequence and structure data that we are now witnessing brings with it a pressing need to bring order to the protein world. Such order enables us to gain insights into the evolution of proteins, their function and the extent to which the functional repertoire can vary across the three kingdoms of life. This has lead to the creation of a wide range of protein family classifications that aim to group proteins based upon their evolutionary relationships.

In this chapter we discuss the approaches and methods that are frequently used in the classification of proteins, with a specific emphasis on the classification of protein domains. The construction of both domain sequence and domain structure databases is considered and we show how the use of domain family annotations to assign structural and functional information is enhancing our understanding of genomes.

Key words Protein domain, Sequence, Structure, Clustering, Classification, Annotation

1 Introduction

The arrival of whole-genome sequencing, heralded by the release of the genome for the first free-living organism, *Haemophilus influenza*, in 1995 [1], promised significant new advances in our understanding of protein evolution and function. Over two decades on we are continuing to understand the full potential of the genome sequencing projects, with over 26000 completed genome projects released to the biological community in 2016 alone [2] (Fig. 1). This rapid increase in genomic sequence data challenges us to unravel the patterns and relationships that underlie the pathways and functions that form genome landscapes.

Detailed comparisons between genomes require the annotation of gene names and functions. Such annotations can be derived from biochemical analysis—a limited approach given the number of sequences we need to functionally classify—or through computational analysis to identify evolutionary relationships. The exploitation of sequence homology for genome annotation is of considerable use because sequence homologues generally share

Jonathan M. Keith (ed.), *Bioinformatics: Volume I: Data, Sequence Analysis, and Evolution*, Methods in Molecular Biology, vol. 1525, DOI 10.1007/978-1-4939-6622-6_7,

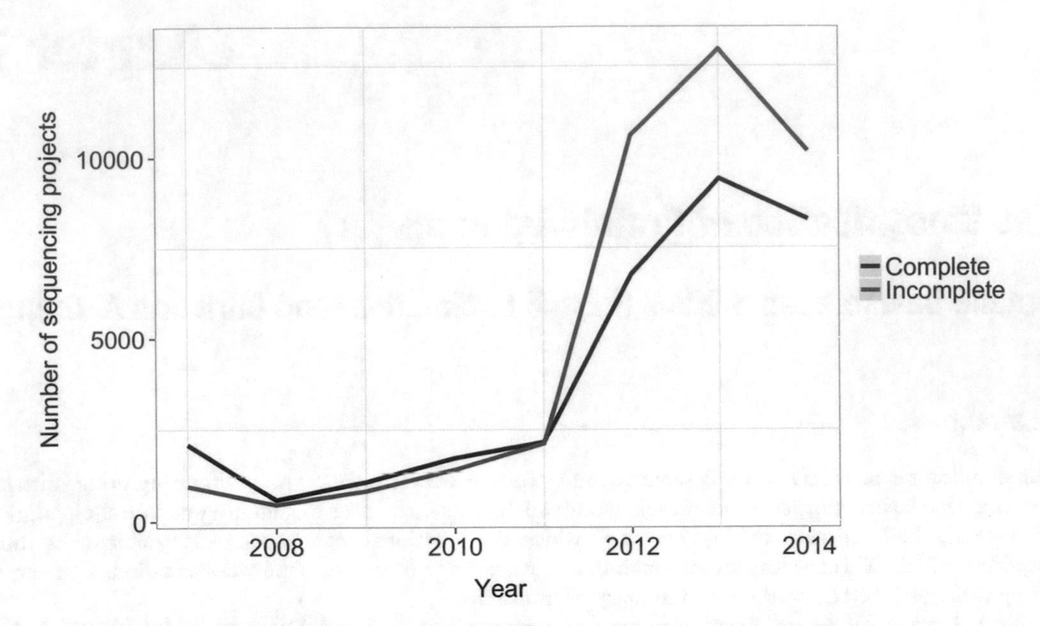

Fig. 1 The number of complete (*black*) and incomplete (*grey*) genome sequencing projects released each year from 2007 to 2016 as reported by the GOLD (Genomes Online Database) resource [2]. As of October 2016, 26402 complete and 14938 genome sequencing projects have been reported

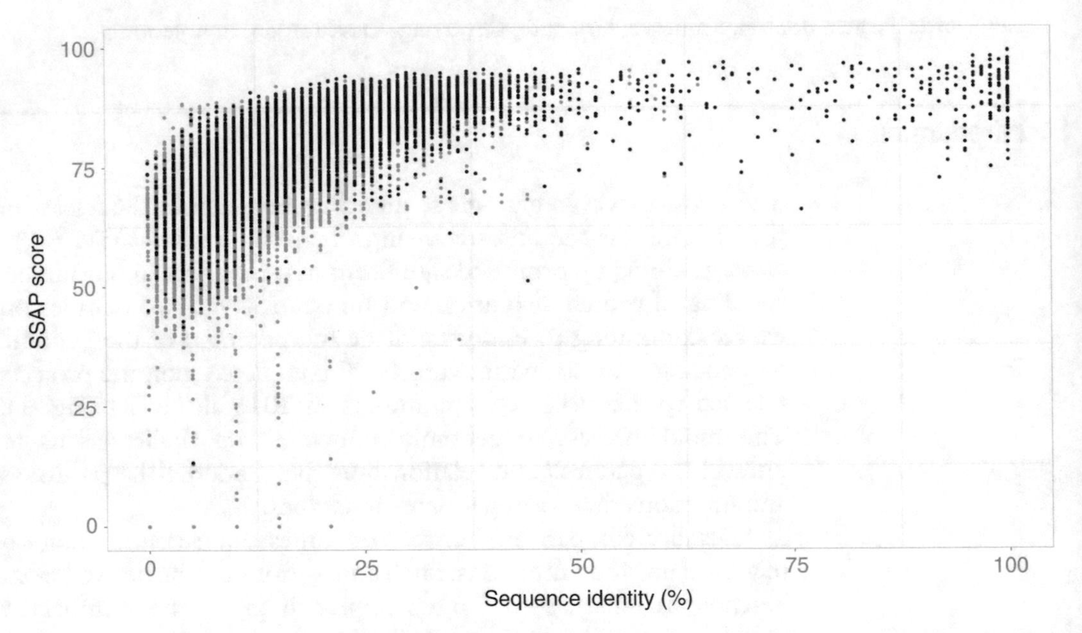

Fig. 2 The correlation between structure similarity and sequence identity for all pairs of homologous domain structures in the CATH domain database. Structural similarity was measured by the SSAP structure comparison algorithm, which returns a score in the range of 0–100 for identical protein structures. *Dark grey circles* represent pairs of domains with the same function, and *light grey* circles represent those with different functions

the same protein fold and often share similarities in their function depending on the degree of relatedness between them (Fig. 2). The diversity of such relationships can be seen within families of proteins that group together protein sequences thought to share a common ancestor. Protein families often exhibit considerable divergence in sequence, structure, and sometimes function. Over the past two decades a varied range of protein family classifications have been developed with a view to transferring functional information from experimentally characterized genes to families of sequence relatives.

In this chapter we discuss the different approaches to classifying proteins, with a particular emphasis on the classification of protein domains into families. We see how the use of domain-centric annotations is pivotal to maximizing our understanding of genomes, and enables experimentally obtained annotation to be imputed to a vast number of protein sequences.

2 What is a Protein Domain?

The protein domain has become a predominant concept in many areas of the biological sciences. Despite this, a single universally accepted definition does not exist, an ambiguity that is partly attributable to the subjective nature of domain assignment. At the sequence level, protein domains are considered to be homologous portions of sequences encoded in different gene contexts that have remained intact through the pathways of evolution. From this perspective, domains are the "evolutionary unit" of protein sequences, and increasing evidence is emerging through the analysis of completed genome data that a large proportion of genes (up to 80 % in eukaryotes) comprise more than one domain [3, 4]. In structural terms, domains are generally observed as local, compact units of structure, with a hydrophobic interior and a hydrophilic exterior forming a globular-like state that cannot be further subdivided. As such, domains may be considered as semi-independent globular folding-units. This property has enabled them to successfully combine with other domains and evolve new functions.

The duplication and recombination of domains has been a staple of evolution: genes have been divided and fused using a repertoire of preexisting components. During the course of evolution, domains derived from a common ancestor have diverged at the sequence level through residue substitution or mutation and by the insertion and deletion of residues, giving rise to families of homologous domain sequences.

Domain assignment provides an approximate view of protein function in terms of a "molecular parts bin": the overall function is a combination of the constituent parts. This view of protein function is sometimes too simplistic, particularly when compared to experimentally derived evidence. Nevertheless, it does provide accurate functional information. Moreover, the classification of domains facilitates study of the evolution of proteins, in particular through the analysis and comparison of domain sequence and structure families.

3 Classification of Domains from Sequence

As sequence databases became more populated in the early 1970s, an increasing amount of research focused on the development of methods to compare and align protein sequences. Needleman and Wunsch [5] devised an elegant solution using dynamic programming algorithms to identify optimal regions of local or global similarity between the sequences of related proteins. More recently, algorithms such as FASTA [6] and BLAST [7] have approximated these approaches in order to make sequence searching several orders of magnitude faster. This has been a significant development, enabling us to work with the ever-increasing collections of sequence data by performing millions of sequence comparisons within reasonable timeframes. Such efforts are essential if we are to identify a significant proportion of the evolutionary relationships within the sequence databases and provide order to these databases through protein clustering and classification.

When sequence comparisons are performed between proteins containing more than one domain, complications can arise when homologous sequence regions are encountered in otherwise non-homologous contexts. This can lead to the assignment of a single function to a multifunctional multidomain protein or to the "chaining problem," which results in unrelated proteins being grouped together by sequence clustering methods [8]. For example, the pairwise comparison of two protein sequences may reveal significant sequence similarity and consequently they may be considered homologous. However, closer inspection of the alignment might reveal that the sequence similarity is only shared over partial regions of each sequence (Fig. 3), suggesting the identification of homologous domains that belong to the same family. Homologous domains like these are frequently found embedded within different proteins with different domain partners and it is this domain recurrence that many domain classification resources exploit when identifying and clustering domain sequences.

3.1 Automatic Domain Sequence Clustering

Automated domain clustering algorithms are used in the construction of many domain sequence classifications to generate an initial clustering of domain-like sequences. This method is often followed by varying levels of expert-driven validation of the proposed domain families.

To implicitly predict domain boundaries from sequence data alone is an immensely challenging problem, though a number of methods have been devised using predicted secondary structure, protein folding or domain-linker patterns recognized through neural networks. Despite extensive research, none of these methods are reliable enough or well enough established to use in large-scale sequence analysis. Instead, most automatic domain clustering

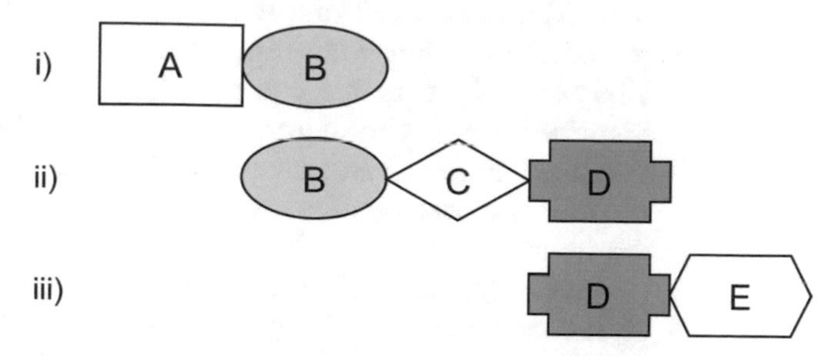

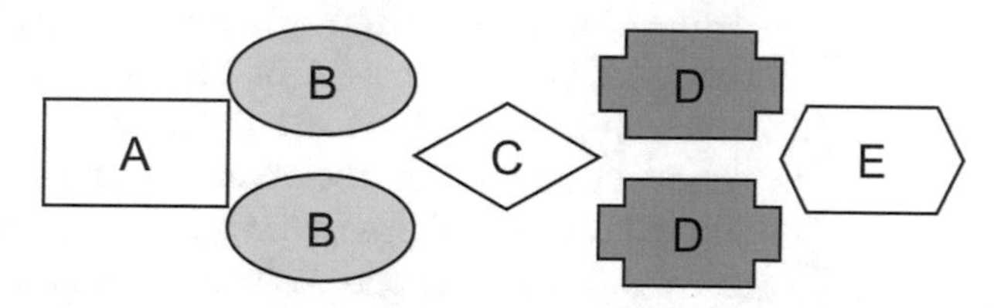

Fig. 3 The domain problem in protein sequence comparison. (**a**) The three sequences (*i, ii,* and *iii*) are incorrectly clustered due to sequence similarity with shared domains (B and D) in sequence ii. (**b**) The identification and excision of domains A to E enables subsequent clustering of the domains into five domain families

methods rely on pairwise alignment information to identify domain sequences and classify them in two main steps. These are, firstly, to divide protein sequences into domain-like regions corresponding to local regions of similarity (i.e., identify recurring domain sequences); and secondly to cluster the sequence fragments. Although conceptually simple, in practice the output of different approaches can vary widely, for example in the number and size of domain families that are generated (*see* **Note 1**). Nevertheless, the clustering of homologous domains remains a rational approach to organizing protein sequence data and many domain clustering approaches have been developed including ProDom [9], EVEREST [10] and ADDA [11].

ProDom is an automated sequence clustering and assignment method that clusters all non-fragment sequences (partial gene sequences that may not correspond to whole domains are excluded) in the UniProtKB (UniProt Knowledgebase) SWISS-PROT and TrEMBL [12] sequence databases. First the database sequences are sorted by size order into a list. The smallest sequence is identified and related sequence regions in the list are identified using PSI-BLAST [13] similarity scores and removed from the list into the current domain cluster. The remaining sequences are once again sorted by size, and the next smallest sequence used as a query to build a new domain cluster. This process is repeated until the supply of sequences in the database is exhausted.

The EVEREST (EVolutionary Ensembles of REcurrent SegmenTs) program [10] uses a combination of machine learning and statistical modeling to interpret the data produced from large-scale all-against-all sequence comparisons. Domain sequences are first identified using the alignment data and clustered into putative domain families. Machine learning methods are then applied to identify the best families, upon which a statistical model is built. An iterative process is then used to scan the models against the original sequence database to recreate a database of domain-like segments which are subsequently re-clustered.

The ADDA algorithm (Automatic Domain Decomposition Algorithm) [11] is another entirely automated approach to domain delineation and clustering with a particular emphasis on the clustering of very large sets of protein sequences. Alignment results from all-against-all BLAST sequence comparisons are used to define a minimal set of domains with as many possible overlapping aligned regions. Pairwise profile-profile comparisons are then used to form links between assigned domains and domain families are generated for sets of domains with links above a given threshold.

3.2 Whole-Chain Sequence Clustering

A number of methods attempt to circumvent the problem of assigning domains by clustering whole-chain protein sequences. For example: CLUSS [14], CLAP [15], FlowerPower from the Phylofacts resource [16], memory-constrained UPGMA from ProtoNet [17], Tribe-MCL [18], UCLUST and USEARCH [19]. CD-HIT [20] is a widely used clustering program that reduces sequence redundancy. It finds the longest sequence in the input and defines it as the first cluster representative. The remaining sequences are ranked in terms of length and the longest to shortest are compared against this first sequence to see whether each is redundant or dissimilar enough to be a new representative [20, 21]. The algorithm, kClust [22] is another popular clustering program that is comparable to CD-HIT and UCLUST in terms of sensitivity, speed, and false discovery rate but differs in that it uses an alignment-free pre-filter and dynamic programming.

Such conservative approaches tend to generate many more sequence families compared to clustering at the domain level, where recurrence of domains leads to a greater reduction in redundancy. However, this cautious approach to clustering sequence space can provide insights into the evolution of domain architectures (the types and orderings of domains along a chain) and associated functions across protein families.

3.3 Families Represented by Multiple Sequence Alignments

The earliest domain family classifications were only able to capture a limited number of evolutionary relationships because the reliability of pairwise sequence comparison quickly descends in the so-called Twilight Zone [23] of sequence similarity (<30 % sequence identity). Comparison of protein structures has shown that evolutionarily related domains can diverge to a point of insignificant sequence similarity—pairwise sequence comparisons do not recognize these relationships even with the use of substitution matrices that allow variation in amino acid residues caused by mutation. The expansion of sequence databases over the past two decades, however, has enlarged the populations of protein families and this has improved homologue recognition in the twilight zone through the derivation of family-specific multiple sequence alignments. Such multiple alignments are an authoritative analytical tool as they create a consensus summary of the domain family, encapsulating the constraints of evolutionary change, and can reveal highly conserved residues that may correspond to function, such as an active site. The unique features that define each domain family can then be captured in the form of patterns, or as profiles represented by Position Specific Scoring Matrices (PSSMs) or hidden Markov models (HMMs).

3.3.1 Patterns

Family-specific patterns are built by automated analysis of multiple sequence alignments in order to characterize highly conserved structural and functional sequence motifs that can be used as a core "fingerprint" to recognize distant homologues. These are usually represented in the form of regular expressions that search for conserved amino acids that show little or no variability across constituent sequences. Patterns are relatively simple to build and tend to represent small contiguous regions of conserved amino acids such as active sites or binding sites rather than providing information across a domain family. They are rigid in their application of pattern matching, which can mean that even small divergences in amino acid type can result in matches to related sequence being missed.

3.3.2 Profiles

Like patterns, profiles can also be automatically built from a multiple sequence alignment, but unlike patterns, they tend to be used to form a consensus view across a domain family. A profile is built to describe the probability of finding a given amino acid at a given location in the sequence relatives of a domain family using position-specific amino acid weightings and gap penalties. These values are stored in a table or matrix and are used to calculate similarity scores between a profile and query sequence. Profiles can be used to represent larger regions of sequence and allow a greater residue divergence in matched sequences in order to identify more divergent family members. A threshold can be calculated for a given set of family sequences to enable the reliable identification and inclusion of new family members into the original set of sequences. Profile methods such as HMMs have been shown to be highly discriminatory in identifying distant homologues when searching the sequence databases [24]. It is also clear that the expansion of diversity within the sequence and domain databases will bring an increase in the sensitivity of profile methods, allowing us to go even further in tracing the evolutionary roots of domain families. The HMMER project [25] comprises a set of widely used programs that use profile HMMs to search for homologues in sequence databases. Profile HMMs are built and then can be searched against a protein sequence database (hmmsearch), or a protein sequence can be searched against a profile HMM database (hmmscan). An HMM-based iterative search against a protein sequence database can be run with jackhmmer, analogous to a PSI-BLAST search. Another method recently developed to find remote homologues is HHblits (HMM-HMM-based lightening-fast iterative sequence search) [26], an extension of HHsearch. The UniProtKB is clustered at 20 % sequence identity using kClust, and each subsequent cluster converted into a HMM. These HMMs differ as they take into account the context of the surrounding amino acids for a particular position in an MSA, which improves the sensitivity and profile HMM quality. Sequences are iteratively searched against the HMM database and added to produce a new profile HMM if they have a significant score, as in jackhmmer.

3.4 Domain Sequence Classifications

A variety of domain sequence databases are now available, most of which, as has been discussed, use patterns or profiles (or in some cases both) to build a library of domain families. Such libraries can often be browsed via the internet or used to annotate sequences over the internet using comparison servers. Some classifications, such as Pfam, provide libraries of HMMs and comparison software to enable the user to generate automated up-to-date genome annotations on their local computer systems. A list of the most popular databases is shown in Table 1.

The consolidation of domain classifications is a logical progression towards a comprehensive classification of all protein domains. However, with so many domain sequence and structure domain classifications, each with their own unique formats and outputs, it can be difficult to choose which one to use or how to meaningfully combine the results from separate sources.

One solution is the manually curated InterPro database [27] (Integration Resource of Protein Families) at the EBI in the UK. This resource integrates 11 of the major protein family classifications and provides regular mappings from these family resources onto primary sequences in Swiss-Prot and TrEMBL. Contributing databases include: CATH-Gene3D [28], HAMAP [29], PANTHER [30], PIRSF [31], Pfam [32], PRINTS [33], ProDom [9], PROSITE [34], SMART [35], SUPERFAMILY [36], and TIGRFAMs [37].

This integration of domain family resources into homologous groups not only builds upon the individual strengths of the component databases, but also provides a measure of objectivity for the

Table 1
Protein domain sequence classifications

Database	Description	URL
ADDA	Automatic domain decomposition and clustering based on a global maximum likelihood model	http://ekhidna.biocenter.helsinki.fi/sqgraph
CDD	Conserved domain database	http://www.ncbi.nlm.nih.gov/Structure/cdd/cdd.shtml
EVEREST	Large-scale automatic determination of domain families	http://www.everest.cs.huji.ac.il
InterPro	Integrated domain family resource	http://www.ebi.ac.uk/interpro/
IProClass	Integrated protein classification database	http://pir.georgetown.edu/iproclass/
Pfam	Database of multiple sequence alignments and hidden Markov models	http://pfam.xfam.org
PIR	Protein Information Resource	http://pir.georgetown.edu/
PRINTS	Database of protein fingerprints based on protein motifs	http://www.bioinf.manchester.ac.uk/dbbrowser/PRINTS/index.php
ProDom	Automatic classification of protein domains	http://prodom.prabi.fr/prodom/current/html/home.php
PROSITE	Database of domain families described by patterns and profiles	http://www.expasy.org/prosite/
ProtoNet	Automated hierarchical classification of UniProtKB database	http://www.protonet.cs.huji.ac.il
SMART	Collection of protein domain families focusing on those that are difficult to detect	http://smart.embl-heidelberg.de
TIGRFAMs	Protein families represented by hidden Markov models	http://www.tigr.org/TIGRFAMs/

database curators, highlighting domain assignment errors or areas where individual classifications lack representation.

Each InterPro entry includes one or more descriptions from the individual member databases that describe an overlapping domain family and an abstract providing annotation to a list of precompiled SWISS-PROT and TrEMBL sequence matches. This match list is also represented in a tabulated form, detailing protein accession numbers and the corresponding match boundaries within the amino acid sequences for each matching InterPro entry. This is complemented by a graphical representation in which each unique InterPro sequence match is split into several lines, allowing the user to view profile matches from the same and other InterPro entries. Proteins can also be viewed in the form of a consensus of domain boundaries from all matches within each entry, thus allowing proteins sharing a common architecture to be grouped.

A selection of the sequence-based domain-based resources (PROSITE, PFAM, and SMART) are described in greater detail below.

3.4.1 PROSITE

PROSITE [34] began as a database of sequence patterns, its underlying principle being that domain families could be characterized by the single most conserved motif observed in a multiple sequence alignment. More recently it has also provided an increasing number of sequence profiles. PROSITE multiple sequence alignments are derived from a number of sources, such as a well-characterized protein family, the literature, sequence searching against SWISS-PROT and TrEMBL and from sequence clustering. PROSITE motifs or patterns are subsequently built to represent highly conserved stretches of contiguous sequence in these alignments. These typically correspond to specific biological regions such as enzyme active sites or substrate binding sites. Accordingly, PROSITE patterns tend to embody short conserved biologically active regions within domain families, rather than representing the domain family over its entire length. Each pattern is manually tested and refined by compiling statistics that reflect how often a certain motif matches sequences in SWISS-PROT. The use of patterns provides a rapid method for database searching, and in cases where global sequence comparison becomes unreliable, these recurring "fingerprints" are often able to identify very remote sequence homologues.

PROSITE also provides a number of profiles to enable the detection of more divergent domain families. Such profiles include a comparison table of position-specific weights representing the frequency distribution of residues across the initial PROSITE multiple sequence alignment. The table of weights is then used to generate a similarity score that describes the alignment between the whole or partial PROSITE profile and a whole or partial sequence. Each pair of amino acids in the alignment is scored based on the probability of a particular type of residue substitution at each position, with scores above a given threshold constituting a true match.

3.4.2 Pfam

Pfam [32] is a comprehensive domain family resource providing a collection of HMMs for the identification of domain families, repeats and motifs. There are two types of families in Pfam: high quality and manually curated Pfam-A families that are represented by the HMMs that cover most of the common protein domain families, and Pfam-B families, which have a higher coverage of sequence space but are less reliable due to being automatically generated. In the latest release of Pfam, version 27.0, new families were generated using data from two main sources: structural domain data from CATH, and human sequences [32]. Each of these domain sequences were iteratively searched against sequence data derived from the UniProtKB with the jackhmmer program to build a seed alignment. The output from this step was then verified to generate the seed alignment for a new Pfam-A family.

The high quality seed alignments are used to build HMMs to which sequences are automatically aligned to generate the final full alignments. If the initial alignments are deemed to be diagnostically unsound the seed is manually checked, and the process repeated until a sound alignment is generated. The parameters that produce the best alignment are saved for each family so that the result can be reproduced. Pfam-B families are created using the ADDA algorithm [38].

3.4.3 SMART

The SMART database (Simple Modular Architecture Research Tool) [35] represents a collection of domain families with an emphasis on those domains that are widespread and difficult to detect. As with Pfam, the SMART database holds manually curated HMMs but unlike Pfam, the "seed" alignments are derived from sequence searching using PSI-BLAST or, where possible, are based on tertiary structure information. An iterative process is used to search for additional relatives for inclusion into each sequence alignment, until no further sequence relatives can be identified. SMART domain families are selected with a particular emphasis on mobile eukaryotic domains and as such are widely found among nuclear, signaling, and extracellular proteins. Annotations detailing function, subcellular localization, phyletic distribution, and tertiary structure are also included within the classification. In general, SMART domains tend to be shorter than structural domains.

3.5 Protein Sequence Classification

There are also a number of databases that classify protein families using whole-protein sequences. The widely used resources: HAMAP, PANTHER, PIRSF, and TIGRFAMs are described further below.

3.5.1 HAMAP

The HAMAP resource (High-quality Automated and Manual Annotation of Proteins) [29] automatically classifies protein sequences into manually curated families using sequence homology information and then functionally annotates them. Family seed profiles are created from reviewed UniProtKB/Swiss-Prot sequences, together with representative sequences covering a broad taxonomic distribution. Iterative BLAST searches are used to find additional homologous sequences, which are verified against family data in other resources including HOGENOM [39], OrthoDB [40], TIGRFAMs, Pfam, and PROSITE. A profile is generated from this seed alignment, which is then scanned against random protein sequences in UniProtKB. HAMAP provides a high-quality automatic pipeline for the annotation of the unreviewed section of the UniProtKB/TrEMBL.

3.5.2 PANTHER

The PANTHER resource [30] annotates protein sequence families with gene and protein functional information to reflect events in gene evolution. Phylogenetic trees are used to infer experimental annotations from a few model organisms with fully sequenced genomes, onto homologues. To create families for a new release, all new protein sequences are scanned against the HMMs from the previous release using InterProScan [41]. Each sequence is assigned to the family with the largest significant score. The CluSTr algorithm is also used to define new families [42]. A phylogenetic tree is built for each family alignment with the GIGA program [43]. Finally, the tree nodes are annotated using three different attributes: protein class membership, GO terms, and subfamily information.

3.5.3 PIRSF

The PIRSF (Protein Information Resource SuperFamily) classification system [44] provides evolutionary relationships for whole proteins using a network structure. Two types of data are provided: preliminary automatically generated sequence clusters, and curated protein families. Protein family members are found within a "homeomorphic" level as they are homologous and homeomorphic, i.e., they have full-length sequence similarity and a common domain architecture. Below this level are subfamily nodes that represent functional subclassification and changes in domain architecture. Above the homeomorphic level are superfamily nodes connecting distantly related families, and orphan proteins that share common domains.

3.5.4 TIGRFAMs

TIGRFAM [37, 45] protein families are composed of protein family alignments and HMMs. The main purpose of this resource is to provide models for functional annotation. Models are provided for superfamilies, subfamilies, and "equivalogs," which are homologous proteins that have performed the same function since their last common ancestor.

4 Classification of Domains from Structure

As we have seen so far, the classification of domains at the sequence level is most commonly based upon the use of sequence comparison methods to identify homologous domains belonging to the same domain family. However, despite advances in sequence comparison algorithms, such as the use of mutation matrices and development of profile-based methods, very remote homologues in the midnight zone (<15 % sequence identity) often remain undetected. Consequently, most sequence-based domain family classifications tend to group closely related sequences sharing significant sequence similarity and possessing similar or identical biological functions.

It is well recognized that there are many distant relationships that can only be identified through structure comparison. This is because protein structure is much more highly conserved through evolution than is its corresponding amino acid sequence. The comparison of proteins at the structure level often allows the recognition of distant ancestral relationships hidden within and beyond the twilight zone of sequence comparison, where sequences share less than 30 % sequence identity (Fig. 2). Under this principle, when sufficient structural data was available, a number of protein structure classifications were developed in the 1990s, each of which aimed to classify the evolutionary relationships between proteins based on their three-dimensional structures. Using structure-based comparisons in a manner comparable to sequence based searching, it becomes possible to traverse the barriers that prevent sequence searches from identifying distant homologues. The ability to classify newly structurally characterized proteins into preexisting and novel structural families has allowed far reaching insights to be gained into the structural evolution of proteins. Whilst the structural core is often highly conserved across most protein families, revealing constraints on secondary structure packing and topology, analysis can also identify considerable structural embellishments that often correspond to changes in domain partners and function.

Like many of the sequence classifications, most structure classifications have also been established at the domain level, *see* Table 2. Each has been constructed using a variety of algorithms and protocols to recognize similarities between proteins. Some groups use all-against-all comparisons to calculate structural relationships with less emphasis on the construction of a formal classification (e.g., the Dali Dictionary) whilst other resources have developed hierarchies based upon these sequence and structure relationships (e.g., the CATH domain database).

4.1 Identification of Domain Boundaries at the Structural Level

Over 40 % of known structures in the Protein Data Bank [46] are multidomain proteins, a percentage that is likely to increase as structure determination methods, such as X-ray crystallography, become better able to characterize large proteins. It is very difficult to reliably assign domain boundaries to distant homologues by sequence-based methods; however, in cases where the three-dimensional structure has been characterized, putative domain boundaries can be delineated by manual inspection through the use of graphical representations of protein structure. Nonetheless, the delineation of domain boundaries by eye is a time consuming process and is not always straightforward, especially for large proteins containing many domains or discontinuous domains in which one domain is interrupted by the insertion of another domain.

The concept of a structural domain was first introduced by Richardson, defining it as a semi-independent globular folding

Table 2
Protein domain structure classifications

Database	Description	URL
CATH-Gene3D	CATH is a hierarchical classification of protein domain structures which clusters proteins at four major levels, **C**lass, **A**rchitecture, **T**opology and **H**omologous superfamily	http://www.cathdb.info
Dali Database	The Dali Database represents a structural classification of recurring protein domains with automated assignment of domain boundaries and clustering	http://ekhidna.biocenter.helsinki.fi/dali/start
ENTREZ/MMDB	MMDB contains precalculated pairwise comparison for each PDB structure. Results are integrated into ENTREZ	http://www.ncbi.nlm.nih.gov/structure
HOMSTRAD	**HOM**ologous **STR**ucture alignment database. Includes annotated structure alignments for homologous families	http://mizuguchilab.org/homstrad/
SCOP	A **S**tructural **C**lassification **O**f **P**roteins. Hierarchical classification of protein structure that is manually curated. The major levels are family, superfamily, fold and class	http://scop.mrc-lmb.cam.ac.uk/scop/
SCOP2	A successor of SCOP that uses complex networks to describe numerous relationship types between protein domains	http://scop2.mrc-lmb.cam.ac.uk
SCOPe	A Structural Classification of Proteins—extended.	http://scop.berkeley.edu
ECOD	Evolutionary Classification of Protein Domains	http://prodata.swmed.edu/ecod/

unit [47]. The following criteria, based upon this premise, are often used to characterize domains:

(a) A compact globular core;

(b) More intra-domain residue contacts than inter-domain contacts;

(c) Secondary structure elements are not shared between domains, most significantly Beta-strands; and

(d) Evidence of domain as an evolutionary unit, such as recurrence in different structural contexts.

The growth in structure data has lead to the development of a variety of computer algorithms that automatically recognize domain boundaries from structural data, each with varying levels of success. Such methods often use a measure of geometric compactness, exploiting the fact that there are more contacts between residues within a domain than between neighboring domains, or searching for hydrophobic clusters that may represent the core of a structural domain. Many of these algorithms perform well on simple multidomain structures in which few residue contacts are found between neighboring domains, though the performance levels tend to be disappointing for more complex structures in which domains are intricately connected. Despite this, their speed (often less than 1 s per structure) does allow a consensus approach to be used for domain assignment, where the results from a series of predictions from different methods are combined, and in cases of disagreement, can be manually validated. To some extent the conflicting results produced by different automated methods is expected if we consider the fact that there is no clear quantitative definition of a domain and the high levels of structural variability in many domain families.

The advent of structure comparison enabled the principle of recurrence to again play a central role in domain definition. The observation of a similar structure in a different context is a powerful indicator of a protein domain, forming the rationale for methods that match domains in newly determined structures against libraries of classified domains. This concept is rigorously applied in the SCOP [48] database, where domains are manually assigned by visually inspection of structures. In other domain classifications such as CATH [49], automated methods are also used to identify putative domains that may not yet have been observed in other contexts. The Dali Database [50] includes an automated algorithm that identifies putative domains using domain recurrence in other structures.

4.2 Methods for Structural Comparison

The use of automated methods for structural comparison of protein domains is essential in the construction of domain structure classifications. Structural comparison and alignment algorithms were first introduced in the early 1970s and methods such as rigid body superposition are still used today for superimposing structures and calculating a similarity measure (root mean square deviation). This is achieved by translation and rotation of structures in space relative to one another in order to minimize the number of non-equivalent residues. Such approaches use dynamic programming, secondary structure alignment and fragment comparison to enable comparison of more distantly related structures in which extensive residue insertions and deletions or shifts in secondary structure orientations have occurred. More recently, some domain structure classifications have employed rapid comparison methods, based on secondary structure, to approximate these approaches (e.g., SEA [51], VAST [52], and CATHEDRAL [53]). This enables a large number of comparisons to be performed that are used to assess the significance of any match via a rigorous statistical analysis. Where necessary, potential relatives can then be subjected to more reliable, albeit more computationally intensive, residue-based comparisons (e.g., SSAP [54] and Dali [55]).

FATCAT (Flexible structural AlignmenT by Chaining Aligned fragment pairs allowing Twists) [56] produces a flexible structural alignment and also accounts for structural rearrangements, which distinguishes it from other methods. In a similar method to DALI, each structure in the pair being compared is broken into fragments before the algorithm builds an alignment. These fragments are compared between each structure and scored to find the most similar aligned fragment pairs (AFPs). The FATCAT algorithm uses dynamic programming to connect AFPs whilst combining gaps and twists between each AFP where applicable.

STRUCTAL [57–59] is an algorithm designed to perform large-scale pairwise structural alignments and was originally applied to aligning members in the same fold groups within SCOP. The structural backbones are aligned using multiple cycles of dynamic programming and least-squares fitting. It has been made with the option to modify the method to take side chain orientation into account when aligning, or even change the cost of opening a gap in the alignment. Multiple alignments can also be generated on a set of related structures.

Differences between methods, and consequently classifications, can arise due to differences in how structural relationships are represented. However, recent attempts to combine sequence- and structure-based domain family resources in the manually curated InterPro database [27] should maximize the number of distant relationships detected.

4.3 Domain Structure Classification Hierarchies

The largest domain structure classification databases are organized on a hierarchical basis corresponding to differing levels of sequence and structure similarity. The terms used in these hierarchies are summarized in Table 3.

At the top level of the hierarchy is domain class, a term that refers to the proportion of residues in a given domain adopting an alpha-helical or beta-strand conformation. This level is usually divided into four classes: mainly alpha, mainly beta, alternating alpha-beta (in which the different secondary structures alternate along the polypeptide chain), and alpha plus beta (in which mainly alpha and mainly beta regions appear more segregated). In the CATH database, these last two classes are merged into a single alpha-beta class as a consequence of the automated assignment of class. CATH also uses a level beneath class classifying the architecture of a given domain according to the arrangement of secondary structures regardless of their connectivity (e.g., barrel-like or layered sandwich). Such a description is also used in the SCOP classification, but it is less formalized, often appearing for a given structural family rather than as a completely separate level in the hierarchy.

Within each class, structures can then be further clustered at the fold (also known as topology) level according to equivalences in the orientation *and* connectivity of their secondary structures. Cases in which domains adopt highly similar folds are often indicative of an evolutionary relationship. However, care must be taken:

Table 3
Overview of hierarchical construction of domain structure classifications

Level of hierarchy	Description
Class	The class of a protein domain reflects the proportion of residues adopting an alpha-helical or beta-strand conformation within the three-dimensional structure. The major classes are mainly alpha, mainly beta, alternating alpha/beta and alpha + beta. In CATH the alpha/beta and alpha + beta classes are merged
Architecture	This is the description of the gross arrangement of secondary structures in three-dimensional space independent of their connectivity
Fold/topology	The gross arrangement of secondary structures in three-dimensional space and the orientation and connectivity between them
Superfamily	A group of proteins whose similarity in structure and function suggests a common evolutionary origin
Family	Proteins clustered into families have clear evolutionary relationships. This generally means that pairwise residue identities between the proteins are 30 % and greater. However, in some cases, similar functions and structure provide sufficient evidence of common descent in the absence of high sequence identity

fold similarity can be observed between two domains that share no significant sequence similarity or features that indicate common functional properties (i.e., are evolutionarily unrelated). As structure databases have grown, an increasing number of domain pairs that adopt similar folds but possess no other characteristics implying homology have been found. Such pairs may consist of two extremely distant relatives from the same evolutionary origins that have diverged far beyond sequence and functional equivalence, or alternatively two domains that have evolved from different ancestors but have converged on the same fold structure (fold analogs).

For this reason, structure relationships must be verified by further evidence such as similar sequence motifs or shared functional characteristics. Such evidence allows fold groups to be further subdivided into broader evolutionary families or superfamilies—a term first coined by Margaret Dayhoff to describe related proteins that have extensively diverged from their common ancestors [60]. Analysis of domain structure classifications shows that some fold groups are particularly highly populated and occur in a diverse range of superfamilies. In fact, nearly 30 % of the superfamilies currently assigned in CATH belong to fewer than tenfold groups. There is also a very biased population of superfamilies, with the top 100 most populated superfamilies accounting for 54 % of the 42 million domain sequences predicted to belong to CATH superfamilies (Fig. 4).

The ten most populated folds, frequently occupied by many domains with no apparent evolutionary relationship, were defined

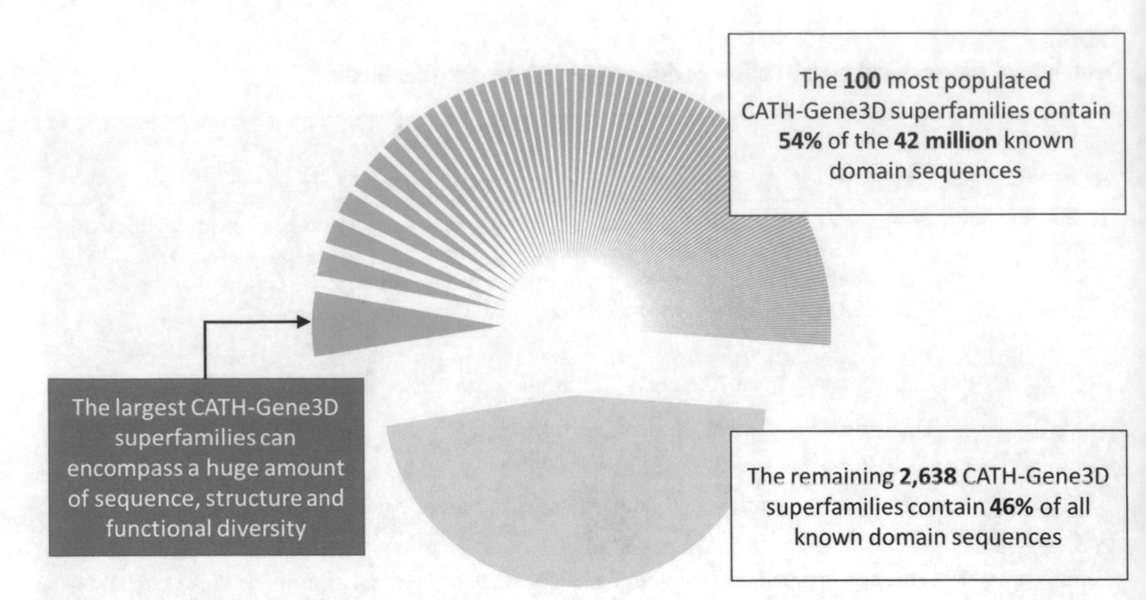

Fig. 4 The 100 largest CATH-Gene3D superfamilies have a biased domain population that contains over half of all known sequence data in the resource

as superfolds by Orengo and coworkers [61]. The occurrence of superfolds suggests that the diversity of possible protein folds in nature is constrained by the need to maintain an optimal packing of secondary structures to preserve a hydrophobic core. Currently, the number of reported folds varies from 981 (SCOP v1.75 database) to 1375 (CATH v4.0 database) due to the different clustering criteria used and the difficulty in distinguishing folds in more continuous regions of fold space. As the amount of structural data has grown, the number of classified homologues has increased and some superfamilies are seen to contain very distant relatives with highly diverse structural folds that no longer appear homologous to the human eye. SCOP2 [62] has recently been developed to redesign the original SCOP hierarchy in order to model such complex relationships (*see* Subheading 4.4 for more details).

Within superfamilies, most classifications subcluster members into families of close relatives based upon similar functional properties. Family clusters are often identified by sequence comparison; for example, in the CATH and Dali databases, close homologues are clustered at 35 % sequence identity. In the SCOP classification, similarity in functional properties is also manually identified and used to generate families. CATH-Gene3D superfamily data is also explicitly subclassified into "functional families," which represent sequences that ideally code for the same, or very similar, function [63]. During subclassification, for each superfamily, domain sequences are subclassified by first producing a hierarchical tree of sequence relatives with a clustering algorithm, GeMMA [64], and then the FunFHMMer algorithm [63] is used to calculate the optimal sets of clusters that should be merged into functional families using specificity-determining position information. The CATH-Gene3D functional families have been shown to perform well in the international CAFA2 competition to predict the protein function of a set of unknown UniProtKB sequences using functional terms from the Gene Ontology.

4.4 Structural Domain Classifications

In this section, the most comprehensive structural domain classifications are briefly discussed: SCOP, SCOP2, CATH, the Dali Database, SCOPe, and ECOD. These represent manual (SCOP and SCOP2), semi-automated (CATH) and automated approaches (the rest of the list) approaches to classification. A more comprehensive list, together with internet links, is also shown in Table 2.

The SCOP database uses an almost entirely manual approach for the assignment of domain boundaries and recognition of structural and functional similarities between proteins to generate superfamilies. This has resulted in an extremely high quality resource even though it requires a significant level of input from the curators.

The SCOP2 prototype was announced in 2014 [62], a new structural classification that succeeds SCOP. SCOP2 similarly aims to classify protein domains based on their structural and evolutionary relationships; however, it uses a directed acyclic graph (DAG) to represent relationships, rather than a hierarchy. With the growth of structural data, the SCOP team found evolutionary relationships to be more complex than first thought, leading to a complete redesign of how the relationships are captured. Structural and evolutionary relationships are now split into two different categories, and are joined by the "protein types" and "evolutionary events" categories [62].

Unlike SCOP, the CATH approach to domain classification aims to automate as many steps as possible, alongside the use of expert manual intervention to differentiate between homologous proteins and those merely sharing a common fold. Domain boundaries in CATH are automatically assigned through the identification of recurrent domains using CATH domain family HMMs, and structure comparison using the CATHEDRAL algorithm. In addition, a consensus method (DBS) is used to predict novel domains (i.e., those that have not been previously observed) directly from structure, with manual validation being applied for particularly difficult domain assignments. Sequence comparison and the CATHEDRAL and SSAP structural comparison algorithms are then used to identify structural and functional relatives within the existing library of CATH domains. Again, manual validation is often used at this stage in order to verify fold and superfamily assignments.

In contrast the Dali Database established by Holm and coworkers uses a completely automated protocol that attempts to provide the user with lists of putative structural neighbors rather than explicitly assigning domains into a hierarchy of fold and superfamilies. The PUU algorithm [65] identifies domain boundaries and assigned domains are clustered using the Dali structure comparison method. The thresholds used for clustering structures are based on Dali Z-scores calculated by scanning new domains against all representative domains in the database. Domains are further grouped according to similarities in functional annotations that are identified automatically using data-mining methods that search text for matching keywords.

The Entrez Global Query cross-database search engine (http://www.ncbi.nlm.nih.gov/gquery/) at the NCBI also provides access to a resource, the CDD (Conserved Domain Database) [66] that, like Dali, uses a "neighborhood" approach for domain classification. It uses the VAST algorithm to identify structural matches. More recently, the RCSB (Research Collaboratory Structural Bioinformatics) PDB has also developed a comparison facility that uses the CE (Combinatorial Extension) [67] and FATCAT [56] algorithms to detect structural relatives (http://www.rcsb.org/pdb/home/home.do#Category-analyze). The PDBe (PDB in Europe) provides the PDBeFold search engine, based upon the SSM (Secondary Structure Matching) algorithm [68] (http://www.ebi.ac.uk/msd-srv/ssm/), which performs either pairwise or multiple structural comparison, and 3D alignment of the protein structures.

In an analogous manner to sequence database searching, some domain structure classifications make it possible to compare a chosen structure against all structures held in the Protein Data Bank. For example, the Dali Database provides an internet-based comparison server (www.ebi.ac.uk/dali) that enables users to scan a given structure against the Dali Database to identify putative relatives.

The CATH database also provides a server that produces a list of structural neighbors identified by one of its automated comparison algorithms, CATHEDRAL, which uses E-values to quantify the degree of similarity (www.cathdb.info).

SCOPe (Structural Classification of Proteins-extended) [69] is a new database resource that extends the SCOP v1.75 database. Automated methods are used to classify additional structures into the SCOP-based hierarchy to try and keep up with the increasing number of deposited PDB structures. Some manual curation is also used [70].

ECOD (Evolutionary Classification of protein Domains) [71] is similar to the other classification methods in that it uses evolutionary information; however, it focuses more on grouping by homologous relationships rather than by topology (i.e., fold). The option to focus more on evolutionary relationships is also captured by SCOP2 through their use of DAGs. As a result, homologous domains that have different topologies can be grouped together and their evolutionary relationships examined.

4.5 Multiple Structural Alignments of Protein Domains

Multiple structural alignments enable detection of conserved structural features across a family that can be encoded in a template. The HOMSTRAD database developed by Blundell and colleagues [72] uses various domain databases including SCOP and Pfam to cluster families of evolutionary relatives that are known to have significant levels of sequence similarity. These clusters are then represented by validated multiple structural alignments for families and superfamilies that can be used to derive substitution matrices or encode conserved structural features in a template in order to identify additional relatives.

In CATH, multiple structure alignments have been generated for each CATH superfamily using the CORA algorithm [73]. Structurally similar groups (SSGs) are identified with multiple structural alignment using the structural comparison algorithm, SSAP [74]. Relatives are grouped using multi-linking clustering if they superpose with an RMSD <9 Å [75].

SSGs from over 270 enzyme superfamilies are used by FunTree [76, 77] to illustrate enzyme function evolution. This is done by depicting the information in phylogenetic tree format. This large-scale analysis provides an overview of how enzyme function changes between structurally similar homologues using Enzyme Commission (EC) [78] information.

To date, multiple structure alignments have not performed as well in generating superfamily or family profiles, as those alignments derived from sequence-based families. However, with the diversity of structural data increasing as a result of the structural genomics initiatives, these approaches should become increasingly useful.

4.6 Consistency of Structure Domain Databases

A comprehensive analysis by Hadley and Jones of the CATH, SCOP and Dali Dictionary classifications revealed that a considerable level of agreement existed between the resources, with 75 % of fold and 80 % of superfamily levels having comparable assignments [79]. Discrepancies tended to be attributable to the different protocols applied to each classification, each of which has a different threshold for the identification of structural relatives and the assignment of fold and superfamily groupings. For example, the manual assessment of structural similarity in SCOP allows a considerable degree of flexibility in the recognition of very distant homologues—a level that is difficult to recapitulate in automatic assignments methods, which must maintain a low assignment error rate.

Collaborations between CATH and SCOP teams in the Genome3D initiative have allowed mapping between the two resources (http://genome3d.eu) to identify equivalent superfamilies. This has helped in the recognition of similar annotations generated by independent structure prediction algorithms, allowing consensus predicted regions to be identified and false positives to be highlighted where algorithms disagree in their predictions.

The most significant recent development in structure comparison has been the improvement to the protocols used both to measure similarities and to assess statistical significance. Although many resources have long exploited Z-scores to highlight the most significant matches to a nonredundant database of structures, recent methods have provided better statistical models by representing the results returned by database scans as extreme value distributions. Match statistics can then be given as an expectation value (E-value) that captures the probability of an unrelated structure returning a particular score within a database of a given size.

The assignment of distinct fold groups is hindered by the recurrence of structural motifs that can be found in nonhomologous structures. Such motifs are often described as super secondary motifs since they comprise two or more secondary structures such as an alpha-beta-motif or alpha-hairpins. The occurrence of such motifs supports the proposal of a fold continuum in which some regions of fold space are densely populated, to the degree that many folds can be linked by common motifs. In such cases the granularity of fold grouping is dependent on the degree of structural overlap required for a significant match and in difficult cases manual (and therefore subjective) validation must be applied. In fact, the structural similarity score between overlapping fold groups can sometimes be as high as structural similarity measures recorded within very diverse superfamilies.

These recurrent structure motifs may well give clues to the origin of protein domains. It is possible that they represent ancient conserved structural cores that have been retained because they are crucial to maintaining both structure and function, whereas the surrounding structure has been modified beyond recognition. Another theory suggests that they represent small gene fragments that have been duplicated and incorporated into nonhomologous contexts [80]. It is therefore a possibility that in the past protein domains were built up from the fusion of short polypeptides, now observed as internal repeats and super secondary motifs.

5 Domain Family Annotation of Genomes

In this section we will consider how the domain classifications can be used to annotate genomes and what these domain-level annotations are beginning to reveal.

The assignment of domain families to individual proteins or complete genomes is typically an automatic procedure, with little or no manual intervention involved. Providing up-to-date and

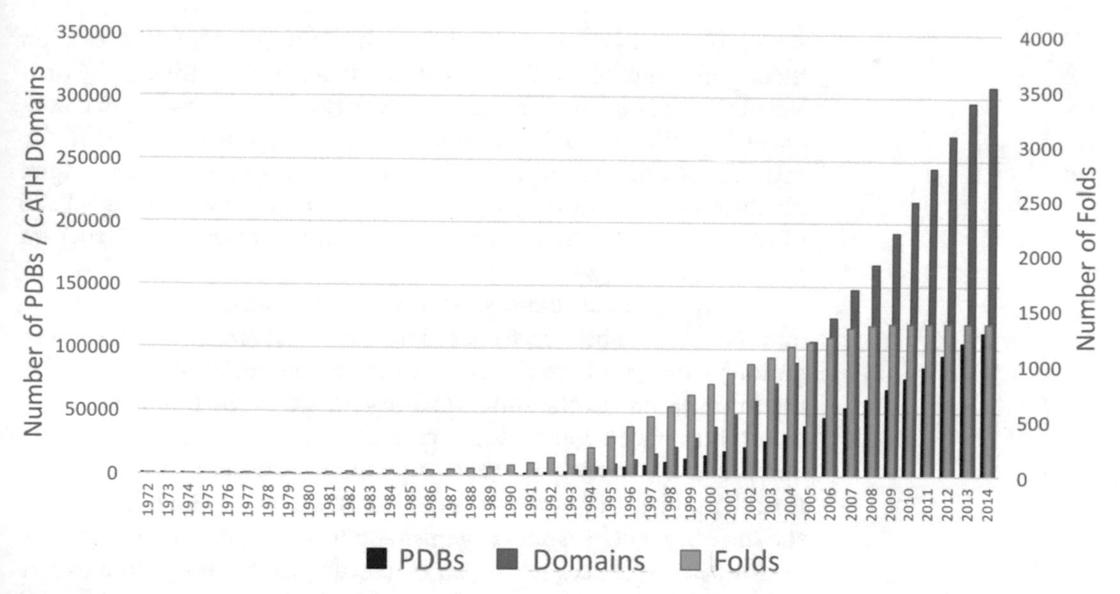

Fig. 5 The growth of structural data in the Protein Data Bank (PDB) and CATH-Gene3D. While the number of PDB structures deposited and domains identified is still increasing annually, the number of new folds classified has leveled off in the last 10 years, showing that CATH-Gene3D has an extensive coverage of structural space

automatic domain-level annotation of genomic data will be essential as genome releases continue to grow. How can we accurately assign domain families to genome sequences on such a large scale? Work pioneered by several groups including Chothia and coworkers [81] used datasets of structurally validated remote homologues to develop benchmarking protocols in order to identify reliable thresholds for accurate homologue detection. Several groups (e.g., SUPERFAMILY [82] and CATH-Gene3D [83]) have used automated HMM assignments to build databases of sequence and structure domain family assignments to the completed genomes.

In comparison to the growth in sequence data, relatively few protein structures have been determined. Despite this, current mappings of structural domains to completed genomes assign over 60 % of genes or partial genes to known structural domain families (e.g., CATH), suggesting that we currently have structural representation for many of the most highly populated domain families in nature [4] (Fig. 5). An additional 20 % of remaining sequences can also be assigned to Pfam domain families, revealing that a significant proportion of genome sequences can now be classified into fewer than 2500 of the largest domain families.

A significant use of domain-level genome annotations has been the identification of structurally uncharacterized superfamilies that are likely to possess a novel fold or function, with the aim that they might be targeted for structural determination. Structural genomics initiatives aimed to characterize a large number of novel structures on a genome scale through the implementation of high throughput methods. These Protein Structure Initiative (PSI) projects initiatives in the USA chose the largest structurally uncharacterized families to help provide structural representatives for the majority of genome sequences. It is interesting to note that, despite structural genomics initiatives targeting such families, only ~15 % of the structures were found to contain novel folds upon structure characterization [84].

Many of the sequences (20–30 %) that cannot be assigned to domain families belong to very small or even single member (singleton) families. Often found in a single species, they may contribute in some way to the functional repertoire of the organism. Each newly sequenced genome generated brings with it new domain families and currently it seems difficult to estimate an upper limit for the number of domain families in nature. It is quite possible that the majority will be small, organism-specific families whilst the bulk of domain sequences within an organism will be assigned to a few thousand characterized domain families.

Another trend that has been revealed from the domain annotation of genomes is the extreme bias in the distribution of families. A very small proportion of families recur extensively in the genomes, with the remaining families occurring only once or a few times within a specific organism or subkingdom. Such a distribution suggests that some domains that duplicated early in evolution were retained because they performed important biochemical activities. Alternatively, some folds have particularly stable structural frameworks that support many diverse sequences, allowing paralogous relatives to tolerate changes in structure that modulate function.

Finally, the analysis of domain architectures and their distribution across genomes [3] has shown that some domain families are partnered with a small number of domains, limited to a kingdom or species, whereas others appear to be much more versatile, combining with a wide range of different partners.

6 Conclusions

There are a large number of high quality protein domain classifications that can be used to provide large-scale automated protein domain family annotations. A challenge remains however in the transfer of useful biological knowledge to protein sequences. Domain classification resources annotate the proteome according to a parts list of component domains rather than protein or gene names. Accordingly, proteome annotation by domain content implies that gene function can be viewed as a generalized function of its component parts. Whilst such predictions cannot be expected to compete with experimentally derived characteristics, they provide the best approach that we have for annotating the majority of functionally uncharacterized genome sequences. The comparison of entire genomes is becoming an important mechanism for progressing beyond the simple cataloging of homologous genes and domains towards an understanding of the biology that underlies the variability within families of protein domains.

7 Notes

1. There are limitations in clustering domains by single or multi-linkage methods. One problem for automated domain classification methods is the choice of hierarchical clustering method used to group related domains. The aim is to compare every domain with all other domains, based on some biologically meaningful similarity score represented as a distance. Domains are then grouped into related clusters based on a distance threshold. In general, two main clustering methods tend to be used; single linkage clustering and multi-linkage clustering. Single linkage clustering defines the distance between any two clusters as the minimum distance between them. Consequently, single-linkage clustering can result in the chaining problem, in which new sequences are assigned to a cluster on the basis of a single member being related (as defined by the similarity threshold), regardless of the positions of the other (unrelated) domains in that cluster. In multi-linkage clustering all members of the cluster must be linked to one another with some minimum similarity. Multi-linkage clustering, in comparison to single linkage, tends to be conservative, producing compact clusters that may have an excessive overlap with other domain clusters. Several other clustering methods are possible, including average linkage clustering, complete linkage clustering. The difficulty in assigning meaningful domain families is that there is no unambiguous measure to evaluate the efficacy of a given method.

References

1. Fleischmann R et al (1995) Whole-genome random sequencing and assembly of Haemophilus influenzae Rd. Science 269:496–512
2. Reddy TBK et al (2015) The Genomes OnLine Database (GOLD) v. 5: a metadata management system based on a four level (meta) genome project classification. Nucleic Acids Res 43:D1099–D1106
3. Vogel C, Bashton M, Kerrison ND, Chothia C, Teichmann SA (2004) Structure, function and evolution of multidomain proteins. Curr Opin Struct Biol 14:208–216
4. Marsden RL, Lee D, Maibaum M, Yeats C, Orengo CA (2006) Comprehensive genome analysis of 203 genomes provides structural genomics with new insights into protein family space. Nucleic Acids Res 34:1066–1080
5. Needleman SB, Wunsch CD (1970) A general method applicable to the search for similarities in the amino acid sequence of two proteins. J Mol Biol 48:443–453
6. Pearson WR, Lipman DJ (1988) Improved tools for biological sequence comparison. Proc Natl Acad Sci U S A 85:2444–2448

7. Altschul SF, Gish W, Miller W, Myers EW, Lipman DJ (1990) Basic local alignment search tool. J Mol Biol 215:403–410
8. Ponting CP (2001) Issues in predicting protein function from sequence. Brief Bioinform 2:19–29
9. Bru C et al (2005) The ProDom database of protein domain families: more emphasis on 3D. Nucleic Acids Res 33:D212–D215
10. Portugaly E, Linial N, Linial M (2007) EVEREST: a collection of evolutionary conserved protein domains. Nucleic Acids Res 35: D241–D246
11. Heger A (2004) ADDA: a domain database with global coverage of the protein universe. Nucleic Acids Res 33:D188–D191
12. The UniProt Consortium (2014) UniProt: a hub for protein information. Nucleic Acids Res 43:D204–D212
13. Altschul SF et al (1997) Gapped BLAST and PSI-BLAST: a new generation of protein database search programs. Nucleic Acids Res 25:3389–3402
14. Kelil A, Wang S, Brzezinski R, Fleury A (2007) CLUSS: clustering of protein sequences based on a new similarity measure. BMC Bioinformatics 8:286
15. Gnanavel M et al (2014) CLAP: a web-server for automatic classification of proteins with special reference to multi-domain proteins. BMC Bioinformatics 15:343
16. Krishnamurthy N, Brown DP, Kirshner D, Sjölander K (2006) PhyloFacts: an online structural phylogenomic encyclopedia for protein functional and structural classification. Genome Biol 7:R83
17. Loewenstein Y, Portugaly E, Fromer M, Linial M (2008) Efficient algorithms for accurate hierarchical clustering of huge datasets: tackling the entire protein space. Bioinformatics 24:i41–i49
18. Enright AJ, Kunin V, Ouzounis CA (2003) Protein families and TRIBES in genome sequence space. Nucleic Acids Res 31:4632–4638
19. Edgar RC (2010) Search and clustering orders of magnitude faster than BLAST. Bioinformatics 26:2460–2461
20. Li W, Godzik A (2006) Cd-hit: a fast program for clustering and comparing large sets of protein or nucleotide sequences. Bioinformatics 22:1658–1659
21. Fu L, Niu B, Zhu Z, Wu S, Li W (2012) CD-HIT: accelerated for clustering the next-generation sequencing data. Bioinformatics 28:3150–3152
22. Hauser M, Mayer CE, Söding J (2013) kClust: fast and sensitive clustering of large protein sequence databases. BMC Bioinformatics 14:248
23. Feng DF, Doolittle RF (1996) Progressive alignment of amino acid sequences and construction of phylogenetic trees from them. Methods Enzymol 266:368–382
24. Eddy SR (1996) Hidden Markov models. Curr Opin Struct Biol 6:361–365
25. Finn RD et al (2015) HMMER web server: 2015 update. Nucleic Acids Res 43:W30–W38
26. Remmert M, Biegert A, Hauser A, Söding J (2012) HHblits: lightning-fast iterative protein sequence searching by HMM-HMM alignment. Nat Methods 9:173–175
27. Mitchell A et al (2015) The InterPro protein families database: the classification resource after 15 years. Nucleic Acids Res 43: D213–D221
28. Sillitoe I et al (2015) CATH: comprehensive structural and functional annotations for genome sequences. Nucleic Acids Res 43: D376–D381
29. Pedruzzi I et al (2014) HAMAP in 2015: updates to the protein family classification and annotation system. Nucleic Acids Res 43: D1064–D1070
30. Mi H, Muruganujan A, Thomas PD (2013) PANTHER in 2013: modeling the evolution of gene function, and other gene attributes, in the context of phylogenetic trees. Nucleic Acids Res 41:D377–D386
31. Nikolskayaw QN, Arighi CN, Huang H, Barker WC, Wu CH (2006) PIRSF family classification system for protein functional and evolutionary analysis. Evol Bioinforma 2:197–209
32. Finn RD et al (2014) Pfam: the protein families database. Nucleic Acids Res 42:D222–D230
33. Attwood TK et al (2012) The PRINTS database: a fine-grained protein sequence annotation and analysis resource—its status in 2012. Database (Oxford) 2012:bas019
34. Sigrist CJA et al (2013) New and continuing developments at PROSITE. Nucleic Acids Res 41:D344–D347
35. Letunic I, Doerks T, Bork P (2015) SMART: recent updates, new developments and status in 2015. Nucleic Acids Res 43:D257–D260
36. Oates ME et al (2015) The SUPERFAMILY 1.75 database in 2014: a doubling of data. Nucleic Acids Res 43:D227–D233
37. Haft DH et al (2013) TIGRFAMs and genome properties in 2013. Nucleic Acids Res 41: D387–D395

38. Heger A, Holm L (2003) Exhaustive enumeration of protein domain families. J Mol Biol 328:749–767
39. Penel S et al (2009) Databases of homologous gene families for comparative genomics. BMC Bioinformatics 10(Suppl 6):S3
40. Kriventseva EV et al (2015) OrthoDB v8: update of the hierarchical catalog of orthologs and the underlying free software. Nucleic Acids Res 43:D250–D256
41. Jones P et al (2014) InterProScan 5: genome-scale protein function classification. Bioinformatics 30:1236–1240
42. Petryszak R, Kretschmann E, Wieser D, Apweiler R (2005) The predictive power of the CluSTr database. Bioinformatics 21:3604–3609
43. Thomas PD (2010) GIGA: a simple, efficient algorithm for gene tree inference in the genomic age. BMC Bioinformatics 11:312
44. Wu CH et al (2004) PIRSF: family classification system at the Protein Information Resource. Nucleic Acids Res 32:D112–D114
45. Haft DH, Selengut JD, White O (2003) The TIGRFAMs database of protein families. Nucleic Acids Res 31:371–373
46. Berman H, Henrick K, Nakamura H (2003) Announcing the worldwide Protein Data Bank. Nat Struct Biol 10:980
47. Richardson JS (1981) The anatomy and taxonomy of protein structure. Adv Protein Chem 34:167–339
48. Murzin A, Brenner S, Hubbard T, Chothia C (1995) SCOP: a structural classification of proteins database for the investigation of sequences and structures. J Mol Biol 247:536–540
49. Orengo CA et al (1997) CATH—a hierarchic classification of protein domain structures. Structure 5:1093–1108
50. Holm L, Sander C (1998) Dictionary of recurrent domains in protein structures. Proteins 33:88–96
51. Sowdhamini R, Rufino SD, Blundell TL (1996) A database of globular protein structural domains: clustering of representative family members into similar folds. Fold Des 1:209–220
52. Gibrat JF, Madej T, Bryant SH (1996) Surprising similarities in structure comparison. Curr Opin Struct Biol 6:377–385
53. Redfern OC, Harrison A, Dallman T, Pearl FMG, Orengo CA (2007) CATHEDRAL: a fast and effective algorithm to predict folds and domain boundaries from multidomain protein structures. PLoS Comput Biol 3:e232
54. Taylor W, Orengo CA (1989) Protein structure alignment. J Mol Biol 208:1–22
55. Holm L, Sander C (1993) Protein structure comparison by alignment of distance matrices. J Mol Biol 233:123–138
56. Ye Y, Godzik A (2003) Flexible structure alignment by chaining aligned fragment pairs allowing twists. Bioinformatics 19:ii246–ii255
57. Subbiah S, Laurents DV, Levitt M (1993) Structural similarity of DNA-binding domains of bacteriophage repressors and the globin core. Curr Biol 3:141–148
58. Gerstein M, Levitt M (1998) Comprehensive assessment of automatic structural alignment against a manual standard, the scop classification of proteins. Protein Sci 7:445–456
59. Kolodny R, Koehl P, Levitt M (2005) Comprehensive evaluation of protein structure alignment methods: scoring by geometric measures. J Mol Biol 346:1173–1188
60. Dayhoff MO (2005) Atlas of protein sequence and structure. Natl. Biomed. Res. Foundation
61. Orengo CA, Jones DT, Thornton JM (1994) Protein superfamilles and domain superfolds. Nature 372:631–634
62. Andreeva A, Howorth D, Chothia C, Kulesha E, Murzin AG (2014) SCOP2 prototype: a new approach to protein structure mining. Nucleic Acids Res 42:D310–D314
63. Das S et al (2015) Functional classification of CATH superfamilies: a domain-based approach for protein function annotation. Bioinformatics 31:3460–3467
64. Lee DA, Rentzsch R, Orengo C (2010) GeMMA: functional subfamily classification within superfamilies of predicted protein structural domains. Nucleic Acids Res 38:720–737
65. Holm L, Sander C (1994) Parser for protein folding units. Proteins 19:256–268
66. Marchler-Bauer A et al (2014) CDD: NCBI's conserved domain database. Nucleic Acids Res 43:D222–D226
67. Shindyalov IN, Bourne PE (1998) Protein structure alignment by incremental combinatorial extension (CE) of the optimal path. Protein Eng 11:739–747
68. Krissinel E, Henrick K (2004) Secondary-structure matching (SSM), a new tool for fast protein structure alignment in three dimensions. Acta Crystallogr D Biol Crystallogr 60:2256–2268
69. Fox NK, Brenner SE, Chandonia J-MM (2014) SCOPe: Structural Classification of Proteins—extended, integrating SCOP and ASTRAL data and classification of new structures. Nucleic Acids Res 42:D304–D309

70. Andreeva A et al (2007) Data growth and its impact on the SCOP database: new developments. Nucleic Acids Res 36:D419–D425
71. Cheng H et al (2014) ECOD: an evolutionary classification of protein domains. PLoS Comput Biol 10:e1003926
72. Sowdhamini R et al (1998) Protein three-dimensional structural databases: domains, structurally aligned homologues and superfamilies. Acta Crystallogr D Biol Crystallogr 54:1168–1177
73. Orengo CA (1999) CORA—topological fingerprints for protein structural families. Protein Sci 8:699–715
74. Orengo CA, Taylor WR (1996) In: Computer methods for macromolecular sequence analysis, vol 266. Elsevier, Amsterdam, pp 617–635
75. Cuff A, Redfern O, Dessailly B, Orengo C (2011) In Protein function prediction for omics era. Springer, Netherlands
76. Furnham N et al (2012) FunTree: a resource for exploring the functional evolution of structurally defined enzyme superfamilies. Nucleic Acids Res 40:D776–D782
77. Furnham N et al (2012) Exploring the evolution of novel enzyme functions within structurally defined protein superfamilies. PLoS Comput Biol 8:e1002403
78. Barrett AJ (1992) Enzyme nomenclature: Recommendations of the Nomenclature Committee of the International Union of Biochemistry and Molecular Biology. Academic, San Diego, CA
79. Hadley C, Jones DT (1999) A systematic comparison of protein structure classifications: SCOP, CATH and FSSP. Structure 7:1099–1112
80. Lupas AN, Ponting CP, Russell RB (2001) On the evolution of protein folds: are similar motifs in different protein folds the result of convergence, insertion, or relics of an ancient peptide world? J Struct Biol 134:191–203
81. Park J et al (1998) Sequence comparisons using multiple sequences detect three times as many remote homologues as pairwise methods. J Mol Biol 284:1201–1210
82. Gough J, Chothia C (2002) SUPERFAMILY: HMMs representing all proteins of known structure. SCOP sequence searches, alignments and genome assignments. Nucleic Acids Res 30:268–272
83. Yeats C et al (2006) Gene3D: modelling protein structure, function and evolution. Nucleic Acids Res 34:D281–D284
84. Todd AE, Marsden RL, Thornton JM, Orengo CA (2005) Progress of structural genomics initiatives: an analysis of solved target structures. J Mol Biol 348:1235–1260

Part II

Sequence Analysis

Chapter 8

Multiple Sequence Alignment

Punto Bawono, Maurits Dijkstra, Walter Pirovano, Anton Feenstra, Sanne Abeln, and Jaap Heringa

Abstract

The increasing importance of Next Generation Sequencing (NGS) techniques has highlighted the key role of multiple sequence alignment (MSA) in comparative structure and function analysis of biological sequences. MSA often leads to fundamental biological insight into sequence–structure–function relationships of nucleotide or protein sequence families. Significant advances have been achieved in this field, and many useful tools have been developed for constructing alignments, although many biological and methodological issues are still open. This chapter first provides some background information and considerations associated with MSA techniques, concentrating on the alignment of protein sequences. Then, a practical overview of currently available methods and a description of their specific advantages and limitations are given, to serve as a helpful guide or starting point for researchers who aim to construct a reliable MSA.

Key words Multiple sequence alignment, Progressive alignment, Dynamic programming, Phylogenetic tree, Amino acid exchange matrix, Sequence profile, Gap penalty

1 Introduction

1.1 Definition and Implementation of an MSA

A multiple sequence alignment (MSA) involves three or more homologous nucleotide or amino acid sequences. An alignment of two sequences is normally referred to as a pairwise alignment. The alignment, whether multiple or pairwise, is obtained by inserting gaps into sequences such that the resulting sequences all have the same length L. Consequently, an alignment of N sequences can be arranged in a matrix of N rows and L columns, in a way that best represents the evolutionary relationships among the sequences.

Organizing sequence data in MSAs can be used to reveal conserved and variable sites within protein families. MSAs can provide essential information on their evolutionary and functional relationships. For this reason, MSAs have become an important prerequisite for virtually all genomic analysis pipelines and many downstream computational modes of analysis of protein families such as homology modeling, secondary structure prediction, and

Jonathan M. Keith (ed.), *Bioinformatics: Volume I: Data, Sequence Analysis, and Evolution*, Methods in Molecular Biology, vol. 1525, DOI 10.1007/978-1-4939-6622-6_8,

phylogenetic reconstruction. They may further be used to derive profiles [1] or hidden Markov models [2, 3] that can be used to scour databases for distantly related members of the family. As the enormous increase of biological sequence data has led to the requirement of large-scale sequence comparison of evolutionarily divergent sets of sequences, the performance and quality of MSA techniques is now more important than ever.

1.2 Reliability and Evolutionary Hypothesis

The automatic generation of an accurate MSA is a computationally complex problem. If we consider the alignment or matching of two or more protein sequences as a series of hypotheses of positional homology, it would obviously be desirable to have a priori knowledge about the evolutionary (and structural) relationships between the sequences considered. Most multiple alignment methods attempt to infer and exploit a notion of such phylogenetic relationships, but they are limited in this regard by the lack of ancestral sequences. Naturally, only observed taxonomic units (OTUs), i.e., present-day sequences, are available. Moreover, when evolutionary distances between the sequences are large, adding to the complexity of the relationships among the homologous sequences, the consistency of the resulting MSA becomes more uncertain (*see* **Note 1**).

When two sequences are compared it is important to consider the evolutionary changes (or sequence edits) that have occurred for the one sequence to be transformed into the other. This is generally done by determining the minimum number of mutations that may have occurred during the evolution of the two sequences. For this purpose several amino acid exchange matrices, such as the PAM [4] and BLOSUM [5] series, have been developed, which estimate evolutionary likelihoods of mutations and conservations of amino acids. The central problem of assembling an MSA is that a compromise must be found between the evolutionarily most likely pairwise alignments between the sequences, and the embedding of these alignments in a final MSA, where changes relative to the pairwise alignments are normally needed to globally optimize the evolutionary model and produce a consistent multiple alignment. However, with increasing numbers of input sequences, a reliable MSA method will be able to compile a biologically more correct alignment, reflecting the evolutionary relationships and sequence conservation more meaningfully. Integrating knowledge over more sequences enables a better estimate of the underlying evolutionary events. As a result, pairwise alignments taken from an MSA are expected to be evolutionarily more relevant than "optimal" pairwise alignments (*see* next section).

1.3 Dynamic Programming

Pairwise alignment can be performed by the dynamic programming (DP) algorithm [6]. A two-dimensional matrix is constructed based on the lengths of the sequences to be aligned, in which each possible alignment is represented by a unique path through the matrix. Using a specific scoring scheme, which defines scores for residue matches, mismatches, and gaps, each position of the matrix is filled. The DP algorithm guarantees that, given a specific scoring scheme, the optimal alignment will be found. Although dynamic programming is an efficient way of aligning sequences, applying the technique to more than two sequences quickly becomes computationally unfeasible. This is due to the fact that the number of comparisons to be made increases exponentially with the number of sequences. Carrillo and Lipman [7] and later on Stoye et al. [8] proposed heuristics to reduce the computational requirements of multidimensional dynamic programming techniques. Nonetheless, computation times required remain prohibitive for all but the smallest sequence sets.

1.4 The Progressive Alignment Protocol

An important breakthrough in multiple sequence alignment has been the introduction of the progressive alignment protocol [9]. The basic idea behind this protocol is the construction of an approximate phylogenetic tree for the query sequences and repeated use of the aforementioned pairwise alignment algorithm. The tree is usually constructed using the scores of all-against-all pairwise alignments across the query sequence set. Then the alignment is build up by progressively adding sequences in the order specified by the tree (Fig. 1), which is therefore referred to as the *guide tree*. In this way, phylogenetic information is incorporated to guide the alignment process, such that sequences and blocks of sequences become aligned successively to produce a final MSA. Fortunately, as the pairwise DP algorithm is only repeated a limited number of times, typically on the order of the square of the number

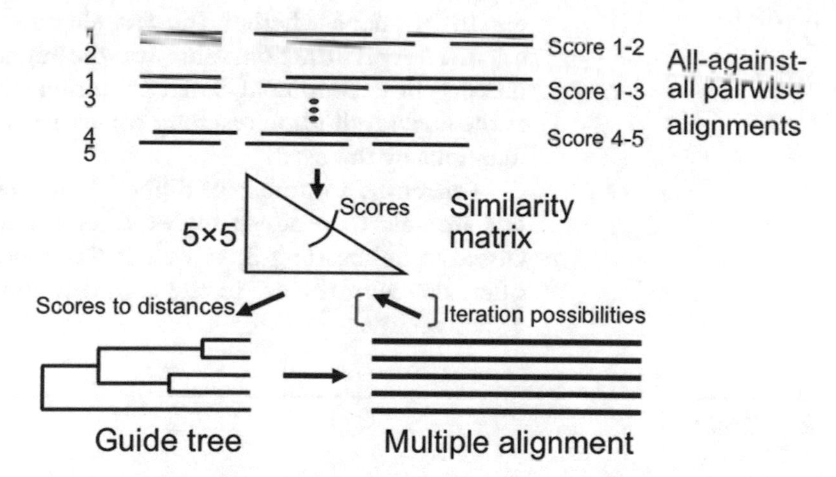

Fig. 1 Schematic representation of the progressive alignment protocol. A similarity (distance) matrix, which contains scores from all pairwise alignments, is used to construct a guide tree. The final alignment is built up progressively following the order of the guide tree. The *black arrow* between brackets indicates possible iterative cycles

of sequences or less, the progressive protocol allows the effective multiple alignment of large numbers of sequences.

However, the obtained accuracy of the final MSA suffers from the so-called greediness of the progressive alignment protocol; that is, alignment errors cannot be repaired and will be propagated into subsequent alignment steps ("Once a gap, always a gap"). In fact, it is only later during the alignment progression that more information from other sequences (e.g., through profile representation) [1] becomes employed in the alignment steps.

1.5 Alignment Iteration

Triggered by the main pitfall of the progressive alignment scenario, some methods try to alleviate the greediness of this strategy by implementing an iterative alignment procedure. Pioneered by Hogeweg and Hesper [10], iterative techniques try to enhance the alignment quality by gleaning increased information from repeated alignment procedures, such that earlier alignments are "corrected" [10, 11]. The idea is to compile an MSA, learn from it, and do it better next time. In this scenario, a previously generated MSA is used for improvement of parameter settings, so that the initial guide tree and consequently the alignment can be optimized. Apart from the guide tree, the alignment procedure itself can also be adapted based on observed features of a preceding MSA. The iterative procedure is terminated whenever a preset maximum number of iterations or convergence is reached. However, depending on the target function of an iterative procedure, it does not always reach convergence, so that a final MSA often depends on the number of iterations set by the user. The alignment scoring function used during progressive alignment can be different from the target function of the iteration process. This means that a decision has to be made whether the last alignment (with the maximal iterative target function value) or the highest scoring alignment that may be encountered earlier on during iteration should be taken as the final result upon reaching convergence or termination of the iterations by the user.

Currently, a number of different progressive alignment methods are able to produce high-quality alignments. These are discussed in Subheading 3, as well as the options and solutions they offer, also with respect to the considerations outlined in the preceding sections.

2 Materials

2.1 Selection of Sequences

Since sequence alignment techniques are based upon a model of divergent evolution, the input of a multiple alignment algorithm should be a set of homologous sequences. Sequences can be retrieved directly from protein sequence databases, but usually a set is created by employing a homology searching technique for a provided query sequence. Widely used programs such as BLAST [12] or FASTA [13] employ carefully crafted heuristics to perform a rapid search over sequence databases and recover putative homologues.

Selected sequences should preferably be orthologous, particularly if speciation events during evolution are to be studied, but in practice it is often difficult to ensure that all sequences have orthologous relationships. It is important to stress that MSA routines will also produce an alignment when given an unrelated set of input sequences; the MSA may appear to have some realistic patterns, but these will be biologically meaningless ("garbage in, garbage out"). For example, it is possible that some columns appear to be well conserved, although in reality no homology exists. Such misinterpretation could well have dramatic consequences for proper conclusions and further analysis modes. Although the development of P- and E-values to estimate the statistical significance of putative homologues found by homology searching techniques limits the chance of false positives, it is entirely possible that essentially non-homologous sequences enter the alignment set, which might confuse the alignment method. A careful selection of the input protein sequences is therefore an essential prerequisite for creating a meaningful alignment.

2.2 Unequal Sequence Lengths: Global and Local Alignment

Query sequence sets comprise sequences that typically will be of unequal length. The extent of such length differences requires a decision whether a *global* or *local* alignment should be performed. A *global* alignment strategy [6] aligns sequences over their entire length. However, many biological sequences are modular and contain shuffled domains [14], which can render a global alignment of two complete sequences meaningless (*see* **Note 2**). Moreover, global alignment can also lead to incorrect alignment when large insertions of gaps are needed, for example, to match two domains A and B in a two-domain protein against the corresponding domains in a three-domain structure ACB. In general, the global alignment strategy is appropriate for sequences of high to medium sequence similarity. At lower sequence identities, the global alignment technique can still be useful provided there is confidence that the sequence set is largely colinear without shuffled sequence motifs or insertions of domains. Whenever such confidence is not present, the *local* alignment technique [15] should be attempted. This technique selects and aligns the most conserved region in either of the sequences and discards the remaining sequence fragments. In cases of medium to low sequence similarity, local alignment is generally the most appropriate approach with which to start the analysis. Techniques have also been developed that use the local alignment technique to iteratively align sequence fragments that remain after previous local alignment (e.g., [16]).

2.3 Type of Alignment

A number of different alignment problems have been identified in the literature. For example, the BAliBASE MSA benchmark database [17] groups these in five basic categories that contain sequence sets comprising the following features:

1. *Equidistant sequences.* Pairwise evolutionary distances between the sequences are approximately the same.
2. *Orphan sequences.* One or more family members of the sequence set are evolutionarily distant from all the others (which can be considered equidistant).
3. *Subfamilies.* Sequences are distributed over two or more divergent subfamilies.
4. *Extensions.* Alignments contain large N- and/or C-terminal gaps.
5. *Insertions.* Alignments have large internal gap insertions.

The preceding classification of alignment problems opens up the possibility of developing different alignment techniques that are optimal for each individual type of problem. Other cases that are challenging for alignment engines include repeats, where different repeat types and copy numbers often lead to incorrect alignment (*see* **Note 3**), and transmembrane segments, where different hydrophobicity patterns confuse the alignment (*see* **Note 4**). However, one would then need a priori knowledge about the alignment problem at hand (*see* **Note 5**), which can be difficult to obtain. A suggestion for investigators is to make a first (quick) multiple alignment using general parameter settings. Often, after this first round, it becomes clear in which problem category the chosen sequence set falls, so that for further alignment parameters can be set accordingly. Alignments can always be manually adjusted by using one of the available alignment editors (*see* **Note 6**). Further, a number of alignment quality measures are available to be used in benchmarking studies in order to choose an alignment technique that performs generally well or in particular cases (*see* **Note 7**).

3 Methods

This section highlights a selection of the most accurate MSA methods to date (Table 1). Each of these follows one or both of two main approaches to address the greediness of the progressive MSA protocol (*see* the preceding): the first is trying to avoid early match errors by using increased information for aligning pairwise sequences; the second is reconsidering alignment results and improving upon these using iterative strategies.

3.1 PRALINE

PRALINE is an online MSA toolkit for protein sequences. It includes a web server offering a wide range of options to optimize the alignment of input sequences, such as global or local

Table 1
Web sites of multiple sequence alignment programs mentioned in this chapter

Name	Web site
PRALINE	http://www.ibi.vu.nl/programs/pralinewww/
MUSCLE	http://www.drive5.com/muscle/
T-Coffee and 3D-Coffee	http://tcoffee.vital-it.ch/apps/tcoffee/index.html
MAFFT	http://mafft.cbrc.jp/alignment/server/
ProbCons	http://probcons.stanford.edu/
Kalign	http://toolkit.tuebingen.mpg.de/kalign
MSAProbs	http://toolkit.tuebingen.mpg.de/msaprobs
Clustal Omega	http://www.ebi.ac.uk/Tools/msa/clustalo/
ProDA	http://proda.stanford.edu/

preprocessing, predicted secondary structure information, sequence motif information and iteration strategies (Fig. 2).

1. *Pre-profile processing options.* Pre-profile processing is an optimization technique used to minimize the incorporation of erroneous information during progressive alignment. The difference between this strategy and the standard global strategy is that the sequences to be aligned are represented by pre-profiles instead of single sequences. Three different options are available: (1) global preprocessing [18, 19], (2) local pre-processing [19], and (3) PSI-Praline [20]. The first two options attempt to maximize the information from each sequence. For each sequence, a pre-profile is built containing information from other sequences in the query set. Under global preprocessing, other sequences can be selected according to a preset minimal pairwise alignment score with the main sequence within each pre-profile. Under local preprocessing, segments of other sequences in the query set are selected based on local alignment scores. The PSI-Praline pre-profile processing strategy employs the PSI-BLAST homology search engine [21] to enrich the information of each of the pre-profiles. Based on a user-specified E-value, PSI-BLAST selects sequence fragments from a large nonredundant sequence database, building more consistent and useful pre-profiles for the alignment. The alignment quality of the PSI-Praline strategy is among the highest in the field [20], but the technique is relatively slow as a PSI-BLAST run needs to be conducted for every sequence in the input set.
2. *DSSP or predicted secondary structure information.* PRALINE currently allows the incorporation of DSSP-defined secondary structure information [22] to guide the alignment. For sequences

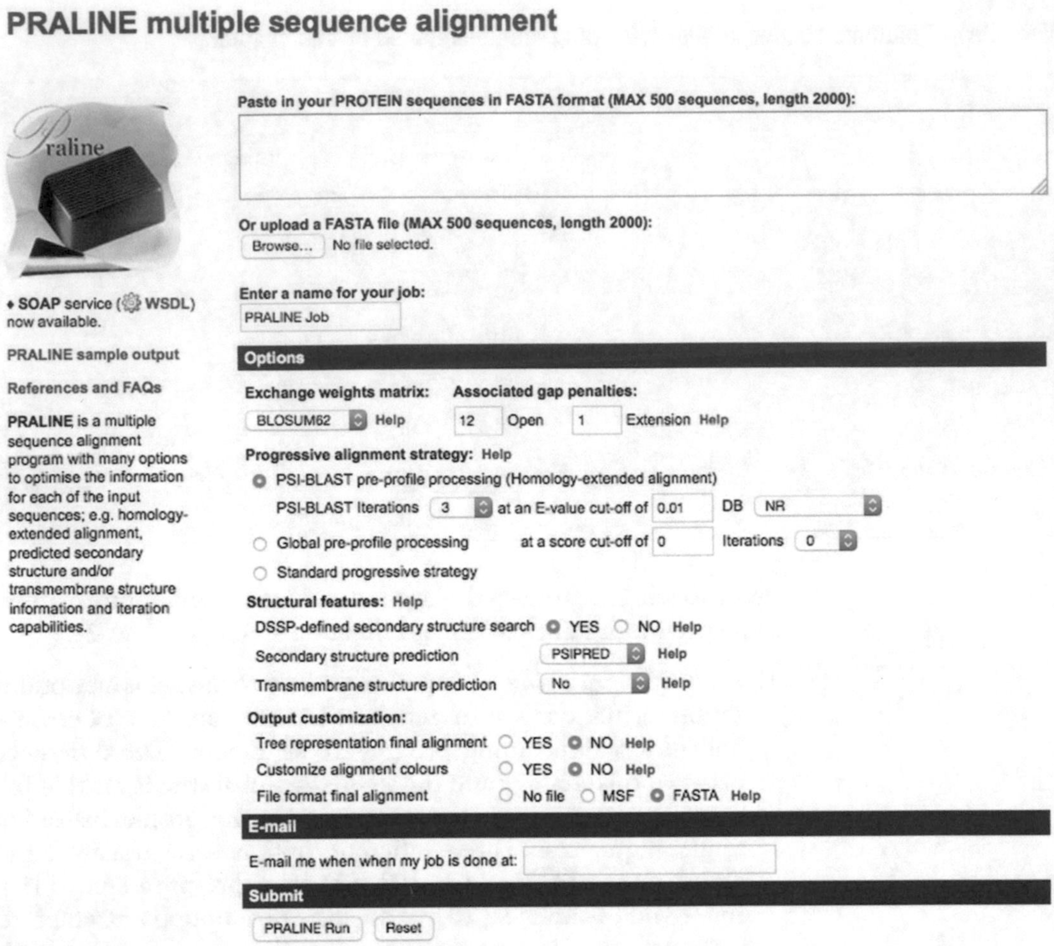

Fig. 2 The PRALINE standard web interface. Protein sequences can be pasted in the upper box in FASTA format or directly uploaded from a file. In addition to using default settings, various alignment strategies can be selected (*see* Subheading 3.1) as well as the desired number of iterations or preprocessing cut-off scores

that do not have a PDB structure, no DSSP file will be available. In such cases, a choice of seven secondary structure prediction methods is provided to determine the putative secondary structure of the sequences. In addition, two different consensus strategies are also included, both relying on the prediction methods PSIPRED [23], PROFsec [24], and YASPIN [25].

3. *Sequence motif information.* Sequence motifs are regions in protein/DNA sequence that possess functional importance. These regions tend to be conserved amongst homologous sequences. However, automatically generating MSAs where motif regions are properly aligned is not easy to achieve. PRALINE provides an option for the user to plug in sequence motif information in the form of a regular expression. This information is then used in

subsequent dynamic programming steps by biasing the residue scoring matrix towards aligning the motif residues. The incorporation of motif information in the alignment process significantly increases the chance that motif residues are aligned, which would otherwise be difficult to accomplish in conventional alignment algorithms, while maintaining overall alignment quality.

4. *Iteration*. For the above global and local preprocessing strategies, iterative optimization is possible. Iteration is based on the consistency of a preceding multiple alignment. Here consistency is defined as the agreement between matched amino acids in the multiple alignment and those in corresponding pairwise alignments. These consistency scores are then fed as weights to a next round of dynamic programming. During iteration, therefore, consistent multiple alignment positions tend to be maintained, whereas inconsistent segments are more likely to become realigned. Iterations are terminated upon reaching convergence or limit cycle (i.e., a number of cyclically recurring multiple alignments). The user can also specify a maximum number of iterations.

3.2 MUSCLE

MUSCLE [26, 27] is multiple alignment software for both nucleotide and protein sequences. It includes an online server, but the user can also choose to download the program and run it locally. The web server performs calculations using predefined default parameters, albeit the program provides a large number of options. MUSCLE is a very fast algorithm, which should be particularly considered when aligning large datasets. The progressive alignment protocol is sped up using a clever pairwise sequence comparison that avoids the slow DP technique for the construction of the so-called guide tree. Because of the computational efficiency gained, MUSCLE by default employs iterative refinement procedures that have been shown to produce high-quality multiple alignments.

1. *Iteration*. The full iteration procedure used by MUSCLE consists of three steps, although only the last can be considered truly iterative.

    (a) In the first step, sequences are clustered according to the number of *k-mers* (contiguous segments of length *k*) that they share using a compressed amino acid alphabet [28]. From this the guide tree is calculated using UPGMA, after which the sequences are progressively aligned following the tree order.

    (b) During the next step the obtained MSA is used to construct a new tree by applying the Kimura distance correction. This step is executed at least twice and can be repeated a number of times until a new tree does not achieve any improvements. As a measure to estimate improvement, the number of internal nodes for which the branching order has changed is recorded. If this number remains constant or increases, the iteration procedure terminates and a final progressive alignment is built for this step.

(c) Finally, the third step involves refinement of the alignment using the now fixed tree-topology. Edges from the tree are deleted in order of decreasing distance from the root. For each subdivision of the tree, the two corresponding profiles are aligned (tree-dependent refinement step). If a resulting alignment has a higher score than the previously retained alignment, the new alignment is taken. Iteration terminates if after traversing all tree edges no new alignment is produced or the user-defined number of iterations has been reached.

2. *Large datasets.* As outlined, one of the most important advantages of MUSCLE is that it is very fast and therefore can handle large datasets in reasonable time. The user can decide for all stages and actions whether to include them, thus making a compromise between speed and accuracy. As an additional option, the user can also define a time range in which the program will select the best solution so far. Another possibility to speed up the program during pairwise k-mer alignment is provided by allowing the user to switch off extending the k-words by dynamic programming (*see* the preceding section). A final option, called "anchor optimization," is designed to reduce computations during tree-dependent refinement by dividing a given alignment in vertical blocks and aligning the associated profiles separately.

The overall quality of MSAs produced by MUSCLE is good, although not exceptional given a number of more recent state-of-the-art techniques (*see below*). A known issue with some input sequence sets is that MUSCLE may produce large erroneous shifts of sequence blocks, which reduces the chance that the user will notice potential locally conserved fragments.

3.3 T-Coffee

The T-Coffee program [29] can also handle both DNA and protein sequences. It includes a web server (following the default settings) as well as an option to download the program. The algorithm derives its sensitivity from combining both local and global alignment techniques. Additionally, transitivity is exploited using triplet alignment information including each possible third sequence. A pairwise alignment is created using a protocol named *matrix extension* that includes the following steps:

1. *Combining local and global alignment.* For each pairwise alignment, the match scores obtained from local and global

alignments are summed, where for every matched residue pair the identity score of the associated (global or local) alignment is taken. For each sequence pair, the ten highest scoring local alignments are compiled using Lalign [30] and a global alignment is calculated using ClustalW [31].

2. *Transitivity.* For each third sequence C relative to a considered sequence pair A and B, the alignments A–C and C–B together constitute an alignment A–B. For each matched residue x in A and y in B, the minimum of the score of the match between residue x in A with residue z in C (alignment A–C) and that of residue z in C with y in B (alignment C–B) is taken; identity scores of associated alignments are taken as in the preceding step and all scores from the direct alignment as well as through all third sequences are summed.

3. For each sequence pair, dynamic programming is performed over the thus extended matrices. Owing to the fact that the signal captured in the extended scores is generally more consistent than noise, the scores of the consistent alignment traces are generally so high that gap penalties can be set to zero, thereby allowing consistent locally aligned fragments to be combined in a single global alignment.

   From the extended alignment scores, a guide tree is calculated using the Neighbor-Joining technique, and sequences are progressively aligned following the dynamic programming protocol. The combined use of local alignment, global alignment, and transitivity effectively alleviates error propagation during progressive alignment. However, the program is constrained by computational demands when aligning larger sets. As a consequence, the T Coffee web server constrains the allowed number of input sequences. The T-Coffee suite provides the following additional options.

4. *Integrating tertiary structures with 3D-Coffee.* A variant of the described protocol, 3D-Coffee [32] allows the inclusion of tertiary structures associated with one or more of the input sequences for guiding the alignment based upon the principle that "Structure is more conserved than sequence." If a partial sequence of a structure is given, the program will only take the corresponding structural fragment into account. The 3D-Coffee web server incorporates two default pairwise structural alignment methods: SAP [33] and FUGUE [34]. The first method is a structure superposition package, which is useful if more than one structure is included. The latter is a threading technique that can improve the multiple alignment process when local structural fragments are available. The advanced interface of the program allows the user to select alternative structural alignment methods.

5. *Accelerating the analyses.* Speed limitations of the T-Coffee program can be partially reduced by running a less demanding version. As an alternative, sequences can be divided into subgroups and aligned separately. To assist in this scenario, the program offers an option to compile a final alignment of these previously aligned subgroups.
6. *Consensus MSA.* A more recent extension is the method M-Coffee [35], which uses the T-Coffee protocol to combine the outputs of other MSA methods into a single consensus MSA.

3.4 MAFFT

The multiple sequence alignment package MAFFT [36, 37] is suited for DNA and protein sequences. MAFFT includes a script and a web server that both incorporate several alignment strategies. An alternative solution is proposed for the construction of the guide tree, which usually requires most computing time in a progressive alignment routine. Instead of performing all-against-all pairwise alignments, Fast Fourier Transformation (FFT) is used to rapidly detect homologous segments. The individual amino acids are characterized using their volume and polarity values, yielding high FFT peaks in a pairwise comparison whenever homologous segments are identified. The segments thus identified are then merged into a final alignment by dynamic programming. Additional iterative refinement processes, in which the scoring system is quickly optimized at each cycle, yield a high overall accuracy of the alignments.

1. *Fast alignment strategies.* Two options are provided for large sequence sets: FFT-NS-1 and FFT-NS-2, both of which follow a strictly progressive protocol. FFT-NS-1 generates a quick and dirty guide tree and compiles a corresponding MSA. If FFT-NS-2 is invoked, it takes the alignment obtained by FFT-NS-1 but now calculates a more reliable guide tree, which is used to compile another MSA.
2. *Iterative strategies.* The user can choose from several iterative approaches. The FFT-NS-i method attempts to further refine the alignment obtained by FFT-NS-2 by realigning subgroups until the maximum weighted sum of pairs (WSP) score [38] is reached. Two more recently included iterative refinement options (MAFFT version 5.66) incorporate local pairwise alignment information into the objective function (sum of the WSP scores). These are L-INS-i and E-INS-i, which use standard affine and generalized affine gap costs [39, 40] for scoring the pairwise comparisons, respectively.
3. *Alignment extension.* Another tool included in the MAFFT alignment package is mafftE. This option enhances the original dimension of the input set by including other homologous sequences, retrieved from the SwissProt database with BLAST [12]. Preferences for the exact number of additional sequences and the e-value can be specified by the user.

3.5 ProbCons

ProbCons [41] is an accurate but slow progressive alignment algorithm for protein sequences. The software can be downloaded but sequences can also be submitted to the ProbCons web server. The method follows the T-Coffee approach in spirit, but implements some of the steps differently. For example, the method uses an alternative scoring system for pairs of aligned sequences. The method starts by using a pair-HMM and expectation maximization (EM) to calculate a posterior probability for each possible residue match within a pairwise comparison. Next, for each pairwise sequence comparison, the alignment that maximizes the "expected accuracy" is determined [42]. In a similar way to the T-Coffee algorithm, information from pairwise alignments is then extended by considering consistency with all possible third "intermediate" sequences. For each pairwise sequence comparison, this leads to a so-called "probabilistic consistency" that is calculated for each aligned residue pair using matrix multiplication. These changed probabilities for matching residue pairs are then used to determine the final pairwise alignment by dynamic programming. Upon construction of a guide tree, a progressive protocol is followed to build the final alignment.

ProbCons allows a few variations of the protocol that the user can decide to adopt:

1. *Consistency replication*. The program allows the user to repeat the probabilistic consistency transformation step, by recalculating all posterior probability matrices. The default setting includes two replications, which can be increased to a maximum of 5.
2. *Iterative refinement*. The program also includes an additional iterative refinement procedure for further improving alignment accuracy. This is based on repeated random subdivision of the alignment in two blocks of sequences and realignment of the associated profiles. The default number of replications is set to 100, but can be changed from 0 to 1000 iterations (for the web server one can select 0, 100, or 500).
3. *Pre-training*. Parameters for the pair-HMM are estimated using unsupervised expectation maximization (EM). Emission probabilities, which reflect substitution scores from the BLOSUM-62 matrix [5], are fixed, whereas gap penalties (transition probabilities) can be trained on the whole set of sequences. The user can specify the number of rounds of EM to be applied on the set of sequences being aligned. The default number of iterations should be followed, unless there is a clear need to optimize gap penalties when considering a particular dataset.

3.6 Kalign

The Kalign algorithm [43] follows the standard progressive alignment strategy for sequence alignment. To gain speed, while not losing too much accuracy, the Kalign method incorporates the Wu–Manber approximate string-matching algorithm [44], which is used to determine local matches and subsequently to calculate the sequence distances. A drawback of using pattern matching techniques is that many spurious local matches may be detected. To address this issue, Kalign incorporates a number of heuristic strategies to filter the matches. This is done to maintain alignment quality and to preserve fast execution times.

Known obstacles for correct alignment are cases with discontinuous alignments requiring large gap regions, for example as a result of a domain being inserted or deleted. To facilitate the insertion of such extensive gap regions, Kalign provides the option to use Wu–Manber matches as anchor points during alignment. It incorporates two extra steps to enable the dynamic programming routine to use the anchor information efficiently:

1. *Consistent match finding*: The largest set of matches is searched that can be included in a single colinear alignment, after which the dynamic programming search matrix is filled with sums of selected matches. Then, dynamic programming is performed over the search matrix without applying gap penalties, so to allow the algorithm to match local segments, even if this requires the insertion of extensive gap regions. As an additional filter, matches occurring on short diagonals (cutoff length is 22) are deleted.
2. *Updating profile match positions*: Early during progressive alignment, the matches found with the Wu–Manber technique are indexed using the sequence positions. Later during progressive alignment, however, these positions must be updated to the appropriate positions in the profiles, because of gap insertion. The updates indexes facilitate rapid computation in a next pairwise alignment step.

The authors compared the speed and accuracy of Kalign to other popular methods, and it turned out to have comparable accuracy to the best other methods on small alignments, but was significantly more accurate when aligning large and evolutionary divergent sequence sets. Overall, the alignment quality is just under the best performers described in this chapter (e.g., MSAProbs and Clustal Omega). The speed of Kalign is about ten times faster than ClustalW.

3.7 MSAProbs

The method MSAProbs [45] reimplements ideas pioneered by T-Coffee and ProbCons. The main innovation is the specific hybrid combination of a pair-HMM and partition function to calculate posterior probabilities, a formalism which has mainly been guided by the objective to attain alignment accuracy of sequences with long N/C-terminal extensions, such as contained in BAliBASE reference 4 alignments (*see above*).

MSAProbs follows the basic progressive alignment protocol to compile multiple protein sequence alignments. The steps taken to arrive at a final MSA include:

1. Calculating all pairwise posterior probability matrices, each representing all possible pairwise residue matches between two aligned sequences or profiles, using both a pair-HMM and a partition function.
2. Calculating a pairwise distance matrix using the posterior probability matrices.
3. Constructing a guide tree from the pairwise distance matrix, and derivation of sequence weights.
4. Performing a weighted probabilistic consistency transformation of all pairwise posterior probability matrices, as inspired by the transitivity protocol of the T-Coffee method (*see* Subheading 3.3).
5. Generating a progressive alignment following the order in the guide tree using the transformed posterior probability matrices in a weighted profile-profile alignment protocol.

To further improve the alignment accuracy, MSAProbs includes an additional iterative refinement step. Iteratively, the MSA is divided randomly in two nonoverlapping subsets, after which a profile–profile alignment of the two subsets is performed. The number of iterations can be set by the user (the default setting is ten iterations).

MSAProbs has been shown to be able to attain statistically significant alignment accuracy improvements over existing state-of-the-art aligners, including ClustalW, MAFFT, MUSCLE, and ProbCons. Furthermore, MSAProbs is optimized for multi-core CPUs by employing a multi-threaded design, leading to significantly reduced execution times, hence allowing efficient alignment of large input sequence sets.

3.8 Clustal Omega

The Clustal Omega alignment engine [46] is the latest version of the widely used Clustal alignment suite. Unlike ClustalW, which employs the standard Dynamic Programming algorithm to align the sequences [31], ClustalOmega incorporates the HHsearch method, which is a HMM profile-profile alignment method, to align the sequences [47]. Similar to MAFFT, Clustal Omega is able to align a large number of sequences (>10,000 sequences) within a reasonable time. The method achieves this by exploiting an algorithm called mBed [48] for rapid generation of a guide tree for the alignment. The mBed technique is able to produce guide trees that are just as accurate as those from conventional methods. The main advantage of mBed, given N input sequences, is that it has a reduced complexity of $O(N \log N)$ as compared to $O(N^2)$, which is common for tree building methods based upon all-against-all sequence comparisons. This dramatic speedup facilitates many more sequences to be aligned, and makes Clustal Omega the fastest option for many large input sets, while its overall alignment accuracy is currently superior.

Clustal Omega provides an additional option for the user to plug in external information in order to improve the alignment. This is done using the "external profile alignment" (EPA) algorithm where the user can provide a profile HMM as an input in addition to the input sequences. The external profile HMM should be derived from sequences that are homologous to the sequences in the input set. The sequences in the input set are then aligned to this external HMM profile to help align them to the rest of the input set. There are a number of databases, such as PFAM, which provide a large collection of HMM profiles. EPA can also be used in a simple iteration scheme. Once an MSA has been made from a set of input sequences, it can be converted into a HMM and used for EPA to help realign the input sequences. This can also be combined with a full recalculation of the guide tree.

4 Notes

1. *Distant sequences*: Although high throughput alignment techniques are now able to make very accurate MSAs, alignment incompatibilities can arise under divergent evolution. In practice, it has been shown that the accuracy of all alignment methods decreases dramatically whenever input sequences share less than 30 % sequence identity [49]. Given this limitation, it is advisable to compile a number of MSAs using different amino acid substitution matrices and alignment tools. Depending on the characteristics of the input sequence set, the accuracy of alignment tools may vary a lot; unfortunately it is not easy to predict a priori which tool will perform best. Among the residue substitution matrices, the PAM [4] and BLOSUM [5] series are the most widely used (especially BLOSUM62). It is helpful to know that higher PAM numbers and low BLOSUM numbers (e.g., PAM250 or BLOSUM45) correspond to exchange matrices that have been designed for the alignment of increasingly divergent sequences, respectively, whereas matrices with lower PAM and higher BLOSUM numbers (e.g., PAM120 or BLOSUM62) are suitable for more closely related sequence sets. Furthermore, it is crucial to attempt different gap penalty values, as these can greatly affect the alignment quality. Gap penalties are an essential part of protein sequence alignment when using dynamic programming. The higher the gap penalties, the stricter

the insertion of gaps into the alignment and consequently the fewer gaps inserted. Gap regions in an MSA often correspond to loop regions in the associated tertiary structures, which are preferentially altered by divergent evolution. Therefore, it may be useful to lower the gap penalty values for more divergent sequence sets, although care should be taken not to deviate too much from the recommended settings. Excessive gap penalty values will enforce a gap-less alignment, whereas low gap penalties will lead to alignments with many gaps, allowing (near) identical amino acids to be matched. In both cases the resulting alignment will be biologically inaccurate. The way in which gap penalties affect the alignment also depends on the residue exchange matrix used. Although recommended combinations of exchange matrices and gap penalties have been described in the literature and most methods include default matrices and gap penalty settings, there is no formal theory yet as to how gap penalties should be chosen given a particular residue exchange matrix. Therefore, gap penalties are set empirically: for example, affine penalty values of 11 (gap opening) and 1 (gap extension) are recommended for BLOSUM62, whereas the suggested values for PAM250 are 10 and 1.

2. *Multi-domain proteins (Dialign, T-Coffee)*: Multi-domain proteins can be a particular challenge for multiple alignment methods. Whenever there has been an evolutionary change in the domain order of the query protein sequences, or if some domains have been inserted or deleted across the sequences, this leads to serious problems for global alignment engines. Global methods are not able to deal with permuted domain orders and normally exploit gap penalty regimes that make it difficult to insert long gaps corresponding to the length of one or more protein domains. For the alignment of multi domain protein sequences, it is advisable to resort to a local multiple alignment method. Alternatively, the T- Coffee [29] and Dialign [50, 51] methods might provide a meaningful alignment of multi-domain proteins, as they are (partly) based on the local alignment technique.

3. *Repeats*: In nature many biologically important protein families contain repeated and/or shuffled domains, thus constituting multi-domain proteins. It is not trivial to reconstruct alignments of repeat proteins using a standard local or global approach; the occurrence of repeats in many sequences can seriously compromise the accuracy of MSA methods, mostly because the techniques are not able to deal with different repeat copy numbers. For sequences with repeats but no rearrangements, the method RAlign applies an alignment algorithm that accounts for repeats and is able to keep track of various repeat types [52]. The method requires the specification of the individual repeats,

which can be obtained by running one of the available repeat detection algorithms, after which a repeat-aware MSA is produced. RAlign produces a global MSA in which information about repeats is merged with similarities deduced from comparing non-repetitious areas within the sequences. The evolutionary model used in the method can cope with translocations that interrupt existing sequence motifs. The method finally compiles a standard global multiple alignment, where shuffled elements remain unaligned in keeping with the global alignment requirement. Although the alignment of repeat sequences can be markedly improved by this method, as compared to methods that are not repeat-aware, it is sensitive to the accuracy of the repeats information provided.

ProDA [53] is an algorithm that was specifically designed for alignment of proteins that contain repeats and/or rearrangements in their domain architecture. In contrast to RAlign, the ProDA method attempts to identify relationships between homologous segments within the same sequence. Briefly, the ProDA algorithm seeks to first compute all-versus-all pairwise local alignments using a pair-HMM approach or the Viterbi algorithm. From these local alignments, likely repeat regions are defined. Subsequently, so-called blocks of alignable sequence fragments are constructed which contain sequences for which at least one region is shared. The sequence blocks are then progressively aligned in a final alignment. An iterative step is implemented to make sure all pairwise alignments are considered. The method was evaluated based on subset 6 of the BAliBASE (2.0) benchmark, which includes sets of proteins with multiple repeats. Further benefits were shown on datasets that include (complex) rearrangements and insertions. The ProDA method is exclusively available as a stand-alone package.

4. *TM regions*: A special class of proteins is comprised of membrane-associated proteins. The regions within such proteins that are inserted in the cell membrane display a profoundly changed hydrophobicity pattern as compared with soluble proteins. Because the scoring schemes (e.g., PAM [4] or BLOSUM [5]) normally used in MSA techniques are derived using sequences of soluble proteins, the alignment methods are in principle not suitable to align membrane-bound protein regions. This means that great care should be taken when using general MSA methods. Fortunately, transmembrane (TM) regions can be reliably recognized using state-of-the-art prediction techniques such as TMHMM [54] or Phobius [55]. Therefore, it may be advisable to mark the putative TM regions across the query sequences, and if their mutual correspondence would be clear, to align the blocks of intervening sequence fragments separately.

5. *Prior knowledge*: In many cases, there is already some prior knowledge about the final alignment. For instance, consider a protein family containing a disulfide bridge between two specific cysteine residues. Given the structural importance of a disulfide bond, constituent Cys residues are generally conserved, so that it is important that the final MSA matches such Cys residues correctly. However, depending on conservation patterns and overall evolutionary distances of the sequences, it can happen that the alignment engine needs special guidance for matching the Cys residues correctly. Currently none of the approaches has a built-in tool to mark particular positions and assign specific parameters for their consistency, although the library structure of the T-Coffee method allows the specification of weights for matching individual amino acids across the input sequences. However, exploiting this possibility can be rather cumbersome. The following suggestions are therefore offered for (partially) resolving this type of problem:
   (a) *Chopping alignments.* Instead of aligning whole sequences, one can decide to chop the alignment in different parts. For example, this could be done if the sequences have some known domains for which the sequence boundaries are known. An added advantage in such cases is that no undesirable overlaps will occur between these pre-marked regions if aligned separately. Finally, the whole alignment can be built by concatenating the aligned blocks. It should be stressed that each of the separate alignment operations is likely to follow a different evolutionary scenario, as for example the guide tree or the additionally homologous background sequences in the PSI-PRALINE protocol can be different in each case. It is entirely possible, however, that these different scenarios reflect true evolutionary differences, such as for instance unequal rates of evolution of the constituent domains.
   (b) *Altering amino acid exchange weights.* Multiple alignment programs make use of amino acid substitution matrices in order to score alignments. Therefore, it is possible to change individual amino acid exchange values in a substitution matrix. Referring to the disulfide example mentioned in the preceding, one could decide to up-weight the substitution score for a cysteine self-conservation. As a result, the alignment will obtain a higher score when cysteines are matched, and as a consequence the method will attempt to create an alignment where this is the case. However, some protein families have a number of known pairs of Cys residues that form disulfide bonds, where mixing up of the Cys residues involved in different disulfide bridges might happen in that Cys residues involved in different

disulfide bonds become aligned at a given single position. To avoid such incorrect matches in the alignment, some programs (e.g., PRALINE) allow the addition of a few extra amino acid designators in the amino acid exchange matrix that can be used to identify Cys residue pairs in a given bond (e.g., J, O, or U). The exchange scores involving these "alternative" Cys residues should be identical to those for the original Cys, except for the cross-scores between the alternative letters for Cys that should be given low (or extreme negative) values to avoid cross alignment. It must be stressed that such alterations are heuristics that can violate the evolutionary model underlying a given residue exchange matrix.

6. *Alignment editors*: A number of multiple alignment editors are available for editing automatically generated alignments, which often can be improved manually. Posterior manual adjustments can be helpful, especially if structural or functional knowledge of the sequence set is at hand. The following editing tools are available:
   (a) *Jalview* (www.jalview.org) [56] is a protein multiple sequence alignment editor written in Java. In addition to a number of editing options, it provides a wide scale of sequence analysis tools, such as sequence conservation, UPGMA, and NJ [57] tree calculation, and removal of redundant sequences. Color schemes may be customized according to amino acid physicochemical properties, similarity to consensus sequence, hydrophobicity, or secondary structure.
   (b) *SeaView* (http://doua.prabi.fr/software/seaview) [58] is a graphical editor suited for Mac, Windows, Unix, and Linux. The program includes a dot-plot routine for pairwise sequence comparison [59] or the ClustalW [31] multiple alignment program to locally improve the alignment and can also perform phylogenetic analyses. Color schemes can be customized.
   (c) *STRAP* (http://www.bioinformatics.org/strap/) [60] is an interactively extendable and scriptable editor program, able to manipulate large protein alignments. The software is written in Java and is compatible with all operating systems. Among the many extra features provided are: enhanced alignment of low-similarity sequences by integrating 3D-structure information, determination of regular expression motifs, and transmembrane and secondary structure predictions.
   (d) *CINEMA* (http://aig.cs.man.ac.uk/research/utopia/cinema/cinema.php) [61] is a Java interactive tool for editing

either nucleotide or amino acid sequences. The flexible editor permits color scheme changes and motif selection. Hydrophobicity patterns can also be viewed. Furthermore, there is an option to load prepared alignments from the PRINTS fingerprint database [62].

7. *Benchmarking*: The main goal of MSA benchmarking is to measure the ability of different MSA methods to reproduce trusted reference alignments [63]. The most widely used approach to benchmark alignment quality is referred to as reference-dependent benchmarking, which aims to compute the similarity between the MSA in question with a gold standard reference alignment of the same sequence set. These reference alignments are normally obtained from a manually (or semiautomatically) curated alignment database, such as BAliBASE [17], OXBENCH [64], or SABmark [65]. The similarity between a reference alignment and the corresponding query alignment can be reflected in a summarizing score such as the column score (CS), the sum-of-pairs (SP) score, the shift score [66] or SPdist score [67]. The CS calculates the query-reference alignment similarity as the fraction of columns in the reference alignment that is exactly reproduced in the query alignment; the SP score determines the similarity by the fraction of aligned residue pairs in the reference alignment that is correctly reproduced in the query alignment. The shift and SPdist scores, on the other hand, also take into account the sequence distance by which a mismatch is displaced, relative to the correctly matched position as given in the reference alignment.

References

1. Gribskov M, McLachlan AD, Eisenberg D (1987) Profile analysis: detection of distantly related proteins. Proc Natl Acad Sci U S A 84:4355–4358
2. Haussler D, Krogh A, Mian IS et al (1993) Protein modeling using hidden Markov models: analysis of globins. In: Proceedings of the Hawaii international conference on system sciences. IEEE Computer Society Press, Los Alamitos, CA
3. Bucher P, Karplus K, Moeri N et al (1996) A flexible motif search technique based on generalized profiles. Comput Chem 20:3–23
4. Dayhoff MO, Schwart RM, Orcutt BC (1978) A model of evolutionary change in proteins. In: Dayhoff M (ed) Atlas of protein sequence and structure. National Biomedical Research Foundation, Washington, DC
5. Henikoff S, Henikoff JG (1992) Amino acid substitution matrices from protein blocks. Proc Natl Acad Sci U S A 89:10915–10919
6. Needleman SB, Wunsch CD (1970) A general method applicable to the search for similarities in the amino acid sequence of two proteins. J Mol Biol 48:443–453
7. Carillo H, Lipman DJ (1988) The multiple sequence alignment problem in biology. SIAM J Appl Math 48:1073–1082
8. Stoye J, Moulton V, Dress AW (1997) DCA: an efficient implementation of the divide-and-conquer approach to simultaneous multiple sequence alignment. Comput Appl Biosci 13:625–626
9. Feng DF, Doolittle RF (1987) Progressive sequence alignment as a prerequisite to correct phylogenetic trees. J Mol Evol 25:351–360
10. Hogeweg P, Hesper B (1984) The alignment of sets of sequences and the construction of phyletic trees: an integrated method. J Mol Evol 20:175–186

11. Gotoh O (1996) Significant improvement in accuracy of multiple protein sequence alignments by iterative refinement as assessed by reference to structural alignments. J Mol Biol 264:823–838
12. Altschul SF, Gish W, Miller W et al (1990) Basic local alignment search tool. J Mol Biol 215:403–410
13. Pearson WR (1990) Rapid and sensitive sequence comparison with FASTP and FASTA. Methods Enzymol 183:63–98
14. Heringa J, Taylor WR (1997) Three-dimensional domain duplication, swapping and stealing. Curr Opin Struct Biol 7:416–421
15. Smith TF, Waterman MS (1981) Identification of common molecular subsequences. J Mol Biol 147:195–197
16. Waterman MS, Eggert M (1987) A new algorithm for best subsequence alignments with application to tRNA-rRNA comparisons. J Mol Biol 197:723–728
17. Thompson JD, Plewniak F, Poch O (1999) BAliBASE: a benchmark alignment database for the evaluation of multiple alignment programs. Bioinformatics 15:87–88
18. Heringa J (1999) Two strategies for sequence comparison: profile-preprocessed and secondary structure-induced multiple alignment. Comput Chem 23:341–364
19. Heringa J (2002) Local weighting schemes for protein multiple sequence alignment. Comput Chem 26:459–477
20. Simossis VA, Heringa J (2005) PRALINE: a multiple sequence alignment toolbox that integrates homology-extended and secondary structure information. Nucleic Acids Res 33: W289–W294
21. Altschul SF, Madden TL, Schaffer AA et al (1997) Gapped BLAST and PSIBLAST: a new generation of protein database search programs. Nucleic Acids Res 25:3389–3402
22. Kabsch W, Sander C (1983) Dictionary of protein secondary structure: pattern recognition of hydrogen-bonded and geometrical features. Biopolymers 22:2577–2637
23. Jones DT (1999) Protein secondary structure prediction based on position-specific scoring matrices. J Mol Biol 292:195–202
24. Rost B, Sander C (1993) Prediction of protein secondary structure at better than 70% accuracy. J Mol Biol 232:584–599
25. Lin K, Simossis VA, Taylor WR et al (2005) A simple and fast secondary structure prediction method using hidden neural networks. Bioinformatics 21:152–159
26. Edgar RC (2004) MUSCLE: a multiple sequence alignment method with reduced time and space complexity. BMC Bioinformatics 5:113
27. Edgar RC (2004) MUSCLE: multiple sequence alignment with high accuracy and high throughput. Nucleic Acids Res 32:1792–1797
28. Edgar RC (2004) Local homology recognition and distance measures in linear time using compressed amino acid alphabets. Nucleic Acids Res 32:380–385
29. Notredame C, Higgins DG, Heringa J (2000) T-Coffee: A novel method for fast and accurate multiple sequence alignment. J Mol Biol 302:205–217
30. Huang X, Miller W (1991) A time-efficient, linear-space local similarity algorithm. Adv Appl Math 12:337–357
31. Thompson JD, Higgins DG, Gibson TJ (1994) CLUSTAL W: improving the sensitivity of progressive multiple sequence alignment through sequence weighting, position-specific gap penalties and weight matrix choice. Nucleic Acids Res 22:4673–4680
32. O'Sullivan O, Suhre K, Abergel C et al (2004) 3DCoffee: combining protein sequences and structures within multiple sequence alignments. J Mol Biol 340:385–395
33. Taylor WR, Orengo CA (1989) Protein structure alignment. J Mol Biol 208:1–22
34. Shi J, Blundell TL, Mizuguchi K (2001) FUGUE: sequence-structure homology recognition using environment-specific substitution tables and structure-dependent gap penalties. J Mol Biol 310:243–257
35. Wallace IM, O'Sullivan O, Higgins DG et al (2006) M-Coffee: combining multiple sequence alignment methods with T-Coffee. Nucleic Acids Res 34:1692–1699
36. Katoh K, Misawa K, Kuma K et al (2002) MAFFT: a novel method for rapid multiple sequence alignment based on fast Fourier transform. Nucleic Acids Res 30:3059–3066
37. Katoh K, Kuma K, Toh H et al (2005) MAFFT version 5: improvement in accuracy of multiple sequence alignment. Nucleic Acids Res 33:511–518
38. Gotoh O (1995) A weighting system and algorithm for aligning many phylogenetically related sequences. Comput Appl Biosci 11:543–551
39. Altschul SF (1998) Generalized affine gap costs for protein sequence alignment. Proteins 32:88–96
40. Zachariah MA, Crooks GE, Holbrook SR et al (2005) A generalized affine gap model significantly improves protein sequence alignment accuracy. Proteins 58:329–338

41. Do CB, Mahabhashyam MS, Brudno M et al (2005) ProbCons: probabilistic consistency-based multiple sequence alignment. Genome Res 15:330–340
42. Holmes I, Durbin R (1998) Dynamic programming alignment accuracy. J Comput Biol 5:493–504
43. Lassmann T, Sonnhammer ELL (2005) Kalign: an accurate and fast multiple sequence alignment algorithm. BMC Bioinformatics 6 (1):298
44. Wu S, Manber U (1992) Fast text searching allowing errors. Commun ACM 35:83–91
45. Liu Y, Schmidt B, Maskell DL (2010) MSA-Probs: multiple sequence alignment based on pair hidden Markov models and partition function posterior probabilities. Bioinformatics 26 (16):1958–1964
46. Sievers F, Wilm A, Dineen D, Li W, Lopez R, McWilliam H, Remmert M, Söding J, Thompson JD, Higgins DG (2011) Fast, scalable generation of high quality protein multiple sequence alignments using Clustal Omega. Mol Syst Biol 7(1):539
47. Söding J (2005) Protein homology detection by HMM–HMM comparison. Bioinformatics 21(7):951–960
48. Blackshields G, Sievers F, Shi W, Wilm A, Higgins DG (2010) Sequence embedding for fast construction of guide trees for multiple sequence alignment. Algorithms Mol Biol 5:21
49. Rost B (1999) Twilight zone of protein sequence alignments. Protein Eng 12:85–94
50. Morgenstern B, Dress A, Werner T (1996) Multiple DNA and protein sequence alignment based on segment-to-segment comparison. Proc Natl Acad Sci U S A 93:12098–12103
51. Morgenstern B (2004) DIALIGN: multiple DNA and protein sequence alignment at BiBiServ. Nucleic Acids Res 32:W33–W36
52. Sammeth M, Heringa J (2006) Global multiple-sequence alignment with repeats. Prot Struct Funct Bioinf 64:263–274
53. Phuong TM, Choung BD, Edgar RC, Batzoglou S (2006) Multiple alignment of protein sequences with repeats and rearrangements. Nucleic Acids Res 34:5932–5942
54. Krogh A, Larsson B, von Heijne G et al (2001) Predicting transmembrane protein topology with a hidden Markov model: application to complete genomes. J Mol Biol 305:567–580
55. Kall L, Krogh A, Sonnhammer EL (2004) A combined transmembrane topology and signal peptide prediction method. J Mol Biol 338:1027–1036
56. Clamp M, Cuff J, Searle SM et al (2004) The Jalview Java alignment editor. Bioinformatics 20:426–427
57. Saitou N, Nei M (1987) The neighbor-joining method: a new method for reconstructing phylogenetic trees. Mol Biol Evol 4:406–425
58. Galtier N, Gouy M, Gautier C (1996) SEA-VIEW and PHYLO_WIN: two graphic tools for sequence alignment and molecular phylogeny. Comput Appl Biosci 12:543–548
59. Li W-H, Graur D (1991) Fundamentals of molecular evolution. Sinauer, Sunderland, MA
60. Gille C, Frommel C (2001) STRAP: editor for STRuctural Alignments of Proteins. Bioinformatics 17:377–378
61. Parry-Smith DJ, Payne AW, Michie AD et al (1998) CINEMA—a novel colour INteractive editor for multiple alignments. Gene 221: GC57–GC63
62. Attwood TK, Beck ME, Bleasby AJ et al (1997) Novel developments with the PRINTS protein fingerprint database. Nucleic Acids Res 25:212–217
63. Golubchik T, Wise MJ, Easteal S, Jermiin LS (2007) Mind the gaps: evidence of bias in estimates of multiple sequence alignments. Mol Biol Evol 24(11):2433–2442
64. Raghava GPS, Searle SMJ, Audley PC, Barber JD, Barton GJ (2003) OXBench: a benchmark for evaluation of protein multiple sequence alignment accuracy. BMC Bioinformatics 4 (1):47
65. Van Walle I, Lasters I, Wyns L (2005) SABmark—a benchmark for sequence alignment that covers the entire known fold space. Bioinformatics 21(7):1267–1268
66. Cline M, Hughey R, Karplus K (2002) Predicting reliable regions in protein sequence alignments. Bioinformatics 18(2):306–314
67. Bawono P, van der Velde A, Abeln S, Heringa J (2015) Quantifying the displacement of mismatches in multiple sequence alignment benchmarks. PLoS ONE 10(5):e0127431

Chapter 9

Large-Scale Sequence Comparison

Devi Lal and Mansi Verma

Abstract

There are millions of sequences deposited in genomic databases, and it is an important task to categorize them according to their structural and functional roles. Sequence comparison is a prerequisite for proper categorization of both DNA and protein sequences, and helps in assigning a putative or hypothetical structure and function to a given sequence. There are various methods available for comparing sequences, alignment being first and foremost for sequences with a small number of base pairs as well as for large-scale genome comparison. Various tools are available for performing pairwise large sequence comparison. The best known tools either perform global alignment or generate local alignments between the two sequences. In this chapter we first provide basic information regarding sequence comparison. This is followed by the description of the PAM and BLOSUM matrices that form the basis of sequence comparison. We also give a practical overview of currently available methods such as BLAST and FASTA, followed by a description and overview of tools available for genome comparison including LAGAN, MumMER, BLASTZ, and AVID.

Key words Homology, Orthologs, Paralogs, Substitutions, Indels, Gap penalty, Conservative substitutions, Dynamic programming algorithm, Heuristic approach, Scoring matrix, Accepted point mutation, BLOcks SUbstitution Matrix, Global and local alignment, *E* value

1 Introduction

With the advent of next-generation sequencing technologies, the number of sequences being added to genomic databases is growing exponentially. For a given set of related biological sequences, the first and foremost step is to compare those sequences. Sequence alignment is an important procedure for comparing two or more than two sequences which are related to each other to investigate their similarity, their patterns of conservation, and the evolutionary relationship shared by these sequences.

1.1 Homology, Similarity, and Identity

When we talk about two sequences we typically want to know how these sequences are related to each other. The terms homology and similarity are often used interchangeably, but these two terms represent different relationships. *Homology* is a general term that is used to describe a shared common evolutionary ancestry [1].

Jonathan M. Keith (ed.), *Bioinformatics: Volume I: Data, Sequence Analysis, and Evolution*, Methods in Molecular Biology, vol. 1525, DOI 10.1007/978-1-4939-6622-6_9, © Springer Science+Business Media New York 2017

Homologous sequences can be further described either as *orthologs* or *paralogs*. *Orthologs* (ortho = exact) are those sequences that are present in different species and have a common ancestry while *paralogs* (para = parallel) are sequences that are present in the same species and have arisen due to gene duplication events. Homology is rather a qualitative term which suggests that two sequences are related but does not specify the degree of relatedness. To describe relatedness quantitatively, the terms *identity* and *similarity* are used. Identity refers to whether two sequences (or parts of sequences) are evolutionarily invariant while similarity acknowledges substitutions that preserve structural or functional roles. For example, substituting the amino acid aspartic acid with glutamic acid is likely to preserve structure and function as both are acidic amino acids. The similarity between two sequences is a score reflecting both the identity and substitutions involving similar bases/amino acids.

1.2 Substitutions and Indels

In order to assess the similarity or identity any two sequences share, these sequences must first be aligned (see also Chapter 8). Alignments of two sequences, referred to as *pairwise alignments*, can be used to compute a similarity score based on the number of matches, mismatches, and *gaps*. The mutations that accumulate in a sequence during evolution are *substitutions*, *insertions*, and *deletions*. Substitutions result from a change in the nucleotide or amino acid sequence. Insertions and deletions (together known as *indels*) denote either addition or removal of residues and are typically represented by a dash. A run of one or more contiguous indels is commonly referred to as a gap in an alignment. Gaps are usually heavily penalized in forming an optimal alignment. The most widely used method for calculation of gap penalties is known as an *affine gap penalty*. In this method two parameters are taken into account: (1) the introduction of a gap and (2) the extension of a gap. The gap score is the sum of the *gap opening penalty* (G) and the *gap extension penalty* (L). For a gap of length n, the gap score will be $G + Ln$. The values for gap opening penalties typically lie in the range 10–15 (a common default is 11) and gap extension penalties are typically 1–2 (default is 1).

The second method for scoring gap penalties is known as *non-affine* or *linear gap penalty*. This method has a fixed penalty to score gaps in an alignment and there is no heavy cost for gap opening as is seen in the affine gap penalty.

In pairwise alignments of proteins (Fig. 1), some aligned residues may be similar (that is, structurally, functionally or biochemically related) but not identical. These similar residues are denoted by a "+" sign in the alignment and are referred to as *conservative substitutions*.

```
Seq1 MSDLDRLASRAAIQDLYSDQLIGVDKRQEGRLASIWWDDAEWTIEGIGTYKGPEGALDLA
     MSDLDRLASRAAIQDLYSD+LI VDKRQEGRLASIWWDDAEWTIEGIGTYKGPEGALDLA
Seq2 MSDLDRLASRAAIQDLYSDKLIAVDKRQEGRLASIWWDDAEWTIEGIGTYKGPEGALDLA

Seq1 NNVLWPMFHETIHYGTNLRLEFVSADKVNGIGDVLCLGNLVEGNQSILIAAVYTNEYERR
     NNVLWPMFHE IHYGTNLRLEFVSADKVNGIGDVL LGNLVEGNQSILIAAV+T+EYERR
Seq2 NNVLWPMFHECIHYGTNLRLEFVSADKVNGIGDVLLLGNLVEGNQSILIAAVFTDEYERR

Seq1 DGVWKLSKLNGCMNYFTPLAGIHFAPPGALLQKS
     DGVW  SK N C NYFTPLAGIHFAPPG     S
Seq2 DGVWKFSKRNACTNYFTPLAGIHFAPPGIHFAPS
```

Fig. 1 Pairwise alignment of two protein sequences indicating the similarities and identities. The similarities are indicated by a "+" sign and identities by the same character between the two sequences in the alignment

2 Materials

2.1 Sequence Selection: DNA or Protein

The comparison of two sequences can be done either at the level of nucleotide or protein. Though it is true that mutations take place in DNA, still comparing two sequences at the level of protein can help in revealing important biological information [2]. Many mutations in DNA are synonymous, that is, they are not reflected at the level of protein and do not lead to any change in the corresponding amino acid. Consequently, for distantly related organisms with low sequence identity at the DNA level, comparison of proteins is preferred.

2.2 Pairwise Alignment and Scoring Matrices

There are various methods available for pairwise alignment. These methods include (i) dot-matrix analysis, (ii) use of a *dynamic programming algorithm* (which depends on the use of scoring matrices for scoring alignment), and (iii) heuristic approaches (word or k-tuple methods).

1. *Dot-matrix analysis*: This is one of the most popular graphical methods of aligning two sequences, first described by Gibbs and McIntyre [3]. The sequences are placed on the *X*- and *Y*-axes of the matrix and a dot is placed wherever a match is found between the two sequences. Diagonal runs of dots are joined to form the alignment, whereas isolated dots are regarded as random matches (Fig. 2). This readily reveals the presence of indels (insertions or deletions) and repeats (directed or inverted).

There are various available tools for dot matrix generation (*see* **Note 1**). The dot matrices give only a graphical representation and do not reveal the similarity score. Therefore other methods for sequence comparison have been developed that depend on scoring any pairwise sequence comparison. The system of scoring implements the *dynamic programming algorithm* (DPA) to yield an optimal alignment between two sequences by breaking an alignment into smaller parts and then joining these subalignments in a sequential manner. Dynamic programming identifies the best

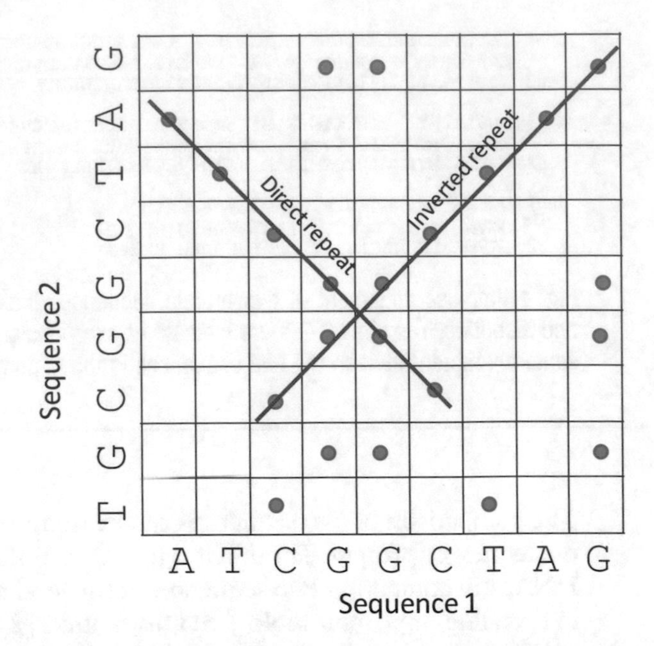

Fig. 2 Diagrammatic representation of dot matrix between sequence 1 "GATCGGCGT" and sequence 2 "TGCGGCTAG". A dot in the box represents a match; a diagonal line connecting the dots represents a sequence alignment whereas dots outside the diagonal represents spurious matches

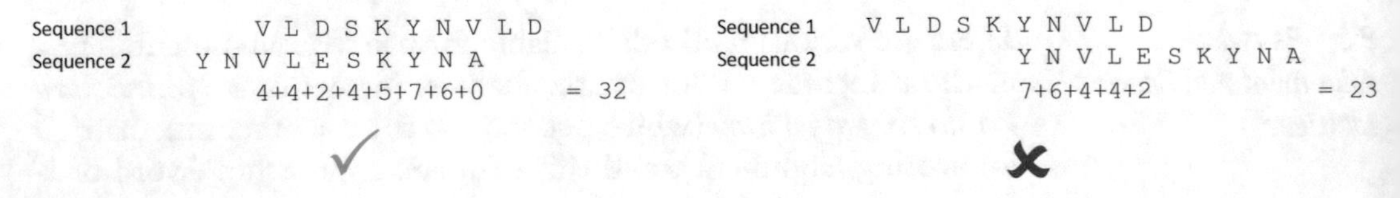

Fig. 3 Alignment of two sequences can be done in many ways, but the one with the highest score is selected by DPA

alignment with highest score among all possible alignments. The dynamic programming algorithm requires *scoring matrices* which are expressed in the form of a 20 by 20 matrix for amino acids and a 4 by 4 matrix for nucleotides. These matrices serve as an evolutionary model and are used to score the substitution of one amino acid by another. For example, the two sequences shown below can be aligned in two different ways (Fig. 3). However, the scores generated are different, and DPA selects the alignment with the higher score (this includes the scores for match, mismatch, and gaps). These scores are generated using the scoring matrix.

A number of scoring matrices are available and the choice of scoring matrix is important as it influences the results obtained.

2.2.1 PAM Matrices

Margaret Dayhoff [4] proposed the first matrices that were used for quantitatively scoring pairwise protein alignments. The Dayhoff model was based on the substitution patterns in a group of proteins. Dayhoff and colleagues compared the sequences of closely related proteins in families that shared more than 85 % sequence similarity. They found 1572 changes in 71 groups of closely related proteins. Based on these results, they constructed a table indicating the frequency of amino acid substitution at a single position that defines the relative mutabilities of amino acids. As the protein sequences used were very closely related, it was expected that the observed mutations would not affect the protein function. The matrices are referred to as *accepted point mutation* (PAM) matrices because these mutations or substitutions were accepted by natural selection.

The replacement of an amino acid by another with similar biochemical properties is sometimes accepted in a protein. These replacements are known as conservative substitutions. For example, replacement of serine with threonine, glutamic acid with aspartic acid, and isoleucine with valine are some of the most common amino acid substitutions that are readily accepted. The less replaced amino acids are those that have some structural or functional role in proteins and their replacement may have some deleterious effects.

Dayhoff and colleagues defined the PAM1 matrix as that which produces 1 accepted point mutation per 100 amino acid residues [4]. By this convention, a PAM0 matrix is the identity matrix, so that no amino acid can change. Since the PAM1 matrix was based on closely related protein sequences that share more than 85 % sequence identity, its use is limited for the protein sequences that are less than 85 % identical. For this, other types of PAM matrices were derived from PAM1 matrix by multiplying PAM1 by itself. PAM100 matrix was derived by multiplying PAM1 by itself 100 times. Similarly, the PAM250 matrix is used for proteins that share about 20 % sequence identity (Fig. 4).

The choice of PAM matrix depends on the relatedness of the protein sequences. If protein sequences are distantly related, a higher value of PAM is used and if they are closely related, a lower value of PAM is considered (Table 1). It is to be noted that the PAM matrices were based on small number of protein sequences available in 1978. With the increase in the number of available sequences, efforts have been made to update these PAM matrices [5, 6]. Moreover, these matrices assume that each site has same mutability. In spite of the limitations that may be present in these matrices, these are still in use and marked an important advancement in understanding the evolutionary relationship between the sequences.

2.2.2 BLOSUM Matrices

Using a slightly different approach, S. Henikoff and J.G. Henikoff [7–9] devised other types of scoring matrices that address the drawbacks in PAM matrices. Their approach was based on the use

A	2																			
R	-2	6																		
N	0	0	2																	
D	0	-1	2	4																
C	-2	-4	-4	-5	12															
Q	0	1	1	2	-5	4														
E	0	-1	1	3	-5	2	4													
G	1	-3	0	1	-3	-1	0	5												
H	-1	2	2	1	-3	3	1	-2	6											
I	-1	-2	-2	-2	-2	-2	-2	-3	-2	5										
L	-2	-3	-3	-4	-6	-2	-3	-4	-2	-2	6									
K	-1	3	1	0	-5	1	0	-2	0	-2	-3	5								
M	-1	0	-2	-3	-5	-1	-2	-3	-2	2	4	0	6							
F	-3	-4	-3	-6	-4	-5	-5	-5	-2	1	2	-5	0	9						
P	1	0	0	-1	-3	0	-1	0	0	-2	-3	-1	-2	-5	6					
S	1	0	1	0	0	-1	0	1	-1	-1	-3	0	-2	-3	1	2				
T	1	-1	0	0	-2	-1	0	0	-1	0	-2	0	-1	-3	9	1	3			
W	-6	2	-4	-7	-8	-5	-7	-7	-3	-5	-2	-3	-4	0	-6	-2	-5	17		
Y	-3	-4	-2	-4	0	-4	-4	-5	0	-1	-1	-4	-2	7	-5	-3	-3	0	10	
V	0	-2	-2	-2	-2	-2	-2	-1	-2	4	2	-2	2	-1	-1	-1	0	-6	-2	4
	A	**R**	**N**	**D**	**C**	**Q**	**E**	**G**	**H**	**I**	**L**	**K**	**M**	**F**	**P**	**S**	**T**	**W**	**Y**	V

Fig. 4 PAM250 matrix, derived by multiplying PAM1 by itself 250 times. This matrix is useful for highly divergent sequences

Table 1
Relationship between PAM matrices and relative sequence identity

PAM	0	30	80	110	200	250
% identity	100	75	50	60	25	20

of the BLOCKS database [8, 10], which consists of large numbers of local multiple alignments (also referred to as blocks) of distantly related proteins. The matrices were named **BLO**cks **SU**bstitution **M**atrix. The aim of BLOSUM matrices was to replace PAM matrices in identifying the distantly related proteins by using much larger set of data as compared to the Dayhoff's data. Blocks represent the alignment without any gap. Henikoff and Henikoff examined about 2000 blocks representing more than 500 groups of related proteins. They replaced those groups of proteins that have similarities higher than the threshold by a single representative or by average weight. The threshold of 62 % produced the most commonly used matrix BLOSUM62 (Fig. 5), which is also the default scoring matrix for BLAST search.

The BLOSUM matrices are thus directly calculated across various evolutionary distances and are based on conserved regions. This makes BLOSUM matrices more sensitive and they can perform better than PAM matrices for local similarity searches [11]. Just like PAM matrices, each BLOSUM matrix is represented by a number that denotes the level of conservation of the protein sequences that were used to derive that particular BLOSUM

	A	R	N	D	C	Q	E	G	H	I	L	K	M	F	P	S	T	W	Y	V
A	**4**	-1	-2	-2	0	-1	-1	0	-2	-1	-1	-1	-1	-2	-1	1	0	-3	-2	0
R	-1	**5**	0	-2	-3	1	0	-2	0	-3	-2	2	-1	-3	-2	-1	-1	-3	-2	-3
N	-2	0	**6**	1	-3	0	0	0	1	-3	-3	0	-2	-3	-2	1	0	-4	-2	-3
D	-2	-2	1	**6**	-3	0	2	-1	-1	-3	-4	-1	-3	-3	-1	0	-1	-4	-3	-3
C	0	-3	-3	-3	**9**	-3	-4	-3	-3	-1	-1	-3	-1	-2	-3	-1	-1	-2	-2	-1
Q	-1	1	0	0	-3	**5**	2	-2	0	-3	-2	1	0	-3	-1	0	-1	-2	-1	-2
E	-1	0	0	2	-4	2	**5**	-2	0	-3	-3	1	-2	-3	-1	0	-1	-3	-2	-2
G	0	-2	0	-1	-3	-2	-2	**6**	-2	-4	-4	-2	-3	-3	-2	0	-2	-2	-3	-3
H	-2	0	1	-1	-3	0	0	-2	**8**	-3	-3	-1	-2	-1	-2	-1	-2	-2	2	-3
I	-1	-3	-3	-3	-1	-3	-3	-4	-3	**4**	2	-3	1	0	-3	-2	-1	-3	-1	3
L	-1	-2	-3	-4	-1	-2	-3	-4	-3	2	**4**	-2	2	0	-3	-2	-1	-2	-1	1
K	-1	2	0	-1	-3	1	1	-2	-1	-3	-2	**5**	-1	-3	-1	0	-1	-3	-2	-2
M	-1	-1	-2	-3	-1	0	-2	-3	-2	1	2	-1	**5**	0	-2	-1	-1	-1	-1	1
F	-2	-3	-3	-3	-2	-3	-3	-3	-1	0	0	-3	0	**6**	-4	-2	-2	1	3	-1
P	-1	-2.	-2	-1	-3	-1	-1	-2	-2	-3	-3	-1	-2	-4	**7**	-1	-1	-4	-3	-2
S	1	-1	1	0	-1	0	0	0	-1	-2	-2	0	-1	-2	-1	**4**	1	-3	-2	-2
T	0	-1	0	-1	-1	-1	-1	-2	-2	-1	-1	-1	-1	-2	-1	1	**5**	-2	-2	0
W	-3	-3	-4	-4	-2	-2	-3	-2	-2	-3	-2	-3	-1	1	-4	-3	-2	**11**	2	-3
Y	-2	-2	-2	-3	-2	-1	-2	-3	2	-1	-1	-2	-1	3	-3	-2	-2	2	**7**	-1
V	0	-3	-3	-3	-1	-2	-2	-3	-3	3	1	-2	1	-1	-2	-2	0	-3	-1	**4**

Fig. 5 BLOSUM62 matrix derived by merging all the proteins that share 62 % or more sequence similarity. It is the most commonly used matrix in BLAST search

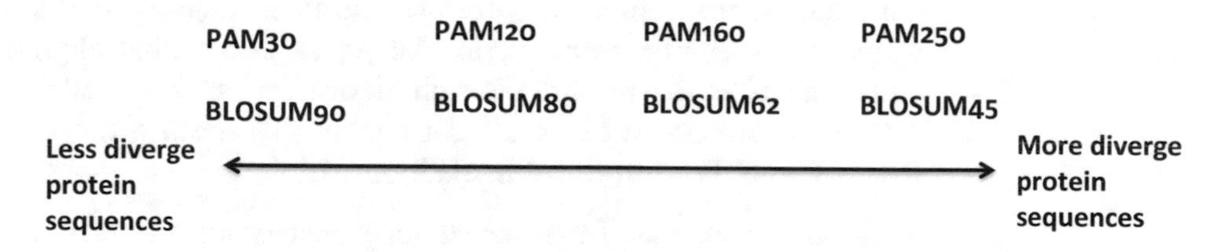

Fig. 6 Relationship between PAM and BLOSUM matrices

matrix. Wheeler [12] tried to relate PAM matrices with BLOSUM matrices and suggested the following relationship (Fig. 6).

2.3 Global and Local Alignment

We now shift our focus to the type of alignments and the algorithms that govern these alignments. There are two main types of alignments: *global*—typified by the Needleman and Wunsch algorithm [13] and *local*—typified by the Smith and Waterman algorithm [14]. Alignment of two sequences over their entire length is known as global alignment. In this type of alignment, both the termini of each sequence participate in alignment, irrespective of matches, mismatches, or gaps. Therefore, such alignments almost always introduce gaps and are preferred for sequences of roughly the same length and high similarity. However, some sequences can show similarity only in some regions (for example, limited to a motif or domain only) and global alignment may misalign these regions of high similarity. In order to align such sequences local alignment is preferred. Local alignment searches for a region of high similarity only, irrespective of the length of the sequence (Fig. 7). For example, if two sequences are of 100 bp length but

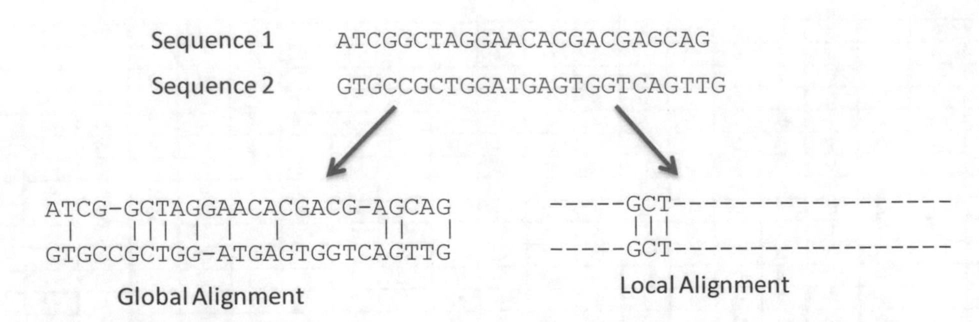

Fig. 7 Global and local alignment: in global alignment, two sequences are aligned over their entire length whereas in local alignment, regions of high similarity are aligned irrespective of the length of the sequence

are highly dissimilar, then local alignment will search for a region of high similarity (even 4–5 bp) generating a small alignment.

2.3.1 Needleman and Wunsch Algorithm: Global Sequence Alignment

Saul Needleman and Christian Wunsch (1970) described one of the important algorithms to align two protein sequences. This algorithm was subsequently modified by Seller [15] and Gotoh [16]. The algorithm tends to produce optimal alignment of two sequences with the introduction of gaps. The global alignment using the Needleman and Wunsch algorithm can be obtained via a three-step process: (1) setting a matrix, (2) scoring the matrix, and (3) identification of optimal alignment.

1. *Setting a matrix*: In order to set a matrix, two sequences are written in two dimensions. The first sequence is written vertically along the y-axis while the second is written horizontally along the x-axis. If the two sequences are identical, then a perfect alignment can be constructed, represented by a diagonal line that extends from top left to bottom right (Fig. 8). Gaps are represented using vertical or horizontal edges (see Fig. 8 c and d respectively).
2. *Scoring the matrix*: Let us consider two sequences of lengths a and b. A scoring matrix of dimensions $a + 1$ and $b + 1$ is created and the upper left cell is assigned the value "0." Subsequent cells across the first row and down the first column will be assigned the terminal gap penalties (Fig. 9a). The scoring system will be +1 for a match, −2 for a mismatch, and −2 for a gap. In each cell the score is derived from the cell diagonally above and to the left ("+/−" score for match/mismatch), the cell to the left ("−"score for gap penalty), and the cell directly above that cell ("−" score for gap penalty) (Fig. 9b). The highest score of these three is then assigned to the cell. Now let us start from the first cell in the example given in Fig. 9. The score derived from the cell diagonally above and to the left will be $0 + 1 = 1$ (+1 for match), the score from the cell to the left will be $-2 - 2 = -4$ (−2 for gap penalty) and the score from the cell directly above

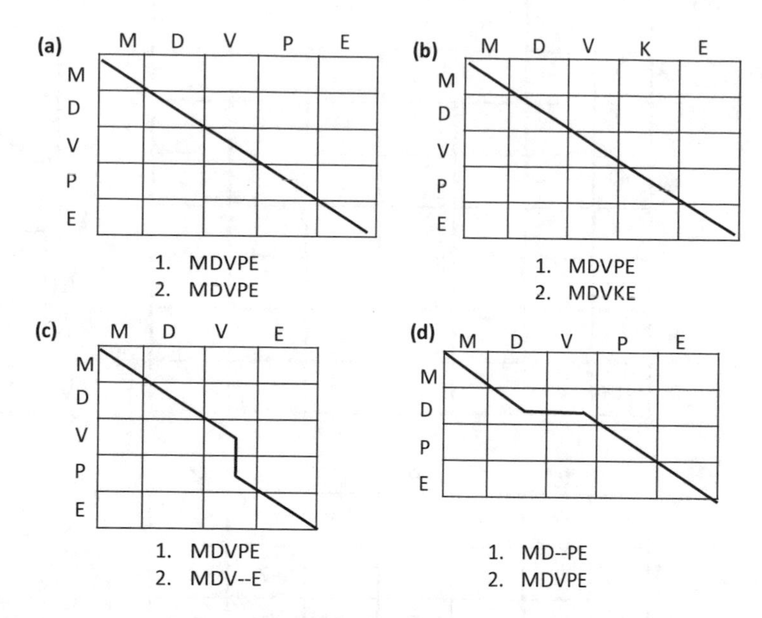

Fig. 8 Representation of a pairwise alignment of two protein sequences using Needleman and Wunsch algorithm. (**a**) Alignment of two identical sequences resulting in a diagonal path, (**b**) alignment of two similar sequences (with a mismatch), (**c**) deletion in sequence 2 resulting in a gap represented by a vertical line, and (**d**) deletion in sequence 1 resulting in a gap represented by a horizontal line

will again be $-2 - 2 = -4$ (-2 for gap penalty). Of these three scores the highest is +1 and therefore the value +1 is put in that cell (Fig. 9c). The scores for other cells are calculated similarly (Fig. 9d).

3. *Identification of optimum alignment*: Once the matrix is completely filled we now need to determine the optimal alignment. The easiest way to do this is by the tracking back process. Start from the last cell, that is, the cell at the extreme lower right. In our example this represents the alignment of two lysine residues with a score of -6. For this cell we can determine from which cell the best score was derived. In this case the best score was derived from the cell diagonally above and to the left. Similarly we can identify the cell from which the highest score for this cell was derived. This will lead to the identification of a path (denoted by arrows and shades (Fig. 9e). This path defines the optimal pairwise alignment. In cases where the score in a cell could have been derived from two or three of the neighboring cells, multiple optimal alignments having the same score can be generated by tracking back from each edge.

(a)

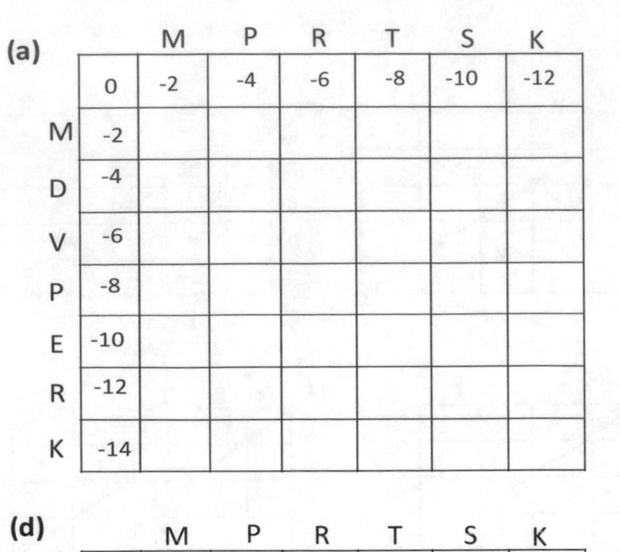

		M	P	R	T	S	K
	0	-2	-4	-6	-8	-10	-12
M	-2						
D	-4						
V	-6						
P	-8						
E	-10						
R	-12						
K	-14						

(b)

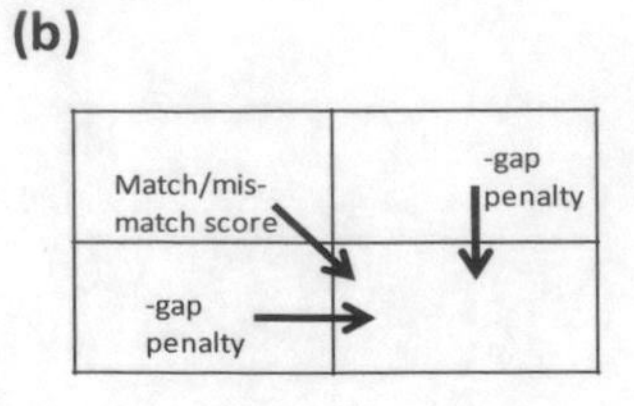

(c)

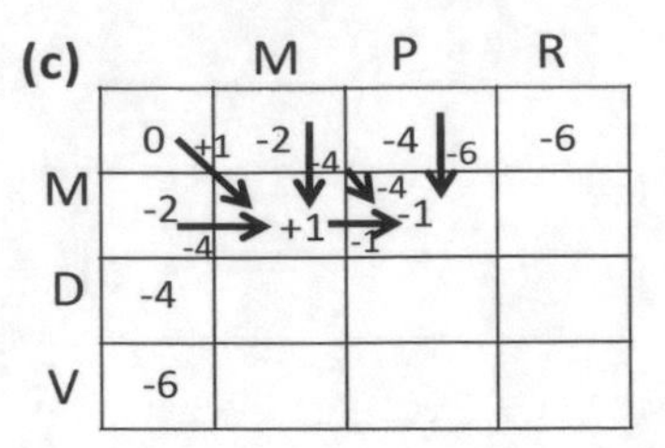

		M	P	R
	0	-2	-4	-6
M	-2	+1	-1	
D	-4			
V	-6			

(d)

		M	P	R	T	S	K
	0	-2	-4	-6	-8	-10	-12
M	-2	+1	-1	-3	-5	-7	-9
D	-4	-1	-1	-3	-5	-7	-9
V	-6	-3	-3	-3	-5	-7	-9
P	-8	-5	-2	-4	-5	-7	-9
E	-10	-7	-4	-4	-6	-7	-9
R	-12	-9	-6	-3	-5	-7	-9
K	-14	-11	-8	-5	-5	-7	-6

(e)

		M	P	R	T	S	K
	0	-2	-4	-6	-8	-10	-12
M	-2	+1	-1	-3	-5	-7	-9
D	-4	-1	-1	-3	-5	-7	-9
V	-6	-3	-3	-3	-5	-7	-9
P	-8	-5	-2	-4	-5	-7	-9
E	-10	-7	-4	-4	-6	-7	-9
R	-12	-9	-6	-3	-5	-7	-9
K	-14	-11	-8	-5	-5	-7	-6

Sequence 1: M D V P E R - - K
Sequence 2: M - - P - R T S K

Fig. 9 Scoring a matrix in Global alignment: (**a**) a matrix of two sequences (**a**, **b**) with a + 1 columns and b + 1 rows is constructed and gap penalties are added in the first row and column. (**b**) The score in a cell is

2.3.2 Smith and Waterman Algorithm: Local Sequence Alignment

The Smith and Waterman algorithm is used to find regions of local similarities and tends to align only part of two sequences without introducing gaps. The matrix construction for the Smith and Waterman Algorithm is similar to the Needleman and Wunsch Algorithm. For two sequences of lengths a and b, a scoring matrix of dimensions $a + 1$ and $b + 1$ is created. However, the scoring in this system is slightly different from the Needleman–Wunsch Algorithm. Here the scores can never be negative.

1. *Setting a matrix*: This step is the same as that of global alignment. The two sequences are written along the x- and y-axes.
2. *Scoring the matrix*: A scoring matrix of dimensions $a + 1$ and $b + 1$ is created and the upper row and first column is assigned the value "0" in each cell. The scoring system will be +1 for a match, $-1/3$ for a mismatch, and -1.3 for a gap. Like global alignment, the score in each cell is derived from the cell diagonally above and to the left ("+/−" score for match/mismatch), the cell to the left ("−" score for gap penalty), and the cell directly above that cell ("−" score for gap penalty), and the highest score out of the three values is then assigned to the cell. However, if the score is negative it is replaced with "0." The score for each cell is obtained as above and the matrix is filled.
3. *Identification of optimum alignment*: This method also uses the tracking back process. But the trace-back procedure starts from the highest score instead of the bottom right cell (Fig. 10); (in this matrix, the highest value is 3 and therefore the trace-back will start in that cell). From this cell, trace-back will proceed in the same way as for global alignment, but will end as soon as a cell containing "0" value is reached. This cell marks the start of alignment.

The Smith–Waterman algorithm and its descendants find the optimal alignment or alignments between two sequences, but like Needleman–Wunsch are rather slow. For the two algorithms that we have just described, the time required is proportional to $a \times b$. These algorithms are too inefficient when a query sequence is to be compared with an entire database. Therefore heuristic approaches have been used to increase the speed of alignment. These heuristic approaches include FASTA and BLAST programs that are based on local alignment but produce alignments more rapidly than the Smith–Waterman algorithm. These programs are described in the next section.

Fig. 9 (Continued) calculated from the upper left cell ("+/−" score for match/mismatch), cell to the left ("−" score for gap penalty), and cell directly above that cell ("−" score for gap penalty). (**c**) The score which is maximum out of these three is put in the cell. (**d**) A complete matrix with overall scores derived as explained. (**e**) Optimal pairwise alignment using the track back process. The cells highlighted are the source of optimum pairwise alignment

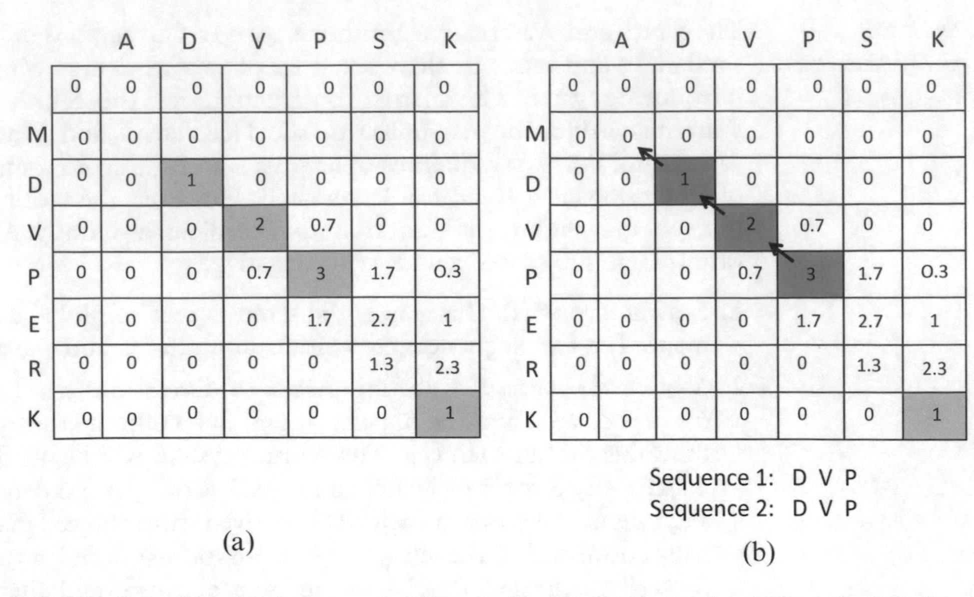

Fig. 10 Scoring a matrix in local alignment. (**a**) The matrix construction and derivation of scores for a cell is similar to global alignment. The scoring system here is +1 for a match, −1/3 for a mismatch, and −1.3 for a gap. But the score cannot be negative and therefore, the lowest score in this matrix is "0." (**b**) The alignment here does not start from the last cell but from the cell with the highest score and is tracked back until a score of "0" is encountered

3 Methods

3.1 BLAST

BLAST stands for Basic Local Alignment Search Tool and was designed by Eugene Myers, Stephen Altschul, Warren Gish, David J. Lipman, and Webb Miler at NIH in 1990 [17]. It is a heuristic method, which is more advance than the Smith–Waterman Algorithm and faster than its counterpart FASTA. This utility helps in searching and comparing biological sequences (nucleotide or protein) to sequence databases. BLAST calculates the statistical significance of matches and generates an overall score based on matches, mismatches, and gaps. The BLAST algorithm relies on *k*-tuples, where *k* is the matching "word length" for searching the query against each database sequence. The default word length for BLAST is 3 for proteins and 11 for nucleic acids. As the name suggests, BLAST explores local alignments, that is, subsequences within the database are compared to the query sequence. A number of databases can be searched using BLAST (Fig. 11). The user needs to enter the query fasta formatted sequence (*see* **Note 2**) in the box provided. Alternatively, the user can also write the accession number of the query (*see* **Note 3**). BLAST can be broadly categorized into basic and specialized BLAST. Basic BLAST includes the

Database	Description
Nucleotide database	
nr	All GenBank, EMBL, DDBJ, PDB sequences
refseq_rna	Reference RNA sequences
refseq_genomic	Genomic sequences from NCBI Reference Sequence Project
chromosome	Complete genomes
est	Expressed sequence tags
gss	Genome Survey Sequence
htgs	Unfinished High Throughput Genomic Sequences
pat	Patent sequences
alu_repeats	Human alu repeat elements
dbsts	Database of Sequence Tag Sites
wgs	Whole Genome Shotgun contigs
TSA	Transcriptome Shotgun Assembly
env_nt	Sequences from environmental samples.
	16S rRNA sequences (from bacteria and archaea)
Protein Sequence database	
nr	non redundant protein sequences
Refseq_proteins	NCBI Protein Reference Sequences
swissprot	SWISS-PROT protein sequence database
pat	Patented protein sequences
pdb	Bank.
env_nr	metagenomics proteins
tsa_nr	Transcriptome Shotgun Assembly proteins

Fig. 11 Nucleotide and protein databases available at BLAST NCBI

variants of BLAST that are used to align either protein or DNA sequences against protein or DNA sequences in the database.

Basic BLAST is further categorized into five sub categories (Fig. 12).

1. Nucleotide BLAST: compares query DNA sequence against the DNA database. Nucleotide BLAST comes with three variants, i.e., blastn, megablast, discontiguous megablast.
2. Protein BLAST: compares query protein sequence against protein database. It has four variants: Blastp, PSI-blast, Phi-blast, and Delta-blast.
3. blastx: translates DNA sequence into all six possible reading frames and then search each translated reading frame against protein database.
4. tblastn: compares protein query against each translated DNA sequence in a database.
5. Tblastx: both query and database DNA are translated and then searches are performed.

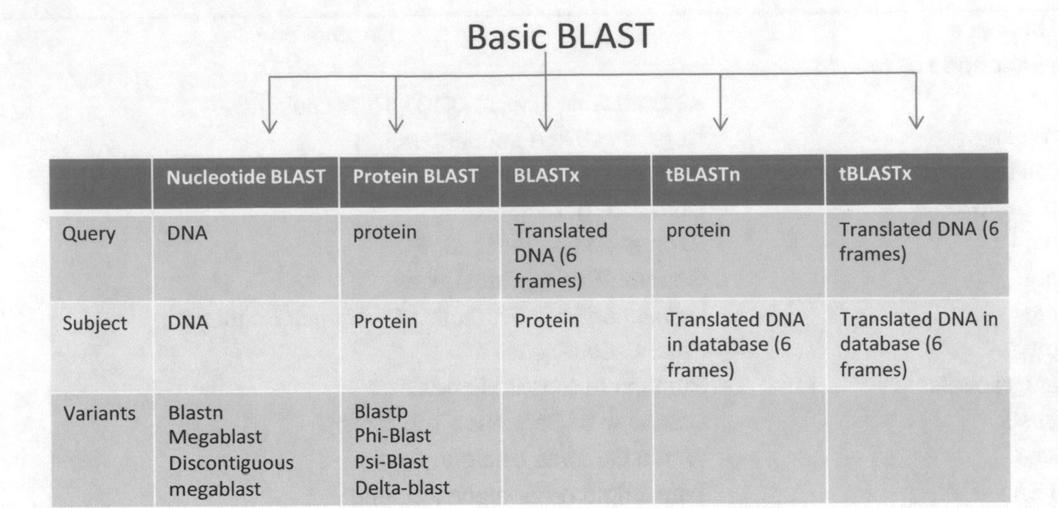

	Nucleotide BLAST	Protein BLAST	BLASTx	tBLASTn	tBLASTx
Query	DNA	protein	Translated DNA (6 frames)	protein	Translated DNA (6 frames)
Subject	DNA	Protein	Protein	Translated DNA in database (6 frames)	Translated DNA in database (6 frames)
Variants	Blastn Megablast Discontiguous megablast	Blastp Phi-Blast Psi-Blast Delta-blast			

Fig. 12 Subcategories of Basic BLAST

3.1.1 Understanding the BLAST Parameters

As already discussed a typical BLAST consists of various parameters that the user has to define. These parameters include Expected threshold (E value), word size, Matrix and gap penalties.

(a) *Expected threshold (E-value)*: An *E-value* represents the expected number of occurrences of a hit in database search purely by chance. Karlin and Altschul [18] defined E-values using a formula also known as the *Karlin–Altschul equation*:

$$E = kmNe^{-\lambda s}$$

where k is the minor constant, m is the query size, N is the total size of the database, S is the score, and λ is a constant used to normalize the score of a high scoring pair. As can be noted from the equation, the value of E decreases exponentially with increase in the score S.

The default threshold E-value in BLAST search is "10." To have a better understanding of E-values, let us consider a case where an alignment has an E value 1e−9 with an alignment score of X bits. This indicates that a score of X bits or better is expected to occur by chance in the database with a probability of 1 in a billion. While defining E-values it is also important to define the scores. Typically there are two types of scores: raw score (S) and bit score (S'). Raw score is calculated using a particular substitution matrix and gap penalty parameters, while bit score is calculated from the raw score after normalizing the variables that define a particular scoring system.
A bit Score (S') is given by:

$$S' = \frac{\lambda S - \ln\ K}{\ln 2}$$

where S is the raw score and λ and K are the statistical parameters of a particular scoring system.

Raw scores are without any units and thus raw scores from different database searches cannot be compared. On the other hand, the bit scores are produced from the raw scores by normalizing the variables and therefore they have standard units, which allows the bit scores from different database searches to be compared.

(b) *Word size*: The BLAST algorithm works by dividing the query into short runs of letters of a particular length, known as ***query words*** or ***k-tuples*** (in case of FASTA). The default word size is 3 for proteins (can be reduced to 2 if the query is a short peptide) and 11 for nucleotides (can be set to 7–15). Lowering the word size results in more accuracy but slower search, while higher word sizes generally match infrequently resulting in a faster search. Let us consider the example given in Fig. 13. The BLAST algorithm will first divide the query into smaller words (of three letters each) and then not only find instances of the first query word (VLD) but also related words where conservative substitutions have taken place (in this case a conservative substitution has taken place in the target

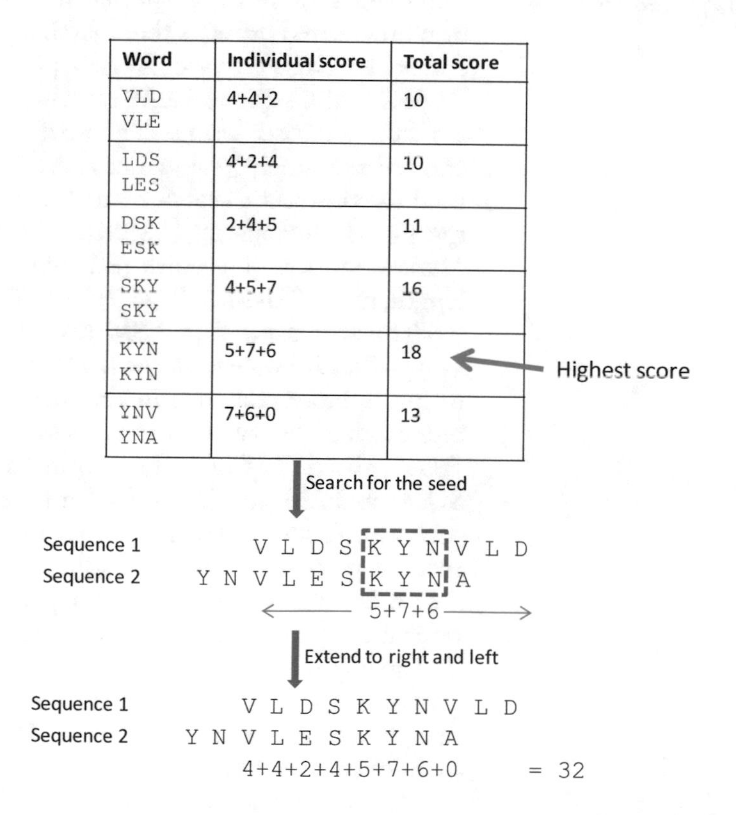

Word	Individual score	Total score
VLD VLE	4+4+2	10
LDS LES	4+2+4	10
DSK ESK	2+4+5	11
SKY SKY	4+5+7	16
KYN KYN	5+7+6	18
YNV YNA	7+6+0	13

Fig. 13 The typical BLAST search begins with query words. (See text for details.)

sequence leading to the replacement of D with E). Similarly, the BLAST algorithm will find instances of the next query word "LDS" and so on. A particular threshold value T is used to detect words that match with the query word. Higher T values will find more exact matches but may overlook some conservative substitutions. The alignment starts with the word that generates the highest score (KYN in this example) and then extends the alignment both to the right and left of the seed.

(c) *Scoring Matrix*: As already mentioned, a number of amino acid substitution matrices are available and the choice depends on the type of search the user is interested in (refer to Subheading 1.4.2 in this chapter). Typically a higher value of PAM is used if the user wants to search highly divergent sequences and a lower value of PAM is used for highly similar sequences. The opposite is true for BLOSUM matrices. Lower values of PAM are also recommended for short query sequences.

For nucleotides, the default values are "+1" for a match and "−2" for a mismatch. Gap penalties are scored differently, but the default is "linear" gap penalty.

3.1.2 PSI-BLAST

PSI-BLAST or position-specific iterated BLAST is used to find distantly related proteins that match a particular protein of interest and is a specialized type of BLAST [19–21]. The initial step in PSI-BLAST is the same as BLASTP where a protein query is used to search a database. Using this initial BLASTP search, PSI-BLAST constructs a multiple-sequence alignment (Fig. 14a), which is then used to construct a search matrix (Fig. 14b) known as a position-specific scoring matrix (PSSM). The PSSMs are also known as Markov models or profiles [22–24]. For a query of a particular length, say L, PSI-BLAST generates a PSSM of dimensions $L \times 20$.

The scores for the PSSM can be either positive or negative values. Positive values denote that this particular amino acid substitution is more common in the alignment, while negative values indicate that the substitution is less common than expected. The PSSMs also differ from other scoring matrices like PAM and BLOSUM; in PAM and BLOSUM matrices a particular amino acid substitution always receives the same score while in PSSMs the score may be different depending on the position. This means that a Ser-Thr substitution at a particular position, say X, may receive a different score than the Ser-Thr substitution at position Y of an alignment.

The PSSM generated and not the initial query is then used as a query to search the database again. The user can decide how many iterations to perform by clicking "*Run PSI-BLAST iteration*."

A different version of BLAST known as RPS-BLAST or Reverse position-specific BLAST can also be accessed at NCBI.

```
                         10        20        30        40        50        60        70        80
                 ....*....|....*....|....*....|....*....|....*....|....*....|....*....|....*....|
3L7Y_A       40 IATDMDGTFLNSKGS-YDhNRFQRILKQLQE--------RDIRFVVASSNPYRQLREHFP-----DCheqlTFVG---EN 102
gi 81747357   6 IYFDIDGTLVPDGVT-LT-NKTIMAIRYLKD--------KNIKIGLATGRNIFFAEYFAKvl--dVDm---PLVC---VN 67
gi 81637281   3 FVFDLDGTICFDGKQ-IP-LNIQHALEQLMQ--------KDHQLIFASARPIRDLLPLLNpkl-----qqaMLIG---GN 64
gi 81759331   5 IATDMDGTFLAEDGT-YNqEQLAALLPKLAE--------KGILFAVSSGRSLLAIDQLFEpfl-----dqiAVIA---EN 67
gi 81538605   8 FATDMDGTFLHNDHS-YNhKKLAEVIKKIQD--------RNLLFAASSGRSLLGLIEVFSey--kDQm---AFVA---EN 70
gi 81726537   9 IASDMDGTLLDQYGR-LD-PEFFDLFLQLEE--------QGILFSAASGRQYYSLRDTFApi--kDRv---LYVA---EN 70
gi 74862782  23 IAIDIDGTLADDTGK-IS-DENLKAIEVCKK--------GGIEIILASGRLHSYAMKMFTne--qIEkykiEKLDgvySH 90
gi 81527515   5 ISVDMDGTFLDGNGE-YNrARFEKIYAELVK--------REIKFIVASGNQYYQLKSFFP-----GKdeelFYVA---EN 67
gi 81437664   6 IALDIDGTVLNHDGHtPS-EGIVAQLARLV---------REHYLVFASGRSTGDTLPVLEslgltQGampqFIIC---CN 72
gi 74861610   6 IALDLDGTILNSNKQ-VS-ENSKRVLQYISKpefknidgEEVLVLLASGRAPYLVSPVEEal--gIDc---YLIG---YN 75
                         90       100       110       120       130       140       150       160
                 ....*....|....*....|....*....|....*....|....*....|....*....|....*....|....*....|
3L7Y_A      103 GANI----ISKNQSLIEVFQQ---REDIASIIYFIEEKYp--qAVIALSGEKKGYLKKG-------------VSENIVKM 160
gi 81747357  68 GAWIvtpkDY--LNISTKYID---FKPQLQLLDVLYKEK----KDFLVYTIDGVYSTSEnqpfykrllsvqeKLKEKKIK 138
gi 81637281  65 GAIT----QVNKQITAQQPIS---SVAFTQIKQWIEEFD----LDYL---ADDLWDYSK-------------------R 110
gi 81759331  68 GSVV----QYHGEILFADMMT---KEQYTEVAKKVLANPhyveTGMVFSGQKAAYILKG-------------ASEEYIQK 127
gi 81538605  71 GGVV----AYKGEILFAKDLT---VAQTQELIDDLQEMPfspkNDYLISGLKGAYYPEG-------------ISKEYLTH 130
gi 81726537  71 GTLV----MYQDKELYSCTIP---KAEVAEIVKAAREIDg---ANIVLCGKRSAYIETH--------------DQQSLEE 126
gi 74862782  91 GAYI----HMKGYDYVYRKFS---YKDLELILFSLGSYNi--lRNAVFLTVDSAYVINDdiklie-eyiytpESEGIISD 160
gi 81527515  68 GAVI----FHQGELLSVNRFD---ERLVQKILRTLIQEYq--dLQVILCGVKSAYLLKA-------------ADPDFKAF 125
gi 81437664  73 GAVTl--kMENNEYVRHRVISfdaSRTLEYIINTLPG------VELAVERCDGVYMHTS-------------------NF 125
gi 74861610  76 GSICfgrkSEGRNTVFSHSID---NSNLKAIFKYVEENN----LFLNIYGDGIVYGIDKpe-----------lAEKPRRYS 138
                        170       180       190       200       210       220       230       240
                 ....*....|....*....|....*....|....*....|....*....|....*....|....*....|....*....|
3L7Y_A      161 LSPFFPVLELVNSFSPLPDERFFKLT-----LQVKEEESAQI--MKAIADYkts---------qrlVGTASGFGYIDIITK 225
gi 81747357 139 SRLVYEMKVNPNLVFYSKQ-KLLKVL-----ICYDNVSEKIK--YENILSNip-----------nlAFSVSQNGIIDIYNG 200
gi 81637281 111 LRQPHSIEHKIDPAKLAQNRPLSAIQhpiktILLNLTQTQFLflKQQLTHLevn---------lieHSEENGRFNLDITAK 182
gi 81759331 128 TKHYYANVKVINGFEDMENDAIFKVS-----TNFTGHTVLEG--SDWLNQAlp-----------yaTAVTTGFDSIDIILK 190
gi 81538605 131 AKLYYPNCQLYHRLDEIDD-KLLKVT-----TNFPEDHVRDC--EQWITDRls-----------fvRATTTGFTSIDIVPN 192
gi 81726537 127 FQKYYHRCETVTDLLEVED-EFIKVA-----ICHFDGSEELL--FPTMNAKfga----------thKVVVSAKIWLDVMNA 189
gi 74862782 161 IEYVKIIDTNYKPILINKIKDIFNIG-----DIVSIEIYDKLypNQDIYSDlfkvlfyelqphykIYIPSSNNKIVLSPI 235
gi 81527515 126 AKKYYFELQEVDSFDVLPDDTFIKFA-----LDVEVAKTGQI--VEDLNQTfag----------eiRAVSSGHGSIDIIIP 189
gi 81437664 126 PLDYLHDKRILSSVSEIVHTPPARVV-----VYEASIDASEI--AKRFESRc--------------INFNNAWGYWDIMPP 185
gi 74861610 139 IMTGATYKLIPSYSTLPEDFTPAKCLi----ILDDDKECDQL--LETMRPLfptlsl---vksncMNKDYKQYYVEFLEH 209
```

(a)

P	C	A	G	I	L	V	M	F	W	P	C	S	T	Y	N	Q	H	K	R	D	E
1	I	-6	-9	6	3	3	-1	2	-7	-8	-6	-7	-3	-2	-8	-8	-8	-8	-8	-8	-8
2	A	5	-6	2	-2	3	-5	5	-6	-7	0	-3	-6	0	-7	-7	-7	-7	-7	-8	-7
3	S	-3	-7	3	1	0	-5	5	-7	-7	4	3	2	-5	-6	-7	-7	-7	-7	-7	-7
4	D	-7	-6	-8	-9	-8	-8	-9	-10	-7	-9	-5	-3	-8	-4	-5	-6	-6	-7	9	-3
5	L	-4	-9	5	4	2	6	-2	-7	-8	-6	-7	-6	-6	-8	-7	-8	-7	-7	-9	-8
6	D	-7	-3	-8	-9	-8	-8	-9	-10	-7	-9	-5	-6	-8	-4	-5	-6	-6	-7	9	-3
7	G	-5	8	-9	-9	-8	-8	-8	-8	-7	-8	-3	-7	-8	-2	-7	-7	-6	-3	-1	-7
8	T	-3	-7	-6	-6	-5	-6	-7	-8	-6	-6	-1	8	-7	-5	-6	-7	-6	-6	-6	-6
9	L	-7	-9	1	6	0	-3	2	-7	-8	-6	-7	-2	-6	-8	-7	-8	-8	-7	-9	-8
10	L	-2	-8	1	6	1	-1	-2	-7	-7	0	-7	-3	-6	-8	-7	-8	-7	-7	-8	-8
11	N	-3	2	4	7	4	7	2	-8	-1	-7	0	2	-6	6	-2	2	-2	1	4	-5
12	[illegible]	2	2	0	0	7	7	7	0	0	-7	3	-3	2	1	0	1	2	-1	3	2
13	D	-4	1	-3	-8	-7	-7	-8	-8	-7	-8	-3	0	-3	5	-1	3	1	-3	5	0
14	K	-6	2	-4	-5	-4	0	-1	0	-3	-8	-1	-6	-6	1	0	6	5	-1	-1	-1
15	K	-4	-7	-3	-4	0	0	-2	-8	-7	-7	1	2	-2	-2	3	1	3	2	-2	1
16	I	-4	-8	6	2	4	0	-5	-7	-1	-6	-7	-3	2	-8	-7	-7	-7	-7	-8	-7
17	S	-5	-4	-3	-1	-3	-6	-7	-7	3	-6	5	2	-6	1	-1	3	-5	-6	0	-5
18	E	-1	-7	-4	-3	-4	-7	-8	-8	4	-8	-1	-4	-3	0	3	0	0	-1	4	3
19	R	-1	-4	-7	-5	-2	-2	1	-7	-7	-8	2	-3	1	-1	2	0	3	3	-3	3
20	T	0	-6	0	-2	-1	-6	1	-8	-7	-6	1	5	-6	5	-6	-2	-6	-6	-3	-6
21	K	0	-7	4	0	1	-5	-1	-8	-7	-7	-6	-3	-6	-6	2	-1	4	3	-7	-2
22	E	0	-7	-2	-3	-4	-1	-7	0	-7	-8	0	-1	-2	0	2	1	0	1	3	4
23	A	6	-2	2	-1	1	-5	-3	-7	-6	-6	-1	2	-2	-3	-3	-7	-6	-7	-7	-4
24	I	-6	-9	6	4	1	0	1	-7	-8	0	-7	-6	0	-8	-8	-8	-8	-8	-9	-8
25	K	0	-4	-1	-2	-6	0	-7	-7	-3	-8	-3	-2	0	-1	3	1	5	3	-2	1
26	K	-1	-7	-7	-7	-2	-2	-3	-7	-7	-8	-2	-2	3	-3	4	-5	4	4	-3	3

(b)

Fig. 14 (**a**) PSI-BLAST works by using the BLASTP search to construct a multiple-sequence alignment (**b**) the alignment is then used to construct a search matrix also known as a position-specific scoring matrix (PSSM). Only a part of an alignment and PSSM is shown here

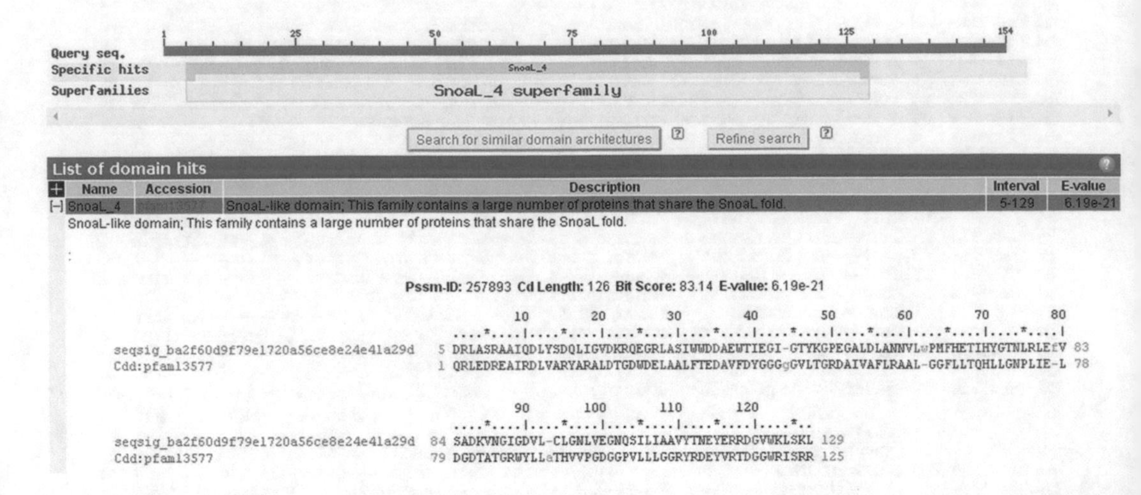

Fig. 15 RPS-BLAST output. The query used in this example is HCH dehydrochlorinase from *Sphingobium indicum* B90A. RPS-BLAST works by searching a protein query against a database of predefined PSSMs. RPS-BLAST can be accessed at CDD at NCBI

The RPS-BLAST program is used to search a protein query against a database of predefined PSSMs. RPS-BLAST is implemented in the Conserved Domain Database (CDD) [25] at NCBI and is generally used to find conserved protein domains in the query. A typical RPS-BLAST search result is shown in Fig. 15 using HCH dehydrochlorinase from *Sphingobium indicum* B90A.

3.1.3 PHI-BLAST

PHI-BLAST (also known as Pattern-Hit Initiated BLAST) is a specialized type of BLAST that allows a user to search a particular pattern or signature present in the query and the database [26]. The pattern or signature can be a stretch of amino acid that may be helpful in identifying a protein family. These patterns or signatures may not be exactly identical. For example, consider the pattern GPX[WF]MX. In this example X denotes any amino acid at positions 3 and 6 of the pattern, while [WF] indicates that the fourth position may have either W or F.

3.2 FASTA

FASTA was the first program developed to address rapid database search for protein or DNA sequences [27–31]. Like BLAST, FASTA also aims to compare a query sequence against the databases. FASTA search begins by looking for matching words called *k-tuples* or *ktup* which can be considered equivalent to the word size of BLAST. The *ktup* length is usually 1–2 for proteins and 4–6 for nucleic acids. Larger values of *ktup* result in a faster search but there may be the possibility of missing similar regions. Once the *ktup* is determined the FASTA program looks for word matches that are close to each other and connects these without introducing any gaps. Once the initial connections are made, an initial score is calculated. In the next step, the FASTA program considers the ten best regions with ungapped pairwise alignments and tries to join them. In the next step, these regions are scored using a modified Smith–Waterman algorithm to come up with the optimal pairwise alignment. The FASTA program can be accessed at http://www.ebi.ac.uk/Tools/sss/fasta/ (*see* **Note 4**).

There are several implementations of the FASTA algorithm [32]. A few of them are:

1. *FASTX*: compares a query DNA sequence to a protein sequence database by translating the query DNA into six reading frames.
2. *TFASTX*: compares a protein sequence to a DNA sequence database by allowing frameshift between codons.
3. *FASTA-pat and FASTA-swap*: these programs are used to compare a query protein sequence against a pattern database. These patterns are characteristics of a protein family.
4. *ssearch*: compares DNA or protein query against a sequence database using Smith–Waterman algorithm.
5. *ggsearch*: compares DNA or protein query against a sequence database using Needleman–Wunsch algorithm.

3.3 Methods and Tools for Large Scale Sequence Comparison

3.3.1 MegaBLAST and Discontiguous MegaBLAST

MegaBLAST is a variation of the BLAST program at NCBI that is optimized for use in aligning large or highly similar DNA queries [33]. MegaBLAST is faster than traditional BLASTN. The speed of MegaBLAST is due to two important changes that have been introduced into the traditional BLASTN program.: (1) The default word size for MegaBLAST is 28 and can be set as high as 64 while the word size for BLASTN is 11. (2) MegaBLAST makes use of a non-affine gap penalty, which means that there is no penalty for gap opening. Figure 16 shows a typical megaBLAST output using human myoglobin coding region.

A variant of MegaBLAST known as Discontiguous MegaBLAST can also be accessed at NCBI. This version has been designed to compare divergent sequences with low sequence identity from different organisms. It uses a discontiguous word approach [34] in which nonconsecutive positions are scanned over longer sequence segments.

3.3.2 BLAT

BLAT or BLAST like Alignment Tool [35] was developed to align large genomic DNA sequences of 95 % and greater similarity. It is quite similar to the MegaBLAST program but different from BLAST due to its high speed. BLAT search is used to find the position of a sequence of interest in a genome. The BLAT query page is shown in Fig. 17. The sequence of interest, which can be DNA, protein, translated RNA or translated DNA, is pasted in the box provided. The user can choose the genomes from the drop down menu and can use them for cross-species analysis. In the following example, we have used myoglobin coding region from

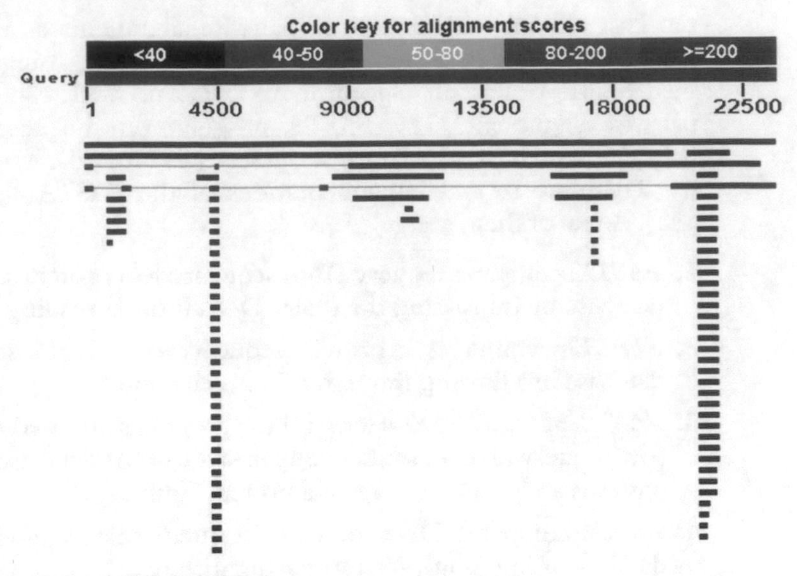

Sequences producing significant alignments:

Select: All None Selected:0

Alignments Download GenBank Graphics Distance tree of results

Description	Max score	Total score	Query cover	E value	Ident	Accession
Homo sapiens myoglobin (MB), RefSeqGene on chromosome 22	43565	46566	100%	0.0	100%	NG_007075.1
Human DNA sequence from clone CITF22-62D4 on chromosome 22, complete sequen	37129	44577	85%	0.0	100%	AL049747.1
Homo sapiens myoglobin transcript variant 1 (MB) gene, complete cds	25787	26788	61%	0.0	99%	DQ003030.1
Human myoglobin gene (exon 1) (and joined CDS)	6913	6913	15%	0.0	99%	X00371.1
Human DNA sequence from clone RP4-569D19 on chromosome 22q13.1, complete se	6621	26811	22%	0.0	100%	AL022334.1
Human myoglobin gene, exon 2	4867	5386	11%	0.0	99%	M14602.1
Human myoglobin gene, exon 1	4691	4691	10%	0.0	99%	M10090.1
Human myoglobin gene (exon 2)	3615	4050	8%	0.0	99%	X00372.1

Fig. 16 A typical Megablast output using the human myoglobin coding region. Megablast is used to search large DNA query sequences rapidly, in part due to the larger word size implemented in Megablast

human. The result from a typical BLAT search is shown in Fig. 18. The results are presented with the identity of each hit along with the position of that particular hit. Details regarding a particular hit can be found by clicking the hyperlink to its left. The matched bases are shown in blue and are capitalized while the unaligned regions are shown in black lower case. The user can also find the distribution of the BLAT hits across the human chromosomes (Fig. 18c).

3.3.3 BLASTZ

BLASTZ is a variation of gapped BLAST [36] that was initially used to align the human and mouse genomes. The method is slightly different from BLASTN, which looks for series of exact matches defined by word size. To begin with, BLASTZ looks for repeat regions in the first genome that are also found in the second genome followed by their removal from both genomes. In the

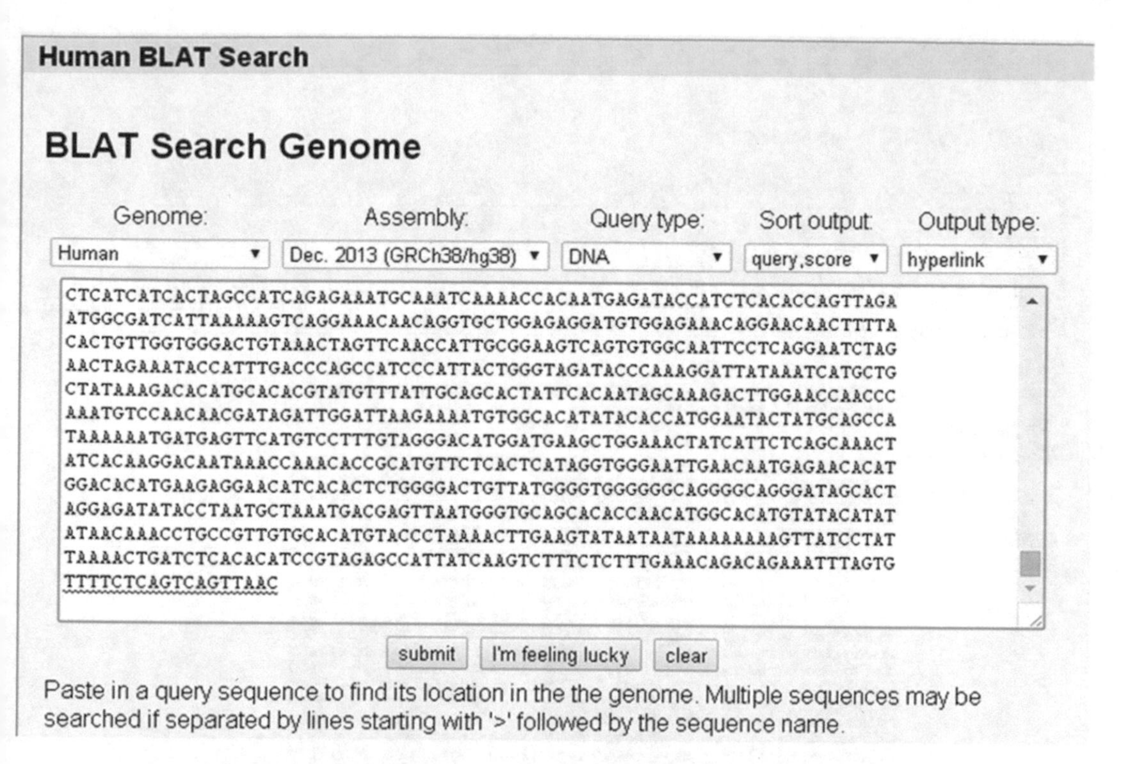

Fig. 17 BLAT homepage. The user can choose the genome against which the query is to be searched using the drop-down menu

next step, it determines initial matches by looking for a string of 19 nucleotides out of which 12 positions have exact matches (for example, a match could correspond to the string 1110100110010101111, where 1 represents a match and 0 a mismatch). This is followed by a gap-free extension until the score reaches a particular threshold. These steps are then repeated and after reaching a particular threshold score, gaps are allowed in the pairwise alignment.

3.3.4 LAGAN

(Limited Area Global Alignment of Nucleotide) is an efficient global alignment tool that is used for pairwise alignment of genomes [37]. LAGAN uses three basic steps to perform global alignment. In the first step, it generates local alignments between the two sequences and selects the best alignment, which serves as an anchor. The use of anchors is important as it limits the search space. LAGAN uses a highly sensitive anchoring algorithm CHAOS [38] which searches for short inexact words instead of longer exact words, making it more efficient in aligning both closely and distantly related organisms. In the next step, a rough global map is constructed using the sets of defined anchors and in the last step a final global map is constructed using the rough map. A variant of LAGAN called Multi-LAGAN is used for aligning multiple genome

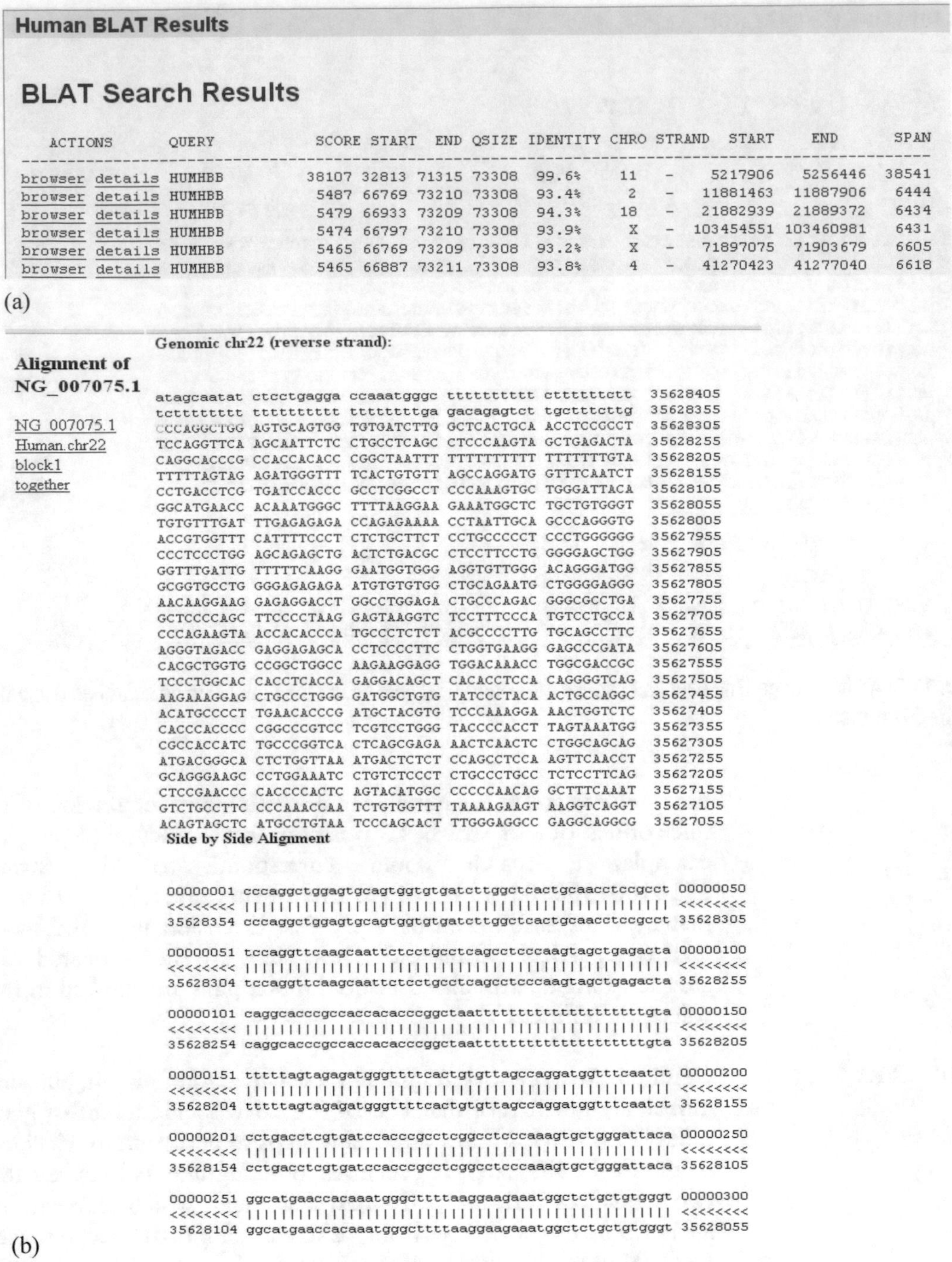

Human BLAT Results

BLAT Search Results

ACTIONS	QUERY	SCORE	START	END	QSIZE	IDENTITY	CHRO	STRAND	START	END	SPAN
browser details	HUMHBB	38107	32813	71315	73308	99.6%	11	-	5217906	5256446	38541
browser details	HUMHBB	5487	66769	73210	73308	93.4%	2	-	11881463	11887906	6444
browser details	HUMHBB	5479	66933	73209	73308	94.3%	18	-	21882939	21889372	6434
browser details	HUMHBB	5474	66797	73210	73308	93.9%	X	-	103454551	103460981	6431
browser details	HUMHBB	5465	66769	73209	73308	93.2%	X	-	71897075	71903679	6605
browser details	HUMHBB	5465	66887	73211	73308	93.8%	4	-	41270423	41277040	6618

(a)

Alignment of NG_007075.1

NG_007075.1
Human.chr22
block1
together

Genomic chr22 (reverse strand):

```
atagcaatat ctcctgagga ccaaatgggc tttttttttt cttttttctt  35628405
tctttttttt tttttttttt ttttttttga gacagagtct tgctttcttg  35628355
CCCAGGCTGG AGTGCAGTGG TGTGATCTTG GCTCACTGCA ACCTCCGCCT  35628305
TCCAGGTTCA AGCAATTCTC CTGCCTCAGC CTCCCAAGTA GCTGAGACTA  35628255
CAGGCACCCG CCACCACACC CGGCTAATTT TTTTTTTTTT TTTTTTTGTA  35628205
TTTTTAGTAG AGATGGGTTT TCACTGTGTT AGCCAGGATG GTTTCAATCT  35628155
CCTGACCTCG TGATCCACCC GCCTCGGCCT CCCAAAGTGC TGGGATTACA  35628105
GGCATGAACC ACAAATGGGC TTTTAAGGAA GAAATGGCTC TGCTGTGGGT  35628055
TGTGTTTGAT TTGAGAGAGA CCAGAGAAAA CCTAATTGCA GCCCAGGGTG  35628005
ACCGTGGTTT CATTTTCCCT CTCTGCTTCT CCTGCCCCCT CCCTGGGGGG  35627955
CCCTCCCTGG AGCAGAGCTG ACTCTGACGC CTCCTTCCTG GGGGAGCTGG  35627905
GGTTTGATTG TTTTTCAAGG GAATGGTGGG AGGTGTTGGG ACAGGGATGG  35627855
GCGGTGCCTG GGGAGAGAGA ATGTGTGTGG CTGCAGAATG CTGGGGAGGG  35627805
AACAAGGAAG GAGAGGACCT GGCCTGGAGA CTGCCCAGAC GGGCGCCTGA  35627755
GCTCCCCAGC TTCCCCTAAG GAGTAAGGTC TCCTTTCCCA TGTCCTGCCA  35627705
CCCAGAAGTA ACCACACCCA TGCCCTCTCT ACGCCCCTTG TGCAGCCTTC  35627655
AGGGTAGACC GAGGAGAGCA CCTCCCCTTC CTGGTGAAGG GAGCCCGATA  35627605
CACGCTGGTG CCGGCTGGCC AAGAAGGAGG TGGACAAACC TGGCGACCGC  35627555
TCCCTGGCAC CACCTCACCA GAGGACAGCT CACACCTCCA CAGGGGTCAG  35627505
AAGAAAGGAG CTGCCCTGGG GATGGCTGTG TATCCCTACA ACTGCCAGGC  35627455
ACATGCCCCT TGAACACCCG ATGCTACGTG TCCCAAAGGA AACTGGTCTC  35627405
CACCCACCCC CGGCCCGTCC TCGTCCTGGG TACCCCACCT TAGTAAATGG  35627355
CGCCACCATC TGCCCGGTCA CTCAGCGAGA AACTCAACTC CTGGCAGCAG  35627305
ATGACGGGCA CTCTGGTTAA ATGACTCTCT CCAGCCTCCA AGTTCAACCT  35627255
GCAGGGAAGC CCTGGAAATC CTGTCTCCCT CTGCCCTGCC TCTCCTTCAG  35627205
CTCCGAACCC CACCCCACTC AGTACATGGC CCCCCAACAG GCTTTCAAAT  35627155
GTCTGCCTTC CCCAAACCAA TCTGTGGTTT TAAAAGAAGT AAGGTCAGGT  35627105
ACAGTAGCTC ATGCCTGTAA TCCCAGCACT TTGGGAGGCC GAGGCAGGCG  35627055
```

Side by Side Alignment

```
00000001 cccaggctggagtgcagtggtgtgatcttggctcactgcaacctccgcct 00000050
<<<<<<<< |||||||||||||||||||||||||||||||||||||||||||||||||| <<<<<<<<
35628354 cccaggctggagtgcagtggtgtgatcttggctcactgcaacctccgcct 35628305

00000051 tccaggttcaagcaattctcctgcctcagcctcccaagtagctgagacta 00000100
<<<<<<<< |||||||||||||||||||||||||||||||||||||||||||||||||| <<<<<<<<
35628304 tccaggttcaagcaattctcctgcctcagcctcccaagtagctgagacta 35628255

00000101 caggcacccgccaccacacccggctaatttttttttttttttttttttgta 00000150
<<<<<<<< |||||||||||||||||||||||||||||||||||||||||||||||||| <<<<<<<<
35628254 caggcacccgccaccacacccggctaatttttttttttttttttttttgta 35628205

00000151 tttttagtagagatgggttttcactgtgttagccaggatggtttcaatct 00000200
<<<<<<<< |||||||||||||||||||||||||||||||||||||||||||||||||| <<<<<<<<
35628204 tttttagtagagatgggttttcactgtgttagccaggatggtttcaatct 35628155

00000201 cctgacctcgtgatccacccgcctcggcctcccaaagtgctgggattaca 00000250
<<<<<<<< |||||||||||||||||||||||||||||||||||||||||||||||||| <<<<<<<<
35628154 cctgacctcgtgatccacccgcctcggcctcccaaagtgctgggattaca 35628105

00000251 ggcatgaaccacaaatgggcttttaaggaagaaatggctctgctgtgggt 00000300
<<<<<<<< |||||||||||||||||||||||||||||||||||||||||||||||||| <<<<<<<<
35628104 ggcatgaaccacaaatgggcttttaaggaagaaatggctctgctgtgggt 35628055
```

(b)

Fig. 18 BLAT output. (**a**) This part of the results shows the number of hits and their details. (**b**) Clicking on details gives additional information about the hit including the alignment. The matched bases are shown in *blue* and are capitalized while the unaligned regions are shown in *black* lower case. (**c**) Distribution of BLAT hits across the human chromosomes. The portion of chromosome 22 that is boxed has the maximum identity with the query

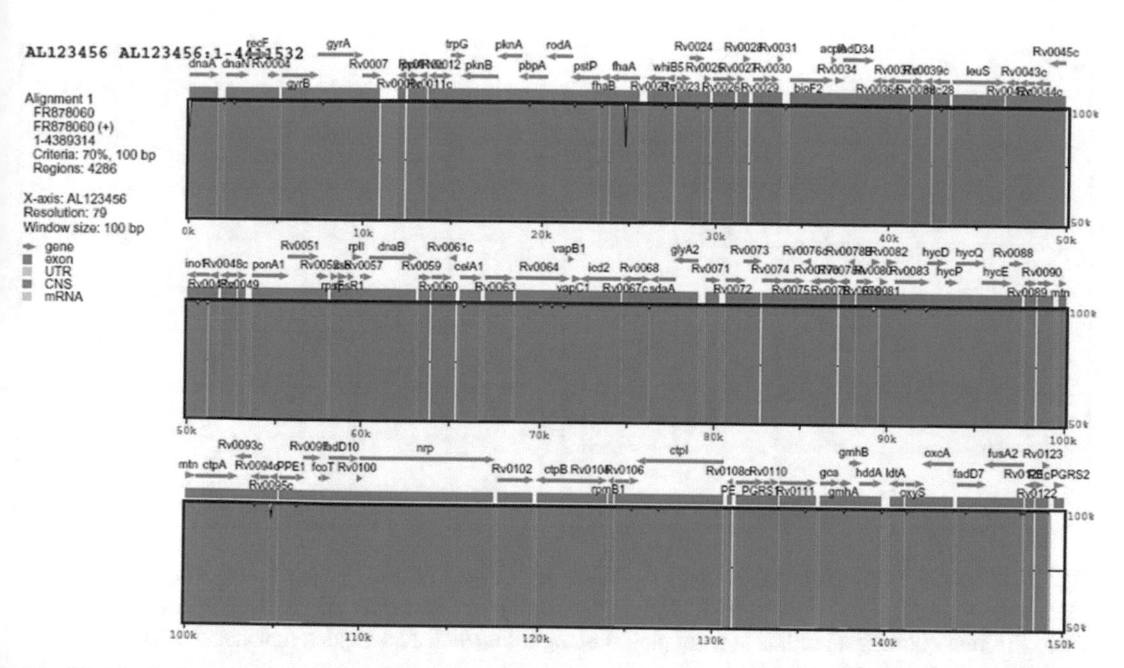

Fig. 19 A typical output of LAGAN showing an alignment between *Mycobacterium tuberculosis* (AL123456) and *Mycobacterium africanum* (FR878060) genomes. The key to the alignment is also given in the *left panel*

sequences. LAGAN, along with AVID, has been implemented in the VISTA browser for large-scale genome comparison. The example in Fig. 19 shows the alignment of two *Mycobacterium* strains using LAGAN implemented in the VISTA browser.

3.3.5 MUMMER

Mummer is a local alignment based tool that is primarily used for aligning whole genomes [39]. Unlike other methods, mummer constructs a suffix tree, a method that is known to find all distinct sub-sequences with high efficiency. Among these sub-sequences, Maximal Unique Matches (MUMs) are selected. These are unique in both the genomes. Any extension of an MUM results in a mismatch. Using these MUMs, the algorithm identifies the longest increasing subsequence (LIS) of MUMs that occur in the same direction in both the genomes. All aligned LIS are then connected by closing gaps between them using Smith–Waterman algorithm. This helps in detecting SNPs, indels, repeats, and polymorphic regions in genomes. The MUMMER package can also align draft genomes with more than 100–1000 contigs using the NUCmer program efficiently. The package can also translate DNA sequences into six frames and generate alignments using the PROmer program. The alignments can be viewed using MUMmer plot (Fig. 20).

3.3.6 AVID

AVID is a global alignment program [40] used to align two genomes. Like BLASTZ, repeat regions are masked but unlike it both

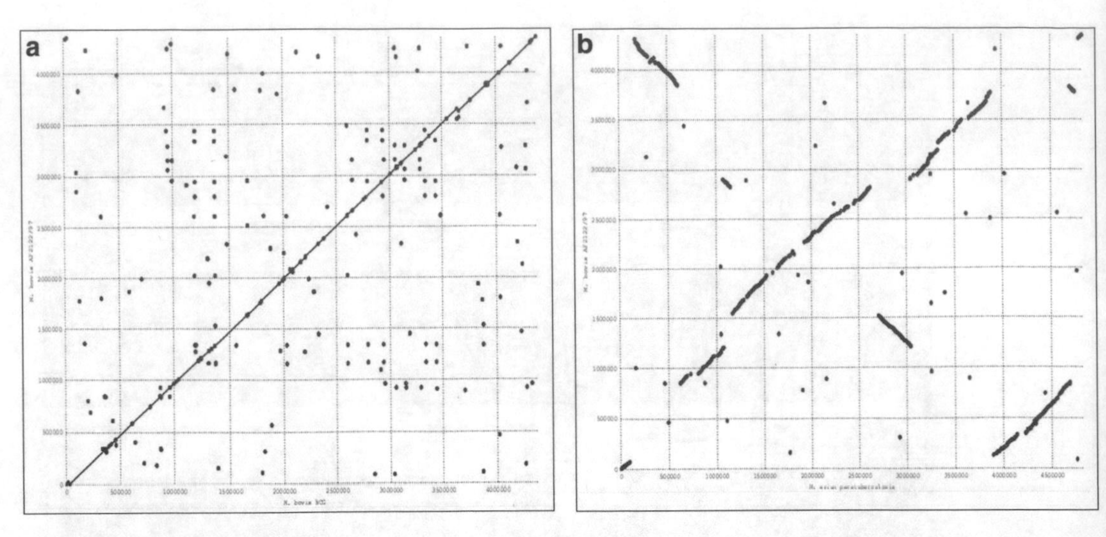

Fig. 20 Whole genome alignments of (**a**) *M. bovis* AF2122/97 (*vertical axis*) and *M. bovis* BCG Pasteur (*horizontal axis*) and (**b**) *M. bovis* AF2122/97 (vertical axis) and *M. avium paratuberculosis* (*horizontal axis*) using NUCmer showing all MUMs in a dot plot. A *straight line* indicates aligned regions. *Blue dots* represent large-scale chromosomal reversals

masked and unmasked regions are used for alignment. Like MUMmer it also relies on the construction of suffix trees to find the initial matches. This is followed by anchoring and aligning the sequences. Here again the anchors are selected using the Smith–Waterman algorithm. After searching for the sub-sequences or matches, anchoring of the sub-sequences (nonoverlapping, non-crossing matches) is done so that "noisy" matches are eliminated. Anchors are then joined to each other using a recursive approach, finally leading to a global alignment. The AVID program provides fast and reliable detection of even weak homologies [40].

3.3.7 Mugsy

Mugsy is a multiple alignment tool that is used to align whole genomes [41]. Mugsy can align multiple genomes without requiring any reference and can identify genetic variations like duplications and rearrangements. Mugsy uses the pairwise aligner NUCmer and an algorithm for identifying locally collinear blocks (LCBs are aligned regions from the genomes under consideration). In the final steps a multiple alignment of LCBs is generated.

3.3.8 WABA

Wobble Aware Bulk Aligner or WABA program is used for large-scale genome alignment [42]. Like BLASTZ, WABA is a variant of gapped BLAST and determines the initial matches by looking for strings of six nucleotides in the pattern 11011011 where 1 should always be a match. WABA is efficient in managing insertions and deletions.

Table 2
Web addresses of the most commonly used tools for large scale genome sequence comparison

Program	Web address	References
BLAT	https://genome.ucsc.edu/cgi-in/hgBlat?command=start	Kent (2002) [35]
BLASTZ	http://www.bx.psu.edu/miller_lab/	Schwartz et al. (2003) [36]
LAGAN	http://lagan.stanford.edu/lagan_web/index.shtml	Brudno et al. (2003) [37]
MUMMER	http://mummer.sourceforge.net/	Delcher et al. (1999) [39]
AVID	http://bio.math.berkeley.edu/avid/	Bray et al. (2003) [40]
WABA	http://users.soe.ucsc.edu/~kent/xenoAli/index.html	Kent and Zahler (2000) [42]
Mugsy	http://mugsy.sourceforge.net/	Angiuoli and Salzberg (2011) [41]
Mauve	http://gel.ahabs.wisc.edu/mauve/download.php	Darling et al. (2004) [43]
Cgaln (Coarse grained alignment)	http://www.iam.u-tokyo.ac.jp/chromosomeinformatics/rnakato/cgaln/	Nakato and Gotoh (2008, 2010) [44, 45]
LAST	http://last.cbrc.jp/	Kielbasa et al. (2011) [46]
Alfresco	http://www.sanger.ac.uk/resources/software/alfresco/	Dalca and Brudno (2008) [47]
M-GCAT	http://alggen.lsi.upc.es/recerca/align/mgcat/	Treangen and Messeguer (2006) [48]

3.3.9 Mauve

Mauve is a multiple genome alignment tool that is efficient in identifying genome rearrangements [43]. Like LAGAN and avid, Mauve depends on anchors to seed the alignment.

Apart from the programs and tools mentioned above, a number of other programs are available that permit users to do large-scale genome comparison. These programs and their web addresses have been summarized in Table 2.

The above-mentioned tools are valuable for generating synteny plots and are helpful in large-scale genome comparisons. There are various other tools that can generate circular maps for genome comparisons, as explained in **Note 5**.

4 Notes

1. Various tools are currently available for generating dot plots. These include Dotter [49], JDotter [50], YASS [51], Dotlet [52], and many more. Dotter is available at Wellcome Trust

(www.sanger.ac.uk/resources/software/seqtools/) and runs on Linux. JDotter is the interactive interface of Dotter (http://athena.bioc.uvic.ca/virology-ca-tools/jdotter/) and can be used for generating dot plots for whole genomes, sub-genomes, or protein alignments. Dotlet (http://www.isrec.isb-sib.ch/java/dotlet/Dotlet.html) is a platform-independent tool that runs on web browsers and is thus easy to use.

YASS (http://bioinfo.lifl.fr/yass/) is a genomic similarity search tool for nucleic acids (DNA/RNA). It is one of the most simple and easy to use tools for comparing two genomes. Users can upload their own file or use inbuilt genomes available for comparison. Users can also specify parameters including gap penalty and threshold *E*-value. The default values for gap penalties are: gap opening -16, gap extension -4, and e-value 10. Dotplot generated using YASS genomic similarity search tool is shown in Fig. 21.

Other programs that construct dot plots of an alignment are BLAST and Mummer (explained in Subheading 3.3.5).

2. FASTA format is the most common type of file format that is used in bioinformatics. In FASTA format a sequence is preceded by a header line beginning with the ">" character and followed by a name or identifier to be associated with that sequence

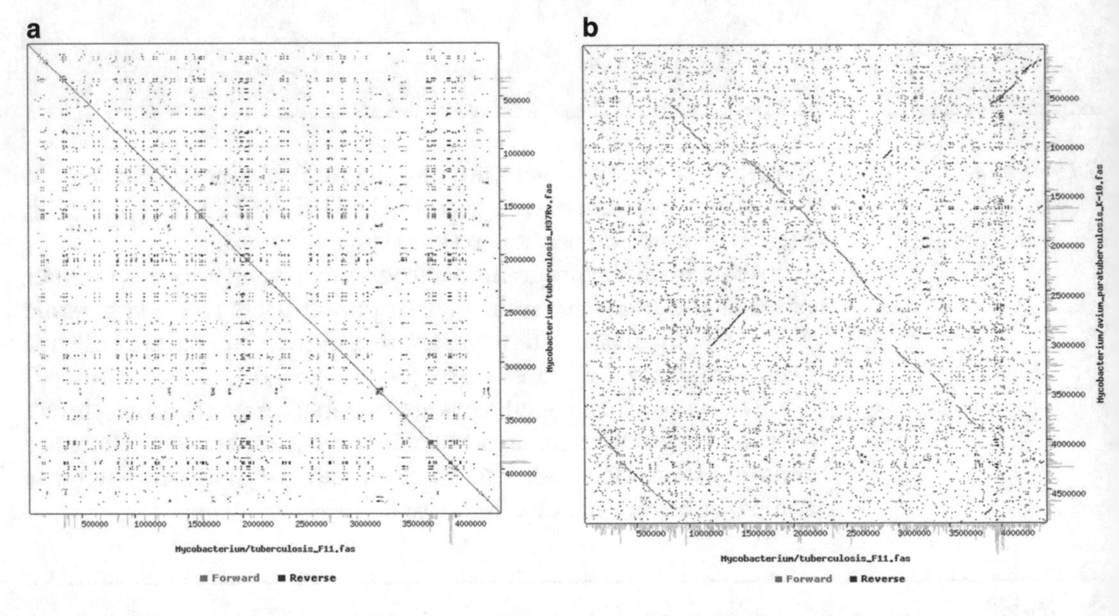

Fig. 21 Dotplot generated using YASS genomic similarity search tool. The figure shows alignment of complete genomes of (**a**) *Mycobacterium tuberculosis* F11 and *Mycobacterium tuberculosis* H37Rv (**b**) *Mycobacterium tuberculosis* F11 and *Mycobacterium paratuberculosis* K10. The *diagonal* shown in (**a**) represents synteny between the two *Mycobacterium* genomes while in (**b**) the two genomes are not in synteny

```
FASTA
>seq1
MSDLDRLASRAAIQDLYSDQLIGVDKRQEGRLASIWWDDAEWTIEGIGTY
>seq2
KGPEGALDLANNVLWPMFHETIHYGTNLRLEFVSADKVNGIGDVLCLGNL
>seq3
VEGNQSILIAAVYTNEYERRDGVWKLSKLNGCMNYFTPLAGIHFAPPGAL

PIR/NBRF
>P1;seq1
seq1
MSDLDRLASRAAIQDLYSDQLIGVDKRQEGRLASIWWDDAEWTIEGIGTY*
>P1;seq2
seq2
KGPEGALDLANNVLWPMFHETIHYGTNLRLEFVSADKVNGIGDVLCLGNL*
>P1;seq3
seq3
VEGNQSILIAAVYTNEYERRDGVWKLSKLNGCMNYFTPLAGIHFAPPGAL*

#NEXUS

begin taxa;
dimensions ntax=4;
taxlabels
seq1
seq2
seq3
seq4
;
end;

begin characters;
dimensions nchar=50;
format datatype=protein gap=-;
matrix
seq1 MSDLDRLASRAAIQDLYSDQLIGVDKRQEGRLASIWWDDAEWTIEGIGTY
seq2 KGPEGALDLANNVLWPMFHETIHYGTNLRLEFVSADKVNGIGDVLCLGNL
seq3 VEGNQSILIAAVYTNEYERRDGVWKLSKLNGCMNYFTPLAGIHFAPPGAL
seq4 LQKS
;
end;

GENBANK
LOCUS      AAR05959       154 aa      linear  BCT 02-APR-2004
DEFINITION LinA [Sphingobium indicum].
ACCESSION  AAR05959
ORGANISM   Sphingobium indicum

ORIGIN
     1 msdldrlasr aaiqdlysdq ligvdkrqeg rlasiwwdda ewtiegigty kgpegaldla
    61 nnvlwpmyhe tihygtnlrl efvsadkvng igdvlclgnl vegnqsilia avytneyerr
   121 dgvwklskln gcmnyftpla gihfappgal lqks
```

Fig. 22 Different file formats that are used in sequence similarity search and retrieval

(Fig. 22). Other types of file formats include PIR/NBRF, NEXUS, and genbank.

3. A number of BLAST servers have been developed and are available freely, but the most widely used is available at the National

Center for Biotechnology Information (NCBI; http://www.ncbi.nlm.nih.gov/). As already mentioned, many variants of BLAST are available and it is the user's choice to choose the BLAST variant depending on the type of search. Users can click on the various hyperlinks available at the BLAST homepage. Let us start a typical BLAST search using BLASTP, a programme that searches a protein database using a protein query.

In our example we have written the accession number for the beta-globin gene of human (P68871) (Fig. 23). Next the user has to define the type of BLAST to be used. The user can also set algorithm parameters like word size (default is 3), expected threshold (default 10), matrix (BLOSUM62 is the default matrix), and gap costs (allows user to define the gap penalty) and the database to which the query sequence is to be aligned.

Clicking on the BLAST hyperlink will submit the query. The first part of the result of a typical BLAST output is shown in Fig. 24a. The results also show the presence of conserved domain(s) in the query sequence (belonging to the globin superfamily). The colors indicate the alignment scores for each hit in the database. Clicking these colored bars will take the user to detailed information regarding that particular hit.

The second part of the BLASTP result shows the identity of each hit (Fig. 24b). The hits are also presented with their accession numbers, % identity, query coverage, *E* value, and the alignment score. The last part of the result gives detailed information regarding each hit that is obtained using BLASTP (Fig. 24c). The details include the identity of the hit and the complete alignment.

4. FASTA program can be used to search protein, nucleotide, genomes or even proteomes. Figure 25 shows the FASTA program homepage. Just like BLAST, the query sequence is pasted in the box provided. The user can change the default settings by clicking the hyperlink "More options." Clicking the "Submit" button will submit the query to the FASTA server. The results of the FASTA search are organized under various categories (Fig. 26). The query used in our example is the HCH dehydrochlorinase from *Sphingobium indicum* B90A. The summary table consists of information including the identity of the hit, score of the alignment, % identities, % positives, and *E*-value. The visual output consists of color codes for the alignment scores for each hit. Functional predictions indicate the conserved domains present.

5. There are various graphical visualization tools that are used for comparing large sequences on different platforms (LINUX, Windows, Mac OS). These visualization software are circos (http://circos.ca/), CGView Comparison Tool (CCT), BRIG, and many more. The CGView Comparison Tool (CCT) is a tool

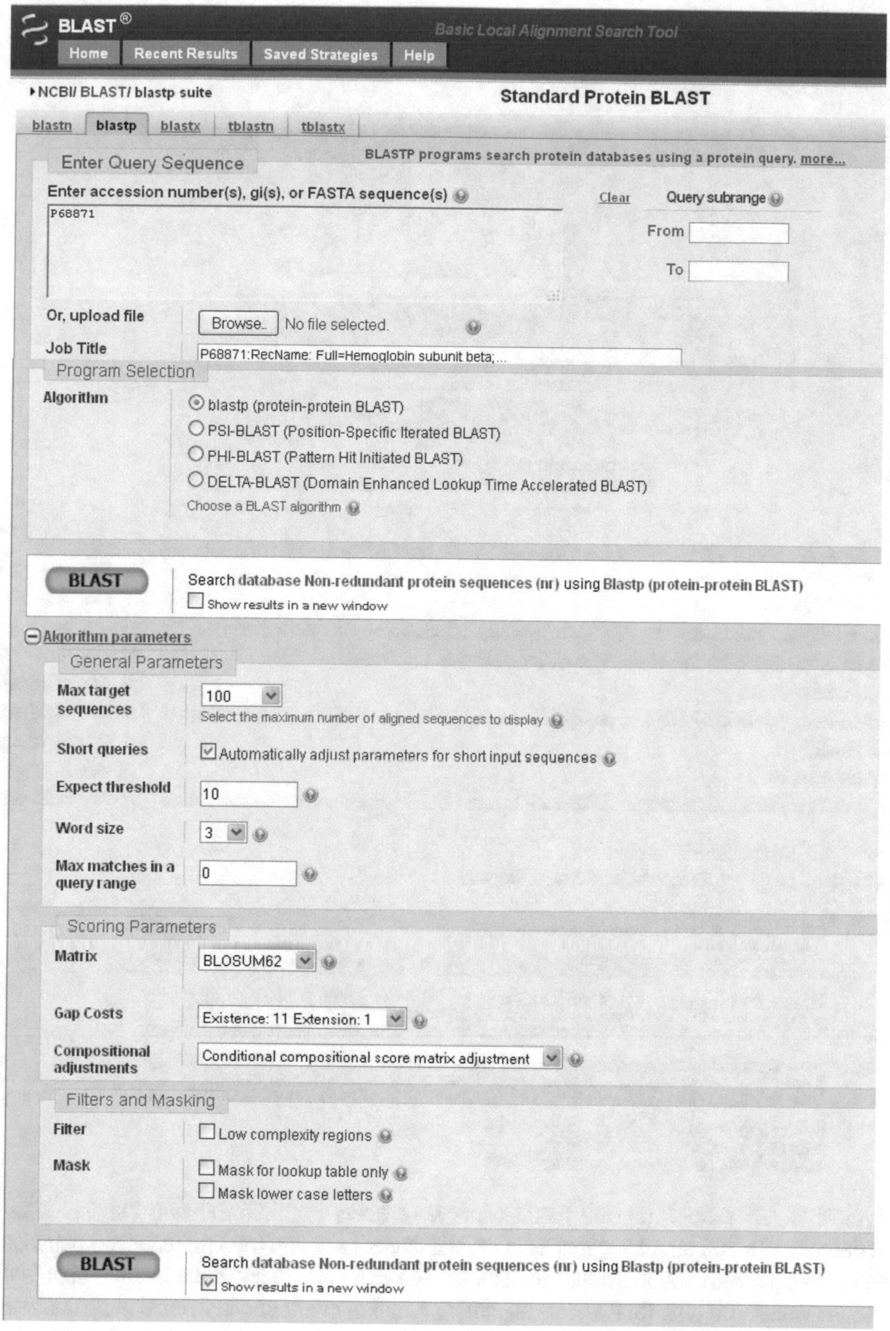

Fig. 23 BLASTp search. In the query box of BLASTp, accession number P68871 for the beta-globin gene of human is typed and the query is searched against the database with default parameters

that is used for comparing small genomes such as bacterial genomes [53] and is freely available at http://stothard.afns.ualberta.ca/downloads/CCT/. It runs on Mac OS X and Linux, and also requires BLAST installation. CCT generates

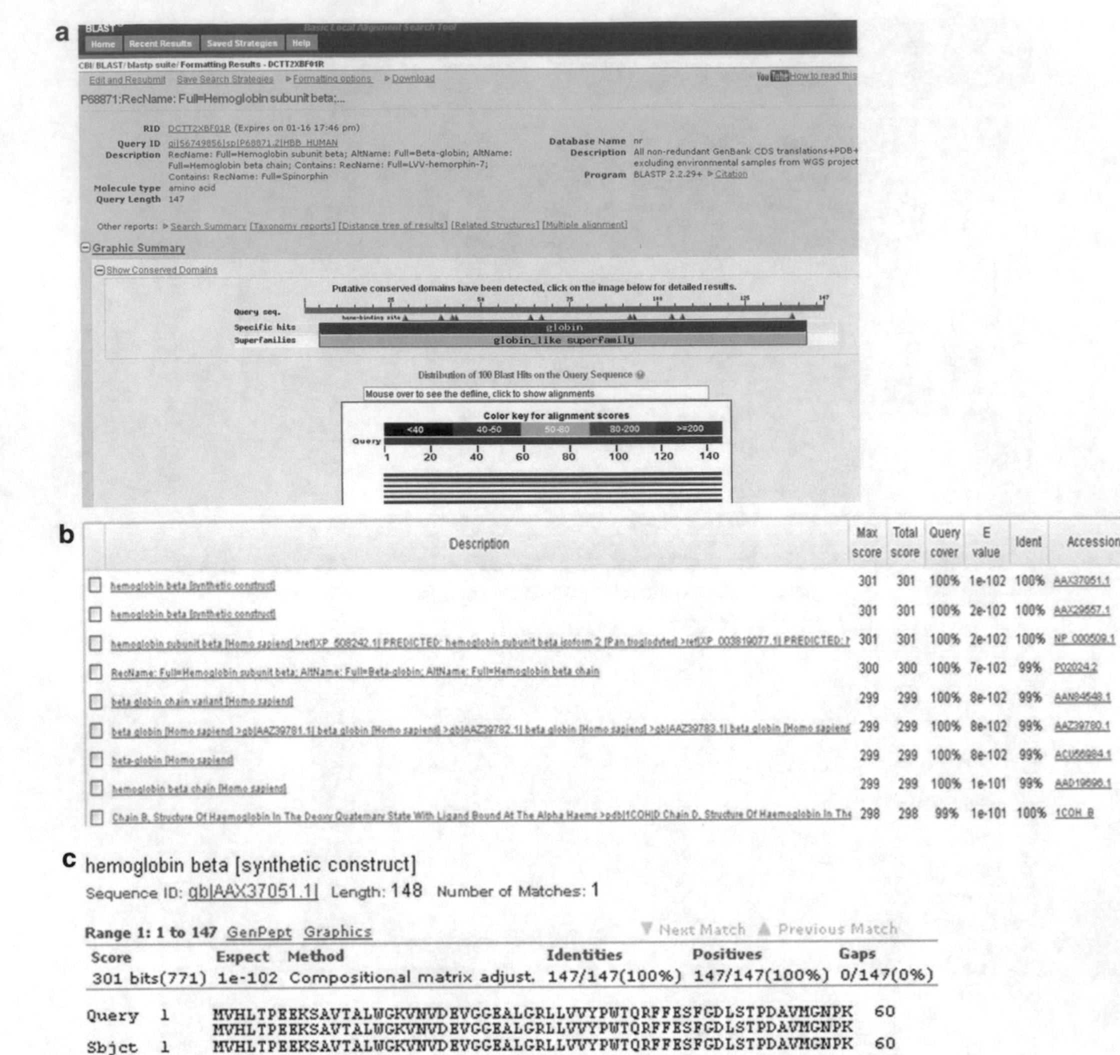

Fig. 24 Typical BLAST output. (**a**) The result window of query P68871 showing the hits with a putative conserved domain with 100 target sequences in the database. Each line (*red*) represents a pairwise alignment of the query sequence with 100 individual sequences. The *color* of this line is dependent upon the alignment score of the individual 100 hits. (**b**) Pairwise alignment result of query and individual subjects presented with their accession numbers, % identity, query coverage, *E*-value, and alignment score. (**c**) Detailed pairwise alignment of query with respective subjects

maps that depict COG (Clusters of Orthologous groups) functional categories for forward and reverse strand coding sequences and sequence similarity rings detected by BLAST. BLAST Ring Image Generator (BRIG) is a tool for comparing multiple prokaryotic genomes [54] and is freely available at

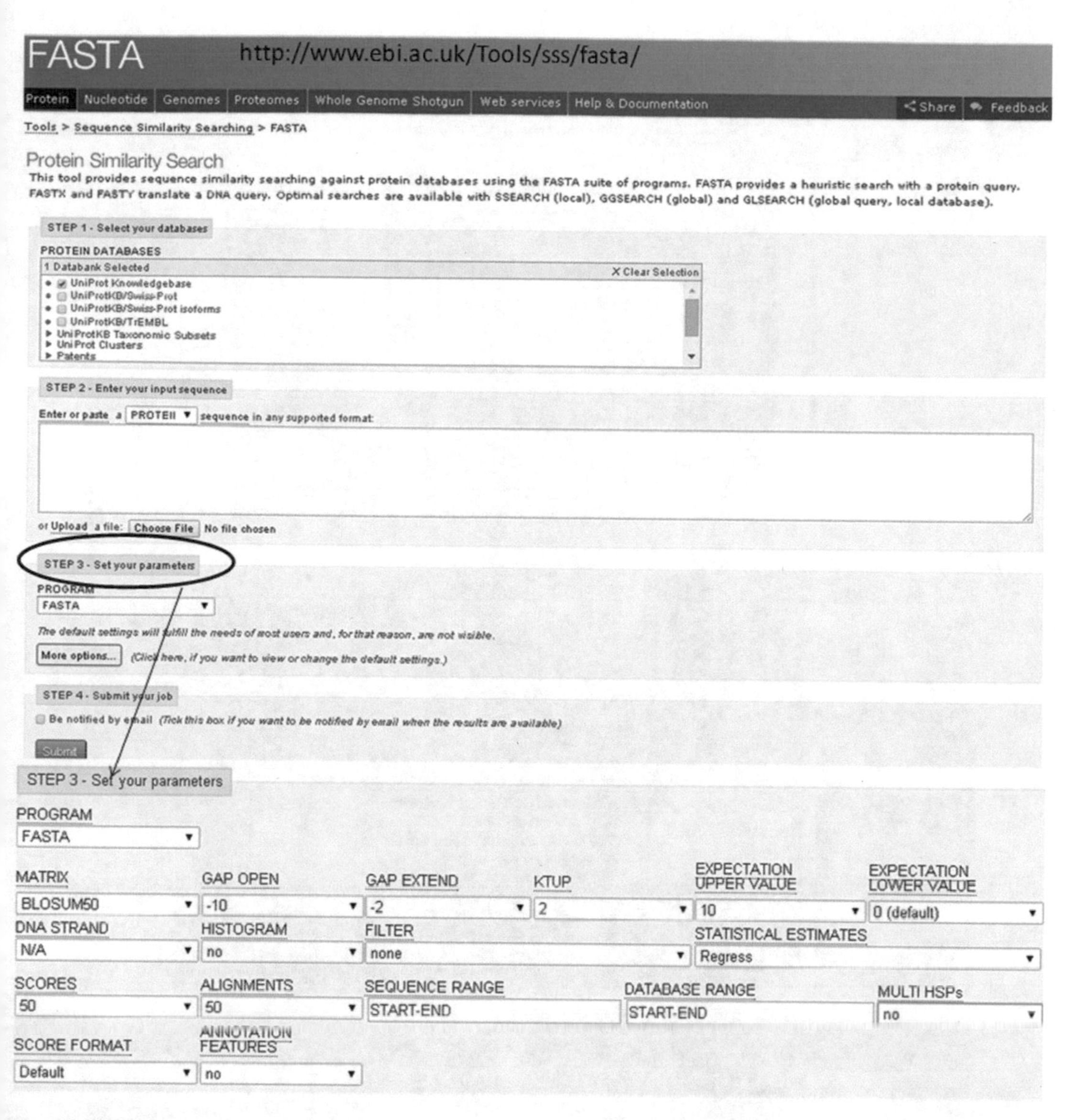

Fig. 25 FASTA homepage for protein query. In *step 1*, the user can choose the database against which the user is interested in aligning the protein. In *step 2*, the query protein sequence is pasted in the fasta format. In *step 3*, the user can select and set various parameters

http://sourceforge.net/projects/brig/. Unlike CCT, it runs on all operating systems. It generates results in the form of circular maps in which each ring represents a BLAST search of one genome each. There is no limit to the number of genomes that can be compared using BRIG.

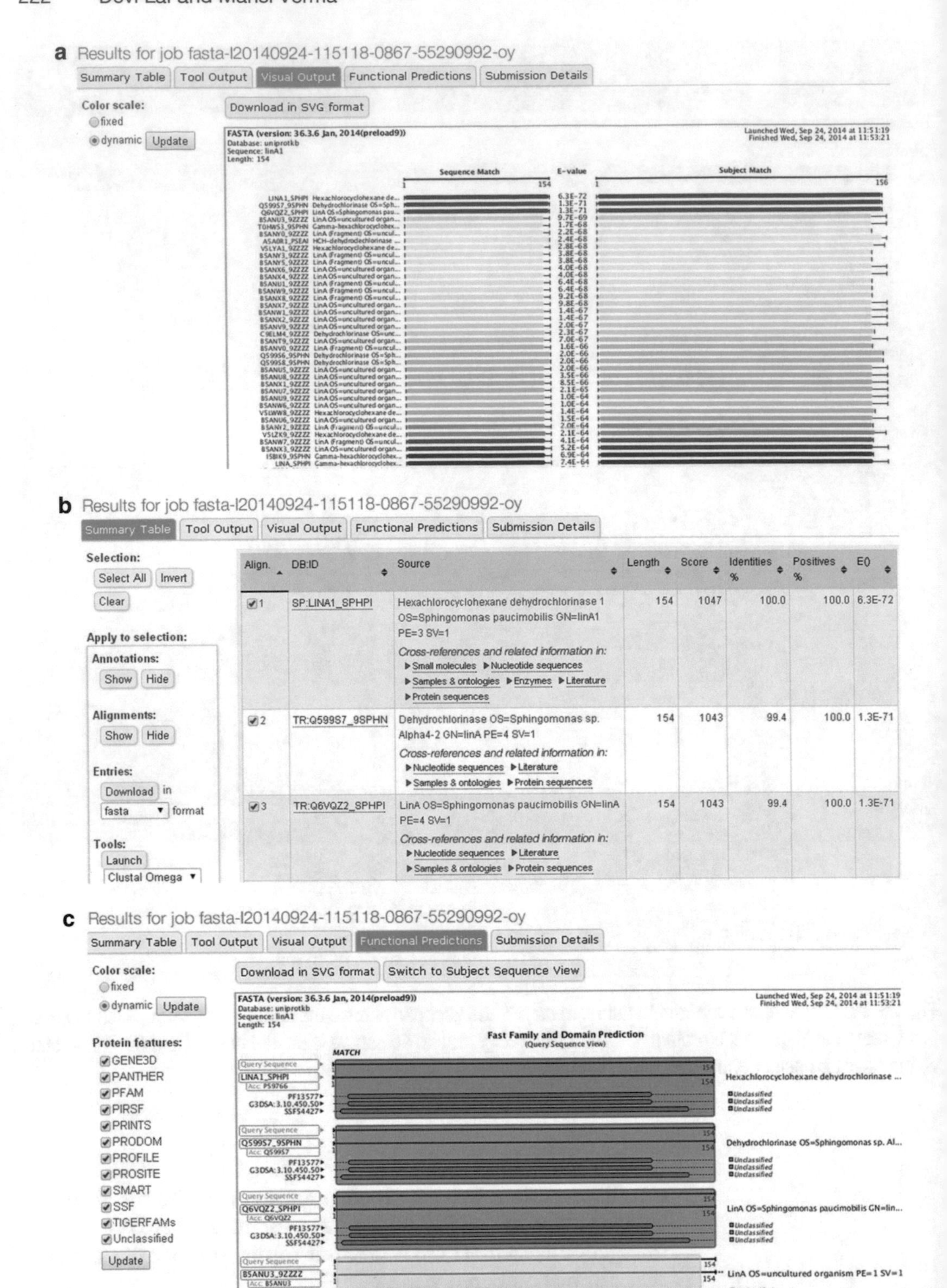

Fig. 26 Typical FASTA search output. Just like the BLAST program, there are various outputs of FASTA: (**a**) visual output giving the color code for the *E*-value of the hits (**b**) summary table that lists all the hits across the database (**c**) functional prediction suggesting the presence of conserved domains in the query protein sequence

References

1. Tautz D (1998) Evolutionary biology. Debatable homologies. Nature 395:17–19
2. Pearson WR (1996) Effective protein sequence comparison. Methods Enzymol 266:227–258
3. Gibbs AJ, McIntyre GA (1970) The diagram, a method for comparing sequences. Its use with amino acid and nucleotide sequences. Eur J Biochem 16:1–11
4. Dayhoff MO, Schwartz RM, Orcutt BC (1978) A model of evolutionary changes in proteins. In: Dayhoff MO (ed) Atlas of protein sequence and structure, vol 5. National Biomedical Research Foundation, Washington, DC, pp 345–352
5. Gonnet GH, Cohen MA, Brenner SA (1992) Exhaustive matching of the entire protein sequence database. Science 256:1443–1445
6. Jones DT, Taylor WR, Thornton JM (1992) The rapid generation of protein mutation data matrices from protein sequences. Cumput Appl Biosci 8:275–282
7. Henikoff S, Henikoff JG (1992) Amino acid substitution matrices from protein blocks. Proc Natl Acad Sci U S A 89:10915–10919
8. Henikoff S, Henikoff JG (1996) Blocks database and its application. Methods Enzymol 266:88–105
9. Henikoff S, Henikoff JG (2000) Amino acid substitution matrices. Adv Protein Chem 54:73–97
10. Henikoff S, Henikoff JG (1991) Automated assembly of protein blocks for database searching. Nucleic Acids Res 19:6565–6572
11. Henikoff S, Henikoff JG (1993) Performance evaluation of amino acid substitution matrices. Proteins Struct Funct Genet 17:49–61
12. Wheeler DG (2003) Selecting the right protein scoring matrix. Curr Protoc Bioinformatics 3.5.1–3.5.6
13. Needleman SB, Wunsch CD (1970) A general method applicable to the search for similarities in amino acid sequence of two proteins. J Mol Biol 48:443–453
14. Smith TF, Waterman MS (1981) Identification of common molecular subsequences. J Mol Biol 147:195–197
15. Sellers PH (1974) On the theory and computation of evolutionary distances. SIAM J Appl Math 26:787–793
16. Gotoh O (1982) An improved algorithm for matching biological sequences. J Mol Biol 162:705–708
17. Altschul SF, Gish W, Miller W, Myers EW, Lipman DJ (1990) Basic local alignment search tool. J Mol Biol 215:403–410
18. Karlin S, Altschul SF (1990) Methods for assessing the statistical significance of molecular sequence features by using general scoring schemes. Proc Natl Acad Sci U S A 87:2264–2268
19. Altschul SF, Madden TL, Schäffer AA, Zhang J, Zhang Z, Miller W, Lipman DJ (1997) Gapped BLAST and PSI-BLAST: a new generation of protein database search programs. Nucleic Acids Res 25:3389–3402
20. Altschul SF, Koonin EV (1998) Iterated profile searches with PSI-BLAST: a tool for discovery in protein databases. Trends Biochem Sci 23:444–447
21. Schaffer AA, Aravind L, Madden TL, Shavirin S, Spouge JL, Wolf YI, Koonin EV, Altschul SF (2001) Improving the accuracy of PSI-BLAST protein database searches with composition based statistics and other refinements. Nucleic Acids Res 29:2994–3005
22. Bucher P, Karplus K, Moeri N, Hofmann K (1996) A flexible motif search technique based on generalized profiles. Comput Chem 20:3–23
23. Staden R (1988) Methods to define and locate patterns of motifs in sequences. Comput Appl Biosci 4:53–60
24. Tatusov RL, Altschul SF, Koonin EV (1994) Detection of conserved segments in proteins: iterative scanning of sequence databases with alignment blocks. Proc Natl Acad Sci U S A 91:12091–12095
25. Marchler-Bauer A, Anderson JB, Chitsaz F, Derbyshire MK, DeWeese-Scott C, Fong JH, Geer LY et al (2009) CDD: specific functional annotation with the Conserved Domain Database. Nucleic Acids Res 37:D205–D210
26. Zhang Z, Schäffer AA, Miller W, Madden TL, Lipman DJ, Koonin EV, Altschul SF (1998) Protein similarity searches using patterns as seeds. Nucleic Acids Res 26:3986–3990
27. Wilbur WJ, Lipman DJ (1983) Rapid similarity searches of nucleic acid and protein data banks. Proc Natl Acad Sci U S A 80:726–730
28. Lipman DJ, Pearson WR (1985) Rapid and sensitive protein similarity searches. Science 227:1435–1441
29. Pearson WR, Lipman DJ (1988) Improved tools for biological sequence comparison. Proc Natl Acad Sci U S A 85:2444–2448

30. Pearson WR (1990) Rapid and sensitive sequence comparison with FASTP and FASTA. Methods Enzymol 183:63–98
31. Pearson WR (2003) Finding protein and nucleotide similarities with FASTA. Curr Protoc Bioinformatics 3.9.1–3.9.23
32. Pearson WR (2000) Flexible sequence similarity searching with the FASTA3 program package. Methods Mol Biol 132:185–219
33. Zhang Z, Schwartz S, Wagner L, Miller WA (2000) A greedy algorithm for aligning DNA sequences. J Comput Biol 7:203–214
34. Ma B, Tromp J, Li M (2002) Patternhunter: faster and more sensitive homology search. Bioinformatics 18:440–445
35. Kent WJ (2002) BLAT-the BLAST like alignment tool. Genome Res 12:656–664
36. Schwartz S, Kent WJ, Smit A, Zhang Z, Baertsch R, Hardison RC, Haussler D, Miller W (2003) Human–mouse alignments with BLASTZ. Genome Res 13:103–107
37. Brudno M, Do CB, Cooper GM, Kim MF, Davydov E, NISC Comparative Sequencing Program, Green ED, Sidow A, Batzoglou S (2003) LAGAN and multi-LAGAN: efficient tools for large-scale multiple alignment of genomic DNA. Genome Res 13:721–731
38. Brudno M, Morgenstern B (2002) Fast and sensitive alignment of large genomic sequences. In: Proceedings IEEE computer society bioinformatics conference, Stanford University, pp 138–147
39. Delcher AL, Kasif S, Fleischmann RD, Peterson J, White O, Salzberg SL (1999) Alignment of whole genomes. Nucleic Acids Res 27:2369–2376
40. Bray N, Dubchak I, Pachter L (2003) AVID: a global alignment program. Genome Res 13:97–102
41. Angiuoli SV, Salzberg SL (2011) Mugsy: fast multiple alignment of closely related whole genome. Bioinformatics 27:334–342
42. Kent WJ, Zahler AM (2000) Conservation, regulation, synteny, and introns in a large-scale *C. briggsae–C. elegans* genomic alignment. Genome Res 10:1115–1125
43. Darling AC, Mau B, Blattner FR, Perna NT (2004) Mauve: multiple alignment of conserved genomic sequence with rearrangements. Genome Res 14:1394–1403
44. Nakato R, Gotoh O (2008) A novel method for reducing computational complexity of whole genome sequence alignment. In Proceedings of the sixth Asia-Pacific bioinformatics conference (APBC2008), pp 101–110
45. Nakato R, Gotoh O (2010) Cgaln: fast and space-efficient whole-genome alignment. BMC Bioinformatics 11:24
46. Kiełbasa SM, Wan R, Sato K, Horton P, Frith MC (2011) Adaptive seeds tame genomic sequence comparison. Genome Res 21:487–493
47. Dalca AV, Brudno M (2008) Fresco: flexible alignment with rectangle scoring schemes. Pac Symp Biocomput 13:3–14
48. Treangen T, Messeguer X (2006) M-GCAT: interactively and efficiently constructing large-scale multiple genome comparison frameworks in closely related species. BMC Bioinformatics 7:433
49. Sonnhammer EL, Durbin R (1995) A dot-matrix program with dynamic threshold control suited for genomic DNA and protein sequence analysis. Gene 167:GC1–GC10
50. Brodie R, Roper RL, Upton C (2004) JDotter: a Java interface to multiple dotplots generated by dotter. Bioinformatics 20:279–281
51. Noe L, Kucherov G (2005) YASS: enhancing the sensitivity of DNA similarity search. Nucleic Acids Res 33:W540–W543
52. Junier T, Pagni M (2000) Dotlet: diagonal plots in a web browser. Bioinformatics 16:178–179
53. Grant JR, Arantes AS, Stothard P (2012) Comparing thousands of circular genomes using the CGView Comparison Tool. BMC Genomics 13:202
54. Alikhan NF, Petty NK, Ben Zakour NL, Beatson SA (2011) BLAST Ring Image Generator (BRIG): simple prokaryote genome comparisons. BMC Genomics 12:402

Chapter 10

Genomic Database Searching

James R.A. Hutchins

Abstract

The availability of reference genome sequences for virtually all species under active research has revolutionized biology. Analyses of genomic variations in many organisms have provided insights into phenotypic traits, evolution and disease, and are transforming medicine. All genomic data from publicly funded projects are freely available in Internet-based databases, for download or searching via genome browsers such as Ensembl, Vega, NCBI's Map Viewer, and the UCSC Genome Browser. These online tools generate interactive graphical outputs of relevant chromosomal regions, showing genes, transcripts, and other genomic landmarks, and epigenetic features mapped by projects such as ENCODE.

This chapter provides a broad overview of the major genomic databases and browsers, and describes various approaches and the latest resources for searching them. Methods are provided for identifying genomic locus and sequence information using gene names or codes, identifiers for DNA and RNA molecules and proteins; also from karyotype bands, chromosomal coordinates, sequences, motifs, and matrix-based patterns. Approaches are also described for batch retrieval of genomic information, performing more complex queries, and analyzing larger sets of experimental data, for example from next-generation sequencing projects.

Key words Bioinformatics, Epigenetics, Genome browsers, Identifiers, Internet-based software, Next-generation sequencing, Motifs, Matrices, Sequences

Abbreviations

API	Application Programming Interface
BED	Browser Extensible Data
BLAST	Basic Local Alignment Search Tool
BLAT	BLAST-Like Alignment Tool
DDBJ	DNA Databank of Japan
EBI	European Bioinformatics Institute
EMBOSS	European Molecular Biology Open Software Suite
ENA	European Nucleotide Archive
ENCODE	Encyclopedia Of DNA Elements
FTP	File Transfer Protocol
GI	GenInfo Identifier
GOLD	Genomes Online Database

Jonathan M. Keith (ed.), *Bioinformatics: Volume I: Data, Sequence Analysis, and Evolution,* Methods in Molecular Biology, vol. 1525, DOI 10.1007/978-1-4939-6622-6_10, © Springer Science+Business Media New York 2017

GRC	Genome Reference Consortium
GUI	Graphical User Interface
HAVANA	Human and Vertebrate Analysis and Annotation
ID	Identifier Code
INSDC	International Nucleotide Sequence Database Collaboration
NCBI	National Center for Biotechnology Information
NGS	Next-Generation Sequencing
PWM	Position Weight Matrix
RegEx	Regular Expression
REST	Representational State Transfer
ROI	Region of Interest
RSAT	Regulatory Sequence Analysis Tools
UCSC	University of California Santa Cruz
URL	Uniform Resource Locator
Vega	Vertebrate Genome Annotation

1 Introduction

The enormous efforts in sequencing, bioinformatics, and annotation during the past four decades have resulted in the generation of reference genomes for thousands of species, revolutionizing biology. Major landmarks in genome sequencing include the first genome, bacteriophage ΦX174, in 1977 [1]; the first organism, *Haemophilus influenzae* in 1995 [2]; the first eukaryote, *Saccharomyces cerevisiae*, in 1996 [3]; and the first multicellular organism, *Caenorhabditis elegans*, in 1998 [4]. The human genome project, launched in 1990, resulted in a "working draft" sequence in 2001 [5, 6], and a "near complete" assembly in 2004 [7]; these are now complemented by sequences for virtually every organism of research interest. Indeed, the Genomes OnLine Database (**GOLD**) [8] records the genome sequencing of over 200,000 organisms, over 13,000 of which are eukaryotes.

For most species, the relationships between the genome and the many biological functions carried out by the organism are still poorly understood. Genomic sequences form the basis for many types of study, including the identification of specific loci bound by particular proteins, those that make contact during chromosome folding, and those forming origins of replication. Genomic information enables in silico prediction of transcripts and polypeptides, providing for comprehensive investigations into the gene expression and proteomic status of cells and tissues, plus genome-wide gene knockdown or gene editing-based functional screens.

Beyond functional studies, genomic sequence data have provided fascinating insights into evolution and species diversity

[9, 10], the emergence of hominids [11], and even the course of European history [12].

For any species, individuals within a population will exhibit genomic variations, and practically none will exactly match the reference genome sequence. Recent advances in next-generation sequencing (NGS) technology have allowed the genomes of many individuals within a population to be sequenced for comparative analysis. The worldwide **1000 Genomes Project** is creating a detailed catalog of human genetic variation [13, 14], whereas the UK-based **100,000 Genomes Project** [15] focuses on genomic differences linked to medical conditions; variation data from both projects are to be made publicly available. Yet even within one individual, or one tissue sample, different cells may exhibit variations in genomic sequence. Technological advances have now enabled genomic sequencing of single cells, revealing much about inter-cell diversity [16, 17].

A major driving force behind the human genome project was the quest to identify genes underlying human diseases, for the development of novel therapies. During the period of the project the field of medical genetics evolved, from using reverse-genetic approaches to identify individual disease-linked genes, to prospects for personalized whole-of-life health care [18–20]. However, early hopes and expectations for genome-based therapies were dampened slightly by the realization that gene regulation, genome organization and the genetic basis of disease are much more complicated than initially thought [21]. Understanding the genetic basis of cancer was the original motivation behind the human genome project [22], and **cancer genomes projects** on both sides of the Atlantic are investigating the relationship between genomic variation and cancers [23], to identify sequence signatures linked to different cancer types [24].

Complementing genomic sequence information are data generated from consortia that attribute evidence-based or computationally predicted functions to segments of the genome, as well as mapping epigenetic modifications in certain cell types. The best known such project, the Encyclopedia of DNA Elements (**ENCODE**), aims to identify all functional elements in the human genome [25]. The parallel **modENCODE** project has similar goals, for the model organisms *C. elegans* and *D. melanogaster* [26, 27], whereas the **GENCODE** project aims "to identify all gene features in the human genome using a combination of computational analysis, manual annotation, and experimental validation" [28]. Data from these projects are also publicly available, and will be complemented by further epigenetic features as collaborative projects currently underway come to fruition [29].

This chapter provides an overview of current methods for searching genomic databases using names and unique identifier codes (IDs) for genes, DNA, or RNA molecules, locus codes,

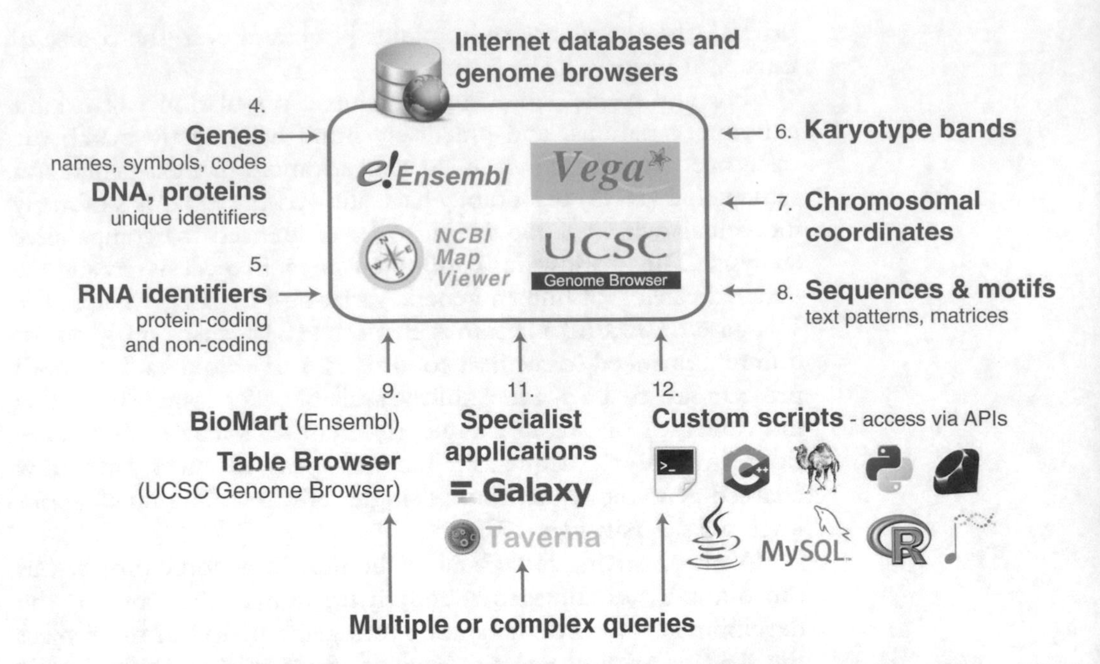

Fig. 1 Overview of approaches for genomic database searching. Genomic data in public databases can be searched using many types of query, including gene names and codes, IDs for DNA, RNA, and protein molecules, karyotype band codes, chromosomal coordinates, sequences, and motifs. A variety of software approaches are possible, including the use of genome browsers, tools such as BioMart and Table Browser, specialist applications such as Galaxy and Taverna, and custom scripts employing APIs, in a variety of languages. Numbers refer to sections within this chapter

chromosomal coordinates, sequences, and motifs (Fig. 1). Separate sets of questions relate to resources and approaches for accessing information about the functions and properties of genes and their products; such methods have been recently described [30–32], and are not covered here.

2 Reference Genomes and Genes

2.1 Reference Genome Sequences

For most experimental organisms, reference genome sequences are the product of collaborative endeavors, with the resulting assemblies being deposited in public databases. But for the human genome project there were two parallel, competing efforts: one led by the publicly funded International Human Genome Sequencing Consortium (IHGSC), and one undertaken by the private company Celera Genomics, with "working draft" genomes from each team reported simultaneously in 2001 [5, 6]. Whereas IHGSC data were immediately made publicly available, Celera data were initially available only via paid subscription, but eventually released to the public, incorporated into GenBank [33]. The IHGSC draft genome underwent refinement, for example by filling in missing sections, to generate a "near complete" reference covering 99 % of the euchromatic genome [7]. Even in their "complete" states, reference genome sequences may still contain a few small gaps and regions of uncertainty. The Genome Reference Consortium (GRC) [34] coordinates the correction, refinement, and occasional release of updated genome assemblies for human, mouse and zebrafish, the latest versions of which are known (at the time of writing) as GRCh38, GRCm38 and GRCz10, respectively; these sequences are the "definitive" ones shared by the major public genomic databases.

2.2 Reference Gene Sets

One issue central to genomic annotation, but which is far from straightforward, is what a gene is, and how a set of genes can be identified within a genome. The definition of a gene has evolved considerably over time, and must now encompass loci corresponding to a range of protein-coding as well as non protein-coding transcripts [35]. Several automated routines have been developed to predict sets of reference genes within genomic sequences, these include Genescan [36], AceView [37], the Mammalian Gene Collection (MGC) [38], Consensus Coding Sequence (CCDS) [39], Ensembl [40], and RefSeq [41]. Complementing automated gene-prediction methods are expert-curated manual annotations, notably the products of the Human and Vertebrate Analysis and Annotation (HAVANA) procedure, available from the Vega database [42].

For the human and mouse genomes, the GENCODE consortium merges Ensembl (automated) and HAVANA (manual) gene annotations to generate a reference gene set that aims to capture the full extent of transcriptional complexity, including pseudogenes, small RNAs and long noncoding RNAs. There are two versions of this gene set: GENCODE Basic and GENCODE Comprehensive. For protein-coding genes the former contains only full-length, protein-coding transcripts, whereas the latter includes a fuller set of variant-length transcripts. The GENCODE Comprehensive gene set in particular gives very good genome coverage [43], and is increasingly being adopted as the standard in the community.

All major genome browsers can be customized to allow the visualization of multiple gene sets aligned in parallel to the reference genome, allowing their outputs to be inspected and compared.

3 Genomic Databases, and Approaches to Searching Them

Genomic reference sequences and related data from major academic consortia are freely available from Internet-based databases. So how can this genomic sequence information be best accessed?

The approach and type of query depends on the starting point, and the question being asked. Genomic database searching can be approached in three main ways:

Firstly, complete genome sequences can be downloaded from Internet-based databases. This gives the user the full freedom and flexibility to search in the manner of their choosing, which may range from using a simple text editor to writing a sophisticated custom program. This approach is also necessary when using software not directly connected to Internet-based databases, but requiring a sequence file to be inputted or uploaded (such as EMBOSS; Subheading 8.2). Genomic sequence files are typically downloadable from the file transfer protocol (FTP) servers of each genomic database, as described in Table 1.

Secondly, genomes can be searched using Web-based genome browsers and other Internet-connected software that employ graphical user interfaces (GUIs). This is the most user-friendly approach to genomic database searching, as genome browsers automatically recognize many types of query, have in-built search routines, and allow searches to be stored and combined. Major Web-based browsers also allow genome searching by incorporating query terms into a Web address (URL). For users with a spreadsheet containing many IDs, links to custom genome searches can thus be automatically created in spreadsheet software, allowing one-click launching of searches in genome browsers (Subheading 9.1).

Thirdly, genomic databases can be searched using custom scripts that execute queries to Internet-based databases by means of their application programming interfaces (APIs). This is ultimately the most flexible and powerful approach, but for those without programming experience this involves a considerable learning curve. Options for this approach are described in Subheading 12.

4 Genomic Databases, and Genome Browser-Based Searches

As the genomic databases and their associated genome browsers are typically accessible from the same Internet home page, they are introduced together. For each browser, the search functionalities and modes of output are briefly described.

4.1 General Features of Genome Browsers

A genome browser produces a graphical representation of a region of interest (ROI) within a selected chromosome of the organism in question, showing annotations of region-specific genomic features, most importantly genes, plus other landmarks such as CpG islands, repeat elements, and single nucleotide polymorphisms (SNPs). Typically a chromosome ideogram appears alongside, showing the

Table 1
Genomic databases and genome browsers

Resource	Website	References
Ensembl home page	http://www.ensembl.org/	[40]
Tutorials:	http://www.ensembl.org/info/website/tutorials/index.html	
Data access (FTP):	http://www.ensembl.org/info/data/ftp/index.html	
Ensembl Genomes	http://ensemblgenomes.org/	[44]
Ensembl Bacteria:	http://bacteria.ensembl.org/	
Ensembl Fungi:	http://fungi.ensembl.org/	
Ensembl Metazoa:	http://metazoa.ensembl.org/	
Ensembl Plants:	http://plants.ensembl.org/	
Ensembl Protists:	http://protists.ensembl.org/	
Vega home page:	http://vega.sanger.ac.uk/	[42]
Documentation:	http://vega.sanger.ac.uk/info/index.html	
Data access (FTP):	ftp://ftp.sanger.ac.uk/pub/vega/	
NCBI resources:		[45, 56]
Genomes	http://www.ncbi.nlm.nih.gov/genome/	
Data access (FTP):	ftp://ftp.ncbi.nlm.nih.gov/genomes/	
Map Viewer:	http://www.ncbi.nlm.nih.gov/mapview/	[57]
Map Viewer Help	http://www.ncbi.nlm.nih.gov/projects/mapview/static/MapViewerHelp.html	
Gene	http://www.ncbi.nlm.nih.gov/gene/	[58]
Viral Genomes:	http://www.ncbi.nlm.nih.gov/genome/viruses/	[59]
UCSC Genome Browser:	https://genome.ucsc.edu/	[51]
User Guide:	http://genome.ucsc.edu/goldenPath/help/hgTracksHelp.html	
Data download:	http://hgdownload.soe.ucsc.edu/downloads.html	

location of the ROI in the context of the whole chromosome and its karyotype banding patterns.

There are three principal ways that a ROI can be selected within a genome browser. Firstly, the name, ID, or sequence of a gene, or DNA, RNA, or protein molecule of interest can be used to perform a genomic database search to identify the corresponding locus. Secondly, an ROI can be selected based on its chromosomal coordinates. Thirdly, a region can be selected by browsing within a selected chromosome: scanning up- and downstream, zooming in and out, or by using a selection tool.

Genomic features relevant to the locus or ROI are shown as a series of linear panels ("tracks") parallel to the reference sequence. In most genome browsers genomic regions and tracks appear horizontally, with the notable exception of NCBI's Map Viewer, where the arrangement appears vertically. Notable features of the major genome browsers are described in the following sections. Detailed descriptions of how to customize the outputs of each browser, and integrate one's own experimental data are beyond the scope of this chapter, and are readily accessible from each browser's online documentation and tutorial documents, listed in Table 1.

4.2 Ensembl

Ensembl [40], a joint initiative between the European Bioinformatics Institute and the Wellcome Trust Sanger Institute, is a resource initially created to allow access to data from the public Human Genome Project. Ensembl now comprises a family of databases providing compiled genomic data for over 80 genome-sequenced vertebrate species, together with gene predictions and corresponding transcript and polypeptide sequences, and various analytical tools.

The main Ensembl resource is complemented by **Ensembl Genomes** [44], a "superfamily" of databases that includes **Ensembl Bacteria** (comprising over 10,000 genomes of bacteria and archaea), **Ensembl Fungi** (comprising genomes of over 40 species, including budding and fission yeasts, and *Aspergillus*), **Ensembl Metazoa** (over 50 invertebrate species, including *Drosophila*, *C. elegans*, and silkworm), **Ensembl Plants** (over 30 species, including *Arabidopsis*, rice, and wheat), and **Ensembl Protists** (over 20 species, including *Dictyostelium*, *Plasmodium*, and *Tetrahymena*).

4.2.1 Searching Ensembl

Starting with a gene name (for human genes, this is referred to as a gene symbol), the genomic locus can be rapidly identified within Ensembl using a simple search, where the relevant organism is also specified. At the Ensembl home page, after "Search:" choose the relevant species from the pull-down menu, enter the gene name into the search box and click "Go". At the search results page, the relevant gene usually appears at the top of the list, and the "Best gene match" is shown at the top-right. Clicking on either of these opens an Ensembl gene page showing information about the gene and a summary of its genomic context.

More straightforward still is searching Ensembl using unique IDs. In addition to its own identifiers, which have the prefix ENS, Ensembl recognizes a wide range of, but not all, popular ID types for genes, nucleic acids and proteins from external databases. For searching using gene IDs, Ensembl recognizes NCBI's GeneID and UniGene [45] codes, as well as several from species-specific databases such as HGNC [46] and FlyBase [47]. For nucleic acids such as mRNAs, Ensembl recognizes IDs from GenBank, RefSeq, European Nucleotide Archive (ENA) [48], DNA Databank of Japan (DDBJ) [49], but notably not NCBI's GenInfo (GI) numbers. For proteins, Ensembl recognizes UniProt [50] and RefSeq IDs, but not NCBI GI numbers.

The genomic locus relevant to the molecule of interest can be located very straightforwardly by directly searching the whole Ensembl database collection using the ID, without the need to specify an organism. From the home page, paste the ID into the search box on the top right ("Search all species..."), and click the magnifying glass icon or press enter. A results page appears, which may list many transcripts, proteins and other DNA segments as well as the gene. To restrict results to the gene, in the top-left of the page under "Restrict category to:", click "Gene". Clicking on the gene name opens the Ensembl gene page.

Genomic database searches can also be initiated by incorporating query terms into URLs, for both the main Ensembl database (*see* **Note 1**) and Ensembl Genomes (*see* **Note 2**). A more advanced approach for retrieving Ensembl genomic data for multiple IDs is the use of the BioMart tool; an example of this is shown in Subheading 9.2.

4.2.2 Navigating and Customizing the Genome Browser

Following an Ensembl gene search, and the user's selection of the gene of interest, a page appears that is rich in information relevant to that gene. This page is organized into tabs, the three main ones being Species, Location, and Gene. The Gene tab is shown as default, displaying summary information about the gene, a table of transcripts, and then a condensed genome viewer zoomed to the full length of the gene. Genes are shown with a color code (red, protein coding; orange, merged Ensembl/Havana; blue, processed transcript; gray, pseudogene). Within a gene, exons are shown as solid blocks, and introns as chevron-shaped connecting lines. To open the full interactive genome browser from the Gene tab, click the "Location" tab at the top.

The Location tab shows the locus in question in its genomic context, in an interactive and customizable view. Within the Location tab there are three main panels: "Chromosome", and under "Region in detail", the 1 MB Region and the Main Panel. The Chromosome panel gives the chromosomal coordinates of the ROI, and a diagram of the chromosome with its banding pattern, with the ROI bounded by a red box. Below this, the 1 MB Region allows the user to see the genes in the vicinity of the current ROI, also bounded by a red box. Genes are color coded by type (e.g., protein coding, or RNA gene), and their orientations indicated next to their names by the > and < symbols. The ROI can be adjusted by either scrolling left and right, or by selecting a specific region by dragging a box, then clicking "Jump to region".

The main panel is the full-featured genome browser, showing tracks corresponding to the ROI. The region shown can be altered by drag-based scrolling left and right, zooming in and out, and can be selected by dragging a box. Clicking a location unoccupied by a track opens a pop-up box that states the location, and allows the user to zoom in or out by a factor of 2, 5, or 10, and to center the view upon that point.

The tracks in the Main Panel are fully user-customizable in terms of which ones are present, their order, and the level of detail visible within each one. By default, the genomic ruler appears at the top, then below that the chromosome bands, forward-strand features, the contig (blue), then reverse-strand features. Towards the bottom, the Regulatory Features track includes promoters, enhancers, and transcription factor binding sites. Underneath is a color legend corresponding to properties of the tracks above.

Each track has to its left a title and a colored handle. Hovering the mouse pointer over the title opens a pop-up box providing information, options for display (expanded or collapsed, with or without labels), and the possibility to remove it. Dragging tracks by their handle allows their vertical order to be changed. Clicking on a track element opens a pop-up window that shows information about that component, and links to the original database records where appropriate.

To add further tracks to the display, click the "Configure this page" link to the left of the browser. In the pop-up window, the active tracks are shown in the main panel, and a large number of possible tracks are shown to the left, under the headings Sequence and assembly, Genes and transcripts, Variation, Somatic mutations, Regulation (which includes numerous sets of ENCODE data), mRNA and protein alignments, ncRNA, Oligo probes, and Information and decorations. Clicking on these allows them to be selected for inclusion in the browser display. For each track on offer, a description is provided as well as options for display. Clicking the tick symbol in the top right closes this window and refreshes the browser display.

4.2.3 Exporting Outputs from Ensembl

Once the search has been performed and relevant information visualized in the browser, the user may wish to export the results, either as an image of the browser output, or as a data file containing the sequence and other features.

To export the current browser output as a publication-quality graphic file, click on the "Export this image" icon within the set of white icons along the left of the blue header bar of the browser panel. A pop-up box allows the graphical output to be saved in high-quality raster (PNG) or vector (PDF, SVG) formats.

To export the sequence and other features in the current view, click the "Export data" link to the left of the browser. In the dialog box, the "output" pull-down menu allows selection of the output format type (BED, FASTA, CSV, TAB, EMBL, GenBank). Below this, options are shown for the features to be included. Clicking "Next >" leads to a page that shows the export format options; clicking on one of these opens a browser window or tab containing the exported data, which can be copied to the clipboard or saved as a file. For a plain-text output of the relevant sequence, select "FASTA sequence" and "Text".

4.3 Vega

The Vertebrate Genome Annotation (**Vega**) database [42] complements Ensembl by providing expert manual annotation of genes (protein-coding, noncoding RNA, processed transcripts, and pseudogenes), currently for five organisms: human, mouse, rat, zebrafish, and pig, with partial annotation for five other vertebrates. The genome annotations in the Vega database are updated by a dedicated team via the HAVANA procedure.

The Vega search and browser interface works almost exactly like Ensembl, and so a separate description of its functionality is not necessary. As with Ensembl, searching the Vega genome database using a molecule's unique ID can be initiated from the search box at the top-right of the home page. Searches can also be launched by incorporating query terms into URLs (*see* **Note 3**). Such searches identify the corresponding Vega gene IDs, which have the prefix OTT, and the corresponding Gene and Location (genome browser) pages.

Manually curated gene information from Vega/HAVANA played an enormously valuable role in enhancing the quality and credibility of the GENCODE gene set; as this information is available from the main Ensembl website, genome searching from the Vega site would only be necessary if a user were interested exclusively in manually annotated genes.

4.4 University of California Santa Cruz (UCSC) Genome Browser

The UCSC Genome Browser [51] is one of the most popular and flexible Web browser-based tools for searching, visualizing, and accessing genomic data. The browser allows access to the genomes of currently over 90 species, the vast majority being metazoans. Access is provided to locus-specific information of many types, including genes, genomic landmarks, gene expression and regulation (including ENCODE project data), epigenetic modifications, comparative genomics, and genomic variations.

A unique feature of the UCSC Genome Browser is the "UCSC genes" gene set [52]. This moderately conservative set of gene predictions is based on RefSeq, GenBank, CCDS, Rfam [53], and tRNAs [54], and comprises both protein-coding and noncoding RNA genes. However in a move towards standardizing on a common gene set within the bioinformatics community, as of July 2015 the UCSC Genome Browser has adopted GENCODE

Comprehensive as the default gene reference for the human genome assembly.

4.4.1 Genome Searching Using the UCSC Genome Browser

Genomes can be searched within the UCSC Genome Browser using gene names, plus many kinds of ID, including those from RefSeq, GenBank, UniProt, and Ensembl—although not all of these ID types are recognized for all organisms. To search for genomic information for a gene or molecule, the starting point is the "Genomes" page. From the pull-down menus "group" and "genome", choose the appropriate entries for the species of interest. Under "assembly", choose the latest version, unless an older one is specifically required. Then under "search term" enter the gene name or molecule ID, and click "submit".

The results page that follows lists entries from nucleic acid databases that correspond to genes showing a match to the search term; each of these is a link to the corresponding genome browser page. Each results entry listed has chromosomal coordinates and an ID, and often a brief title, which may include the gene name. Matches are categorized according to the type of gene set, such as UCSC, RefSeq, or GENCODE, as appropriate. Matches to genomes of other organisms also sometimes appear, listed separately. Whereas searches using unique IDs usually produce one hit per gene set, searches using gene names may also generate several "off-target" hits due to non-exact text matching (e.g., "CDK1" also matches "CDK10"), thus the user must exercise caution when choosing a match from the results list. One useful approach is that when searching using gene names, these may be automatically recognized before "submit" is clicked, in the form of a drop-down menu of gene matches; clicking on the appropriate gene match takes the user to the correct genome browser page.

The UCSC Genome Browser can also be searched by incorporating query terms into a URL (*see* **Note 4**). A more sophisticated method of searching the UCSC Genome Browser for multiple IDs is provided by the site's Table Browser tool; an example of this is provided in Subheading 9.3.

4.4.2 Genome Browser Navigation and Customization

The main genome browser page features a menu bar along the top, then navigation controls allowing the user to move the ROI left or right; zoom in by a factor of 1.5, 3, 10, or right in to the base-pair level; and zoom out by a factor or 1.5, 3, 10 and 100. Below this appear the coordinates of the current ROI and its length, plus an additional search box. Below this, a chromosomal ideogram with banding patterns is shown, with a red line or box indicating the location of the current ROI.

Below this is the main browser panel, showing the tracks. The uppermost track, "Base Position", shows by default a genome ruler and a scale bar. Navigation left and right can be performed by dragging any part of the panel except the Base Position track. To zoom in to a specific desired region, drag a box over the Base

Position track, then in the "Drag-and-select" pop-up box click "Zoom In".

The series of tracks presented in parallel to the Base Position track displays a variety of pertinent genomic features within the ROI. The tracks shown as default depend on the species; for the human genome these include contig clones, GENCODE, RefSeq, and GenBank transcripts, ESTs, ENCODE project data (histone H3K27 acetylation, DNase I hypersensitivity, transcription factor ChIP), alignment and conservation with other vertebrate genomes, common SNPs, and interspersed repeat sequences. For gene tracks, exons are shown as solid blocks, with introns as connecting lines with a repeated arrow pattern indicating the gene's directionality.

Each track has a short name to its left, a brief explanation text above, and a handle (small gray box) to its far left. Right-clicking the handle opens a pop-up menu showing viewing density options: hide, dense, squish, pack and full. In "dense", features are collapsed onto one line, "squish" and "pack" show features closely packed together, "full" shows each feature on a separate line, and "hide" removes the track altogether. Where features are collapsed in "dense" viewing mode, a single click within the track converts it to the "pack" format. Dragging tracks vertically by their handles enables their order to be changed, whereas a single click on a handle opens a window giving a full description of the track and the origin of the data within. A single click on a feature within the track opens a window showing full information about the relevant data.

Below the main panel are listed a large number of additional sources of genome annotations available as supplementary tracks, by which the user can customize the browser. Annotations are categorized as follows: Mapping and Sequencing, Genes and Gene Predictions, including protein features and domains from Pfam [55], and UniProt [50], Phenotype and Literature, mRNA and EST, Expression, Regulation, Comparative Genomics, hominids, Variations, and Repeats. Each feature is listed together with its viewing status: most are "hide" by default, and can be altered from the pull-down menu.

4.4.3 Exporting Outputs from the UCSC Genome Browser

To export the current graphical output of the genome browser as a publication-quality vector file, go to the top menu, select "View" then "PDF/PS". This opens a page called "PDF Output", where the current browser graphic or chromosome ideogram can be downloaded in PDF or EPS formats.

To obtain genomic sequence corresponding to the current browser view, in the top menu, click "View" then "DNA". The "Get DNA in Window" page opens, with the relevant genomic coordinates already filled in under "Position". Under "Sequence Retrieval Region Options:", select: "Add 1 extra bases upstream (5′) and 0 extra downstream (3′)." Under "Sequence Formatting Options:" choose upper or lower case as desired. Click "get DNA".

A new page appears containing the relevant DNA sequence in FASTA format.

4.5 NCBI Genome, Map Viewer and Gene

The National Center for Biotechnology Information (NCBI) has several tools for the retrieval and navigation of genomic and gene-based data [45, 56].

The **NCBI Genome** [45, 56, 57] home page provides links to NCBI genome-related resources. The search box can be used to query by species name, leading to a page summarizing genomic information for that organism. However this is not the place to search for genomic information based on gene names or codes, or molecule IDs.

Map Viewer [45, 56, 57] is the NCBI's main Web-based genome browser tool. The home page is a search interface listing sequenced organisms by taxonomic classification.

4.5.1 Searching Map Viewer

Performing a genomic database search from the Map Viewer home page proceeds by first selecting an organism (after "Search:"), entering a query term (after "for:"), then clicking "Go". The range of ID types recognized is somewhat narrower than those of Ensembl and the UCSC Genome Browser. Gene names and NCBI IDs from RefSeq, GenBank, and UniGene are recognized, but not NCBI GI numbers nor external IDs from Ensembl or UniProt.

Following a database search, a results page is produced that shows a karyotype of the organism in question. For a successful search that matches the query to a single genomic locus, its location on the corresponding chromosome is indicated by a red line. When a search matches several loci, the positions of each are indicated on their respective chromosomes. Where the search term contains degeneracy (for example "CDK" matches all members of the cyclin-dependent kinase family), this provides a useful overview of the distribution of matches across the whole genome. Below the karyotype is a table listing all the matching nucleotide entries, listed under the headings Chromosome, Assembly, Match, Map element, Type and Maps. For many genes there may be matches to dozens of transcript entries in the NCBI nucleotide database, and so finding whole gene hits among these may require some effort. Fortunately, a Quick Filter option to the right of the table allows the user to narrow down the hits to only Gene or RefSeq entries. A more extensive range of filters and search options are available by clicking the "Advanced Search" button. Here, the search can be narrowed down to the chromosome, assembly, type of mapped object, and map name. The query is entered in the "Search for" box, and the search initiated by clicking "Find". Genomic database searching using Map Viewer can also be performed by incorporating query terms into a URL (*see* **Note 5**).

From the results page, clicking an ID link under the "Map element" header opens the Map Viewer genome browser to show the relevant ROI and features within.

4.5.2 Genome Browser Navigation and Customization

The NCBI Map Viewer browser differs from browsers described previously in that tracks appear vertically rather than horizontally, with a vertical ideogram of the relevant chromosome to the far left. Each track has a short title above it; hovering the mouse pointer over this opens a pop-up box providing further information. In Map Viewer one track is designated as the "Master Map"; this appears as the rightmost track, with its title in red. Next to each track title are two small buttons: a right-facing arrow and an "X". Clicking the former makes that track the Master Map; clicking the latter removes the track from the display. Each element within the vertical track is connected by a gray line to a horizontal textual annotation. For the Master Map track these annotations appear larger, and where appropriate further information is provided in a table to the right of the graphical panel.

Navigating the Map Viewer by zooming and scrolling is possible, but not with the same instantaneous interactivity as Ensembl, Vega or the UCSC Genome Browser. Zooming in and out can be achieved by clicking on a small panel to the left of the main graphic. Clicking on a track opens a pop-up window with options for zooming and re-scaling to 10 Mb, 1 Mb, 100 kb, and 10 kb. Scrolling can be performed by clicking on the blue up or down arrowheads at either end of the Master Map track.

Where genes are shown in a track, the exons are shown as solid blocks, and introns as thin lines. The track called "Genes_seq" presents a flattened view of all exons that can be spliced together in various ways. It is advantageous to have Genes_seq as the Master Map, as the resulting table to the right of the graphical output contains a wealth of useful information, including gene descriptions, plus links to numerous external databases of gene and protein function.

Clicking the "Maps & Options" button opens a window that gives the user control over the tracks displayed, and their order. The window has two sections: the Available Tracks section on the left shows a variety of additional data types, including genes from different sources, transcripts, clones, STSs, CpG islands and contigs. Clicking the "+" next to a feature adds it to the Tracks Displayed panel on the right, which lists currently displayed tracks and allows their order to be changed. Here, highlighting an "R" icon next to a track adds a genomic ruler to its left in the genome browser. Clicking "OK" closes the Maps & Options window and updates the graphical output.

4.5.3 Export from Map Viewer

Options for exporting Map Viewer's graphical output appear fairly limited. One could take the approach of clicking the right mouse button over the browser graphic, choosing Save Image As. . ., and saving in the PNG raster format, however this does not capture annotations to the Master Map track, nor any of the table to its right. So the leading option at present is to perform a screenshot or selective screen grab. In either case, smaller textual elements and finer lines are of low resolution, appearing pixilated.

In contrast, Map Viewer's options for exporting genomic and associated data shown in the browser are very comprehensive. Below the graphical panel is a section called Summary of Maps, which lists each of the tracks (here called maps), and summarizes their contents. Next to each map summary are two links. Clicking the Table View link opens a page that tabulates all of the features for each track, with genomic coordinates and other relevant data; this can be downloaded as a tabulated text file. Clicking the Download/Sequence/Evidence link opens a window that allows the corresponding sequence to be saved in GenBank or FASTA formats.

4.5.4 NCBI Gene

Whereas the **NCBI Gene** database [58] focuses on genes rather than genomic regions, its search and genome browsing functionality complements and in some cases supersedes that of Map Viewer. NCBI Gene can be searched using all standard NCBI IDs, including nucleotide and protein GI numbers, which should be prefaced "GI:". Some external IDs are also recognized, such as Ensembl genes and UniProt proteins. URL-based searching is also possible (*see* **Note 6**).

The NCBI Gene results page lists possible gene hits, tabulated as Name/Gene ID, Description, Location, and Aliases. Clicking a gene name link opens the relevant gene page, which (in "Full Report" format) hosts a wealth of information about the gene, in 12 sections. The "Genomic context" section summarizes information about the locus, with a graphical overview of the gene's relationship to neighboring genes. The "Genomic regions, transcripts, and products" section contains a horizontally oriented interactive genome browser that in some ways surpasses the usability of the Map Viewer (to which there is a link to the right); for example this browser supports drag-based scrolling and ROI selection, and export of the graphic in high-quality PDF format. Thus, in cases where a certain ID is not recognized by Map Viewer, then a search at NCBI Gene, and the use of its horizontal browser or the link to Map Viewer is one solution to access genomic locus information.

4.6 Viral Genomes

The NCBI Genomes database contains genomic information from over 4500 viruses, bacteriophages, viroids, archaeal phages, and virophages, with a dedicated search portal, **NCBI Viral Genomes**

[59]. The home page allows for a NCBI Genomes search, which can be initiated using a virus name or RefSeq ID as a query. Searches can also be incorporated into a URL, thus:

http://www.ncbi.nlm.nih.gov/genome/?term=**PhiX174**

http://www.ncbi.nlm.nih.gov/genome/?term=**NC_001422**

The results page shows a list of virus hits; clicking on one shows a summary of genomic information for that virus, including an overview of the whole genome. Clicking on a link under either the RefSeq or INSDC headings opens a page showing collected information about the genome, including translated open reading frames, and the whole genome sequence. Alternatively, clicking on the "Gene" link under the "Related Information" heading to the right of the page lists the genes, from the NCBI Gene database, that are encoded by the viral genome. Clicking on a gene name link opens the NCBI Gene page; under the "Genomic regions, transcripts, and products" section, the whole viral genome can be inspected in the interactive genome browser. From here, the sequence of the ROI can be exported in FASTA or GenBank formats, or the graphical output exported as a PDF file.

4.7 Stand-Alone Genome Browsers

The genome browsers described above are Web-based, and allow both retrieval and display of genomic data. Stand-alone genome browser applications, not being restricted to Web-browser functionality and speed, are generally more flexible and responsive. However a time-consuming step is the import to local storage of a whole reference genome's worth of data, which can take more than a gigabyte of memory. The strong point of these tools is, rather than genomic database searching, their advanced abilities to display and integrate multiple datasets, including the user's own quantitative experimental data, relative to a reference genome. Several such browsers are available (and in some cases functionally overlap with software described in Subheading 11); three popular ones are described here.

The **Integrated Genome Browser** (IGB, http://bioviz.org/igb/) [60] is a desktop application that allows the visualization of genes, genomic features and experimental data aligned to reference genomes. Genomic sequences can be loaded in to the program from a remote server, and a certain amount of genomic querying is also possible. IGB provides an interactive environment for the exploration of genomic data, with a powerful and user-friendly GUI. It is very flexible regarding options for data upload, and can handle multiple sets of quantitative data, which can be visualized in parallel and customized in terms of their visual aspect, to generate high-quality figures.

The **Integrative Genomics Viewer** (IGV, https://www.broadinstitute.org/igv/) [61] is a desktop application whose primary function is the visualization of large-scale genomic datasets to allow their exploration and interpretation. Experimental data, as well as known features such as genes, are shown aligned to a reference genome, a variety of which are held on the IGV server. Its strengths are its flexibility, in accepting a large number of different data file types, in particular NGS data, and its interactive zooming, allowing the display of experimental data across a wide range of scales, from the whole chromosome to the base-pair resolution, which it achieves by a process it calls "data tiling".

The **Savant Browser** (http://www.savantbrowser.com) [62] was designed primarily for the analysis of NGS data. Its aims are to provide a single platform to allow automated procedures such as read mapping, and visualization (genome browsing) to be used together, so that users can seamlessly inspect and perform computation on their data, iteratively refining their analyses. The software features a colorful and interactive GUI, and several modes of visual display, for example graphs showing the frequency of genetic variants across datasets. The core program is enhanced by plugins created by the user community that allow additional analyses and visualization modes for specific experimental purposes.

5 Genomic Database Searching with RNA Identifiers

The vital roles of mRNAs, tRNAs and rRNAs in expressing the genetic material have been known for about half a century, and are a cornerstone of molecular biology. In the last two decades the discovery and characterization of several classes of noncoding RNA (ncRNA) in a wide range of organisms has changed our view of how gene expression is regulated, and indeed has necessitated an update to the very definition of a gene [35].

A large variety of nomenclature has arisen for the different classes of ncRNAs, and a concerted effort is underway to standardize and assign gene names to the corresponding loci [63]. For each class of RNA, "expert" databases have been developed to allow access to sequence and functional data for their respective molecules [64]. Resources complementing these databases include **RNAcentral** [65], a meta-database that collates and compares RNA species from expert databases, and **Rfam** [53], which classifies RNA molecules into families. However these two resources emphasize the commonality and evolutionary similarity between RNA molecules, rather than providing efficient means for researchers to "drill down" to the genomic origins of an RNA molecule of interest.

In this section, approaches for genomic database searching based on four major classes of RNA are presented. URL-based searching is possible for all the databases described below; *see* **Note 7** for examples.

5.1 Messenger RNAs (mRNAs)

For several decades, sequences of protein-coding mRNAs have been gathered in public nucleotide databases. The repositories GenBank, ENA, and DDBJ collaborate and exchange sequence data within the International Nucleotide Sequence Database Collaboration (**INSDC**), forming a collective resource commonly referred to as DDBJ/EMBL/GenBank [66]. In addition, mRNA entries are present in the RefSeq, Ensembl and Vega databases. Database entries may comprise mRNAs whose existence is confirmed experimentally, as well as transcripts predicted by computational analysis of genome sequences. For most genes several mRNA species (database entries) correspond to a gene or genomic locus, due to the presence of alternative transcriptional start and stop sites, and the effects of alternative splicing.

Fortunately, due to data sharing and integration efforts, mRNA IDs from any of these databases mentioned above are recognized by the four main genome browsers (although Map Viewer seems not to recognize Vega IDs). Therefore genomic database searching from standard mRNA IDs can be performed using the user's favorite genome browser, as described in the previous sections.

5.2 MicroRNAs (miRNAs)

miRNAs are 19–25 nucleotide single-stranded ncRNAs that bind to 3′-untranslated regions (3′-UTRs) of target mRNAs, destabilizing or inhibiting their translation [67]. miRNA genes encode primary transcripts known as pri-miRNAs, which are processed into short stem-loop structures called pre-miRNAs, then modified into mature single-stranded miRNAs.

The nomenclature for miRNAs is complicated by the presence of colloquial and official names and IDs for genes and miRNAs at different stages of processing, as well as database accession codes. As many of these terms are unrecognized or misinterpreted by genome browser databases, the most effective means of accessing reliable information is from the expert database **miRBase** [68], which recognizes all miRNA IDs, and documents the relationships between them.

To search miRBase from the home page, paste an ID into the search box at the top right, and click "Search". The Search Results page lists the number of hits for each class of miRNA. Clicking on an entry under the Accession or ID columns opens a page of information. If the query is a mature miRNA, then the user should click the entry following the "Stem-loop" heading to open the corresponding page. In the Stem Loop Sequence page, under the "Genome context" heading, a link is provided showing the genomic coordinates; clicking this opens an Ensembl genome browser window at the locus corresponding to the miRNA of interest.

5.3 Piwi-Interacting RNAs (piRNAs)

piRNAs are a class of small ncRNAs of length range 26–32 bases that bind to Piwi and other proteins of the Argonaute family. Roles for piRNAs have been found in transposon silencing, and epigenetic and post-transcriptional regulation of gene expression [69, 70]. piRNAs are transcribed from cluster regions on chromosomes, trimming events then generating a population of RNAs of heterogeneous length.

A variety of nomenclature exists for piRNAs. An individual piRNA molecule may have an abbreviated name (e.g., piR-30025), and a GenBank accession code (e.g., DQ569913). For human piRNA clusters, official gene symbols take the form *PIRC#* [63], but these terms appear not to be shared with other organisms. There are three expert piRNA databases: **piRNABank** database [71], **piRBase** [72], and **piRNAQuest** [73], each with its own nomenclature.

Thus the approach to perform a genomic database search for a piRNA depends on the type and origin of the ID, which the researcher should check carefully. A complication to be aware of is that a piRNA sequence may match to more than one genomic locus. The piRNAs terms mentioned above are not currently recognized by Ensembl, Vega or NCBI's Map Viewer. The UCSC Genome Browser recognizes abbreviated names and GenBank IDs, but not IDs from the expert databases.

Where the IDs are of the form generated by the expert databases, genomic information is available by searching these resources. The following examples are for human piRNAs. IDs from piRNABank take the form hsa_piR_000001. Searching for this term, along with species and assembly, leads to a record page. Under "Genomic position" the chromosomal coordinates are provided, as a Web link to the locus in Ensembl. piRBase IDs take the form piR-hsa-237. Here, the record page contains a "Location" section, with chromosomal coordinates as a Web link. Clicking on this opens an embedded version of the UCSC Genome Browser displaying the relevant location. IDs from piRNAQuest take the form hsa_piRNA_6754. A piRNA record page shows a list of matching loci for that piRNA, each one being a Web link to open a page showing the relevant sequence within an embedded JBrowse Genome Browser [74].

5.4 Long Noncoding RNAs (lncRNAs)

lncRNAs are a class of ncRNA ranging in length from 200 bases to more than 15 kb. lncRNAs may be transcribed antisense to protein coding genes, within introns, overlapping genes, or outside of genes, and are believed to have roles in the regulation of gene expression, especially during development [75, 76]. In terms of standardized gene nomenclature, lncRNA gene symbols consist of either function-based abbreviations, existing protein-coding gene symbols with the suffices *-AS*, *-IT*, or *-OT*, or symbols starting *LINC* for Long Intergenic Non-protein Coding RNAs [77].

The efforts of the GENCODE consortium in including lncRNAs in their comprehensive gene set means that these entries are present and searchable from Ensembl and the UCSC Genome

Browser. Several specialist lncRNAs databases exist [78], the resource of reference for generating gene symbols being **lncRNAdb** [79], which in addition to genomic locus data provides information about function and expression of individual lncRNAs. Searching lncRNAdb can be performed from the search box on the home page, using any term as a query, plus optionally the species. Despite gene name standardization efforts, lncRNAdb does not at present include or recognize all official lncRNA gene symbols. The record page for an individual lncRNA lists many characteristics of the lncRNA. Under the "Species" heading, next to the species of interest genomic coordinates are shown; this is a Web link to the corresponding locus in the UCSC Genome Browser.

6 Genome Searching Using Karyotype Bands

One of the first methods of characterizing and subdividing chromosomes was the use of karyotypes and banding patterns, and this remains a cornerstone of cytogenetics today [80]. Chromosomal band patterns define a standard nomenclature for chromosomal loci, for example the human gene *TP53* (coding for the p53 tumor suppressor), is found at the locus 17p13.1, representing chromosome 17, short arm, region 1, band 3, sub-band 1.

Genomic database searches can be performed using such locus codes, enabling genes in the region to be identified, and the complete sequence to be accessed. This is probably easiest performed using the UCSC Genome Browser. Taking the example of human *TP53*, with the genome selected as "human", the term "17p13.1" can be entered as the search term, and click "submit". The equivalent operation can be performed directly from a URL-based search, thus:

http://genome.ucsc.edu/cgi-bin/hgTracks?db=hg18&position =**17p13.1**

In the resulting output, the band 17p13.1 is highlighted on the chromosomal ideogram by a red box, and the main browser panel below shows genes and other features within that region.

7 Genome Searching Using Chromosomal Coordinates

A more precise specification of a genomic region is provided by chromosomal coordinates. These take the form of the chromosome number followed by a colon, then the start and end positions on that chromosome separated by a hyphen, e.g., 17:7661779-7687550. Essential supplementary information to coordinates includes the relevant species and genome build or version number (e.g., GRCh38 or mm10). When a sequence output is desired, an additional parameter is the sense of the strand, usually denoted by another colon then "+" or "−".

Chromosomal coordinates can be used directly as the query term to search most genome browsers, to identify the relevant region of the genome, and features within. The species and genome version should first be chosen (the latest version usually being selected as default), then the coordinates for example "17:7661779-7687550" can be used as the query term. One exception is NCBI's Map Viewer, which does not recognize queries of this kind from its search box.

For all four major Web-based genome browsers, a locus of interest can be chosen by configuring the URL. Examples are shown below:

Ensembl: http://www.ensembl.org/**Homo_sapiens**/Location/View?r=**17:7661779-7687550**

Vega: http://vega.sanger.ac.uk/**Homo_sapiens**/Location/View?r=**17:7661779-7687550**

UCSC Genome Browser: http://genome.ucsc.edu/cgi-bin/hgTracks?db=**hg38**&position=chr**17:7661779-7687550**

NCBI Map Viewer: http://www.ncbi.nlm.nih.gov/projects/mapview/maps.cgi?TAXID=**9606**&CHR=**17**&BEG=**7661779**&END=**7687550**

A method for retrieving multiple sequences from a list of chromosomal coordinates is provided in Subheading 9.3.

8 Sequence, Motif and Matrix-Based Genome Searching

A researcher may wish to search a genome of interest using one or more DNA sequences, for example from clones obtained experimentally during a genetic screen. Researchers may also be interested to determine the locations and distribution of short sequence elements, termed motifs; this can be performed using pattern-matching and matrix methods. Approaches for performing these types of search are described, with relevant links provided in Table 2.

8.1 Sequence-Based Searches

Several algorithms exist to search a database for entries that exhibit homology to a given sequence of interest (the "query sequence"); the best-known is probably **BLAST** (Basic Local Alignment Search Tool) [81]. This can be applied to whole genomic database searching, however the alternative algorithm **BLAT** (BLAST-Like Alignment Tool) [82] is faster, and has been adopted as the default

Table 2
Software utilities for searching genomes using sequences, motifs and matrices

Utility	Website	References
Sequence-based		
Ensembl BLAST/BLAT:	http://www.ensembl.org/Multi/Tools/Blast	[40]
Vega BLAST/BLAT:	http://vega.sanger.ac.uk/Multi/Tools/Blast	[42]
NCBI Genome BLAST:	http://blast.ncbi.nlm.nih.gov/Blast.cgi	[45, 56]
UCSC BLAT:	https://genome.ucsc.edu/cgi-bin/hgBlat	[51]
Motif-based		
MEME FIMO:	http://meme-suite.org/tools/fimo	[93]
RSAT DNA Pattern:	http://rsat.sb-roscoff.fr/genome-scale-dna-pattern_form.cgi	[94]
EMBOSS DREG:	http://emboss.bioinformatics.nl/cgi-bin/emboss/dreg	[95]
Matrix-based		
MATCH:	http://www.gene-regulation.com/pub/programs.html#match	[97]
TRANSFAC:	http://www.gene-regulation.com/pub/databases.html	[98]
ModuleMaster:	http://www.ra.cs.uni-tuebingen.de/software/ModuleMaster/	[99]
RSAT Matrix Scan:	http://rsat.ulb.ac.be/rsat/matrix-scan-quick_form.cgi	[94, 100]

method by Ensembl, Vega and the UCSC Genome Browser. The query length ranges for these two algorithms are similar: for BLAST it is 30 nucleotides and upwards, whereas BLAT accepts queries of 20–25,000 nucleotides. The following paragraphs describe methods employing these algorithms for searching genomic databases using a query sequence, which it is assumed the user has in plain-text format, copied to the computer's clipboard.

Ensembl's BLAST/BLAT tool has the advantage that multiple DNA or protein sequences can be searched against the genomes of several organisms from a single operation. From the home page, the "BLAST/BLAT" link at the top opens a search page. In the "Sequence data" window, paste in the query sequence(s). Up to 30 sequence queries can be entered here, in a multi-sequence FASTA format (where each sequence has a distinct header line beginning with >), and will be searched as separate jobs with the same parameters. These include the type of sequence (DNA or protein), the relevant species (several can be chosen), the type of database (choose DNA for genome searches), the search tool (BLAT, BLASTN, or TBLASTX for DNA queries; TBLASTN for protein), and sensitivity. Under "Configuration options", other parameters can be adjusted, including maximum number of hits to report, maximum E-value, gap penalties, and the option of filtering low complexity regions. Clicking "Run" starts the search; a table appears listing the status of the jobs submitted, each labeled "Running" if underway, or "Done: n hits found" if completed. Clicking the "View Results" link next to a finished job opens a results page that tabulates the hits: their genomic locations, overlapping genes, and alignment and percentage identity scores, with links to view the alignment. Each Genomic Location entry is a link that opens the genome browser, showing the query matched to the relevant genome. An alternative means of accessing these hits is provided by a karyotype diagram below, which shows matching regions highlighted by red boxes and arrows; clicking an arrow opens a pop-up box containing information and links about the matches.

Within the **UCSC Genome Browser**, searching genomic databases with sequences using BLAT is accessible via the menu bar at the top of the page, under >Tools >Blat. In the BLAT Search Genome page, choose the species and genomc assembly, and paste in the query sequence (DNA or protein). Multiple sequence queries can be entered in a multi-sequence FASTA format. The search is initiated by clicking "submit". When the query is a single sequence that the user is confident will map to a unique genomic locus, one option is the "I'm feeling lucky" button, which bypasses the results page, going straight to a browser showing the best match from the BLAT search. Following a standard submission, the BLAT Search Results page lists the set of matches found for each query, showing the genomic coordinates of the match and the percentage identity. In addition there are two hyperlinks: "details" opens a page showing the alignment of bases for that match. The "browser" link opens the genome browser, where the query sequence appears as a blue track alongside the genome, annotated as "Your Sequence from Blat Search" and labeled on the left with the FASTA heading or "YourSeq".

Searching the **NCBI**'s collection of genomic databases using DNA or protein sequence-based queries is performed from the BLAST page (http://blast.ncbi.nlm.nih.gov/Blast.cgi). More search parameters are available here than on the Ensembl or UCSC sites, but BLAT is not offered. Under the "BLAST Assembled Genomes" heading, first choose the organism by entering a species name or taxonomic ID, or by clicking a link to the right. In the relevant species page, choose the type of BLAST search by selecting a tab across the top (blastn, blastp, blastx, tblastn, or tblastx; for a standard nucleotide-based search, choose blastn). Under "Enter Query Sequence", paste in the sequence(s) to be searched. Multiple sequence queries can be entered in a multi-sequence FASTA format. The database selection can be left in its default state ("Genome..."). Advanced search options are available by clicking "Algorithm parameters", but for unique-sequence searches this is usually not necessary. Clicking "BLAST" runs the search, and upon completion a results page shows details of query-genome matches identified. Where more than one query sequence was submitted, a pull-down menu after "Results for:" allows selection of results for each query separately. The "Alignment" section shows alignments between the query sequence and subject (genome sequence), in order of decreasing score. To the right of each alignment, under "Related Information" is a Web link to Map Viewer, allowing the user to view the corresponding locus in that genome browser.

8.2 Motif-Based Searches

The specificity of many factors that interact with and modulate the genome is owed to the recognition of short stretches of DNA sequence, termed motifs. The specific recognition of DNA at motifs drives and regulates many fundamental nuclear processes, notably gene expression [83], and has been studied mechanistically in much detail [84, 85]. Sequence motifs may also be used to predict the formation of unusual DNA secondary structures [86].

A motif may comprise a single exact sequence, for example the restriction endonuclease *Eco*RI cuts at the palendromic motif G↓AATTC, where ↓ indicates the cleavage site [87]. Alternatively a motif may contain some degeneracy (variability in the bases acceptable at certain positions); such is the case for the transcription factor p53 [88]. Some motifs are even more flexible, involving repeated segments and variable-length gaps, as is the case for those predicting G-quadruplex structures, found in telomeric regions and 5′ to the majority of metazoan origins of replication [89–91].

So how can motifs, including those containing degeneracy and variable gaps, be expressed and used for genomic database searching? One means is the **Regular Expression** (RegEx), a standardized format for representing text patterns, popular in the computing world (for more information, *see* http://www.regular-expressions.info). When applied to DNA sequences, the four bases appear simply as the letters A, C, G and T, a set of letters within square brackets represents the possible bases at a certain position, and numbers in curly brackets refer to the number of times that part of the pattern appears. Taking G-quadruplexes as an example, a simplified motif for sequences likely to form these structures can be represented using standard biochemical nomenclature like this (where N represents any of the four standard DNA bases A, C, G or T):

$G_3\text{-}N_{1\text{-}7}\text{-}G_3\text{-}N_{1\text{-}7}\text{-}G_3\text{-}N_{1\text{-}7}\text{-}G_3$

Using RegEx nomenclature, this same pattern can be represented thus:

G{3}[ACGT]{1,7}G{3}[ACGT]{1,7}G{3}[ACGT]{1,7}G{3}

where [ACGT] represents "any standard DNA base", {3} means "exactly three times", and {1,7} means "between one and seven times".

Searching genomic databases using RegEx-based motifs requires software that accepts this type of query format. Although none of the main genome browsers can do this, other tools are available that can be used to perform such operations. In each case, the output of these programs is a table that includes the start and end coordinates of each match, plus the matching sequence. These genomic coordinates can be used to configure custom genome browser URLs to open and inspect the loci of interest.

The **MEME** suite of motif-based sequence analysis tools [92] includes a program called **FIMO** ("Find Individual Motif Occurrences") [93] that can scan genomic databases for motifs in RegEx format. Under "Input the motifs", choose "Type in motifs", and under "Databases" select "Ensembl Genomes". One limitation of FIMO is that this motif has to be of a fixed length; no curly brackets are allowed.

The Regulatory Sequence Analysis Tools (**RSAT**) suite provides several DNA sequence analysis utilities [94]. Within this, the **Genome-scale DNA pattern** tool searches for matches between a motif and a genomic sequence of interest. Motifs can be in RegEx format, with "N" (for any base) being an optional extra character, and curly brackets are allowed.

An additional program that can search sequences using variable-length RegEx-based motifs is the **DREG** routine from the European Molecular Biology Open Software Suite (**EMBOSS**) [95]. Here, full sequence files need to be downloaded from genomic databases, then uploaded to the program. As genomic sequences are typically available as single-chromosome files, searching the human genome would require 24 separate analyses.

8.3 Matrix-Based Searches

Genome searching using RegEx-based motifs allows for degeneracy at certain positions, but does not allow information about preferences for certain bases over others at each position to be taken into account. This information can be expressed by a Position Weight Matrix (PWM), also known as a Position-Specific Scoring Matrix (PSSM). Here, typically experimental data are used to generate a table (matrix) that gives a score for the likelihood of occurrence of each base at each position within the motif. A genomic sequence is then searched using the matrix, the quality of each motif match being assessed and given a score [96].

Software routines that can perform PWM-based genome searches include the **MATCH** algorithm [97], which works together with the **TRANSFAC** database of transcription factors, their binding sites, and nucleotide distributions [98]. Multiple MATCH searches can be launched from and managed by the stand-alone program **ModuleMaster** [99]. The **MEME** suite contains programs that can run matrix-based sequence searches, including **FIMO**, plus **GLAM2Scan**, which accepts gapped motifs. The **RSAT** suite contains a program called **Matrix-Scan** [100], which performs a similar functionality.

9 Performing Multiple Genomic Database Searches

Subheadings 4–7 described methods to perform one-at-a-time searches of genomic databases. But often researchers have a list or table of IDs or chromosomal coordinates, and may wish to launch genome browser windows at will from selected entries in the table, or retrieve genomic data for the whole list. Here, three methods are described for performing such searches. The methods assume the user has a personal computer running Microsoft Windows, Apple OS X, or a Linux-based operating system, a standards-compliant Web browser such as Mozilla Firefox or Google Chrome, a text editor, and spreadsheet software such as Microsoft Excel or Calc from the free Apache OpenOffice suite.

9.1 Creating a Hyperlinked ID Table

Where a researcher has a list of gene names or molecule IDs in a table, they may wish to perform genomic database searches with these IDs using one or more genome browsers. A useful approach to this is to create a series of Web links beside each ID, allowing one-click direct searching within the browsers. All major spreadsheet software packages incorporate a HYPERLINK function; the following is a method that employs this to generate links referring to IDs within the spreadsheet, creating a hyperlinked ID table (Fig. 2).

1. *In spreadsheet software*—open the file containing the list of IDs. For RefSeq IDs (e.g., NM_000546.5), the decimal points and version-number suffixes should be removed, as these can cause recognition problems, notably in Ensembl. This can be performed in the spreadsheet software by selecting the relevant IDs, then using Find and Replace to convert ".*" (without the quotes) to "" (nothing). For the purposes of this exercise, the top-most ID is in cell A2. If the column to the right of the IDs is not blank, insert an additional column there.
2. *Obtain the URL*—identify the direct-search URL of the genome browser you wish to use (e.g., from **Notes 1–7**). Copy the URL, and paste it into a text editor.
3. *Re-format the URL to create a HYPERLINK function*—replacing the part of the URL specific to the ID by a cell reference (here, A2), along the lines of this example:

   This URL: **http://www.ensembl.org/human/Location/View?g=**NM_000546

   rearranges to:
   =HYPERLINK("**http://www.ensembl.org/human/Location/View?g=**"&A2,"Ensembl")
   Note:,"Ensembl" (with a comma) in Excel, but;"Ensembl" (with a semicolon) in Calc.

Copy this HYPERLINK function to the clipboard.

4. *Add the HYPERLINKs*—paste the copied HYPERLINK function into cell B2. Press return to activate the hyperlink. Using the handle at the bottom-right corner of cell B2, drag the formula down to the bottom of the list of IDs. The hyperlink functions will be updated so that each refers to the ID in the same row.

9.2 Creating an Annotated ID Table

Also starting with a table of IDs, a researcher may wish to supplement the list with relevant data such as gene names, genomic coordinates, and even brief functional descriptions, to generate an annotated ID table (Fig. 2). One method for performing this, using the BioMart facility of Ensembl [101], is described here.

1. *In spreadsheet software*—open the file containing the list of IDs. For RefSeq IDs, decimal points and version-number suffixes must be removed, as described in Subheading 9.1. Select the IDs, and copy them to the clipboard.
2. *Go to Ensembl BioMart*—in a Web browser, go to http://www.ensembl.org/biomart/.
3. *Inputting the IDs*—from the "CHOOSE DATABASE" pull-down menu choose "Ensembl Genes". From the "CHOOSE DATASET" menu choose the relevant organism and genome, e.g., "Homo sapiens genes (GRCh38.p3)". Click "Filters" on the left, then "[+]" next to "GENE:" to expand the options. Check the box next to "Input external references ID list limit [Max 500 advised]". Choose the relevant type of ID from the pull-down menu; an example is shown for each ID type, for example "RefSeq mRNA ID(s) [e.g., NM_001195597]". In the box under the ID type, paste in the list of IDs.
4. *Selecting output options*—click "Attributes" on the left, then "[+]" next to "GENE:" to expand the options. Ensembl Gene ID and Ensembl Transcript ID are usually selected by default; uncheck "Ensembl Transcript ID" to restrict the output to genes. Select additional output parameters by checking the relevant boxes, for example the following (in this order): Associated Gene Name, Chromosome Name, Gene Start (bp), Gene End (bp), Strand, Band, Description.
5. *Generating and exporting the results table*—click the "Results" button on the top left of the page. An HTML table appears listing the Ensembl genes corresponding to the input IDs, and further attributes. At "Export all results to" choose "File", and "XLS" from the menus. Check the box for "Unique results only", then click "Go". BioMart will export an Excel spreadsheet file for the computer to download that contains the results table with attributes as hyperlinks.

ID Table

Genes, or DNA, RNA or protein molecules.

RefSeq
NM_000546.5
NM_006325.3
NM_006231.3
NM_005030.3
NM_001071775.2
NM_001114121.2
NM_030928.3
NM_001786.4
NM_001287582.1
NM_000038.5

Hyperlinked ID Table

Additional columns containing hyperlinks allow one-click direct searching in genome browsers.

RefSeq	Ensembl	Vega	NCBI	UCSC
NM_000546	Ensembl	Vega	NCBI	UCSC
NM_006325	Ensembl	Vega	NCBI	UCSC
NM_006231	Ensembl	Vega	NCBI	UCSC
NM_005030	Ensembl	Vega	NCBI	UCSC
NM_001071775	Ensembl	Vega	NCBI	UCSC
NM_001114121	Ensembl	Vega	NCBI	UCSC
NM_030928	Ensembl	Vega	NCBI	UCSC
NM_001786	Ensembl	Vega	NCBI	UCSC
NM_001287582	Ensembl	Vega	NCBI	UCSC
NM_000038	Ensembl	Vega	NCBI	UCSC

e!Ensembl

Vega

NCBI Map Viewer

UCSC Genome Browser

Annotated ID Table

Additional columns provide further information about each molecule: gene codes and names, genomic coordinates, and brief descriptions.

RefSeq	Ensembl Gene ID	Gene Name	Chr	Gene Start	Gene End	Strand	Band	Description
NM_000546	ENSG00000141510	TP53	17	7565097	7590856	-1	p13.1	tumor protein p53
NM_006325	ENSG00000132341	RAN	12	131356424	131362223	1	q24.33	RAN, member RAS oncogene family
NM_006231	ENSG00000177084	POLE	12	133200348	133263951	-1	q24.33	polymerase (DNA directed), epsilon
NM_005030	ENSG00000166851	PLK1	16	23688977	23701688	1	p12.2	polo-like kinase 1
NM_001071775	ENSG00000204899	MZT1	13	73282495	73301825	-1	q22.1	mitotic spindle organizing protein 1
NM_001114121	ENSG00000149554	CHEK1	11	125495036	125546150	1	q24.2	checkpoint kinase 1
NM_030928	ENSG00000167513	CDT1	16	88869621	88875666	1	q24.3	chromatin licensing & DNA replication factor 1
NM_001786	ENSG00000170312	CDK1	10	62538089	62554610	1	q21.2	cyclin-dependent kinase 1
NM_001287582	ENSG00000158402	CDC25C	5	137620954	137674044	-1	q31.2	cell division cycle 25C
NM_000038	ENSG00000134982	APC	5	112043195	112181936	1	q22.2	adenomatous polyposis coli

Gene data link to Ensembl gene pages

Genomic coordinates link to Ensembl locus pages

Fig. 2 Two approaches to genomic data retrieval from a list of identifiers. Starting with a spreadsheet table containing DNA, RNA, or protein identifiers (ID Table; shown here are RefSeq transcript IDs), two methods are described for obtaining relevant genomic information. The Hyperlinked ID Table (method in Subheading 9.1) contains multiple Web links, providing one-click access to relevant entries within multiple genome browsers. The Annotated ID Table (method in Subheading 9.2) contains gene names and genomic locus information for each entry, in this case obtained from Ensembl

9.3 Batch Retrieval of Sequences from Multiple Genomic Coordinates

A researcher may have a set of genomic coordinates, for example as an output from a bioinformatic program, and may wish to obtain the corresponding DNA sequences. The following is a method for performing this, using the UCSC Genome Browser's Table Browser tool.

1. *In spreadsheet software*—open the file containing the genomic coordinates. Use the editing capabilities of the spreadsheet software to rearrange the coordinates into the three-column format shown below (these are the three obligatory columns of the Browser Extensible Data (BED) format). Note for this procedure the "chr" prefix to the chromosome number is obligatory, and must be added if not already present.

Chromosome	Start	End
chr1	3444690	3446689
chr1	3669682	3671681
chr1	3669950	3671949
chr1	3670295	3672294

Select the desired three-column region (not including the column headers), and copy this to the clipboard.

2. *In a Web browser*—go to the UCSC Genome Browser Web page: https://genome.ucsc.edu/. In the Genome Browser Gateway page, choose the species from the "group" and "genome" pull-down menus. Choose the genome version from the "assembly" pull-down menu. Click "add custom tracks" (if for the first time) or "manage custom tracks" (on subsequent occasions).
3. *Add Custom Tracks page*—in the window "Paste URLs or data:" paste in the columns copied from the spreadsheet. Click "Submit".
4. *Manage Custom Tracks page*—in the pull-down menu after "view in", choose "Table Browser", then click "go".
5. *Table Browser page*—in the "group" menu select "Custom Tracks", after "region" select "genome", in the "output format" menu select "sequence". Click "get output".
6. *User Track Genomic Sequence page*—under "Sequence Retrieval Region Options", add **1** extra bases upstream (5′) and **0** extra downstream (3′). Choose "All upper case." or "All lower case." (as desired). Click "get sequence". A new page will appear containing the sequences in FASTA format. The sequences can be saved from the Web browser as a plain-text file, or copied and pasted into another application.

10 Genomic Database Searching Using NGS Data

The advent of next-generation sequencing has revolutionized molecular biology in the past decade by providing a quantitative means to sequence-analyze populations of DNA or RNA molecules obtained from a wide variety of experimental conditions [102–104]. Primary data obtained from such analyses are typically in the form of hundreds of millions of sequence "reads", typically of a length range 25–100 bases, although some technologies produce reads that are much longer.

The database searching task is to identify where within a reference genome each of these millions of reads matches, in a relatively short time. An algorithm suitable for mapping NGS data should therefore be: (1) able to map short reads to whole genomes, (2) fast, (3) accurate. For this, algorithms such as BLAST and BLAT are no longer suitable, and several new ones have been developed. Among the best known such algorithms ("mappers") are the Short Oligonucleotide Alignment Program (SOAP) [105], Mapping and Assembly with Qualities (MAQ) [106], Bowtie [107], and the Burrows-Wheeler Alignment tool (BWA) [108]. The latest generation mappers, which offer greatly enhanced speed and precision, include Stampy [109], Bowtie2 [110], BWA-MEM [111], NextGenMap [112], Hive-hexagon [113], and MOSAIK [114].

Factors for making decisions about which mapper to use include: (1) suitability for the read length of the experiment, (2) whether it will accept the relevant data format, and can produce output files suitable for downstream processing, (3) accuracy, (4) ability to accommodate gaps, (5) speed, (6) memory usage. No attempt here is made to evaluate mappers; readers are directed to the following recent reviews [115–118]. Further online information to guide users regarding mappers can be found at the following Internet sites:

SEQanswers [119] http://seqanswers.com/

Wikipedia: http://en.wikipedia.org/wiki/List_of_sequence_alignment_software#Short-Read_Sequence_Alignment

NGS data mappers are typically incorporated into custom routines and run on a high-performance workstation or cluster, but recent developments and trends in NGS analysis include the incorporation of mapping algorithms into user-friendly workflow software such as Galaxy [120, 121] (Subheading 11.2), and taking advantage of the processing power of cloud-based servers [122].

11 Complex Searches of Genomic Databases

For most straightforward genomic database searching tasks, the utilities and approaches described in the previous sections are sufficient. But where researchers wish to combine searches, filter results, perform more sophisticated analyses and create reproducible workflows, freely available specialist applications are available that offer powerful functionality with GUIs. These main approaches and software tools are described only briefly, as for each one extensive documentation and tutorials are readily accessible.

11.1 UCSC Table Browser

As shown in the example in Subheading 9.3, the UCSC Genome Browser contains a Table Browser tool that provides a powerful and flexible means of performing genomic database searches. Multiple genomic queries can be performed, for example starting with a set of gene IDs the corresponding genomic coordinates, transcription start and end sites can rapidly be obtained, as indeed can any of the data stored in the UCSC Genome Browser. Flexible options are offered for formatting the results, for example exons and introns can be listed separately, or appear in upper or lower case as desired. The output can be as FASTA lists, or table formats suitable for other applications and spreadsheets.

Options exist to perform more complex queries. Filters can be applied to searches, for example to restrict the output to matches from a certain chromosome or set of genomic regions. Tables can be stored, and pairs of tables combined through a union or intersection (for example, to show which genes are present in both tables, or in one table but not the other), generating a single output. For further information on this functionality see the tutorial publication [123], and the Table Browser User's Guide: http://genome.ucsc.edu/goldenPath/help/hgTablesHelp.html.

11.2 Galaxy

Galaxy (https://www.galaxyproject.org) [124, 125] is a scientific workflow, data integration and analysis platform, and probably the best-known GUI-based program for performing complex genomic database searches and analyses. It exists as a Web browser-based version and as a stand-alone application. The main window contains three vertical panels: to the left, a Tools panel listing utilities for obtaining data and performing analyses, in the center the main workspace, and to the right a History panel showing jobs completed or underway. The Get Data tool allows data to be uploaded from a local file or from Internet-based databases, including the UCSC Genome Browser and the BioMart portal [126], which provides access to Ensembl, Vega, and some species-specific databases, among others. Data import features almost seamless integration with these external servers: the user is taken to the external website where the search is performed, the retrieved data being automatically imported into their Galaxy workspace. Once data are imported, tools can be applied including filtering, annotation, statistical analysis and graphing, as well as routines including the EMBOSS suite. Galaxy is also designed to handle and analyze NGS data [121], being able to recognize various industry-standard data formats, and to apply search algorithms including BWA, Bowtie2, and others. Operations can be linked together as linear workflows, allowing for transparent and reproducible bioinformatic analyses. Galaxy features extensive documentation, video tutorials, and a large support community.

11.3 NCBI Genome Workbench

The NCBI Genome Workbench (http://www.ncbi.nlm.nih.gov/tools/gbench/) is a standalone application for Windows, Mac or Linux. It provides for a large variety of analytical functions, such as genomic database searching, and the import of the user's own experimental data as well as from public databases. Tools are included for the construction and rendering of analyses with graphical outputs, such as multiple sequence alignments, dot-matrix plots, and phylogenetic trees. Graphical outputs can be exported in a high-quality PDF format. Accompanying the software is its own macro language, which enables aspects of its functionality to be customized and reproduced.

11.4 Taverna

Taverna (http://www.taverna.org.uk/) [127] is a software environment for creating, managing and executing scientific and analytical workflows. Taverna software comprises a desktop application and a Web server. Workflows may integrate routines from a wide variety of Web services (a list is maintained at https://www.biocatalogue.org/), including all major genomic databases. Complex workflows, including loops, can be generated; these can be visualized as a flowchart, managed and edited. Workflows can be stored, reused and shared, a public repository of workflows being kept at http://www.myexperiment.org/. A utility called Taverna 2-Galaxy allows the automatic generation of Galaxy tools from Taverna workflows. Taverna is transitioning to being a project of the Apache Software Foundation, so future releases of the software will be hosted by Apache.

12 Genome Searching Using Application Programming Interfaces

For researchers with computer programming experience, the most powerful and flexible approach to searching genomic databases is to write or adapt dedicated programs (scripts) that incorporate commands that query Internet-based databases using application programming interfaces (APIs). Here, queries can be performed, information retrieved, stored and used for downstream operations, allowing the creation of an analytical pipeline. There is no common or unified approach for this, due to the different architectures of the

Table 3
Application Programming Interface (API) information

Utility	Website	References
Bio toolkits*:		
Open Bioinformatics Foundation:	http://www.open-bio.org/wiki/Main_Page	
OBF Projects:	http://www.open-bio.org/wiki/Projects	[128]
Ensembl		[40]
Perl API Documentation:	http://www.ensembl.org/info/docs/api/index.html	[129]
REST API Endpoints:	http://rest.ensembl.org/	[130]
UCSC Genome Browser		[51]
Kent Source Tree:	http://hgdownload.soe.ucsc.edu/admin/exe/	
MySQL access	http://genome.ucsc.edu/goldenpath/help/mysql.html	
Ruby API information:	https://rubygems.org/gems/bio-ucsc-api/	[131]
NCBI databases		[45]
E-Utilities help:	http://eutils.ncbi.nlm.nih.gov/	
Ebot:	http://www.ncbi.nlm.nih.gov/Class/PowerTools/eutils/ebot/ebot.cgi	
Entrez Direct guide:	http://www.ncbi.nlm.nih.gov/books/NBK179288/	
Bioconductor		
Bioconductor home page:	http://bioconductor.org	[134]
Bioconductor packages:	http://bioconductor.org/packages/release/	

different databases, and the programming languages involved. Below is a brief overview of programming-based methods and resource-specific APIs available to facilitate genomic database searching; links to information and resources are provided in Table 3. Detailed descriptions and specific methods are beyond the scope of this chapter; documentation, tutorials and examples for each API or resource are provided by their respective servers.

12.1 Bio* toolkits

For each of the major programming languages used in bioinformatics, over the past two decades large sets of open-source routines for performing operations including genomic database searches have been developed by the scientific community and released for public use. Starting with BioPerl, a family of routine sets now includes Biopython, BioRuby, BioJava, and others. These are collectively known as Bio* toolkits [128], and are supported by the Open Bioinformatics Foundation. Each of these toolkits provides routines for searching Internet-based genomic databases, plus methods for interconversion, comparison, and analysis of the outputs.

12.2 Ensembl

The main API for Ensembl is based on the Perl language [129], and depends in part on BioPerl. The API allows searching of the Ensembl database, and retrieval of genomic segments, genes, transcripts, and proteins, their comparison, and provides a set of analytical tools. A recent development for Ensembl is the development of an API that adopts a REpresentational State Transfer (REST) architectural style [130]. These "RESTful" API commands allow genomic database queries to be made using the HTTP protocol, thus making data access possible in any programming language for which HTTP libraries are available (this includes virtually all major languages, e.g., Perl, Python, Ruby, Java), or by using a program capable of handling the HTTP protocol (e.g., Curl, Wget). An online tool is also provided that generates example scripts for several languages.

12.3 The UCSC Genome Browser

The UCSC Genome Browser offers a set of utilities known as the **Kent Source Tree**, based on the C programming language. This comprises nearly 300 command-line applications for Linux and UNIX platforms, providing a wide range of functionalities interacting with the UCSC Genome Browser database, including data retrieval, genomic searching using BLAT and pattern-finding routines. Additionally, a MySQL database of genomic data is maintained, allowing queries to be performed via MySQL commands. Complementing this, but created independently, is an API to the UCSC Genome Browser written in and for the Ruby language [131]. Taking the form of a BioRuby plugin, this API allows access to the main genomic databases, using a dynamic framework to cope with the complex tables in which the UCSC Genome Browser holds its data.

12.4 NCBI Genome Resources

The NCBI offers access to their collection of databases via the Entrez Programming Utilities (**E-Utilities**), a suite of nine programs that support a uniform set of parameters used to search, link and download data [45, 132]. The functionalities of E-Utilities can be accessed from any programming language capable of handling the HTTP protocol; query results are typically returned in the extensible markup language (XML) format. E-Utility commands can be linked together within a script to generate an analysis pipeline or application. To help with this, the NCBI's **Ebot** tool guides the user step by step in generating a Perl script implementing an E-Utility pipeline. An additional recent facility is Entrez Direct (**EDirect**) [45, 133], an advanced method for accessing the NCBI's set of interconnected databases using UNIX terminal command-line arguments, which can be combined to build multi-step queries.

12.5 Bioconductor

Bioconductor [134] is a project that provides a wealth of resources for facilitating biological data analysis, mostly within the statistical programming language R. It consists of several hundred interoperable routines known as packages, each accompanied by a document known as a vignette that provides a textual, task-oriented description of its functionality. Among Bioconductor packages are routines for searching and retrieving data from genomic databases, and communication with genome browsers. Additional packages provide sophisticated tools for the analysis of large sets of experimental data, and powerful and flexible methods for visualization and graphical representation of quantitative data.

13 Conclusions and Perspectives

The genomic revolution that has transformed biology is progressing at full speed and is continuing to advance the fields of biology and medicine. The genomes of thousands of organisms and pathogen strains have now been sequenced, with data released into public repositories. Developments in NGS technology have caused an explosion in genomic data production, and have enabled new biological questions to be explored. Personalized genomes, and the fields of large-population genomics and single-cell genomics have become a reality.

Also accompanying the genomic data boom in recent years have been welcome developments such as the sharing of nucleotide data by the INSDC, standardization of reference genomes by the GRC, and the wider adoption of GENCODE gene sets by genome browsers.

Data repositories and search routines written to allow access to human genome project data have shown remarkable flexibility in incorporating not just new volumes of data but very heterogeneous types of data into their repertoires. What started as simple search portals have developed into sophisticated, interactive, but mostly very user friendly software tools, exhibiting innovative means of integrating and visualizing genomic data sets.

For genomic database searching, virtually all of the basic queries one could imagine regarding the genomic loci of genes, transcript or proteins of interest, or about features in a genomic region of interest can now be posed and rapidly answered through GUI-based Web tools such as Ensembl and the UCSC Genome Browser. For more complex genomic queries and workflows, software such as Galaxy and Taverna is available that combines and integrates diverse analyses; the evolution of these packages has benefited greatly from input from a wide community of users. For those performing searches via custom scripts, libraries of routines and APIs such as Ensembl's RESTful set now allow genomic database searching using virtually any platform or programming language.

Central to the success of the genomics revolution has been a philosophy of openness, involving the early public release of data, sharing of libraries and routines, and the open-source nature of algorithms and software. Databases are only as accessible as their means to search them, and for genomic databases and browsers this crucial aspect has been greatly enhanced by extensive documentation, open-access publications, online tutorials, webinars, public events, and the like that accompany them. The usability of resources is continuing to improve thanks to the work of curators and developers, and their willingness to incorporate features through user feedback and suggestions.

Among the wider community of molecular biologists and bioinformaticians there is a notable spirit of collegiality and mutual support, for example through the **Biostars** question and answer website (http://www.biostars.org/) [135]. For genomic database searching, this spirit of openness, information sharing and assistance has proven to be crucial for clarity of navigation in an otherwise bewilderingly complex terrain, and hopefully will continue in the future.

14 Notes

1. Ensembl databases can be searched by incorporating query terms into a URL. The following are some examples:

   Gene search (any term, any species): http://www.ensembl.org/Multi/Search/Results?facet_feature_type=Gene;q=**Mztl**

   Gene page (Ensembl ID, any species): http://www.ensembl.org/Gene/Summary?g=**ENSLACG00000022197**

   Gene page (gene name/symbol, stating species): http://www.ensembl.org/**Homo_sapiens**/Gene/Summary?g=**TP53**

   Gene page (gene ID, stating species): http://www.ensembl.org/**Mus_musculus**/Gene/Summary?g=**MGI:98834**

   Genomic locus view (any term, stating species): http://www.ensembl.org/**Rattus_norvegicus**/Location/View?g=**BC167758**

   Genomic locus view (Ensembl ID, stating species): http://www.ensembl.org/**Danio_rerio**/Location/View?g=**ENSDARG00000035559**

2. Ensembl Genomes can be searched by incorporating query terms into a URL. The following are some examples:

   Bacteria search (any species, any term): http://bacteria.ensembl.org/Multi/Search/Results?q=**Escherichia+coli+dnaA**

   Bacteria genomic locus view (stating species, any term): http://bacteria.ensembl.org/**geobacillus_stearothermophilus/Gene/Summary?g=GT94_09925**

   Protists search (any species, any term): http://protists.ensembl.org/Multi/Search/Results?q=**PF11_0183**

Protists genomic locus view (stating species, any term): http://protists.ensembl.org/**Dictyostelium_discoideum_ax4**/Location/View?g=**ranA**

Fungi search (any species, any term): http://fungi.ensembl.org/Multi/Search/Results?q=AN9504.2

Fungi genomic locus view (stating species, any term): http://fungi.ensembl.org/**Schizosaccharomyces_pombe**/Location/View?g=**mzt1**

Plants search (any species, any term): http://plants.ensembl.org/Multi/Search/Results?q=**AT5G24280**

Plants genomic locus view (stating species, any term): http://plants.ensembl.org/**Arabidopsis_thaliana**/Location/View?g=**GMI1**

Metazoa search (any species, any term): http://metazoa.ensembl.org/Multi/Search/Results?q=**WBGene00004953**
Metazoa genomic locus view (stating species, any term): http://metazoa.ensembl.org/**Drosophila_melanogaster**/Location/View?g=**stg**

3. The Vega database can be searched by incorporating query terms into a URL. The following are some examples:

   Gene search (any species, any term): http://vega.sanger.ac.uk/Multi/Search/Results?facet_feature_type=Gene;q=**P04637**

   Gene page (any species, Vega ID): http://vega.sanger.ac.uk/Gene/Summary?g=**OTTHUMG00000162125**

   Gene page (any species, gene ID): http://vega.sanger.ac.uk/**Mus_musculus**/Location/View?g=**MGI:98834**

   Genomic locus view (stating species, gene symbol): http://vega.sanger.ac.uk/**Homo_sapiens**/Location/View?g=**TP53**

   Genomic locus view (stating species, any ID): http://vega.sanger.ac.uk/**Mouse**/Location/View?g=**P02340**

4. The UCSC Genome Browser can be searched by incorporating query terms into the URL. In all cases the species needs to be stated using a common name or abbreviation. The following are some examples:

   Gene name: http://genome.ucsc.edu/cgi-bin/hgTracks?org=**Rat**&position=**Chek1**

   NCBI GeneID: https://genome.ucsc.edu/cgi-bin/hgTracks?org=**Human**&position=**3643**

   RefSeq nucleotide ID: http://genome.ucsc.edu/cgi-bin/hgTracks?org=**D.+melanogaster**&position=**NM_057663**

   RefSeq protein ID: https://genome.ucsc.edu/cgi-bin/hgTracks?org=**rat**&position=**NP_001278730**

UniProt ID: http://genome.ucsc.edu/cgi-bin/hgTracks?org=**Mouse**&position=**P02340**

5. NCBI Map Viewer can be searched by incorporating query terms into a URL. For example:

   Via NCBI GeneID: http://www.ncbi.nlm.nih.gov/mapview/map_search.cgi?direct=on&idtype=gene&id=**54801**

   For the following examples, the taxonomic ID ("taxid") relevant to the species needs to be included. For more information about taxids, *see* http://www.ncbi.nlm.nih.gov/taxonomy. The following are some example searches for specific species:

   Gene symbol (human): http://www.ncbi.nlm.nih.gov/projects/mapview/map_search.cgi?taxid=**9606**&query=**ANAPC16**

   Gene name (*S. pombe*): http://www.ncbi.nlm.nih.gov/projects/mapview/map_search.cgi?taxid=**284812**&query=**mztl**

   RefSeq nucleotide (*Drosophila*): http://www.ncbi.nlm.nih.gov/mapview/map_search.cgi?taxid=**7227**&query=**NM_057663.4**

   Genomes can be searched using degenerate term(s). This example retrieves and displays the genomic locations of MAP kinase (Mapk) genes within the mouse genome: http://www.ncbi.nlm.nih.gov/mapview/map_search.cgi?taxid=**10090**&query=**Mapk**

6. NCBI Gene can be searched by incorporating query terms into a URL. The following are some examples:

   Any term(s), e.g., species and gene name: http://www.ncbi.nlm.nih.gov/gene/?term=**chicken+CHEK1**

   Gene code: http://www.ncbi.nlm.nih.gov/gene/?term=**MGI:98834**

   RefSeq nucleotide ID: http://www.ncbi.nlm.nih.gov/gene/?term=**NM_001170407**

   Protein GI number: http://www.ncbi.nlm.nih.gov/gene/?term=**GI:115392148**

   UniProt ID: http://www.ncbi.nlm.nih.gov/gene/?term=**Q9H410**

7. For RNA molecules, several genomic databases can be searched by incorporating query terms into URLs. Searching mRNAs is covered by **Notes 1–6**. The following examples are for miRNAs, piRNAs and lncRNAs:

   miRNA, stem-loop: http://www.mirbase.org/cgi-bin/mirna_entry.pl?acc=**mmu-mir-122**

   miRNA, stem-loop: http://www.mirbase.org/cgi-bin/mirna_entry.pl?acc=**MI0000256**

   miRNA, mature: http://www.mirbase.org/cgi-bin/query.pl?terms=**mmu-miR-122-5p**

miRNA, mature: http://www.mirbase.org/cgi-bin/mature.pl?mature_acc=**MIMAT0000246**

piRNA (piRNABank): http://pirnabank.ibab.ac.in/cgi-bin/accession.cgi?accession=**hsa_piR_000001**&acc_organism=Human+%28NCBI+Build+37%29

piRNA (piRBase): http://www.regulatoryrna.org/database/piRNA/pirna.php?name=**piR-hsa-237**

piRNA (piRNAQuest): http://bicresources.jcbose.ac.in/zhumur/pirnaquest/search_pirna.php?organism=Human+build+37.3%2F+hg19&pirna_id=**hsa_piRNA_6754**

lncRNA (lncRNAdb): http://lncrnadb.com/search/?q=**ncRNA-a1** or http://lncrnadb.com/**LINC00568/**

lncRNA (UCSC Genome Browser): https://genome.ucsc.edu/cgi-bin/hgTracks?org=Human&position=**LINC00568**

lncRNA (Ensembl): http://www.ensembl.org/Human/Search/Results?facet_feature_type=Gene;q=**LINC00568**

Acknowledgements

I would like to thank the numerous developers and support staff of genomic databases who provided valuable information during the researching and writing of this chapter. Grateful thanks also go to colleagues past and present who provided helpful information and advice. During the preparation of this chapter I worked in the laboratory of Dr. M. Méchali, whom I gratefully acknowledge for his guidance and support. I was supported financially by La Fondation pour la Recherche Médicale (FRM), and by the Centre National de la Recherche Scientifique (CNRS).

References

1. Sanger F, Air GM, Barrell BG et al (1977) Nucleotide sequence of bacteriophage phi X174 DNA. Nature 265:687–695
2. Fleischmann RD, Adams MD, White O et al (1995) Whole-genome random sequencing and assembly of Haemophilus influenzae Rd. Science 269:496–512
3. Johnston M (1996) The complete code for a eukaryotic cell. Genome sequencing. Curr Biol 6:500–503
4. C. elegans Sequencing Consortium (1998) Genome sequence of the nematode C. elegans: a platform for investigating biology. Science 282:2012–2018
5. Lander ES, Linton LM, Birren B et al (2001) Initial sequencing and analysis of the human genome. Nature 409:860–921
6. Venter JC, Adams MD, Myers EW et al (2001) The sequence of the human genome. Science 291:1304–1351
7. IHGSC (2004) Finishing the euchromatic sequence of the human genome. Nature 431:931–945
8. Reddy TB, Thomas AD, Stamatis D et al (2015) The Genomes OnLine Database (GOLD) v. 5: a metadata management system based on a four level (meta)genome project classification. Nucleic Acids Res 43:D1099–D1106

9. Warren WC, Hillier LW, Marshall Graves JA et al (2008) Genome analysis of the platypus reveals unique signatures of evolution. Nature 453:175–183
10. Amemiya CT, Alfoldi J, Lee AP et al (2013) The African coelacanth genome provides insights into tetrapod evolution. Nature 496:311–316
11. Prüfer K, Racimo F, Patterson N et al (2014) The complete genome sequence of a Neanderthal from the Altai Mountains. Nature 505:43–49
12. King TE, Fortes GG, Balaresque P et al (2014) Identification of the remains of King Richard III. Nat Commun 5:5631
13. Abecasis GR, Altshuler D, Auton A et al (2010) A map of human genome variation from population-scale sequencing. Nature 467:1061–1073
14. Abecasis GR, Auton A, Brooks LD et al (2012) An integrated map of genetic variation from 1,092 human genomes. Nature 491:56–65
15. Torjesen I (2013) Genomes of 100,000 people will be sequenced to create an open access research resource. BMJ 347:f6690
16. Baslan T, Hicks J (2014) Single cell sequencing approaches for complex biological systems. Curr Opin Genet Dev 26C:59–65
17. Liang J, Cai W, Sun Z (2014) Single-cell sequencing technologies: current and future. J Genet Genomics = Yi Chuan Xue Bao 41:513–528
18. Dykes CW (1996) Genes, disease and medicine. Br J Clin Pharmacol 42:683–695
19. Chan IS, Ginsburg GS (2011) Personalized medicine: progress and promise. Annu Rev Genomics Hum Genet 12:217–244
20. Bauer DC, Gaff C, Dinger ME et al (2014) Genomics and personalised whole-of-life healthcare. Trends Mol Med 20(9):479–486
21. Check Hayden E (2010) Human genome at ten: life is complicated. Nature 464:664–667
22. Dulbecco R (1986) A turning point in cancer research: sequencing the human genome. Science 231:1055–1056
23. International Cancer Genome Consortium, Hudson TJ, Anderson W et al (2010) International network of cancer genome projects. Nature 464, 993–998
24. Alexandrov LB, Stratton MR (2014) Mutational signatures: the patterns of somatic mutations hidden in cancer genomes. Curr Opin Genet Dev 24C:52–60
25. Hoffman MM, Ernst J, Wilder SP et al (2013) Integrative annotation of chromatin elements from ENCODE data. Nucleic Acids Res 41:827–841
26. modEncode Consortium, Roy S, Ernst J et al (2010) Identification of functional elements and regulatory circuits by Drosophila modENCODE. Science 330:1787–1797
27. Gerstein MB, Lu ZJ, Van Nostrand EL et al (2010) Integrative analysis of the Caenorhabditis elegans genome by the modENCODE project. Science 330:1775–1787
28. Harrow J, Frankish A, Gonzalez JM et al (2012) GENCODE: the reference human genome annotation for The ENCODE Project. Genome Res 22:1760–1774
29. Almouzni G, Altucci L, Amati B et al (2014) Relationship between genome and epigenome—challenges and requirements for future research. BMC Genomics 15:487
30. Hériché JK (2014) Systematic cell phenotyping. In: Hancock JM (ed) Phenomics. CRC Press, Boca Raton, FL, pp 86–110
31. Hutchins JRA (2014) What's that gene (or protein)? Online resources for exploring functions of genes, transcripts, and proteins. Mol Biol Cell 25:1187–1201
32. Schmidt A, Forne I, Imhof A (2014) Bioinformatic analysis of proteomics data. BMC Syst Biol 8(Suppl 2):S3
33. Kaiser J (2005) Genomics. Celera to end subscriptions and give data to public GenBank. Science 308:775
34. Church DM, Schneider VA, Graves T et al (2011) Modernizing reference genome assemblies. PLoS Biol 9:e1001091
35. Gerstein MB, Bruce C, Rozowsky JS et al (2007) What is a gene, post-ENCODE? History and updated definition. Genome Res 17:669–681
36. Burge C, Karlin S (1997) Prediction of complete gene structures in human genomic DNA. J Mol Biol 268:78–94
37. Thierry-Mieg D, Thierry-Mieg J (2006) AceView: a comprehensive cDNA-supported gene and transcripts annotation. Genome Biol 7(Suppl 1):S12.1–S12.14
38. MGC Project Team, Temple G, Gerhard DS et al (2009) The completion of the Mammalian Gene Collection (MGC). Genome Res 19:2324–2333
39. Farrell CM, O'Leary NA, Harte RA et al (2014) Current status and new features of

the Consensus Coding Sequence database. Nucleic Acids Res 42:D865–D872

40. Cunningham F, Amode MR, Barrell D et al (2015) Ensembl 2015. Nucleic Acids Res 43: D662–D669
41. Pruitt KD, Brown GR, Hiatt SM et al (2014) RefSeq: an update on mammalian reference sequences. Nucleic Acids Res 42:D756–D763
42. Harrow JL, Steward CA, Frankish A et al (2014) The Vertebrate Genome Annotation browser 10 years on. Nucleic Acids Res 42: D771–D779
43. Frankish A, Uszczynska B, Ritchie GR et al (2015) Comparison of GENCODE and RefSeq gene annotation and the impact of reference geneset on variant effect prediction. BMC Genomics 16(Suppl 8):S2
44. Kersey PJ, Allen JE, Christensen M et al (2014) Ensembl Genomes 2013: scaling up access to genome-wide data. Nucleic Acids Res 42:D546–D552
45. NCBI Resource Coordinators (2015) Database resources of the National Center for Biotechnology Information. Nucleic Acids Res 43:D6–D17
46. Gray KA, Yates B, Seal RL et al (2015) Genenames.org: the HGNC resources in 2015. Nucleic Acids Res 43:D1079–D1085
47. dos Santos G, Schroeder AJ, Goodman JL et al (2015) FlyBase: introduction of the Drosophila melanogaster Release 6 reference genome assembly and large-scale migration of genome annotations. Nucleic Acids Res 43:D690–D697
48. Silvester N, Alako B, Amid C et al (2015) Content discovery and retrieval services at the European Nucleotide Archive. Nucleic Acids Res 43:D23–D29
49. Kodama Y, Mashima J, Kosuge T et al (2015) The DDBJ Japanese Genotype-phenotype Archive for genetic and phenotypic human data. Nucleic Acids Res 43:D18–D22
50. UniProt Consortium (2015) UniProt: a hub for protein information. Nucleic Acids Res 43:D204–D212
51. Rosenbloom KR, Armstrong J, Barber GP et al (2015) The UCSC Genome Browser database: 2015 update. Nucleic Acids Res 43:D670–D681
52. Hsu F, Kent WJ, Clawson H et al (2006) The UCSC known genes. Bioinformatics 22:1036–1046
53. Nawrocki EP, Burge SW, Bateman A et al (2015) Rfam 12.0: updates to the RNA families database. Nucleic Acids Res 43: D130–D137
54. Chan PP, Lowe TM (2009) GtRNAdb: a database of transfer RNA genes detected in genomic sequence. Nucleic Acids Res 37: D93–D97
55. Punta M, Coggill PC, Eberhardt RY et al (2012) The Pfam protein families database. Nucleic Acids Res 40:D290–D301
56. Tatusova T (2010) Genomic databases and resources at the National Center for Biotechnology Information. Methods Mol Biol 609:17–44
57. Wolfsberg TG (2011) Using the NCBI Map Viewer to browse genomic sequence data. Curr Protoc Hum Genet. Chapter 18. Unit 18.15
58. Brown GR, Hem V, Katz KS et al (2015) Gene: a gene-centered information resource at NCBI. Nucleic Acids Res 43:D36–D42
59. Brister JR, Ako-Adjei D, Bao Y et al (2015) NCBI viral genomes resource. Nucleic Acids Res 43:D571–D577
60. Nicol JW, Helt GA, Blanchard SG Jr et al (2009) The Integrated Genome Browser: free software for distribution and exploration of genome-scale datasets. Bioinformatics 25:2730–2731
61. Thorvaldsdottir H, Robinson JT, Mesirov JP (2013) Integrative Genomics Viewer (IGV): high-performance genomics data visualization and exploration. Brief Bioinform 14:178–192
62. Fiume M, Smith EJ, Brook A et al (2012) Savant Genome Browser 2: visualization and analysis for population-scale genomics. Nucleic Acids Res 40:W615–W621
63. Wright MW, Bruford EA (2011) Naming 'junk': human non-protein coding RNA (ncRNA) gene nomenclature. Hum Genomics 5:90–98
64. Agirre E, Eyras E (2011) Databases and resources for human small non-coding RNAs. Hum Genomics 5:192–199
65. The RNAcentral Consortium (2015) RNAcentral: an international database of ncRNA sequences. Nucleic Acids Res 43:D123–D129
66. Nakamura Y, Cochrane G, Karsch-Mizrachi I (2013) The International Nucleotide Sequence Database Collaboration. Nucleic Acids Res 41:D21–D24
67. Ameres SL, Zamore PD (2013) Diversifying microRNA sequence and function. Nat Rev Mol Cell Biol 14:475–488
68. Kozomara A, Griffiths-Jones S (2014) miRBase: annotating high confidence microRNAs using deep sequencing data. Nucleic Acids Res 42:D68–D73

69. Mani SR, Juliano CE (2013) Untangling the web: the diverse functions of the PIWI/piRNA pathway. Mol Reprod Dev 80:632–664
70. Peng JC, Lin H (2013) Beyond transposons: the epigenetic and somatic functions of the Piwi-piRNA mechanism. Curr Opin Cell Biol 25:190–194
71. Sai Lakshmi S, Agrawal S (2008) piRNABank: a web resource on classified and clustered Piwi-interacting RNAs. Nucleic Acids Res 36:D173–D177
72. Zhang P, Si X, Skogerbo G et al (2014) piRBase: a web resource assisting piRNA functional study. Database (Oxford) 2014, bau110
73. Sarkar A, Maji RK, Saha S et al (2014) piRNAQuest: searching the piRNAome for silencers. BMC Genomics 15:555
74. Skinner ME, Uzilov AV, Stein LD et al (2009) JBrowse: a next-generation genome browser. Genome Res 19:1630–1638
75. Kung JT, Colognori D, Lee JT (2013) Long noncoding RNAs: past, present, and future. Genetics 193:651–669
76. Bonasio R, Shiekhattar R (2014) Regulation of transcription by long noncoding RNAs. Annu Rev Genet 48:433–455
77. Wright MW (2014) A short guide to long non-coding RNA gene nomenclature. Hum Genomics 8:7
78. Fritah S, Niclou SP, Azuaje F (2014) Databases for lncRNAs: a comparative evaluation of emerging tools. RNA 20:1655–1665
79. Quek XC, Thomson DW, Maag JL et al (2015) lncRNAdb v2.0: expanding the reference database for functional long noncoding RNAs. Nucleic Acids Res 43: D168–D173
80. Craig JM, Bickmore WA (1993) Chromosome bands—flavours to savour. Bioessays 15:349–354
81. Altschul SF, Gish W, Miller W et al (1990) Basic local alignment search tool. J Mol Biol 215:403–410
82. Kent WJ (2002) BLAT—the BLAST-like alignment tool. Genome Res 12:656–664
83. Jacox E, Elnitski L (2008) Finding occurrences of relevant functional elements in genomic signatures. Int J Comput Sci 2:599–606
84. Brennan RG, Matthews BW (1989) Structural basis of DNA-protein recognition. Trends Biochem Sci 14:286–290
85. Hudson WH, Ortlund EA (2014) The structure, function and evolution of proteins that bind DNA and RNA. Nat Rev Mol Cell Biol 15:749–760
86. Wells RD (1988) Unusual DNA structures. J Biol Chem 263:1095–1098
87. Hedgpeth J, Goodman HM, Boyer HW (1972) DNA nucleotide sequence restricted by the RI endonuclease. Proc Natl Acad Sci U S A 69:3448–3452
88. Wei CL, Wu Q, Vega VB et al (2006) A global map of p53 transcription-factor binding sites in the human genome. Cell 124:207–219
89. Mergny JL (2012) Alternative DNA structures: G4 DNA in cells: itae missa est? Nat Chem Biol 8:225–226
90. Giraldo R, Suzuki M, Chapman L et al (1994) Promotion of parallel DNA quadruplexes by a yeast telomere binding protein: a circular dichroism study. Proc Natl Acad Sci U S A 91:7658–7662
91. Cayrou C, Coulombe P, Puy A et al (2012) New insights into replication origin characteristics in metazoans. Cell Cycle 11:658–667
92. Brown P, Baxter L, Hickman R et al (2013) MEME-LaB: motif analysis in clusters. Bioinformatics 29:1696–1697
93. Grant CE, Bailey TL, Noble WS (2011) FIMO: scanning for occurrences of a given motif. Bioinformatics 27:1017–1018
94. Medina-Rivera A, Defrance M, Sand O et al (2015) RSAT 2015: regulatory sequence analysis tools. Nucleic Acids Res 43: W50–W56
95. Rice P, Longden I, Bleasby A (2000) EMBOSS: the European Molecular Biology Open Software Suite. Trends Genet 16:276–277
96. Stormo GD, Zhao Y (2010) Determining the specificity of protein-DNA interactions. Nat Rev Genet 11:751–760
97. Kel AE, Gossling E, Reuter I et al (2003) MATCH: A tool for searching transcription factor binding sites in DNA sequences. Nucleic Acids Res 31:3576–3579
98. Wingender E (2008) The TRANSFAC project as an example of framework technology that supports the analysis of genomic regulation. Brief Bioinform 9:326–332
99. Wrzodek C, Schroder A, Drager A et al (2010) ModuleMaster: a new tool to decipher transcriptional regulatory networks. Biosystems 99:79–81
100. Turatsinze JV, Thomas-Chollier M, Defrance M et al (2008) Using RSAT to scan genome sequences for transcription factor binding sites and cis-regulatory modules. Nat Protoc 3:1578–1588

101. Kinsella RJ, Kahari A, Haider S et al (2011) Ensembl BioMarts: a hub for data retrieval across taxonomic space. Database (Oxford) 2011, bar030
102. Metzker ML (2010) Sequencing technologies—the next generation. Nat Rev Genet 11:31–46
103. Niedringhaus TP, Milanova D, Kerby MB et al (2011) Landscape of next-generation sequencing technologies. Anal Chem 83:4327–4341
104. Ozsolak F, Milos PM (2011) RNA sequencing: advances, challenges and opportunities. Nat Rev Genet 12:87–98
105. Li R, Li Y, Kristiansen K et al (2008) SOAP: short oligonucleotide alignment program. Bioinformatics 24:713–714
106. Li H, Ruan J, Durbin R (2008) Mapping short DNA sequencing reads and calling variants using mapping quality scores. Genome Res 18:1851–1858
107. Langmead B, Trapnell C, Pop M et al (2009) Ultrafast and memory-efficient alignment of short DNA sequences to the human genome. Genome Biol 10:R25
108. Li H, Durbin R (2009) Fast and accurate short read alignment with Burrows-Wheeler transform. Bioinformatics 25:1754–1760
109. Lunter G, Goodson M (2011) Stampy: a statistical algorithm for sensitive and fast mapping of Illumina sequence reads. Genome Res 21:936–939
110. Langmead B, Salzberg SL (2012) Fast gapped-read alignment with Bowtie 2. Nat Methods 9:357–359
111. Li H (2013) Aligning sequence reads, clone sequences and assembly contigs with BWA-MEM. arXiv preprint arXiv:1303.3997
112. Sedlazeck FJ, Rescheneder P, von Haeseler A (2013) NextGenMap: fast and accurate read mapping in highly polymorphic genomes. Bioinformatics 29:2790–2791
113. Santana-Quintero L, Dingerdissen H, Thierry-Mieg J et al (2014) HIVE-hexagon: high-performance, parallelized sequence alignment for next-generation sequencing data analysis. PLoS One 9:e99033
114. Lee WP, Stromberg MP, Ward A et al (2014) MOSAIK: a hash-based algorithm for accurate next-generation sequencing short-read mapping. PLoS One 9:e90581
115. Fonseca NA, Rung J, Brazma A et al (2012) Tools for mapping high-throughput sequencing data. Bioinformatics 28:3169–3177
116. Lindner R, Friedel CC (2012) A comprehensive evaluation of alignment algorithms in the context of RNA-seq. PLoS One 7:e52403
117. Buermans HP, den Dunnen JT (2014) Next generation sequencing technology: advances and applications. Biochim Biophys Acta 1842:1932–1941
118. van Dijk EL, Auger H, Jaszczyszyn Y et al (2014) Ten years of next-generation sequencing technology. Trends Genet 30:418–426
119. Li JW, Schmieder R, Ward RM et al (2012) SEQanswers: an open access community for collaboratively decoding genomes. Bioinformatics 28:1272–1273
120. Scholtalbers J, Rossler J, Sorn P et al (2013) Galaxy LIMS for next-generation sequencing. Bioinformatics 29:1233–1234
121. Blankenberg D, Hillman-Jackson J (2014) Analysis of next-generation sequencing data using galaxy. Methods Mol Biol 1150:21–43
122. Liu B, Madduri RK, Sotomayor B et al (2014) Cloud-based bioinformatics workflow platform for large-scale next-generation sequencing analyses. J Biomed Inform 49:119–133
123. Zweig AS, Karolchik D, Kuhn RM et al (2008) UCSC genome browser tutorial. Genomics 92:75–84
124. Goecks J, Nekrutenko A, Taylor J (2010) Galaxy: a comprehensive approach for supporting accessible, reproducible, and transparent computational research in the life sciences. Genome Biol 11:R86
125. Hillman-Jackson J, Clements D, Blankenberg D et al (2012) Using Galaxy to perform large-scale interactive data analyses. Curr Protoc Bioinformatics Chapter 10, Unit 10.15
126. Smedley D, Haider S, Durinck S et al (2015) The BioMart community portal: an innovative alternative to large, centralized data repositories. Nucleic Acids Res 43:W589–W598
127. Wolstencroft K, Haines R, Fellows D et al (2013) The Taverna workflow suite: designing and executing workflows of Web Services on the desktop, web or in the cloud. Nucleic Acids Res 41:W557–W561
128. Mangalam H (2002) The Bio* toolkits—a brief overview. Brief Bioinform 3:296–302
129. Stabenau A, McVicker G, Melsopp C et al (2004) The Ensembl core software libraries. Genome Res 14:929–933
130. Yates A, Beal K, Keenan S et al (2014) The Ensembl REST API: Ensembl data for any language. Bioinformatics 31(1):143–145
131. Mishima H, Aerts J, Katayama T et al (2012) The Ruby UCSC API: accessing the UCSC

genome database using Ruby. BMC Bioinformatics 13:240

132. Sayers E (2013) Entrez programming utilities help [Internet]. National Center for Biotechnology Information (US), Bethesda, MD. http://www.ncbi.nlm.nih.gov/books/NBK25497/
133. Kans J (2014) Entrez programming utilities help [Internet]. National Center for Biotechnology Information (US), Bethesda, MD. http://www.ncbi.nlm.nih.gov/books/NBK179288/
134. Huber W, Carey VJ, Gentleman R et al (2015) Orchestrating high-throughput genomic analysis with Bioconductor. Nat Methods 12:115–121
135. Parnell LD, Lindenbaum P, Shameer K et al (2011) BioStar: an online question & answer resource for the bioinformatics community. PLoS Comput Biol 7:e1002216

Chapter 11

Finding Genes in Genome Sequence

Alice Carolyn McHardy and Andreas Kloetgen

Abstract

Gene finding is the process of identifying genome sequence regions representing stretches of DNA that encode biologically active products, such as proteins or functional noncoding RNAs. As this is usually the first step in the analysis of any novel genomic sequence or resequenced sample of well-known organisms, it is a very important issue, as all downstream analyses depend on the results. This chapter describes the biological basis for gene finding, and the programs and computational approaches that are available for the automated identification of protein-coding genes. For bacterial, archaeal, and eukaryotic genomes, as well as for multi-species sequence data originating from environmental community studies, the state of the art in automated gene finding is described.

Key words Gene prediction, Gene finding, Genomic sequence, Protein-coding sequences, Next-generation sequencing, Environmental sequence samples

1 Introduction

The coding regions of a genome contain the instructions to build functional proteins. This fact gives rise to several characteristic features that can universally be used for their identification. First, if transcript or protein sequences are already known for an organism, these can be used to determine the location of the corresponding genes in the genomic sequence. In recent years, cost-effective next-generation sequencing (NGS) techniques have become widely available, which can produce tens to hundreds of million sequence reads per run [1] and can be used to sequence complementary DNA (cDNA) generated from RNA transcripts. For instance, the database GenBank contains more than 62 million assembled sequences obtained from transcriptome shotgun assembly studies in its release version 205, one of the most rapidly growing divisions of GenBank [2]. Although this approach seems fairly straightforward, the complex structure of genes, along with alternative splicing events, makes this a nontrivial task for eukaryotic organisms. Second, natural selection acting on the encoded protein product restricts the rate of

Jonathan M. Keith (ed.), *Bioinformatics: Volume I: Data, Sequence Analysis, and Evolution*, Methods in Molecular Biology, vol. 1525, DOI 10.1007/978-1-4939-6622-6_11, © Springer Science+Business Media New York 2017

mutation in coding sequences (CDSs) compared to nonfunctional genomic DNA. Thus, many CDSs can be identified based on statistically significant sequence similarities that they share with evolutionally related proteins from other organisms. Selection is also evident in genome sequence comparisons between closely related species, where stretches of more highly conserved sequences correspond to the functionally active regions in the genome. Here, protein-coding genes in particular can be identified by their characteristic three-periodic pattern of conservation. Due to the degeneracy of the genetic code, different nucleotides at the third codon position often encode the same amino acid, thereby allowing for alterations at this codon position without changing the encoded protein. Approaches which use information about expressed sequences or sequence conservation are called "extrinsic," as they require additional knowledge besides the genomic sequence of the organism being analyzed. One can also take the "intrinsic" approach to gene identification, which is based on the evaluation of characteristic differences between coding and noncoding genomic sequence regions. In particular, there are characteristic differences in the distribution of short DNA oligomers between coding and noncoding regions (often called codon bias). One biological reason for these differences is the influence of "translational selection" on the use of synonymous codons and codon combinations in protein-coding genes. Synonymous codons which encode the same amino acid are not used with equal frequency in CDSs. Instead, there is a genome-wide preference in many organisms for the codons that are read by the more highly expressed tRNAs during the translation process, which increases translational efficiency. This effect is especially pronounced for highly expressed genes [3–7]. Many measures are available for characterizing the codon bias and expression preferences for a particular genome, such as a log-likelihood score or the MinMax measure [8, 9]. There is also evidence that codon combinations that are prone to initiating frameshift errors during the translation process tend to be avoided [10]. Another influence is evident in GC-rich genomes, where the genome-wide tendencies towards high GC usage establish themselves in the form of a three-periodic skew towards high GC content in the generally more flexible third codon position (*see* **Note 1**). In the early 1990s, Fickett systematically evaluated the suitability of a large variety of intrinsic coding measures to discriminate between CDSs and noncoding sequences that had been proposed at that time and found that simply counting DNA oligomers is the most effective method [11].

2 Methods

2.1 Gene Finding in Bacteria and Archaea

Finding genes in genome sequences is a simpler task in bacteria and archaea than it is in eukaryotic organisms. First, the gene structure is less complex: a protein-coding gene corresponds to a single open reading frame (ORF) in the genome sequence that begins with a start and ends with a stop codon. The protein-coding ORFs (ORFs that are transcribed and translated in vivo) are commonly referred to as CDSs. The translation start site is defined by a ribosome binding site that is typically located about 3–16 bp upstream of the start codon. Second, more than 90 % of the genome sequence is coding, as opposed to higher organisms, where vast stretches of noncoding DNA exist.

For a gene finding program, the task is (1) to discriminate between the coding and noncoding ORFs (nORFs) in the genome, as there are usually many more ORFs than CDSs in the sequence [12], and (2) to identify the correct start codon for those genes, as these can also appear internally in a gene sequence or even upstream of the actual translation start site.

Many programs make use of both extrinsic and intrinsic sources of information to predict protein-coding genes, as they complement each other very well. External evidence of evolutionary conservation provides strong evidence for the presence of a biologically active gene, but genes without (known) homologs cannot be detected in this way. By contrast, intrinsic methods do not need external knowledge; however, an accurate model of oligonucleotide frequencies for CDS and noncoding sequence regions is required.

A wide variety of techniques are used to solve the gene finding problem (Table 1). These include probabilistic methods such as hidden Markov models (HMMs) [13–15] or interpolated context models [16], and machine learning techniques for supervised or unsupervised classification, such as support vector machines (SVMs) [17] and self-organizing maps [18]. Most gene finders apply their classification models locally to individual sequence fragments (using a sliding window approach) or ORFs in the sequence. In the "global" classification approach implemented with the HMM architecture of GeneMark.hmm, a maximum likelihood parse is derived for the complete genomic sequence, to either coding or noncoding states. An interesting property of this HMM technique is that the search for the optimal model and the creation of the final prediction occur simultaneously; after iteratively training the optimal model, it uses the maximum likelihood sequence parse thus found, which is also the most likely assignment of genes to the sequence.

Another important issue in gene prediction relates to the fact that many bacterial and archaeal genomes exhibit a considerable portion of sequence that is "atypical" in terms of sequence composition compared to other parts. This, in combination with the other properties of such regions, usually indicates a foreign origin and acquisition of the respective region by horizontal gene transfer [19]. To enable the accurate detection of genes based on sequence composition in such regions, some gene finding programs distinguish between "atypical" CDSs and nORFs. This comparison is

Table 1
Bacterial and archaeal gene finders

Program	I[a]	E[b]	Method	URL (if available)
BDGF [25]	–	+	Classifications based on universal CDS-specific usage of short amino acid "seqlets"	
Critica [26]	+	+	Comparative analysis is based on amino acid sequence similarity to other species	http://www.ttaxus.com/software.html
EasyGene [14]	+	+	Uses HMMs. Model training is based on BLAST-derived "reliable" genes	http://www.cbs.dtu.dk/services/EasyGene
GeneMark.hmm/S [13, 15]	+	–	Uses HMMs.	http://Opal.biology.gatech.edu/GeneMark
Gismo [17]	+	+	Uses SVMs. Model training is based on "reliable" genes found with PFAM protein domain HMMs. *See* Fig. 2 for a schematic version of the program output	
Orpheus [27]	+	+	Seed and extend	
Reganor [28]	+	+	Uses Glimmer and Critica	
Yacop [29]	+	+	Utilizes Glimmer, Critica and Orpheus	http://gobics.de/tech/yacop.php
ZCurve [30]	+	–	Uses the "Z-transform" of DNA as the information source for classification	http://tubic.tju.edu.cn/Zcurve_B
RescueNet [18]	+	+	Unsupervised discovery of multiple gene classes using a self-organizing map. No exact start/stop prediction	http://bioinf.nuigalway.ie/RescueNet

[a]Uses intrinsic evidence
[b]Uses extrinsic evidence

based on the input features by providing a state for genes with atypical sequence properties within the HMM architecture or by the use of techniques such as SVMs, where classifiers can be created that are also able to distinguish between atypical CDSs and nORFs. With even more generalized approaches, the unsupervised discovery of characteristic CDS classes in the input dataset initiates the gene finding procedure [18, 20].

The following sequence of steps is often employed by a bacterial and archaeal gene finder (*see also* Fig. 1):

1. In the initial phase, an intrinsic classifier for discriminating between nORFs and CDSs is learnt. This is often initiated with similarity searches of the genomic sequences against protein or DNA sequence databases, or a search for motifs of protein domains. The information generated hereby can be used for a partial labeling of the sequence into nORFs and CDSs, and also

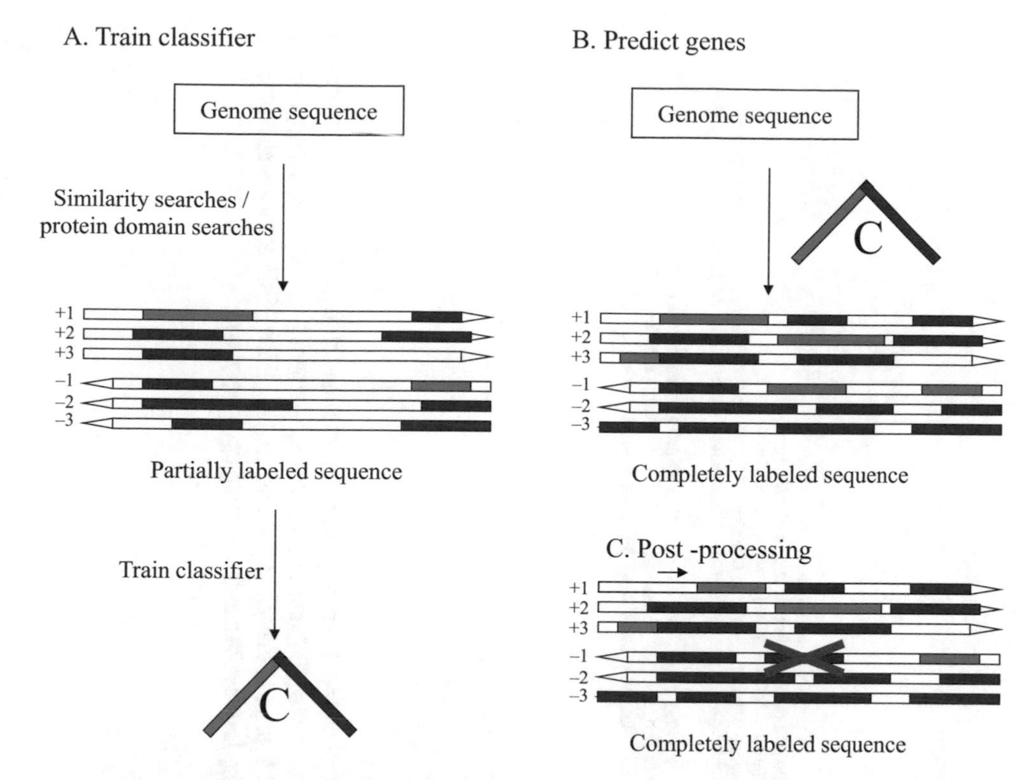

Fig. 1 Overview of a sequence of steps employed by a bacterial and archaeal gene finder. In (**a**), the sequence is initially searched for regions which exhibit significant conservation on amino acid level relative to other protein-coding regions or which show motifs of protein domains. By extending such regions to a start and stop codon, a partial labeling of the genome sequence into coding regions (*light gray*) and noncoding ORFs (nORFs), which significantly overlap with such coding sequences in another frame (*dark gray*), can be obtained. The labeled parts can be used as training sequences to derive the vectors of intrinsic sequence features for training a binary classifier. In (**b**), the classifier is applied to classify all ORFs above a certain length in the sequence as either coding sequences (CDSs) or nORFs. In the post-processing phase (**c**), the start positions of the predicted CDSs are reassigned with the help of translation start site models, and conflicts between neighboring predictions are resolved

to generate reliable training data for the classifier. ORFs supported by external sequence similarities can be used as training material for the protein-coding genes, and ORFs that significantly overlap with such regions (which are located in their "shadow") as training material for the nORF class.

2. In the prediction phase, the classifier is applied to identify the protein-coding genes of the sequence.
3. In the post-processing phase, the start positions of predicted CDSs are relocated to the most likely translation start site using translation start site models. Such models usually incorporate information about the ribosome binding site and are applied to search for the start codon with the strongest upstream signal for any given gene. Programs which have been specifically designed

```
##gff-version 2
##date 2006-03-21

##source-version Parser::GFF 1.2
36521 . CDS 381 82 -0.262462 . -1 rbs "381" ; reliable "0"
36521 . CDS 1353 1967 1.00021 . 3 rbs "1353" ; evalue "5.3e-18" ; description "Molybdopterin oxidoreductase Fe4S4 do" ; reliable "1"
36521 . CDS 1980 4457 1.06453 . 3 rbs "1980" ; evalue "6e-22" ; description "Molybdopterin oxidoreductase" ; reliable "1"
36521 . CDS 4454 5446 0.420936 . 2 rbs "4454" ; evalue "2.8e-05" ; description "4Fe-4S binding domain" ; reliable "1"
36521 . CDS 5424 6110 0.105856 . 3 rbs "5424" ; comment "overlapping" ; evalue "0.0041" ; description "Cytochrome b561 family" ; reliable
36521 . CDS 5519 7069 1.38546 . 2 rbs "5519" ; comment "overlapping" ; evalue "1.7e-39" ; description "Protein involved in formate dehydrogen" ; reliable "1"
36521 . CDS 7318 6782 -0.536073 . -2 rbs "7318" ; discard "1" ; reliable "0"
36521 . CDS 7074 8474 1.0716 . 3 rbs "7074" ; comment "overlapping" ; evalue "3.7e-255" ; description "L-seryl-tRNA selenium transferase" ; reliable "1"
36521 . CDS 8302 9255 1.08006 . 1 rbs "8302" ; comment "overlapping" ; evalue "4.5e-10" ; description "Sel1 repeat" ; reliable "1"
36521 . CDS 9252 11252 1.00039 . 3 rbs "9252" ; evalue "4.4e-30" ; description "Elongation factor Tu GTP binding doma" ; reliable "1"
36521 . CDS 11453 11776 -0.0957853 . 2 rbs "11453" ; reliable "0"
36521 . CDS 11506 12063 -0.338037 . 1 rbs "11506" ; reliable "0"
36521 . CDS 12026 11763 -0.596191 . -3 rbs "12026" ; discard "1" ; reliable "0"
36521 . CDS 12546 12259 0.233941 . -1 rbs "12546" ; reliable "1"
36521 . CDS 12754 13161 -0.579761 . 1 rbs "12754" ; reliable "0"
36521 . CDS 13227 13700 -0.00217621 . 3 rbs "13227" ; reliable "0"
36521 . CDS 13718 14068 -0.564218 . 2 rbs "13718" ; reliable "0"
36521 . CDS 15094 14450 1.0002 . -2 rbs "15094" ; evalue "0.0034" ; description "Response regulator receiver domain" ; reliable "1"
36521 . CDS 15753 15358 -0.525293 . -1 rbs "15753" ; reliable "0"
36521 . CDS 15727 16209 0.247978 . 1 rbs "15727" ; reliable "1"
36521 . CDS 16418 16251 -0.466092 . -3 rbs "16418" ; reliable "0" "1"
```

Fig. 2 Output of the program GISMO (Table 1). The output is in GFF format. For each prediction, the contig name, the start and stop positions, the support vector machine (SVM) score and the reading frame are given. The position of a ribosome binding site (RBS) and a confidence assignment (*1* for high confidence; *0* for low confidence) are also given. If a prediction has a protein domain match in PFAM, the e-value of that hit and the description of the PFAM entry are also reported. Predictions that were discarded in the post-processing phase (removal of overlapping low-confidence predictions) are labeled as discarded ("1")

for this task are available [21–24]. The abovementioned HMM-based gene finders deviate from this procedure, as they incorporate the ribosome binding site signal directly into the HMM model and identify the optimal start sites simultaneously with the overall optimal sequence parse. Conflicts of overlapping predictions that could not be solved by relocation of the start sites are resolved by the removal of the "weaker" prediction, where this call can be based on intrinsic and on extrinsic information generated in Phase 1.

2.2 Gene Finding in Environmental Sequence Samples

The application of sequencing techniques to DNA samples obtained directly from microbial communities inhabiting a certain environment has spawned the field of environmental or community genomics, also called metagenomics [31]. The field has delivered insights into numerous questions which cannot be addressed by the sequencing of individual lab-cultivated organisms. According to estimates, less than 1 % of all microorganisms can be grown in pure culture with standard techniques. As a result of this, our current knowledge of bacterial and archaeal biology, and also the sequenced genomes, exhibit a strong bias towards four phyla which contain many cultivable organisms. By bypassing the need for pure culture, environmental genomics allows the discovery of novel organisms and protein-coding genes that could not have been obtained otherwise. NGS technology has made an essential contribution to the dissemination of metagenomics, as it enables the deep sequencing of non-cultivable organisms from a single environmental sample [32]. These studies also increase our understanding of the processes and interactions that shape the metabolism, structure, function, and evolution of a microbial community seen as a whole [33]. Metagenomics has broadened the knowledge on bacterial and archaeal phyla in recent years [34] and has increased the number of known bacterial and archaeal phyla to 60, with a much higher number of expected phyla [33, 35]. Evidence for novel phyla can be obtained by the analysis of microbial communities. Sequences coding for marker genes like 16S rRNAs that cannot be assigned unambiguously to any known phyla are likely to originate from organisms of a novel phylum, which is also called a "candidate phylum" [36, 37].

Because microbial genomes with low abundance in an environmental sample are still difficult to identify and reconstruct from metagenomic data, the sequencing of isolated single cells is an alternative approach [38]. Nevertheless, single-cell sequencing also still faces technical challenges, such as amplification biases and errors, which complicate genome recovery [39].

Sequences from an environmental sample can sometimes represent a very large number of organisms, each of which is represented approximately in proportion to its abundance in the habitat, as NGS protocols in general do not require targeted cloning and amplification. For environments with up to medium organismic complexity (with several 100 community members), enough sequence is often sampled to allow the reconstruction of (nearly) complete genomes for the more abundant organisms in a sample [40, 41]. In addition, numerous short fragments and unassembled singleton reads are generated from many less abundant organisms. The metagenome data create challenges at all levels for the bioinformatics tools that have been established in genome sequence analysis [42, 43], including the gene finding programs currently available. First, the short sequences that are created frequently contain truncated and frameshifted genes, which many of the standard gene finders have not been designed to identify. Second, the sequence properties of both CDSs and noncoding sequences can vary considerably between genomes [44], which might cause difficulties in creating an appropriate intrinsic model (*see* **Note 2**). During recent years, gene finding programs specifically designed for application on metagenome sequence samples have become available (Table 2), which allows identification of truncated and frameshifted genes with good specificity [45, 46].

The recovery of genes from a complex metagenome sample is still a difficult task. On the one hand, samples from an environment can be prepared with different DNA preparation methods and the combination of sequencing results can refine the read assembly and gene prediction [47]. On the other hand, multiple algorithms can be applied to the metagenome sequences to overcome individual issues within the algorithms for specific sample properties, thereby complementing each other in gene predictions [48]. This is called the "consensus approach."

Table 2
Metagenomic gene finders

Tool	Method	URL
MetaGeneMark [49]	Update of GeneMark.hmm with improved model parameters for metagenomic samples	http://ergatis.diagcomputing.org/cgi/documentation.cgi?article=components&page=metagenemark
MetaGUN [50]	SVM-based. Phylogenetic binning and assignment of protein sequences to each bin	ftp://162.105.160.21/pub/metagun/
MOCAT [51]	Pipeline for metagenome samples including gene prediction. Uses Prodigal and MetaGeneMark	http://vm-lux.embl.de/~kultima/MOCAT/
FragGeneScan [46]	HMM-based. Combines sequencing error models with codon usage	http://sourceforge.net/projects/fraggenescan/files/latest/download
Prodigal [52]	Log-likelihood coding statistics trained from data	http://compbio.ornl.gov/prodigal/

2.3 Gene Finding in Eukaryotes

Gene prediction for bacterial and archaeal genomes has reached high levels of accuracy and is more accurate than that for eukaryotic organisms (*see* **Note 3**). However, newer programs for eukaryotic datasets have reached more than 90 % specificity and 80 % sensitivity, at least for exon identification in compact eukaryotic genomes [53] (*see also* Subheading 2.3.4). Nevertheless, the process of gene prediction in eukaryotic genomes is still a complex and challenging problem for several reasons. First, only a small fraction of a eukaryotic genome sequence corresponds to protein-encoding exons, which are embedded in vast amounts of noncoding sequence. Second, the gene structure is complex (Fig. 3).

The CDS encoding the final protein product can be located discontinuously in two or more exonic regions, which are sometimes very short and are separated from each other by an intron sequence. The junctions of exon–intron boundaries are characterized by splice sites on the initial transcript, which guide intron removal in fabrication of the ripe mRNA transcript at the spliceosome. Additional signal sequences, such as an adenylation signal, are found in proximity to the transcript ends. These determine a cleavage site corresponding to the end of the ripe transcript to which a polyadenylation (polyA) tail is added for stability. The issue is further complicated by the fact that genes can have alternative splice and polyA sites, as well as alternative translation and transcription initiation sites. Third, due to the massive sequencing requirements, additional eukaryotic genomes which can be used to study sequence conservation are becoming available more slowly than their bacterial and archaeal counterparts. So far, 283 eukaryotic genomes have been finished compared to 241 archaeal and 5,047 bacterial genomes in the current GOLD release v6 [54].

The complex organization of eukaryotic genes makes determination of the correct gene structure the most difficult problem in eukaryotic gene prediction. Signals of functional sites are very informative—for instance, splice site signals are the best means of locating exon–intron boundaries. Methods which are designed to identify these or other functional signals such as promoter or polyA sites are generally referred to as "signal sensors." Methods that classify genomic sequences into coding or noncoding content are called "content sensors." Content sensors can use all of the above-mentioned sources of information for gene identification—

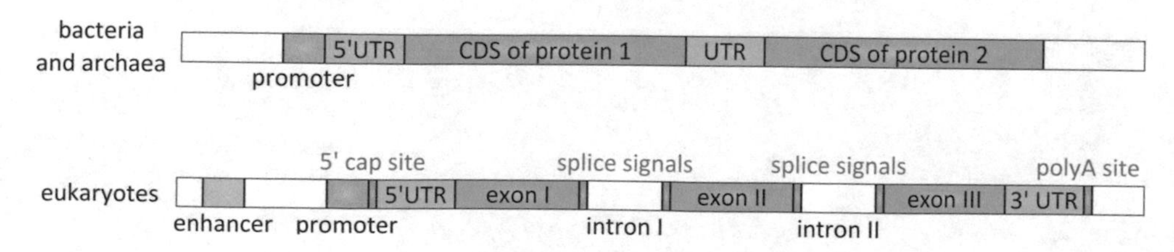

Fig. 3 Comparison of bacterial and archaeal vs. eukaryote gene structure

extrinsic information such as transcript or protein sequences, comparisons to genomic sequences from other organisms or intrinsic information about CDS-specific properties. Extrinsic evidence such as the sequences of known proteins and sequenced cDNA libraries tends to produce reliable predictions, but it is biased towards genes which are highly and ubiquitously expressed. Programs for de novo gene identification are also categorized as intrinsic "one genome" approaches, or "dual" or "multi-genome" approaches that use comparative information from other genomes [55]. Some gene finders use HMMs for the task of "decoding" the unknown gene structure from the sequence, which allows the combination of content and signal sensor modules into a single, coherent probabilistic model with biological meaning. Typically, these models include states for exons, introns, splice sites, polyA signals, start codons, stop codons, and intergenic regions, as well as an additional model for single exon genes. A similar approach is used by SVM classifiers that incorporate both intrinsic and extrinsic information into their model. Another statistical learning method known as conditional random fields (CRFs) has recently been implemented in several gene finding programs (Table 3) that combine extrinsic and intrinsic sources of information into a single model. *See* [55–57] for recent reviews of the programs and techniques used in the field.

Table 3
Eukaryotic gene finders

Tool	Method	URL (if available)
CONRAD [58]	Semi-Markov CRF method combining extrinsic and intrinsic information. The underlying model can easily be exchanged to tackle other problems	http://www.broadinstitute.org/annotation/conrad/
CONTRAST [59]	CRF-based. Combines local classifiers with the global gene structure model	http://contra.stanford.edu/contrast/
eCRAIG [60]	CRF-based	
mGene [61]	Structural HMM combined with discrimination training techniques similar to SVMs	http://mgene.org/
GeneMark.hmm [15]	HMM-based. Iteratively trains and improves the model in an unsupervised manner	http://www.genepro.com/Manuals/PrGM/PrGM_usage.aspx
Augustus [62]	General HMM	http://bioinf.uni-greifswald.de/webaugustus/
SNAP [63]	Semi-HMM	http://korflab.ucdavis.edu/software.html
GIIRA [64]	Only uses extrinsic evidence of RNA-seq data and reassigns ambiguous reads	https://sourceforge.net/projects/giira/

2.3.1 RNA-seq

The sequencing of an organism's entire transcriptome using reverse transcription and NGS technologies generates data known as RNA-seq. It allows us to identify novel transcripts, rare or alternatively spliced variants, transcription initiation sites and gene fusions [65–67]. RNA-seq data also approximate transcript abundances accurately, as the genome sequence coverage correlates with this. The analysis of RNA-seq data comes with its own difficulties, which have been addressed as far as possible in newer programs. The main problem can be that sequence assemblers that were initially designed for DNA reads make assumptions about similarity in sequencing coverage (except for repetitive regions) throughout the entire genome or assume that both strands are sequenced simultaneously [65, 67], both of which are incorrect for RNA-seq data. An overview of the first steps of de novo assembly and annotation of an eukaryote genome using NGS data and including RNA-seq data as evidence for gene annotation is given in [68].

2.3.2 Gene Prediction and Annotation Pipelines

Not only gene finding programs but also entire computational pipelines for data analysis, including gene prediction and annotation, are of great importance for genome projects (*see* **Note 4**). A brief description of the widely used Ensembl gene prediction and annotation pipeline [69] and its updates [70] is therefore given at this point. The complete pipeline involves a wide variety of programs (Fig. 4). In this pipeline, repetitive elements are initially identified and masked to remove them from the analyzed input.

1. The sequences of known proteins from the organism are mapped to a location on the genomic sequence. Local alignment programs are used for a prior reduction of the sequence search space, and the HMM-based GeneWise program [71] is used for the final sequence alignment.
2. An ab initio predictor such as Genscan [72] is run. For predictions confirmed by the presence of homologs, these homologs are aligned to the genome using GeneWise, as before.
3. Simultaneously with **step 1**, Exonerate [73] is used to align known cDNA sequences to the genome sequence. If RNA-seq data are integrated directly, gaps in the protein-coding models identified in **step 1** are filled in or added if they are completely missing.
4. The candidates found via this procedure are merged to create consensus transcripts with 3′ untranslated regions (UTRs), CDSs, and 5′ UTRs.
5. Redundant transcripts are merged and genes are identified, which, by Ensembl's definition, correspond to sets of transcripts with overlapping exons.

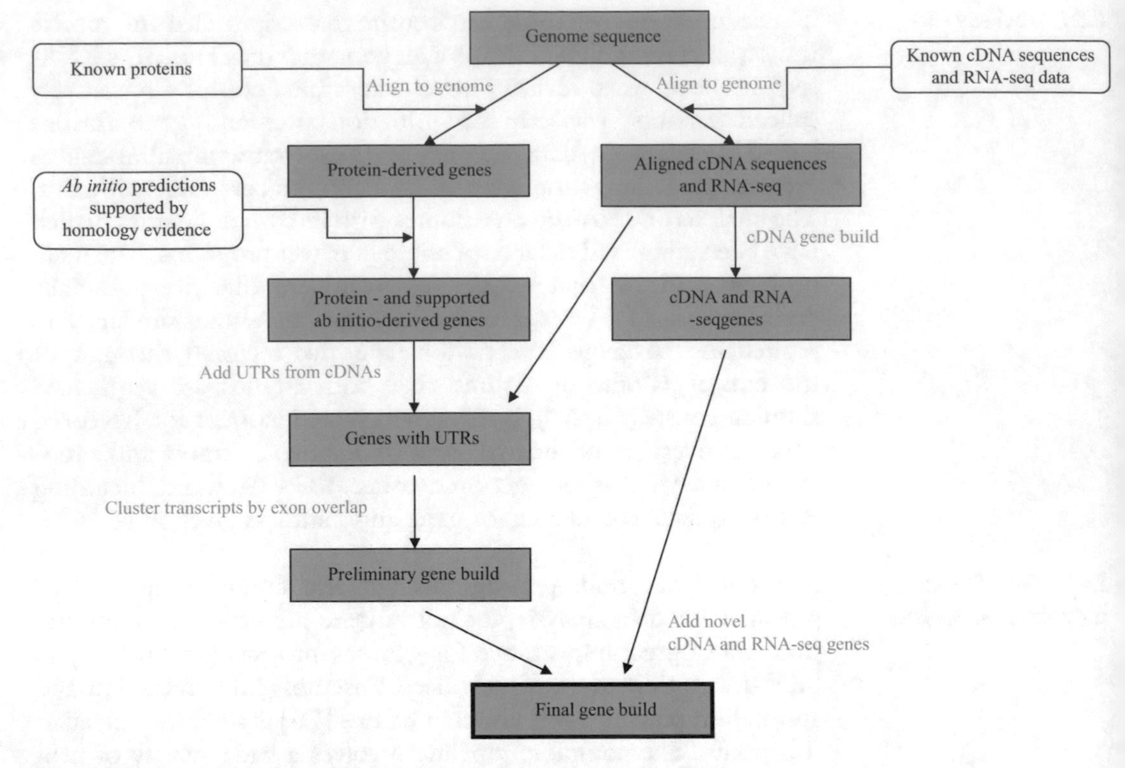

Fig. 4 Overview of the Ensembl pipeline for eukaryotic gene prediction

6. Novel cDNA genes which do not match with any exons of the protein-coding genes identified so far are added to create the final gene set.

The Ensembl pipeline leans strongly towards producing a specific prediction with few false positives rather than a sensitive prediction with few false negatives. Every gene prediction that is produced is supported by direct extrinsic evidence, such as transcript sequences, known protein sequences from the organism's proteome, or known protein sequences of related organisms. It is important to note that for the genomes of different organisms, this procedure is applied with slight variations, depending on the resources available. Since 2012, supporting evidence provided by RNA-seq data has been integrated into the process. For some species, the additional data were directly considered in the annotation process, whereas for others, the results of the standard pipeline were updated based on RNA-seq evidence [70]. For the latter approach, an update script was developed and applied to the zebrafish genome [74]. Other annotation pipelines that are very similar to Ensembl exist, such as the UCSC known genes [75] and the publicly available tool Maker [76]. These also use extrinsic evidence from protein or transcript sequences and subsequently remove redundant predictions.

2.3.3 De Novo Prediction in Novel Genomes

Traditionally, the models of eukaryotic de novo gene finding programs are trained in a supervised manner, in order to learn the optimal, species-specific parameters for splice sites, intron lengths and content sensor models. Their complexity requires large and reliable training sets, which can best be derived by mapping transcript sequence information to the genome. In cases where sufficient information is not available, models that have been created for other species can be used, although this can deliver suboptimal results. The programs SNAP [63] and Genemark.hmm ES-3.0 [77] employ two different techniques to circumvent this problem, and allow high-quality gene prediction for genomes where little reliable annotation is available. SNAP uses an iterative "bootstrap" procedure, in which the results initially derived by application of models from other organisms are used in an iterative manner for model training and prediction refinement. Genemark.hmm derives its model in a completely unsupervised manner from the unannotated genomic sequence, starting with broadly set generic model parameters, which are iteratively refined based on the prediction/training set obtained from the raw genomic sequence.

2.3.4 Accuracy Assessment and Refining Annotation

The increasing availability of high-coverage genomic sequence for groups of related eukaryotic organisms has resulted in significant advances in de novo gene finding accuracy (*see* **Note 5**). To assess the accuracy of gene finders, they are applied to already annotated genomes and the results are compared with the real annotations on different levels: base-pair level, exon level and gene (structure) level, with increasing stringency. On the base-pair level, only the differentiation between coding and noncoding nucleotides has to be correct for the given gene annotations. On the more stringent exon level, start and end coordinates for exons must be specified correctly. For the higher and more stringent levels, like the gene level, accuracy is measured by the correct identification of translation start and end sites, and all exon–intron boundaries for at least one transcript of a particular gene. For compact eukaryotic genomes, multi-genome approaches achieve up to 99 % specificity and 93 % sensitivity in de novo prediction on the base-pair level, 91 % specificity and 83 % sensitivity on the exon level, and 80 % specificity and 57 % sensitivity on complete gene structures [53]. Experimental follow-up experiments led to the biological verification of 42–75 % of novel predictions selected for testing [61, 78, 79]. However, performance is still markedly lower for the mammalian genomes because of their larger fractions of noncoding sequences and pseudogenes.

For the human genome, the ENCODE (Encyclopedia of DNA Elements) project was launched in September 2003, which ultimately aims to produce an accurate and experimentally verified annotation of all functional elements in the human genome. In the pilot phase, the goal was to produce an accurate, experimentally verified annotation for 30 megabases (1 %) of the human genome and to delineate a high-throughput strategy that would be suitable for application to the complete genome sequence.

As part of this effort, the EGASP (ENCODE Genome Annotation Assessment Project) '05 was organized [80], which brought together more than 20 teams working on computational gene prediction. Evaluation of the different programs on the high-quality annotation map showed, not surprisingly, that the extrinsic programs which use very similar information as the human annotators perform best. Of the de novo prediction programs, those including comparative genomic analyses came in second; the programs making predictions based solely on intrinsic genomic evidence came last. Comparisons with existing annotation and the subsequent experimental evaluation of several hundred de novo predictions showed that only very few human genes are not detected by computational means, but even for the best programs, the accuracy on gene structure level was only about 50 % [81]. Of the experimentally validated de novo predictions, only a few (3.2 %) turned out to correspond to real, previously unknown exons.

The results of different functional analyses within ENCODE were made available, such as information on SNP–disease associations, DNA–protein interaction sites, biochemical RNA- or chromatin-associated events, and more [82]. The database and results are still updated frequently to increase knowledge on the human genome. So far, 2886 experiments have been published by the ENCODE consortium and are available for access [83]. After successful completion of the ENCODE pilot phase, another consortium started to annotate all genes and gene variants for the human genome, a project called GENCODE [84]. Its current version (V19) includes 20,345 protein-coding genes. The GENCODE gene annotations were automatically created with the Ensembl pipeline (including supporting evidence) and curated manually afterwards.

2.3.5 Performance Estimates on Nematode Genomes: The nGASP Competition

To estimate the performance of a wide range of eukaryotic gene finders on less complex nematode genomes, the "nematode Genome Annotation Assessment Project" (nGASP) was launched [53]. The *Caenorhabditis elegans* genome is well studied and annotated, but genomes of other *Caenorhabditis* species were missing accurate annotations. Overall, 17 groups from around the world participated in the nGASP competition. The resulting predictions were evaluated using experimentally confirmed annotations. Overall, "combiner" algorithms, which also use the multi-genome approach for de novo prediction, performed best. The best programs reached more than 91 % specificity and 83 % sensitivity on the exon level, but 80 % sensitivity and 57 % specificity on complete gene structures, showing that these gene finders still have issues, particularly with genes with an unusually large number of exons or exons of unusual length.

The authors of the subsequently published program eCRAIG [60] used the data from the nGASP competition for their evaluation to show its superior accuracy. In contrast to the low specificity of 57 % for the predictions of the best program on the gene level for the *C. elegans* genome seen in the nGASP competition, eCRAIG showed a specificity of 66.8 % with a simultaneously high sensitivity of 82.7 %.

3 Conclusions

Automated gene finding in genome sequences has a long history since its beginning in the late 1980s and can be considered to be one of the more mature fields in bioinformatics with many sophisticated programs available for data analysis. Nowadays high levels of accuracy have been reached for gene finding in bacterial and archaeal genomes as well as in eukaryotic genomes of lower complexity. Nevertheless, for applications such as the analysis of metagenome sequence samples and the exact identification of gene structures in eukaryotes of a higher complexity, improvements are still required. The recent development of RNA-seq techniques has led to the discovery of many new genes in higher eukaryotes. As these have an essential contribution to gene finding especially in newly sequenced genomes, RNA-seq data should always be considered for such sequencing projects in the near future if possible.

4 Notes

1. The periodic skew towards high GC content in the third codon position of the CDSs in GC-rich genomes can be visualized with "frame plots," where the phase-specific GC-content of the sequence (averaged over a sliding sequence window) is visualized in three different curves (Fig. 5). Such plots are commonly used by annotators for start site annotation and are part of many genome annotation packages.
2. Currently, genes in metagenome samples in some studies are identified based on sequence similarities only, which leaves genes that do not exhibit similarities to known genes (or to other genes found in the sample) undiscovered. By translating the consensus read sequences into all six reading frames, genes are identified by comparing the translated sequences with databases

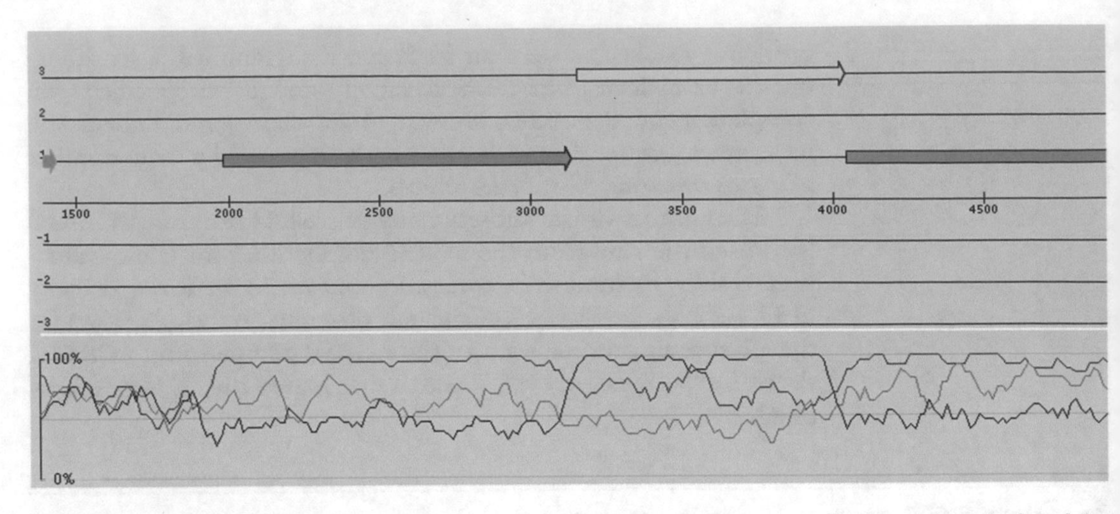

Fig. 5 The location of genes in GC-rich genomes is indicated by the frame-specific GC content. A plot of the frame-specific GC content was computed for a sliding window 26 bp in size that was moved by a step size of 5 across the sequence. The GC content is plotted in the lower panel for the three frames—Frame 3, Frame 2 and Frame 1. The *upper panel* shows the location of annotated protein-coding genes in the genome (*arrows*)

of protein sequences. The advantage is that functional annotations are given directly and thereby functional predictions for the organisms can be made [85]. An alternative procedure is de novo prediction by creating a universal intrinsic model of bacterial and archaeal sequence composition from known genomic sequences. This prevents a bias in the model towards the dominating organisms of a metagenome community, but could still be biased towards the phyla that have been characterized so far. Such gene predictions can be partially confirmed by sequence similarity comparisons to genes identified from other metagenome samples.

3. The number of completely sequenced bacterial and archaeal organisms now includes several thousand different isolates. The existence of such a large and diverse data set allows a thorough assessment of gene finding accuracy across the wide variety of sequenced organisms. For several programs, evaluations on more than 100 genome sequences have been undertaken. Prediction of CDSs for bacterial and archaeal genomes has generally become very accurate. The gene finder Reganor [28], which incorporates evidence from the programs Glimmer and Critica, reproduces bacterial annotations with 95 % specificity and 98 % sensitivity in its identification of "certain" genes that are supported by external evidence. The SVM-based gene finder GISMO [17] and the HMM-based EasyGene [86] reach sensitivity levels of up to 99 % and display high levels of specificity. Automated gene prediction has become so accurate that a

considerable fraction of the additional "false positive" predictions according to the manually compiled annotations seem to represent real genes that have not been annotated [87] (*see* **Note 5**). Programs for start site identification, such as TICO [23] and GS-Finder [21], have also been found to perform with more than 90 % accuracy. However, due to the currently limited numbers of CDSs with experimentally verified N-termini, it is too early to generalize this observation. A challenge that remains to be addressed by bacterial and archaeal gene finders is the correct identification of "short" genes with fewer than 100 codons. Evidence indicates that the automatic identification as well as the current annotation for such short genes is incomplete and need to be improved [28, 86, 88].

4. Gene finding, gene prediction, or gene annotation is often used synonymously in the literature, although the meaning is not always the same. Some gene finders also report structural annotation information such as the location of splice sites, 5′ and 3′ UTRs or polyA signals, whereas others only predict the genomic location of the entire transcript without further information. Additionally, "combiner" algorithms or annotation pipelines such as the Ensembl pipeline also return functional annotations for some genes.
5. Annotations are not perfect and are not always supported by experimental evidence, so accuracy estimates obtained with this standard of truth can be questionable. Accordingly, claims that additional predictions might correspond to real but currently missing genes could be correct, just as genes that were predicted in accordance with the annotation might possibly be false positives in both cases. Thus accuracy estimates should generally be based on further evidence that supports or refutes the predictions. A strong indicator of a valid prediction, for instance, is its location in a cluster of genes with homologs in a similar arrangement in related genomes [89].

Acknowledgments

The authors thank Lutz Krause, Alan Grossfield, and Augustine Tsai for their comments. A.K. acknowledges the support by the Düsseldorf School of Oncology (funded by the Comprehensive Cancer Center Düsseldorf/Deutsche Krebshilfe and the Medical Faculty HHU Düsseldorf).

References

1. Metzker ML (2010) Sequencing technologies—the next generation. Nat Rev Genet 11:31–46
2. Benson DA, Cavanaugh M, Clark K, Karsch-Mizrachi I, Lipman DJ, Ostell J, Sayers EW (2013) GenBank. Nucleic Acids Res 41: D36–D42
3. Dong H, Nilsson L, Kurland CG (1996) Co-variation of tRNA abundance and codon usage in *Escherichia coli* at different growth rates. J Mol Biol 260:649–663
4. Ikemura T (1981) Correlation between the abundance of *Escherichia coli* transfer RNAs and the occurrence of the respective codons in its protein genes: a proposal for a synonymous codon choice that is optimal for the *E. coli* translational system. J Mol Biol 151:389–409
5. Sharp PM, Bailes E, Grocock RJ, Peden JF, Sockett RE (2005) Variation in the strength of selected codon usage bias among bacteria. Nucleic Acids Res 33:1141–1153
6. Rocha EP (2004) Codon usage bias from tRNA's point of view: redundancy, specialization, and efficient decoding for translation optimization. Genome Res 14:2279–2286
7. Wallace EW, Airoldi EM, Drummond DA (2013) Estimating selection on synonymous codon usage from noisy experimental data. Mol Biol Evol 30:1438–1453
8. McHardy AC, Pühler A, Kalinowski J, Meyer F (2004) Comparing expression level-dependent features in codon usage with protein abundance: an analysis of 'predictive proteomics'. Proteomics 4:46–58
9. Saunders R, Deane CM (2010) Synonymous codon usage influences the local protein structure observed. Nucleic Acids Res 38:6719–6728
10. Hooper SD, Berg OG (2000) Gradients in nucleotide and codon usage along *Escherichia coli* genes. Nucleic Acids Res 28:3517–3523
11. Fickett JW, Tung CS (1992) Assessment of protein coding measures. Nucleic Acids Res 20:6441–6450
12. Hayashi T, Makino K, Ohnishi M, Kurokawa K, Ishii K, Yokoyama K, Han CG, Ohtsubo E, Nakayama K, Murata T et al (2001) Complete genome sequence of enterohemorrhagic *Escherichia coli* O157:H7 and genomic comparison with a laboratory strain K-12. DNA Res 8:11–22
13. Besemer J, Lomsadze A, Borodovsky M (2001) GeneMarkS: a self-training method for prediction of gene starts in microbial genomes. Implications for finding sequence motifs in regulatory regions. Nucleic Acids Res 29:2607–2618
14. Larsen TS, Krogh A (2003) EasyGene—a prokaryotic gene finder that ranks ORFs by statistical significance. BMC Bioinformatics 4:21
15. Lukashin AV, Borodovsky M (1998) GeneMark.hmm: new solutions for gene finding. Nucleic Acids Res 26:1107–1115
16. Delcher AL, Harmon D, Kasif S, White O, Salzberg SL (1999) Improved microbial gene identification with GLIMMER. Nucleic Acids Res 27:4636–4641
17. Krause L, McHardy AC, Nattkemper TW, Pühler A, Stoye J, Meyer F (2007) GISMO—gene identification using a support vector machine for ORF classification. Nucleic Acids Res 35:540–549
18. Mahony S, McInerney JO, Smith TJ, Golden A (2004) Gene prediction using the Self-Organizing Map: automatic generation of multiple gene models. BMC Bioinformatics 5:23
19. Ochman H, Lawrence JG, Groisman EA (2000) Lateral gene transfer and the nature of bacterial innovation. Nature 405:299–304
20. Hayes WS, Borodovsky M (1998) How to interpret an anonymous bacterial genome: machine learning approach to gene identification. Genome Res 8:1154–1171
21. Ou HY, Guo FB, Zhang CT (2004) GS-Finder: a program to find bacterial gene start sites with a self-training method. Int J Biochem Cell Biol 36:535–544
22. Suzek BE, Ermolaeva MD, Schreiber M, Salzberg SL (2001) A probabilistic method for identifying start codons in bacterial genomes. Bioinformatics 17:1123–1130
23. Tech M, Pfeifer N, Morgenstern B, Meinicke P (2005) TICO: a tool for improving predictions of prokaryotic translation initiation sites. Bioinformatics 21:3568–3569
24. Zhu HQ, Hu GQ, Ouyang ZQ, Wang J, She ZS (2004) Accuracy improvement for identifying translation initiation sites in microbial genomes. Bioinformatics 20:3308–3317
25. Shibuya T, Rigoutsos I (2002) Dictionary-driven prokaryotic gene finding. Nucleic Acids Res 30:2710–2725
26. Badger JH, Olsen GJ (1999) CRITICA: coding region identification tool invoking comparative analysis. Mol Biol Evol 16:512–524
27. Frishman D, Mironov A, Mewes HW, Gelfand M (1998) Combining diverse evidence for gene

recognition in completely sequenced bacterial genomes. Nucleic Acids Res 26:2941–2947
28. McHardy AC, Goesmann A, Puhler A, Meyer F (2004) Development of joint application strategies for two microbial gene finders. Bioinformatics 20:1622–1631
29. Tech M, Merkl R (2003) YACOP: enhanced gene prediction obtained by a combination of existing methods. In Silico Biol 3:441–451
30. Guo FB, Ou HY, Zhang CT (2003) ZCURVE: a new system for recognizing protein-coding genes in bacterial and archaeal genomes. Nucleic Acids Res 31:1780–1789
31. Venter JC, Remington K, Heidelberg JF, Halpern AL, Rusch D, Eisen JA, Wu D, Paulsen I, Nelson KE, Nelson W et al (2004) Environmental genome shotgun sequencing of the Sargasso Sea. Science 304:66–74
32. Wu D, Hugenholtz P, Mavromatis K, Pukall R, Dalin E, Ivanova NN, Kunin V, Goodwin L, Wu M, Tindall BJ (2009) A phylogeny-driven genomic encyclopaedia of bacteria and archaea. Nature 462:1056–1060
33. Walker A (2014) Adding genomic 'foliage' to the tree of life. Nat Rev Microbiol 12:78
34. Hugenholtz P (2002) Exploring prokaryotic diversity in the genomic era. Genome Biol 3:0003.1–0003.8
35. Kantor RS, Wrighton KC, Handley KM, Sharon I, Hug LA, Castelle CJ, Thomas BC, Banfield JF (2013) Small genomes and sparse metabolisms of sediment-associated bacteria from four candidate phyla. MBio 4: e00708–e00713
36. Harris JK, Caporaso JG, Walker JJ, Spear JR, Gold NJ, Robertson CE, Hugenholtz P, Goodrich J, McDonald D, Knights D (2012) Phylogenetic stratigraphy in the Guerrero Negro hypersaline microbial mat. ISME J 7:50–60
37. Ley RE, Harris JK, Wilcox J, Spear JR, Miller SR, Bebout BM, Maresca JA, Bryant DA, Sogin ML, Pace NR (2006) Unexpected diversity and complexity of the Guerrero Negro hypersaline microbial mat. Appl Environ Microbiol 72:3685–3695
38. Rinke C, Schwientek P, Sczyrba A, Ivanova NN, Anderson IJ, Cheng J-F, Darling A, Malfatti S, Swan BK, Gies EA (2013) Insights into the phylogeny and coding potential of microbial dark matter. Nature 499:431–437
39. Ning L, Liu G, Li G, Hou Y, Tong Y, He J (2014) Current challenges in the bioinformatics of single cell genomics. Front Oncol 4:7
40. Pope P, Smith W, Denman S, Tringe S, Barry K, Hugenholtz P, McSweeney C, McHardy A, Morrison M (2011) Isolation of Succinivibrionaceae implicated in low methane emissions from Tammar wallabies. Science 333:646–648
41. Iverson V, Morris RM, Frazar CD, Berthiaume CT, Morales RL, Armbrust EV (2012) Untangling genomes from metagenomes: revealing an uncultured class of marine Euryarchaeota. Science 335:587–590
42. Chen K, Pachter L (2005) Bioinformatics for whole-genome shotgun sequencing of microbial communities. PLoS Comput Biol 1:106–112
43. Scholz MB, Lo C-C, Chain PS (2012) Next generation sequencing and bioinformatic bottlenecks: the current state of metagenomic data analysis. Curr Opin Biotechnol 23:9–15
44. Sandberg R, Branden CI, Ernberg I, Coster J (2003) Quantifying the species-specificity in genomic signatures, synonymous codon choice, amino acid usage and G + C content. Gene 311:35–42
45. Krause L, Diaz NN, Bartels D, Edwards RA, Puhler A, Rohwer F, Meyer F, Stoye J (2006) Finding novel genes in bacterial communities isolated from the environment. Bioinformatics 22:e281–e289
46. Rho M, Tang H, Ye Y (2010) FragGeneScan: predicting genes in short and error-prone reads. Nucleic Acids Res 38:e191
47. Albertsen M, Hugenholtz P, Skarshewski A, Nielsen KL, Tyson GW, Nielsen PH (2013) Genome sequences of rare, uncultured bacteria obtained by differential coverage binning of multiple metagenomes. Nat Biotechnol 31:533–538
48. Yok NG, Rosen GL (2011) Combining gene prediction methods to improve metagenomic gene annotation. BMC Bioinformatics 12:20
49. Zhu W, Lomsadze A, Borodovsky M (2010) Ab initio gene identification in metagenomic sequences. Nucleic Acids Res 38:e132
50. Liu Y, Guo J, Hu G, Zhu H (2013) Gene prediction in metagenomic fragments based on the SVM algorithm. BMC Bioinformatics 14:S12
51. Kultima JR, Sunagawa S, Li J, Chen W, Chen H, Mende DR, Arumugam M, Pan Q, Liu B, Qin J (2012) MOCAT: a metagenomics assembly and gene prediction toolkit. PLoS One 7: e47656
52. Hyatt D, Chen G-L, LoCascio PF, Land ML, Larimer FW, Hauser LJ (2010) Prodigal: prokaryotic gene recognition and translation initiation site identification. BMC Bioinformatics 11:119
53. Coghlan A, Fiedler TJ, McKay SJ, Flicek P, Harris TW, Blasiar D, Stein LD (2008)

nGASP—the nematode genome annotation assessment project. BMC Bioinformatics 9:549
54. Reddy TBK, Thomas A, Stamatis D, Bertsch J, Isbandi M, Jansson J, Mallajosyula J, Pagani I, Lobos E, Kyrpides N (2015) The Genomes OnLine Database (GOLD) v. 5: a metadata management system based on a four level (meta)genome project classification. *Nucleic Acids Res.* 43:D1099–1106
55. Brent MR, Guigo R (2004) Recent advances in gene structure prediction. Curr Opin Struct Biol 14:264–272
56. Brent MR (2008) Steady progress and recent breakthroughs in the accuracy of automated genome annotation. Nat Rev Genet 9:62–73
57. Sleator RD (2010) An overview of the current status of eukaryote gene prediction strategies. Gene 461:1–4
58. DeCaprio D, Vinson JP, Pearson MD, Montgomery P, Doherty M, Galagan JE (2007) Conrad: gene prediction using conditional random fields. Genome Res 17:1389–1398
59. Gross SS, Do CB, Sirota M, Batzoglou S (2007) CONTRAST: a discriminative, phylogeny-free approach to multiple informant *de novo* gene prediction. Genome Biol 8:R269
60. Bernal A, Crammer K, Pereira F (2012) Automated gene-model curation using global discriminative learning. Bioinformatics 28:1571–1578
61. Schweikert G, Zien A, Zeller G, Behr J, Dieterich C, Ong CS, Philips P, De Bona F, Hartmann L, Bohlen A (2009) mGene: accurate SVM-based gene finding with an application to nematode genomes. Genome Res 19:2133–2143
62. Stanke M, Diekhans M, Baertsch R, Haussler D (2008) Using native and syntenically mapped cDNA alignments to improve de novo gene finding. Bioinformatics 24:637–644
63. Korf I (2004) Gene finding in novel genomes. BMC Bioinformatics 5:59
64. Zickmann F, Lindner MS, Renard BY (2013) GIIRA–RNA-Seq driven gene finding incorporating ambiguous reads. Bioinformatics 30:606–613
65. Martin JA, Wang Z (2011) Next-generation transcriptome assembly. Nat Rev Genet 12:671–682
66. Wang Z, Gerstein M, Snyder M (2009) RNA-Seq: a revolutionary tool for transcriptomics. Nat Rev Genet 10:57–63
67. Ozsolak F, Milos PM (2011) RNA sequencing: advances, challenges and opportunities. Nat Rev Genet 12:87–98
68. Yandell M, Ence D (2012) A beginner's guide to eukaryotic genome annotation. Nat Rev Genet 13:329–342
69. Curwen V, Eyras E, Andrews TD, Clarke L, Mongin E, Searle SM, Clamp M (2004) The Ensembl automatic gene annotation system. Genome Res 14:942–950
70. Flicek P, Ahmed I, Amode MR, Barrell D, Beal K, Brent S, Carvalho-Silva D, Clapham P, Coates G, Fairley S (2013) Ensembl 2013. Nucleic Acids Res 41:D48–D55
71. Birney E, Clamp M, Durbin R (2004) GeneWise and Genomewise. Genome Res 14:988–995
72. Burge C, Karlin S (1997) Prediction of complete gene structures in human genomic DNA. J Mol Biol 268:78–94
73. Slater GS, Birney E (2005) Automated generation of heuristics for biological sequence comparison. BMC Bioinformatics 6:31
74. Collins JE, White S, Searle SM, Stemple DL (2012) Incorporating RNA-seq data into the zebrafish Ensembl genebuild. Genome Res 22:2067–2078
75. Hsu F, Kent WJ, Clawson H, Kuhn RM, Diekhans M, Haussler D (2006) The UCSC known genes. Bioinformatics 22:1036–1046
76. Cantarel BL, Korf I, Robb SM, Parra G, Ross E, Moore B, Holt C, Alvarado AS, Yandell M (2008) MAKER: an easy-to-use annotation pipeline designed for emerging model organism genomes. Genome Res 18:188–196
77. Lomsadze A, Ter-Hovhannisyan V, Chernoff YO, Borodovsky M (2005) Gene identification in novel eukaryotic genomes by self-training algorithm. Nucleic Acids Res 33:6494–6506
78. Tenney AE, Brown RH, Vaske C, Lodge JK, Doering TL, Brent MR (2004) Gene prediction and verification in a compact genome with numerous small introns. Genome Res 14:2330–2335
79. Wei C, Lamesch P, Arumugam M, Rosenberg J, Hu P, Vidal M, Brent MR (2005) Closing in on the *C. elegans* ORFeome by cloning TWINSCAN predictions. Genome Res 15:577–582
80. Guigo R, Reese MG (2005) EGASP: collaboration through competition to find human genes. Nat Methods 2:575–577
81. Guigo R, Flicek P, Abril JF, Reymond A, Lagarde J, Denoeud F, Antonarakis S, Ashburner M, Bajic VB, Birney E et al (2006) EGASP: the human ENCODE genome annotation assessment project. Genome Biol 7(Suppl 1):S2
82. ENCODE Project Consortium (2012) An integrated encyclopedia of DNA elements in the human genome. Nature 489:57–74
83. Rosenbloom KR, Sloan CA, Malladi VS, Dreszer TR, Learned K, Kirkup VM, Wong MC, Maddren M, Fang R, Heitner SG (2013) ENCODE data in the UCSC genome browser: year 5 update. Nucleic Acids Res 41:D56–D63

84. Harrow J, Frankish A, Gonzalez JM, Tapanari E, Diekhans M, Kokocinski F, Aken BL, Barrell D, Zadissa A, Searle S (2012) GENCODE: the reference human genome annotation for the ENCODE project. Genome Res 22:1760–1774
85. Sharpton TJ (2014) An introduction to the analysis of shotgun metagenomic data. Front Plant Sci 5:209
86. Nielsen P, Krogh A (2005) Large-scale prokaryotic gene prediction and comparison to genome annotation. Bioinformatics 21:4322–4329
87. Linke B, McHardy AC, Krause L, Neuwege H, Meyer F (2006) REGANOR: a gene prediction server for prokaryotic genomes and a database of high quality gene predictions for prokaryotes. Appl Bioinformatics 5:193–198
88. Warren AS, Archuleta J, Feng W-C, Setubal JC (2010) Missing genes in the annotation of prokaryotic genomes. BMC Bioinformatics 11:131
89. Osterman A, Overbeek R (2003) Missing genes in metabolic pathways: a comparative genomics approach. Curr Opin Chem Biol 7:238–251

Chapter 12

Sequence Segmentation with changeptGUI

Edward Tasker and Jonathan M. Keith

Abstract

Many biological sequences have a segmental structure that can provide valuable clues to their content, structure, and function. The program `changept` is a tool for investigating the segmental structure of a sequence, and can also be applied to multiple sequences in parallel to identify a common segmental structure, thus providing a method for integrating multiple data types to identify functional elements in genomes. In the previous edition of this book, a command line interface for `changept` is described. Here we present a graphical user interface for this package, called `changeptGUI`. This interface also includes tools for pre- and post-processing of data and results to facilitate investigation of the number and characteristics of segment classes.

Key words Multiple change-point analysis, Genome segmentation, Functional element discovery, Model selection

1 Introduction

Understanding how biological function is encoded in DNA sequences is an important goal of bioinformatics. One technique for investigating this is to separate the sequence into compositionally homogenous blocks—a technique called "sequence segmentation." The purpose of segmenting DNA sequences is to facilitate the annotation of genomes by associating the characteristics of defined segments with biological functions. Sequence segmentation can be applied to the DNA sequence itself or to other profiles over genomic positions, such as information obtained from DNA sequence alignments (e.g., [1–4]) or epigenetic markers [5]. There is an increasing focus on incorporating multiple data-types in genomic segmentation [6].

The original version of this chapter was revised. An erratum to this chapter can be found at DOI 10.1007/978-1-4939-6622-6_19

Electronic supplementary material: The online version of this chapter (doi:10.1007/978-1-4939-6622-6_12) contains supplementary material, which is available to authorized users.

Jonathan M. Keith (ed.), *Bioinformatics: Volume I: Data, Sequence Analysis, and Evolution*, Methods in Molecular Biology, vol. 1525, DOI 10.1007/978-1-4939-6622-6_12, © Springer Science+Business Media New York 2017

Algama and Keith [7] provide a review of sequence segmentation techniques in the field of bioinformatics in greater detail than is described here: this introduction briefly summarizes the review. There are four main types of technique for sequence segmentation: sliding window analysis, recursive segmentation algorithms, hidden Markov models (HMM), and multiple change-point analysis.

Sliding window techniques were common in the earlier days of bioinformatics and are a relatively simple technique that involves averaging for a chosen property of the sequence over a "window" of fixed size, then "sliding" this window along the sequence and identifying sharp changes in the average. The main drawback of this technique is that the scale of features which can be identified is dependent on the fixed window size and not solely determined by the data; large window sizes tend to 'blur' sharp changes in the property of interest and thus lead to the identification of larger segments; using smaller windows can reduce this problem, but this decreases the signal-to-noise ratio and will lead to the identification of smaller segments, some of which may be artifactual. Although the sliding window technique is not often used in this simple form, more complex variations of this technique still suffer from the need to choose a "window" size.

Recursive segmentation algorithms identify segments by recursively optimizing the positions of segment boundaries based on predefined measures of sequence composition. The algorithms are repeated until statistically significant segment boundaries can no longer be identified.

HMMs are widely applied in probability and statistics and can be used for a wide variety of purposes. Sequence segmentation is modeled with HMMs by assuming the observed sequence is generated by a Markov process. Segment boundaries are defined when the process transitions from one hidden state to another; the sequence of hidden states is also modeled by a Markov process. The *order* of the HMM is the number of preceding sequence positions required to condition the Markov process. Both the *order* and the number of hidden states will generally be unknown a priori and will thus need to be specified. Some approaches however will incorporate inference about the *order* and number of classes as part of the technique. Sequence segmentation using HMMs can be modeled within a Bayesian framework, which allows for the estimation of model parameters in the form of probability distributions and thus provides robust estimates of uncertainty regarding the number and position of change-points. A challenge of Bayesian HMMs, however, is that they can be computationally intensive and thus the application to genome-wide segmentation may be infeasible without simplifying heuristics.

Sequence segmentation by change-point analysis (the term "change-point" refers to segment boundaries), despite developing separately from HMMs, can be thought of as a "zeroth-order"

HMM. The model is "zeroth-order" because the general assumption is that there is no Markov dependence for the observed sequence or the sequence of hidden states. Change-point models thus have fewer parameters than HMMs, which is an advantage when applying the techniques to genome-wide segmentation, particularly within a Bayesian framework, due to the reduced computational burden.

2 Change-Point Analysis

2.1 Software for Sequence Segmentation: changept

Liu and Lawrence [8] introduced a Bayesian multiple change-point model for sequence segmentation in 1999. The Bayesian model has since been extended by Keith and coworkers [1–4, 9] and along with the development of an efficient technique for Gibbs sampling from a distribution with varying dimension [10], has been encoded as the C program `changept`, which samples parameter estimates from the underlying change-point model given a target sequence. This chapter describes some of the practical issues involved in implementing `changept` and the associated graphical user interface (GUI) `changeptGUI`. An attractive feature of `changept` is that it is not only a sequence segmentation algorithm, it also includes the classification of segments into groups that share similar sequence characteristics (these classes can be considered similar to the hidden states in HMMs). This is consistent with the modular nature of DNA functionality; functional elements within DNA sequence, for example transcription factor binding sites (TFBS), often have defined boundaries (segmentation) and similar functional elements will often have similar sequence characteristics (classification).

As is typical of Bayesian methodologies, `changept` does not merely generate a single segmentation, optimized according to some scoring function. Rather, it generates multiple segmentations, sampled from a posterior distribution over the space of all possible segmentations. Markov chain Monte Carlo (MCMC) simulation is applied to estimate the posterior probabilities of the underlying model parameters.

`Changept` estimates, for each genomic position, the probability that the given genomic position belongs to a given segment class for each of the segment classes. As seen later, it is these probabilities in association with the estimates of the segment class characteristics that will be used to guide biological insight. The ability to estimate probabilities in this way derives from the Bayesian modeling framework. However, `changept` is relatively fast compared to alternative Bayesian genomic segmentation methods, and can feasibly be applied to whole eukaryotic genomes. Although a full description of the Bayesian model and sampling algorithm is beyond the scope of this chapter, a few brief explanations are necessary.

2.2 The Bayesian Segmentation and Classification Model (BSCM): A Model for `changept`

It is convenient to think about the observed sequence as a random vector which has been generated given parameters of an underlying model. The sequence is presumed to be drawn from a probability distribution, where the probability distribution is a function of the parameters in the model. The model incorporates three main concepts: *change-points*, *segments*, and *segment classes*.

For a sequence which contains characters from an *alphabet* of any given size (D), a ***segment*** is a region of the sequence within which the probability of observing each of the D characters is consistent throughout the segment. ***Change-points*** are positions in the sequence at which the characters either side of the *change-point* belong to different segments. Each segment in the sequence belongs to one of a fixed number (T) of ***segment classes***.

Individual *segments* are each defined by a vector (θ) that contains the probability of observing each of the D characters in that segment. The θ probability vectors are assumed to be drawn from a mixture of Dirichlet distributions, where the number of components is the number of *segment classes* (T). Each *segment class* is distinguished by a vector, $\alpha^{(j)}$ (with D entries), which is the parameter vector for one of the Dirichlet distributions in the mixture. The mixture proportions are represented by the vector $\pi = (\pi_1, \ldots, \pi_T)$. `Changept` samples from the posterior distribution: the positions of K change-points, the $\alpha^{(j)}$ vectors and the π vector. A full description of the distribution can be found in the supplementary materials of [2].

2.3 MCMC Simulation

Two features of MCMC algorithms that need to be explained here are the burn-in phase and subsampling. MCMC methods involve a Markov chain for which the limiting distribution is the distribution from which one wishes to sample, in this case a posterior distribution over the space of segmentations and classifications. However, the chain approaches this distribution asymptotically, and thus the elements generated early in the chain are not typical and need to be discarded. This early phase of sampling is known as burn-in. Even after burn-in, it is an inefficient (and often infeasible) use of disk space to record all of the segmentations generated by the algorithm. The algorithm therefore asks the user to specify a sampling block length. The chain of segmentations will be divided into blocks of this length and only one element in each block will be recorded for future processing.

Estimates of the $\alpha^{(j)}$ and π vectors are calculated by averaging the values over post burn-in samples. The weights of the D components in the $\alpha^{(j)}$ vectors can be considered as the frequency of each of the D characters in the given segment class. As we see in more detail later this is the feature which distinguishes each segment class.

The probability that a position in the sequence belongs to a given segment class is estimated by first calculating the probability of the position belonging to each class under the posterior distribution for the values of a given sample. This probability is then averaged over each of the post burn-in samples. Again, we see later how these probabilities are used to generate a *map* for the posterior estimate of the segmentation and classification of the given sequence.

3 Changept Application: The GUI and a Worked Example

Changept and the changeptGUI have been developed with the goal of estimating parameters of the segmentation and classification model for DNA sequence alignments. The changeptGUI provides the user a seamless and convenient interface for implementing changept with a given sequence or DNA sequence alignment. Currently changeptGUI is only available for use in a Microsoft Windows operating system. In order to run the changeptGUI, changeptGUI14.exe needs to be located in the same directory as changept-multiseq.exe and readcp-multiseq.exe (the executables and source code are available at LINK). Changept can also be run from the command line; for guidelines on how to use changept from the command line see the "Sequence segmentation" chapter from the first edition of this book.

As an example, throughout this chapter the application of changept is demonstrated using the changeptGUI for the segmentation of the alignment of the mouse genome (mm10) to the human Y chromosome (hg19) (available for download at http://hgdownload.soe.ucsc.edu/goldenPath/hg19/vsMm10/) [11, 12]. Changept need not be exclusively used for this purpose, and can be used for any discrete character sequence, including nonbiological sequences. For the given sequence being modeled, the final output of changeptGUI will be a map of posterior estimates of segments and segment classifications along with estimates of the mixture proportions and character frequencies for each segment class.

3.1 The Input Sequence(s)

Changept, and the underlying BSCM, operates on a sequence composed of letters from an arbitrary alphabet; when using changept, the sequence(s) for which the model parameters are being estimated (which is referred to as the *input sequence*) must be in a single line of a text file.

The changeptGUI is capable of taking as input: a single sequence, parallel sequences, or a DNA sequence alignment in *axt* format (description at https://genome.ucsc.edu/goldenPath/help/axt.html). If using an alignment in *axt* format, the GUI will convert the alignment into a single input sequence before running the changept algorithm. The process of converting the alignment is described here; if another alignment format were to be used, a similar procedure would need to be performed by the user.

3.1.1 The Input Sequence Format

Although the input sequence can be composed of a relatively arbitrary alphabet, the following guidelines should be used:

1. The alphabet should be made up of only alphanumeric characters, i.e., a–Z and 0–9.
2. Capitalized letters will be considered as different characters from lower-case letters.
3. The characters I,J,K,L,M,N,O (all capitals) will be ignored by *changept*. That is, the characters either side of these special characters will be considered as adjacent by *changept*. These characters may be necessary in the sequence as place-holders when there is no appropriate information for that position, for example: when there are gaps in DNA sequence alignment.
4. The character '#' is used to mark the location of fixed change-points; in every sample generated by *changept*, positions marked with '#' will be considered as a change-point. This is useful when concatenating alignment blocks into a single input sequence; it would not make sense to allow the *changept* algorithm to consider two points of the genome that are not adjacent to each other as part of the same segment.

3.1.2 Alignment Encoding

The conversion of a DNA sequence alignment to a single sequence will be referred to as an alignment *encoding*. For a 2-species DNA sequence alignment, the GUI is capable of performing three different encodings: *conservation*, *bi-directional*, or *full*. An alphabet with more characters allows for more information retained in the input sequence; however, a larger alphabet leads to a longer running time of *changept*. When deciding the appropriate encoding, the user should consider the trade-off between information retained and speed. Figure 1 depicts each of the three encodings.

1. The **conservation** encoding assigns the character '1' to positions in the alignment with matches and '0' to positions with mismatches. Gaps in the alignment relative to the reference

```
Alignment
Species 1:          AAAACCCCC-GGGGTTTT
Species 2:          ACGTACGT-AACGTACGT

Encoding
Conservation:       10000100IJ00100001
Bi-directional:     abcdefghIJhgfedcba
Full:               abcdefghIJijklmnop
```

Fig. 1 A demonstration of the three single-sequence encodings available when using the GUI for the segmentation of a two-way DNA sequence alignment in *ax* format

species are assigned the character 'I', and gaps relative to the second species are assigned a 'J'.

2. The **bi-directional** encoding conserves both match/mismatch information and DNA sequence information. The encoding allows for the fact that DNA is a double stranded molecule in which functional elements can be coded for on either strand. It also allows for the fact that when a whole genome sequence alignment is performed the ordering of the reference genome is kept constant, however alignment blocks from the aligning species may be in either direction. Considering this, the alignment of a particular sequence can be considered as equivalent to the alignment of the complement of that sequence.
3. The **full** encoding preserves fully both sequences in the alignment. Each possible alignment pairing is assigned a unique character from 'a'–'p'.

3.1.3 Recording Genomic Position Information: `input.log`

When generating the final output (the *segmentation map*), in order for the `changeptGUI` to relate the position of segments in the model to genomic coordinates, an `input.log` file is automatically generated (for the *axt* input), which records the genomic positions of the alignment blocks. A user who is generating their own input sequence that they want to be related to genomic coordinates will need to generate their own `input.log` in the following format:

1. The file should be in a tab separated format, where each line contains the genomic coordinates of the ends of the alignment blocks for each species, the coordinates should be 1-based and inclusive at both ends (the same as for *axt* format).
2. On each line, separated by tabs should be the following pieces of information: the alignment block number; chromosome for the reference species; starting position of the alignment block for the reference species; ending position of the alignment block for the reference species; the strand of the alignment block for the reference species (either + or −); chromosome for the aligned species; starting position of the alignment block for the aligned species; ending position of the alignment block for the aligned species; the strand of the alignment block for the aligned species (either + or −).

For an input that is not in *axt* format, if an `input.log` file is not provided then the final output positions will be relative to positions in the input sequence (as opposed to genomic coordinates).

3.1.4 Example: Alignment Encoding

The following figures demonstrate the encoding of an alignment block from the alignment of the mouse genome to the human Y chromosome; the example shows alignment block 16 using the ***bi-directional*** encoding. Figure 2 shows the bi-directional encoding for alignment block 17 and the corresponding `input.log` entry.

```
Alignment block 17
 16 chrY 252611 252675 chr5 109552634 109552702 + 2462
 TACGTACCGTGTGACTGCTCCTGAGA----AGATCCTGTCTATCATCTTGGTAGAAAGGGCTGGAAAGG
 TGCTCACTGGGTGACAGCACCGGAGAGAGAAGACGCAGTCTATCATCCAGGAAGAGATGGCTGCAAGGG

Bi-directional encoding
#acfecafhfbfafafdffdffbfafaJJJJafacgfdfafaaafaafcdffdafacaefffafgaacff#

Input.log entry
 17 chrY 252611 252675 + chr5 109552634 109552702 +
```

Fig. 2 A demonstration of the *bi-directional* encoding and the `input.log` line entry using alignment block 17 of the alignment of the Mouse genome to the Human Y chromosome as an example. The top shows the alignment block entry in *axt* format, the middle shows the *bi-directional* encoding and the bottom shows the `input.log` line entry

```
Species 1: AAAACCCCC-GGGGTTTT
Species 2: ACGTACGT-AACGTACGT
Encoding:  aaaaccccIJccccaaaa
           10000100IJ00100001
```

Fig. 3 A demonstration of a 2-sequence encoding for a two-way DNA sequence alignment. The top sequence of the encoding captures the GC content of species 1 and the bottom sequence of the encoding captures the matches/mismatches in the alignment

Notice that the encoded alignment is bookended by the '#' character. For the full alignment, the alignment blocks are concatenated and are separated by '#' characters (note that the first and last character of the input sequence should not be a '#' character, they should only be used to **separate** segments).

3.1.5 Parallel Input Sequences

The BSCM can be extended to multiple parallel sequences. Each sequence may be composed of its own alphabet; the alphabets of the multiple sequences need not be the same. The model assumes that the positions of change-points in each of the sequences are the same, and that corresponding segments are always allocated to the same class. However, the corresponding segments from each of the parallel sequences have distinct character frequencies.

The guidelines for constructing multiple sequences to be run through `changept` in parallel are the same as for a single sequence, with the additional comment that care should be taken to ensure that the sequences are the same length, that the positions of any '#' characters are the same and that the positions of any of the special characters I,J,K,L,M,N,O are the same in each parallel sequence.

There is a conceptual difference between segmenting parallel sequences and segmenting a single sequence encoding multiple sequences. Parallel segments are assumed to be generated independently, conditional on the character frequencies in each sequence. To illustrate this, consider the two-sequence encoding shown in Fig. 3 and how it differs from the *bi-directional* encoding in Fig. 1.

For the first encoding, an 'a' is assigned where there is either an 'A' or 'T' in the reference species, and a 'c' is assigned where there is a 'C' or a 'G' in the reference species. The second sequence is the *conservation* encoding. When these two sequences are segmented in parallel, the probability of a match occurring at a position is the same for positions encoded by 'a' as for positions encoded by 'c' within the same segment. The bi-directional encoding, in contrast, allows for matches to be more frequent for 'a' positions than for 'c' positions within the same segment.

3.2 Running `changept` Using the `changeptGUI`

3.2.1 Steps for a Sequence Segmentation Project Using `changeptGUI`

1. Begin project.
2. Load input.
   (a) Choose encoding (for DNA sequence alignment).
   (b) Load `input.log` (optional).
3. Choose the number of segment classes.
4. Choose the sampling parameters: Number of samples and sampling block size.
5. Run `changept`.
6. Assess convergence.
   (a) Run additional samples if needed.
7. Model selection.
8. Generate segment class parameter estimates.
9. Generate the *segmentation map*.

3.2.2 Beginning the Project

Upon opening `changeptGUI` the user will be prompted to either open an existing project or begin a new project. When beginning a new project, the user will be prompted to give the project a name and specify whether or not the input will be a DNA sequence alignment (in *axt* format). A subdirectory called "`ProjectName`" will be created which will store the output files and the file "`projectname.cpsegpro`" which is the file used to load existing projects.

3.2.3 Loading the Input Sequence

If the user is running `changept` for a DNA sequence alignment (in *axt* format), the user can load the alignment file(s) by clicking the "load alignment file(s)" button. The user will then need to select the desired encoding for the alignment (*see* Subheading 3.1.2). If multiple alignment files are selected then the alignment blocks from all selected files will be concatenated into a single input sequence.

If the user is not using a DNA sequence alignment as input then they will be able to load the input sequence by clicking the "load input sequence(s)" button (*see* Subheading 3.1.1). If multiple files are chosen then `changeptGUI` will assume the user is running `changept` for parallel sequences (note `changept` can be run with a maximum of ten parallel sequences). If the input sequence(s) are representative of genomic positions then a manually created `input.log` file can be loaded by clicking the "load input log" button (*see* Subheading 3.1.3).

3.2.4 Choosing Input Parameters

In order to run `changept`, the user needs to specify three parameters. Firstly, the user will need to specify the number of segment classes in the model. In general, the appropriate number of segment classes will not be known in advance. Consequently, `changept` should be run for a range of models and then the appropriate model chosen using model selection criteria (*see* Subheading 3.3). The other two parameters required are the sampling block size and number of samples. There is considerable latitude in the choice of these parameters, but the following guidelines can be used.

3.2.5 Sampling Block Size

The sampling block size is the number of samples that `changept` generates for each sample that is output to file. For a sampling block size of say 10,000, `changept` generates 10,000 samples, then 1 of the 10,000 samples is chosen uniformly and randomly to output to file. The sampling block size should be large enough so that consecutive post burn-in samples appear relatively independent. A rough guideline for the sampling block size is ~10 % of the length of the input sequence, or ~10 times the average number of change points. For large sequences (e.g., whole chromosomes) 10 % of the sequence length may be too long and too time consuming (the larger the sampling block size the longer the sampler will take to produce the desired number of samples). In this case `changept` can be run for a small number of iterations (until convergence first occurs) with a relatively small sampling block size, then extended with a sampling block size of ~10 times the average number of change-points.

3.2.6 Number of Samples

This is the number of samples that *changept* will output to file. The number should be large enough so that the sampler can both converge and provide sufficient post-burn-in samples from which to estimate model parameters. At least 500 post burn-in samples are recommended. It is recommended to initially underestimate the number of samples needed. If burn-in is not achieved or the number of post-burn-in samples is smaller than desired, additional samples can then be generated. The sampler generates a Markov chain, thus the last sample from the initial run can be used as the starting point for a new run, without requiring a second period of burn-in. After the `changept` algorithm has run for the initial number of samples, the run can be extended from the "Run Segmentation" tab in the `changeptGUI`.

3.2.7 Assessing Convergence

Once changept has finished running, the "Model Analysis" tab can be used to inspect the output. The first thing that should be checked is that the sampler has converged; each model will need to be checked individually. Models with fewer segment classes will have fewer parameters to estimate and will generally converge faster than models with more segment classes. To check for convergence, go to the "Model Analysis" tab and select a model from the drop-down menu, then click on the "Sample plots" interface. The "Sample plots" interface can be used to plot model parameter values from each of the samples generated by changept. The fist plot to generate should be a plot of the log-likelihood over all samples. If this plot has not converged then changept should be run for more samples. It can be the case that the log-likelihood has converged but the sampler has not yet converged for the other parameters. The next plots that should be checked for convergence are the mixture proportions for each segment class and then the character frequencies for each segment class. If the sampler has not converged it can be set to run again for a larger number of samples.

3.2.8 Example: Running changept

The alignment of the mouse genome to the human Y chromosome in *axt* format was loaded and changept was run using the *bi-directional* encoding for models with the number of classes ranging from 2–10. The input sequence length is 2,524,681 (ignoring gaps). Changept was initially run for 1000 samples with a sampling block size of 10,000. Figure 4 (top) shows the log-likelihood over the 1000 samples for the 7-class model. The sampler converges in log-likelihood after a relatively small number of samples (less than 50). Figure 4 (middle and bottom) show the mixture proportions and the *conservation rate* respectively (*see* **Note 1**) for each segment class in the 7-class model across the 1000 samples. Figure 4 shows that the sampler has converged in the 1000 samples. However, convergence in mixture proportions and character frequencies takes longer than for log-likelihood.

As is expected for a sequence of that length, a sampling block size of 10,000 is too small. This is apparent in Fig. 4 (middle and bottom), as there is an observable level of correlation between the values in consecutive samples. Using the "All models" selection in the drop-down menu, the average number of change-points can be calculated using the last 500 samples (after convergence). The number of change-points is approximately 15,000, which suggests that a good choice for the sampling block size is approximately 150,000. Thus, for each model changept was run for an additional 500 samples using a sampling block size of 150,000. Figure 5 shows the mixture proportions for the 1500 samples after running for an additional 500 samples. Figure 5 indicates that for the last 500 samples, with the sampling block size of 150,000, there is less correlation between consecutive samples. A similar observation can be made for each of the models (from 2–10 segment classes); thus

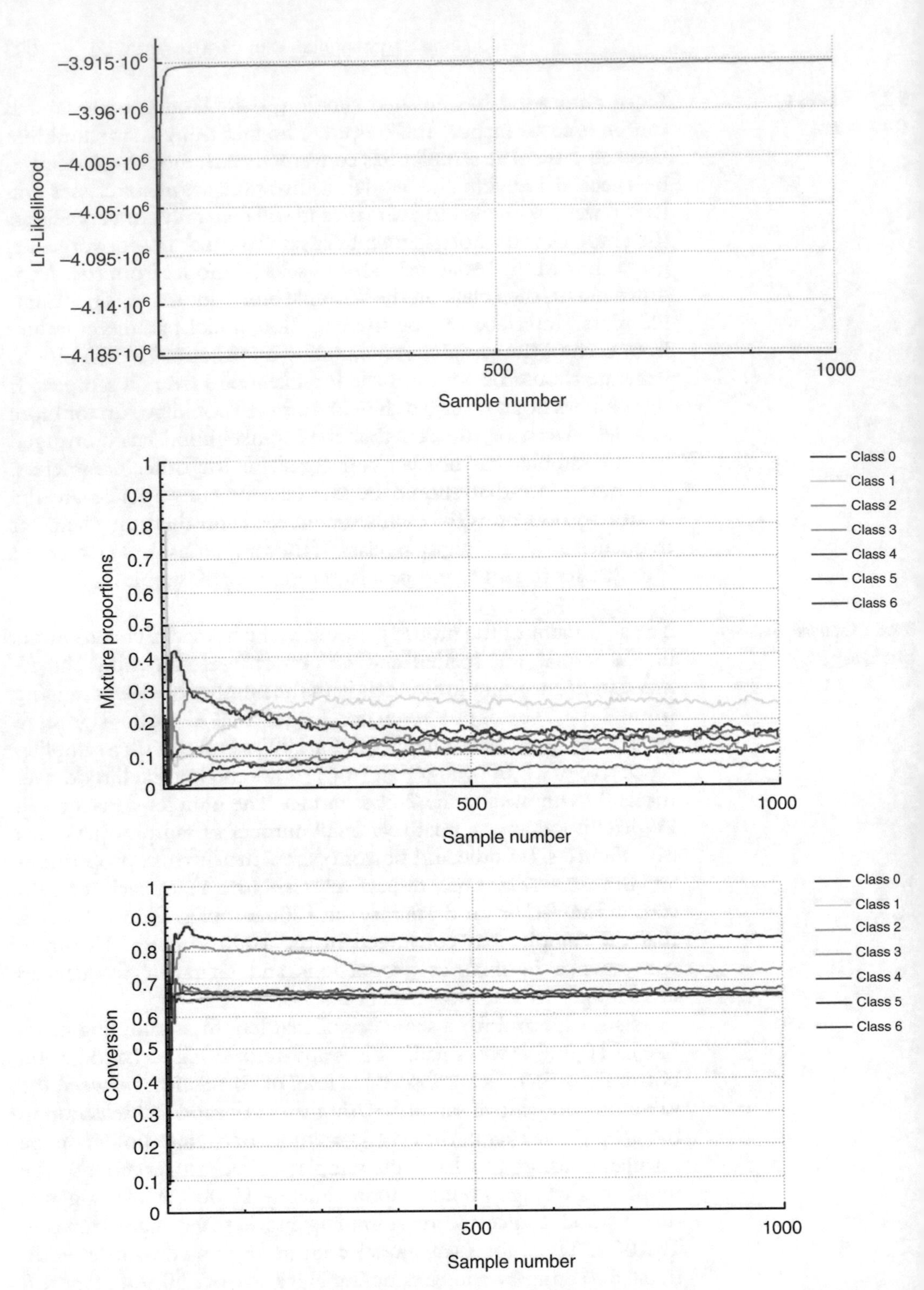

Fig. 4 All three images are plots generated using the "Sample plots" interface of `changeptGUI` depicting values from 1000 samples output by `changept` for the 7-class segmentation of the alignment of the Mouse genome to the Human Y chromosome. (*Top*) A plot of the log-likelihood for each sample. (*Middle*) A plot of the mixture proportions for each of the seven classes for each sample. (*Bottom*) A plot of the conservation rate (*see* **Note 1**) for each of the seven classes for each sample

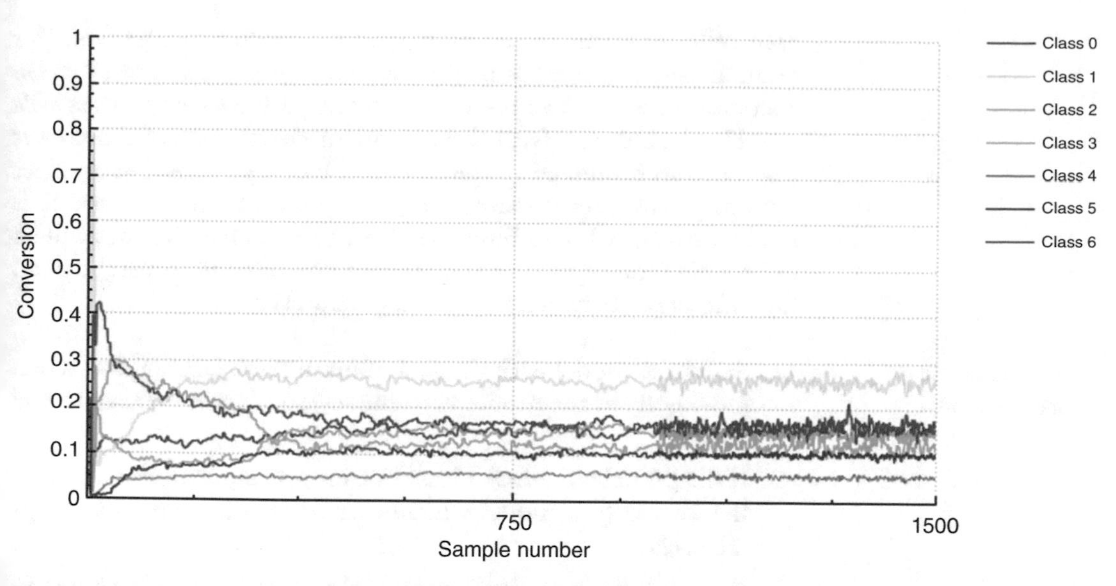

Fig. 5 A plot of the mixture proportions for the 1500 samples generated by `changept` for the 7-class segmentation of the alignment of the Mouse genome to the Human Y chromosome. The first 1000 samples were output by `changept` in an initial run using a sampling block size of 10,000 and the second 500 samples were output using a sampling block size of 150,000. The image is generated using the "Sample plots" interface in the "Model analysis" tab of the `changeptGUI`

the additional 500 samples with a sampling block size of 150,000 are used to make each of the estimates in the remaining steps.

3.3 Model Selection

3.3.1 Information Criteria

As already stated, generally `changept` will be run for multiple models with different numbers of segment classes. The appropriate model can then be selected using *information criteria*; a model with a lower value of information criteria will be preferred to a model with a higher value. A full discussion of information criteria is beyond the scope of this chapter, a discussion of the use of information criteria for model selection when using `changept` can be found in Ref. [3]. `ChangeptGUI` can calculate estimates of three different information criteria. In the "Model Analysis" tab, under the drop-down menu select "All models", the user will then be prompted to specify a number of samples to use for generating the "All models summary" (*see* **Note 2**). The GUI will then show properties for each of the models including the estimates of the information criteria DICV, AIC, and BIC. As discussed in Ref. [3], the DICV estimate is generally preferred for model selection.

There are two more considerations that need to be made for model selection. Both relate to the idea that the selected model should be informative to the user. The first consideration is that a model which contains an *empty* segment class is not informative; an *empty* segment class is one for which the mixture proportion for that class is zero or close to zero. The second consideration is that segment classes should differ significantly (and informatively) from each other in terms of the estimates of character frequencies. For example: say *changept* is being used to segment the *conservation* encoding of a DNA sequence alignment, a binary sequences (0's and 1's). In a model with three segment classes, one of the classes may have an estimated frequency for 1's of say 55 %, and another class may have an estimated frequency of 56 %. In this case it is unlikely that this 1 % difference between the segment classes will be informative for anyone trying to gain any biological insight from the differences between the two segment classes.

3.3.2 Guidelines for Model Selection

1. Select the model with the *first* minimum for DICV (*first*, when moving from the model with the fewest segment classes to the most).
2. If the model contains at least one empty segment class, choose the next largest model which does not contain an empty segment class.
3. If the model contains segment classes that contain character frequency estimates which are indistinguishable, then choose the next largest model for which all segment classes are distinguishable.

3.3.3 Segment Class Parameter Estimates

The estimates of mixture proportions and character frequencies for each class can be calculated using the "Model parameters" interface in the "Model Analysis" tab. Similarly for the "All models summary", the user will be prompted to specify the number of samples to use to generate the parameter estimates (*see* **Note 2**). Once the number of samples to use for parameter estimates has been chosen, the GUI will calculate the estimates and display them in both tabular and graphical displays.

3.3.4 Example Continued: Model Selection and Segment Class Parameter Estimates

`Changept` was run for models with 2–10 segment classes. As explained in the previous example section, each model was run for 1500 samples and in each case the last 500 samples were appropriate to use for subsequent parameter estimates. Figure 6 shows the plot of the estimates of DICV for each model (2–10 segment classes).

Following the guidelines for model selection (*see* Subheading 3.3.2) the first step is to identify the model with the *first* minimum in DICV, which as Fig. 6 shows is the 7-class model. The next step is to check whether the 7-class model contains *empty* segments: to do this, select the 7-class model from the drop-down menu in the "Model Analysis" tab, then select the "Model parameters" interface. Figure 7 shows the parameter estimates for the 7-class model using 500 post burn-in samples to generate the estimates.

The segment class with the smallest mixture proportion is class 4 with 5.47 %, which would generally not be considered *empty*.

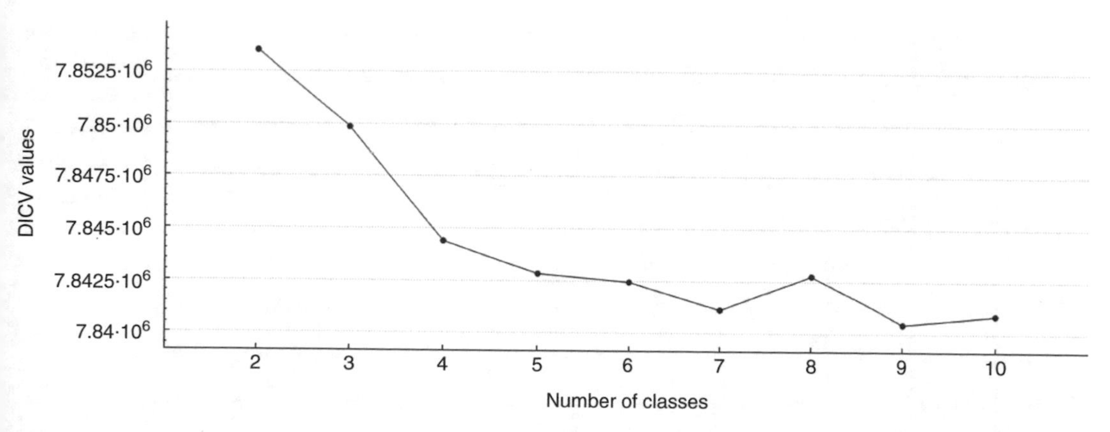

Fig. 6 A plot of DICV estimates calculated using the last 500 samples (of 1500) for each model (2–10 classes) for the segmentation of the alignment of the Mouse genome to the Human Y chromosome. The image is generated by selecting "All models" from the drop-down menu in the "Model analysis" tab of the `changeptGUI`

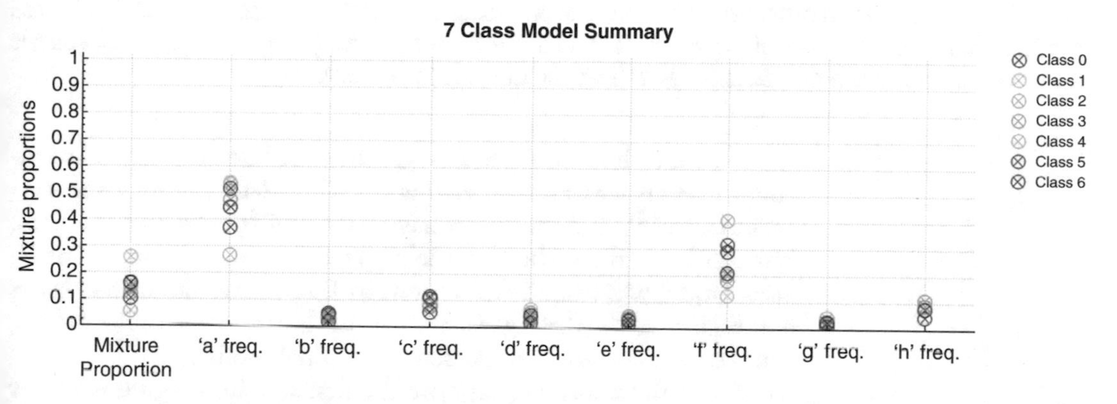

Fig. 7 A plot of the point estimates of parameters (including mixture proportions and character frequencies) for the 7-class segmentation of the alignment of the Mouse genome to the Human Y chromosome. The image is generated using the "Parameter estimates" interface in the "Model analysis" tab of the `changeptGUI`

Note that what is considered *empty* is a subjective judgement and will depend on what the user determines to be informative; reasonable limits might be below 1 % or below 0.5 % mixture proportions. The decision for the limit might also depend on the length of the sequence being studied. The next step is to check whether each segment class can be differentiated based on character frequencies. For characters 'a' and 'f' it appears that each class has a relative unique and informative point of difference. Figure 8 shows more clearly how each class can be distinguished; the plot shows the *conservation rate* on the vertical axis and *GC content* on the horizontal axis (*see* **Note 1**) for each of the seven segment classes across the 500 post burn-in samples.

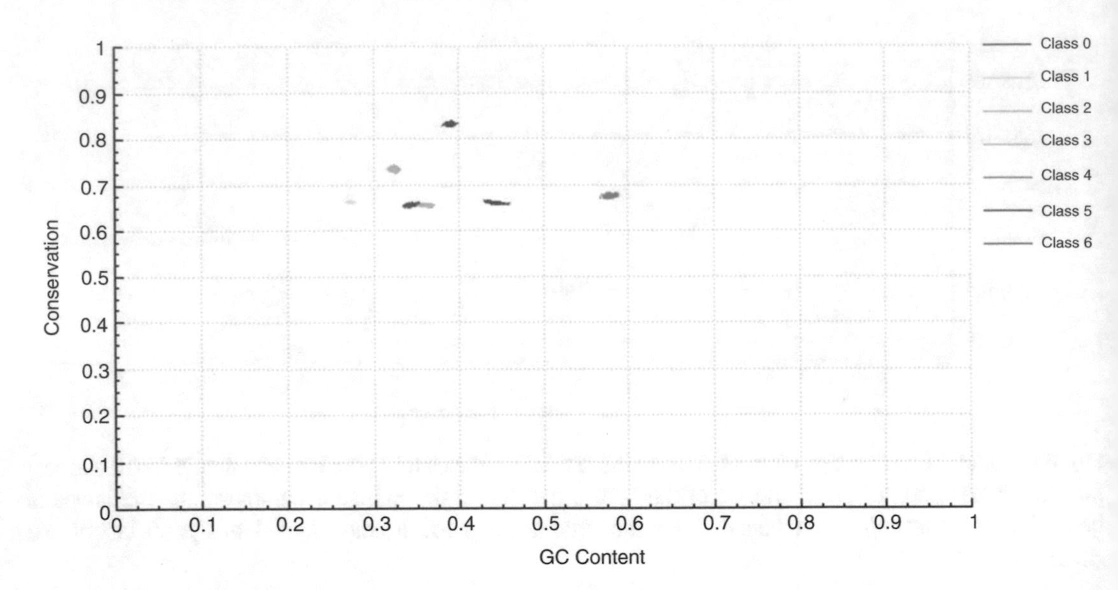

Fig. 8 A plot depicting parameter values from the last 500 samples (of 1500) for each class from the 7-class segmentation of the alignment of the Mouse genome to the Human Y chromosome. The *vertical axis* shows the conservation rate and the *horizontal axis* shows the GC content (*see* **Note 1**)

Figure 8 shows how each segment class has a distinct relationship between *conservation rate* and *GC content*. Each of the segment classes 0, 1, 3, 4, and 6 have very similar conservation rates, but can be distinguished by their respective GC contents. The 7-class model was the *first* minimum in DICV, had no *empty* classes and each class was distinguished from each other class in the model, thus the 7-class model is chosen as the final model.

For arguments sake, suppose the first step in the guidelines for model selection had been to select the model with the minimum DICV over the range of models (which would still be a reasonable and justifiable guideline). The model selected would then have been the 9-class model (*see* Fig. 6). The next step would then have been to check for *empty* segment classes; Class 7 in the 9-class model has a very low mixture proportion. In the tabular display of parameter estimates we can see that the mixture proportion for Class 7 is 0.35 %. Using a threshold of 0.5 %, class 7 would be considered as *empty*, and thus we would need to select a model with fewer segment classes. The next lowest DICV value is for the 10-class model, which also contains an empty class, then the next lowest DICV value again is the 7-class model and thus we would have ultimately arrived at the same decision, selecting the 7-class model.

3.4 The Segmentation Map

3.4.1 Generating Map of Segment Positions

Once the model (the number of segment classes) has been chosen, the *segmentation map* of segment positions for each segment class can be generated for the chosen model. It should be emphasized again that the map of segment positions is based on estimates of the probability of each position being assigned to each segment class (given by the posterior distribution). These estimates are calculated by averaging (over post burn-in samples) the posterior probability of each position belonging to each segment class.

The user will need to specify a threshold probability that will define whether a position in the sequence belongs to a segment class or not. For a given threshold, a position will be assigned as belonging to a given segment class if the estimate of the probability of belonging to that segment class is above the threshold. The minimum threshold that can be used is 0.50; this ensures that any position can be assigned to at most one segment class. The threshold 0.50 can be viewed as a natural choice, as it means that if a position is assigned to a given segment class, it is more likely to belong to that class than not. For the chosen threshold probability, segments are defined as a contiguous run of positions which are each assigned to the same segment class. The user can specify a minimum segment length and also whether gaps should be allowed in segments (reminder: the special characters I,J,K,L,M,N,O represent gaps in the input sequence). If the user allows gaps within segments then the user must also specify the maximum proportion of the segment length which consists of gap characters, and also the maximum number of consecutive gap characters within a segment.

To generate the map, go to the "Segmentation Map" tab and select the chosen model from the drop-down menu. In order to generate the map the user needs to specify which segment classes should be included in the map, the number of burn-in samples, the threshold probability for segments and how gaps are treated (as discussed above). Once the parameters have been chosen, the map of segments can be generated by clicking the "Generate segmentation map" button.

The GUI will calculate the posterior probabilities and average them over the post burn-in samples. The segment positions will then be calculated based on the threshold probabilities. Once segment positions have been calculated, they will be output to file in *bed* format (description at https://genome.ucsc.edu/FAQ/FAQformat.html#format1). A separate file will be created for each segment class (and for each species if the input was from a DNA sequence alignment). If the input was an *axt* formatted DNA sequence alignment or an *input.log* file was loaded (*see* Subheading 3.1.3) then the bed file will contain the segment positions in genomic coordinates for the relevant species in the alignment. Otherwise, if the input was not a DNA sequence alignment and no *input.log* file was loaded, the segment positions referred to in the *bed* files will be relative to the input sequence (*see* **Note 3**).

3.4.2 The Segmentation Viewer

In addition to the segmentation map being saved in *bed* files, the map is also able to be viewed in the "Segmentation viewer" interface in the "Segmentation Map" tab. Bed files are a convenient

Fig. 9 The segmentation map generated using `changeptGUI` for the 7-class segmentation of the alignment of the Mouse genome to the Human Y chromosome. The map is visualized using the "Segmentation viewer" interface of the GUI; at the top is the segmentation map with positions of segments color-coded by the class to which they belong; below the segmentation map are two tracks which have been loaded from *bed* formatted files. The middle track shows regions with annotated genes, and the bottom track shows the regions which were aligned

format for further investigation and processing and for viewing in the UCSC genome browser [11, 12]. The "Segmentation viewer" function of the GUI is not intended to have the full functionality of the UCSC genome browser, rather it is a convenient first step in understanding and visualizing the output of `changept`. The user is also able to load in other bed files so that the segmentation map can be viewed in the context of any other features of interest to the user.

3.4.3 Example Continued: Segmentation Map

The 7-class model was chosen for the segmentation of the alignment of the mouse genome to the human Y chromosome (*see* Subheading 3.3.4). The *segmentation map* was generated for the 7-class model with a burn-in of 1000 samples using a probability threshold of 0.5, a minimum segment length of 1, allowing for up to ten consecutive gaps within a segment and no more than 50 % of the segment length coming from gaps (these are the default parameters for the GUI). Figure 9 shows, for the first 10,000,000 bases of the human Y chromosome, the *segmentation map* as displayed in the "Segmentation viewer" interface. The map is displayed with the addition of tracks showing the regions of the chromosome which were aligned and the regions which contain genes (as annotated in "ensGene.txt.gz" available at http://hgdownload.soe.ucsc.edu/goldenPath/hg19/database/).

4 Notes

1. When using the "Sample plots" interface, the character frequencies can be plotted across the samples generated by `changept`. If the user selects multiple characters then the

frequencies of each character selected will be summed. For example, when using the *bi-directional* encoding the characters 'a' and 'f' can be selected simultaneously as a measure of the "conservation rate" within the given segment class. This is because both characters represent matches in the alignment: an 'a' represents a match of either an A or T in the alignment; an 'f' represents a match of either a G or a C. Similarly, when using the *bi-directional* encoding, the combination of the characters 'e', 'f', 'g', and 'h' is a measure of "GC content" in the reference species (*see* Subheading 3.1.2).

2. In the "Model analysis" tab, when calculating the "All models summary" and the "Parameter estimates" the user is prompted to choose a number of samples to be used. The user needs to select a number that ensures only post burn-in samples are included. For example, if `changept` was run for 1000 samples and convergence occurred (at least) before the 200th sample, the user could specify to use up to 800 samples.
3. When `changept` is not used for the segmentation of a DNA sequence alignment and no *input.log* file is loaded the output files describing the segmentation map are still in *bed* format with the following considerations. The first column will be a reference to which *block* of the input sequence the segment is in, this differs from the standard *bed* format in which the first column usually refers to the chromosome. The term *block* is used here to refer to the section of sequence which is contained between two fixed change-points. For example consider a simplified input sequence: `00111000111#000111000#000111000`. A segment that includes the three 1's in the second block would have an entry in the *bed* file: "`2 3 6`", i.e., the "`2`" in the first column refers to the second *block* in the input sequence.

References

1. Keith JM, Adams P, Stephen S, Mattick JS (2008) Delineating slowly and rapidly evolving fractions of the Drosophila genome. J Comput Biol 15(4):407–430
2. Oldmeadow C, Mengersen K, Mattick JS, Keith JM (2010) Multiple evolutionary rate classes in animal genome evolution. Mol Biol Evol 27(4):942–953
3. Oldmeadow C, Keith JM (2011) Model selection in Bayesian segmentation of multiple DNA alignments. Bioinformatics 27(5):604–610
4. Algama M, Oldmeadow C, Tasker E, Mengersen K, Keith JM (2014) Drosophila 3′ UTRs are more complex than protein-coding sequences. PLoS One 9(5):e97336
5. Hoffman MM, Buske OJ, Wang J, Weng Z, Bilmes JA, Noble WS (2012) Unsupervised pattern discovery in human chromatin structure through genomic segmentation. Nat Methods 9:473–476
6. Hoffman MM, Buske OJ, Bilmes JA, Noble WS (2011) Segway: simultaneous segmentation of multiple functional genomics data sets with heterogeneous patterns of missing data. http://noble.gs.washington.edu/proj/segway/manuscript/temposegment.nips09.hoffman.pdf

7. Algama M, Keith JM (2014) Investigating genomic structure using *changept*: a Bayesian segmentation model. Comput Struct Biotechnol J 10:107–115
8. Liu JS, Lawrence CE (1999) Bayesian inference on biopolymer models. Bioinformatics 15:38–52
9. Keith MJ (2006) Segmenting eukaryotic genomes with the generalized Gibbs sampler. J Comput Biol 13(7):1369–1383
10. Keith MJ, Kroese DP, Bryant D (2004) A generalised Markov sampler. Methodol Comput Appl Probab 6:29–53
11. Karolchik D, Baertsch R, Diekhans M, Furey TS, Hinrichs A et al (2003) The UCSC genome browser database. Nucleic Acids Res 31(1):51–54
12. Fujita PA, Rhead B, Zweig AS, Hinrichs AS, Karolchik D et al (2011) The UCSC genome browser database: update 2011. Nucleic Acids Res 39:D876–D882

Part III

Phylogenetics and Evolution

Chapter 13

Measuring Natural Selection

Anders Gonçalves da Silva

Abstract

In this chapter, I review the basic algorithm underlying the CODEML model implemented in the software package PAML. This is intended as a companion to the software's manual, and a primer to the extensive literature available on CODEML. At the end of this chapter, I hope that you will be able to understand enough of how CODEML operates to plan your own analyses.

Key words Natural selection, CODEML, PAML, Codon substitution model, dN/dS ratio, Site models, Branch models, Branch-Site models

1 Introduction

Natural selection is one of the principal mechanisms for evolutionary change. Since its formal description by Charles Darwin and Alfred Wallace [1], many have sought to observe natural selection in nature [2]. However, its trans-generational effects mean it is difficult to observe natural selection in nature over the course of a normal human life-time for all but the most rapidly breeding organisms or the longest of studies [3]. The recent explosion of genetic data, fueled by the technological innovations in DNA sequencing technology, has led to a renewed interest in natural selection [4], and how to identify its signatures in genomic data.

Here, I review the approach most commonly employed to identify signatures of natural selection in coding sequence data (i.e., sequence data from genome regions that code for proteins). It is important to note that any evolutionary approach for detecting signatures of natural selection will only identify candidate markers or codons that are *likely* to be under natural selection. Further experimental or observational work is necessary to demonstrate that such sites are indeed subjected to natural selection [5].

Jonathan M. Keith (ed.), *Bioinformatics: Volume I: Data, Sequence Analysis, and Evolution,* Methods in Molecular Biology, vol. 1525, DOI 10.1007/978-1-4939-6622-6_13,

2 Expected Signatures of Natural Selection in Coding Sequences

A coding sequence produces a protein. Thus the DNA sequence can be grouped in non-overlapping sets of three nucleotides called codons, each of which code for an amino acid (Table 1). There are six ways we can group a DNA sequence into sets of three non-overlapping bases, three in one direction (starting from the first, second, or third base, respectively), and three in the reverse-complement direction. The correct reading frame is usually indicated by a start codon (almost always "ATG," which codes for the amino acid methionine), and continues until a stop codon (which in the standard code is: "TAG," "TAA," or "TGA").

The DNA sequence is generally thought to be under natural selection only indirectly, through the protein it encodes [6, 7]. This is because the protein is the molecule that is usually directly involved in biological processes, and thus subject to natural selection. For instance, if a mutation occurs in the coding sequence that causes the amino acid sequence of the protein to change (called non-synonymous mutations), the change in the nucleotide itself is not likely to affect the biological process. However, the amino acid change might cause significant disruption of the biological process, and thus lead to lower fitness (ability to survive and reproduce) of the carrier organism. Thus, such a mutation would be quickly eliminated from a population. On the other hand, a mutation that does not cause a change in the amino acid (called synonymous mutations) would not suffer the same fitness consequence. As such a mutation would not be under natural selection, its frequency in a population would only be a function of luck.

Because of this indirect pathway of how natural selection acts on the DNA sequence, we can build an expectation about patterns we might observe when natural selection is occurring. Thus, changes that are synonymous are expected to be largely ignored by natural selection. These provide us with a null model, of what happens when natural selection is not happening. We can then contrast this with the pattern at non-synonymous sites to measure any potential effects of natural selection. This is usually achieved by taking the ratio of non-synonymous (usually represented as dN) to synonymous (dS) changes across a set of sequences. If dN/dS (often called ω) is significantly smaller than 1, this usually means that non-synonymous changes cause a significant fitness disadvantage, leading to purifying selection; if it is close to 1, then non-synonymous changes are behaving like synonymous changes and likely do not have a high fitness impact, and thus changes are neutral with respect to selection; and if > 1, then the amino acid change has had a positive effect on fitness, and there is thus positive selection.

To identify signatures of natural selection in coding sequences we require data on more than one organism. This is because we can only identify mutations once we compare two or more DNA sequences. It is also required that the organisms be separated long enough in evolutionary time for a number of synonymous and non-synonymous mutations to occur.

In the following sections, I will review how to naively estimate dN and dS from a set of aligned codons at several sequences. I will then review the model implemented in the software `CODEML` of the `PAML` package [8, 9] (the software can be obtained at http://abacus.gene.ucl.ac.uk/software/paml.html). The goal is to provide readers with a companion to the software's manual, and a primer to the extensive literature available on `CODEML` and all its possible model iterations. The manual provides a good overview of how to run different models, and is a good source for the relevant literature. However, I have found that the manual is largely lacking on guidance that will allow the user to fully understand the output file. Thus, I have put an emphasis on how to interpret various portions of the output file, compiling information gleaned from various sources (most of them blogs and forums) and from my own experiments with the software. At the end of this chapter, I hope that you will be able to understand enough of how `CODEML` operates to plan your own analyses.

3 Three Steps to dN/dS

The estimator of ω in `CODEML` is based on three steps: (1) count synonymous/non-synonymous sites across the data (*see* Subheading 3.1); (2) count the number of synonymous/non-synonymous differences in the data (*see* Subheading 3.2); and (3) correct for number of multiple mutation steps connecting the observed data (*see* Subheading 3.3). We need both number of sites (**step 1**) and number of differences (**step 2**) because non-synonymous sites are far greater in number than synonymous sites. We thus need to normalize the observed changes (**step 2**) by the total amount of possible changes (**step 1**) in order for dN/dS to be meaningful [10, 11].

Thus, at heart of the `CODEML` models are algorithms designed to obtain estimates of the expected number of synonymous sites (S) per codon and the expected number of synonymous differences (S_d) among sequences (as we will see below, once S and S_d are obtained, estimates of expected number of non-synonymous sites (N) per codon, and the expected number of non-synonymous differences (N_d) among sequences arise naturally).

3.1 Step 1: Counting Synonymous (S) and Non-synonymous (N) Sites

To understand how these parameters come together, let us work through a simple example. Let us say we have the same codon from two separate sequences, both code for the amino acid Isoleucine (I), but using different codons:

```
Seq1 ATT
Seq2 ATC
```

The naive calculation of the number of synonymous sites for *Seq1* would be to examine all the possible changes at each site that lead to a synonymous substitution. Isoleucine, in the universal code, is coded for three separate codons (**ATT**, **ATC**, and **ATA**, Table 1). For positions 1 and 2, any change would lead to a change in the coded amino acid, and thus NO synonymous mutations are possible at these sites. At the third position, however, 2 out of the possible 3 changes lead to a codon that also codes for Isoleucine. Thus, the number of synonymous sites at the codon **ATT** is

$$S_{ATT} = \frac{0}{3} + \frac{0}{3} + \frac{2}{3}$$

$$S_{ATT} = \frac{2}{3}$$

And, by contrast, the number of non-synonymous sites is

$$N_{ATT} = \frac{3}{3} + \frac{3}{3} + \frac{1}{3}$$

$$N_{ATT} = \frac{7}{3}$$

For many codons, S can be estimated as the sum over codons, and N is defined as $N = 3 * r - S$, where r is the number of codons in the sequence. We can then take the average per sequence to obtain an expectation of S and N for the data-set. At this point it should be clear why the number of non-synonymous sites is much larger than the number of synonymous sites.

3.2 Step 2: Counting Synonymous (S) and Non-synonymous (N) Differences

We can then estimate the number of synonymous differences per synonymous site S_d and the number of non-synonymous differences per non-synonymous site N_d across the data-set. For *Seq1* and *Seq2* above, the calculation is straightforward. There is a single synonymous difference out of a single observed difference. Thus, $S_d = 1$ and $N_d = 0$.

The calculation becomes a little more involved if there are additional observed changes. For instance, in the example given by Yang and Nielsen [12], we have

```
Seq1 TTA
Seq2 CTC
```

Both these codons code for Leucine (L). There are a total of two differences, however, we do not know the order of the changes. If we start from sequence 1, the T → C mutation could have happened first, followed by the A → C, or vice-versa. Because we do not know, we must calculate S_d and N_d across both pathways. If we suppose T → C happened first, that would lead to a codon CTA, which also codes for L, which in turn would suffer an A → C mutation, leading to CTC. In this case, we would have two synonymous mutations in a total of two changes. If the order were reversed, starting with the A → C mutation, this would lead to a codon TTC, which codes for Phenylalanine (F), this would be followed by a T → C mutation at the first codon position, leading to the CTC codon, which now codes for L. Thus, in this pathway, we have two non-synonymous changes over a total of two possible changes. Across both pathways, therefore, we have two synonymous changes across four possible changes:

$$S_d = \frac{2+0}{4}$$
$$S_d = \frac{1}{2}$$

If changes occurred across all three positions, then there would be a total of six possible pathways. Again, the total S_d is obtained by summing across r codon sites that are being compared. We can then obtain the proportion of synonymous changes per synonymous site (ρ_S), and the proportion of non-synonymous changes per non-synonymous site (ρ_N), by dividing S_d by S, and N_d by N, respectively.

$$\rho_S = \frac{S_d}{S}$$
$$\rho_N = \frac{N_d}{N}$$

3.3 Step 3: Correcting for Multiple (Latent) Mutational Events to Obtain dN and dS

The proportions ρ_S and ρ_N are uncorrected estimates of dS and dN, respectively—uncorrected for multiple, latent, and mutation events. The way to correct these estimates is by employing a nucleotide substitution model that estimates the number of mutation events given the amount of divergence between sequences. The simplest model is the Jukes and Cantor [13] model, which would apply the following corrections:

$$d_S = \frac{-3*ln\left(1-\frac{4*\rho_s}{3}\right)}{4}$$
$$d_N = \frac{-3*ln\left(1-\frac{4\rho_N}{3}\right)}{4}$$

If we carry the example started in **step 1** above all the way through, we have

$$S = \frac{2}{3}$$

$$S_d = \frac{1}{1}$$

$$\rho_S = \frac{1}{\frac{2}{3}}$$

$$= \frac{3}{2}$$

$$d_S = -1$$

$$N = \frac{7}{3}$$

$$N_d = \frac{0}{1}$$

$$\rho_N = \frac{0}{\frac{7}{3}}$$

$$= 0$$

$$d_N = 0$$

$$\omega = \frac{d_N}{d_S}$$

$$= \frac{0}{-1}$$

$$= 0$$

The result is not surprising. Our data contained a single codon and two sequences. A single synonymous event occurred, and thus the dN/dS ratio should be zero. The surprising element was that the number of synonymous mutations per synonymous site (dS) was estimated at −1 after correction for multiple mutational events. It is a condition of the Jukes and Cantor model that the value in the natural logarithm always be positive. This serves to illustrate, albeit in a rather extreme way, how having too little data can have a detrimental effect on the estimates obtained with CODEML.

4 Moving Away from the Naive Model

The issue with the naive model is that it assumes that all nucleotide changes have equal weights. One instance in which this assumption is obviously violated relates to how often we expect a transition (A <-> C or T <-> G mutations) relative to a transversion (all four other possible mutations). By chance, we would expect one half transitions for every transversion. Empirical observations, however, demonstrate that the actual ratio of transitions to transversions (κ) is often larger than one, and sometimes is as great as 40 [14]. However, before we delve into the violations that affect how different parameters are treated, I first want to quickly review how they are put together into a mathematical framework [12, 15–17].

4.1 The Markov Process

While the data unit of concern is usually a sequence of nucleic acids or amino acids, it is easier to model the evolutionary process at the single base/amino acid level, and assume that each is independent from the other. It is known that this assumption is violated, but it provides mathematical convenience and the results are often reasonable. With this simplification in mind, we can model the probability of change at a single site, and multiply through all the observed sites to obtain the likelihood of the data.

When we think of the evolutionary process forward in time, we can imagine a single nucleotide that may or may not change over a specific time period, t. Let us say that the probability of it changing in an infinitesimal time dt is udt. If change does occur, it may take on any of the three other possible nucleotides (in which case the identity of the base changes at that site). It is also convenient to allow the possibility it may take on the same base it had before (in this case change has occurred but no difference is observed). On the other hand, with probability $1-udt$ the base will not change. With this simple model, we can derive the instantaneous probability of observing base j some small time dt in the future given that we now have base i [15]:

$$P_{ij}(dt) = (1 - udt)\delta_{ij} + udt\pi_j$$

The expression to the right of the plus sign gives us the joint probability of change occurring (with probability udt) and that the change was to base j (with probability π_j). In the expression to the left of the plus sign, $\delta_{ij} = 1$ if $i = j$ and zero otherwise. Thus, if after dt has passed, and we still observe i, two events are possible and we need to sum over both: (1) the i we now observe is the same as the i we observed dt time ago, which happens with probability $(1-udt)$; or (2) the i we now observe is a new i resulting from a change from the previous i, which happens with probability $udt\,\pi_i$.

It follows [15] that for any t:

$$P_{ij}(t) = e^{-ut}\delta_{ij} + (1 - e^{-ut})\pi_j$$

In this expression, e^{-ut} is now the probability that no change occurs. This equation describes a Markov process, in which the state of the base at time t in the future is independent of past states, given the current state. This property allows one to model the Markov process along a tree, starting from the tips, and working towards the root (Fig. 1). The likelihood of the tree for a particular site can be computed by the product of the probabilities of changes along the tree. The trick to obtaining these probabilities lies in that the observed tips have likelihood 1 for the observed base, and zero for all others. The likelihood of an unobserved internal node k, that has decendents i and j, is obtained by calculating the joint probability of i changing to k and of j changing to k, as we move down the tree. But, because we don't know k, we must sum over all possible k.

$$L_k = \sum_k \left(\sum_i P_{ki}(v_i) * L_i \right) \left(\sum_j P_{kj}(v_j) * L_j \right)$$

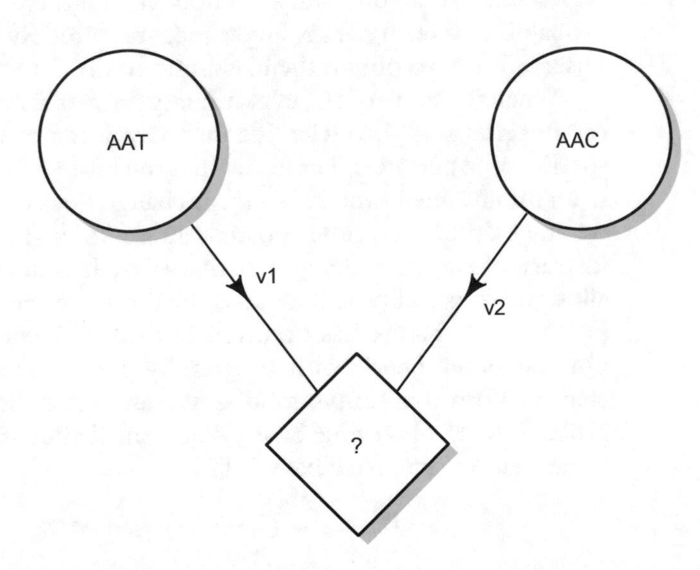

Fig. 1 Calculating the likelihood of a tree can be achieved by collapsing nodes starting at the tips, and working towards the root, calculating the likelihood of each internal node. Here, we present a single example, that can be easily expanded to a whole tree by following the same principles. The *circles* represent observed data at a single site. One sequence has codon "AAT" while the other has "AAC." These sequences are connected to their ancestral node (*diamond*) through branches of length "v1" and "v2," respectively. We do not know the codon of the ancestral node (*diamond*). Any of the 61 sense codons are plausible. If we fix the unknown ancestral state at "AAT," for instance, it is possible to calculate the joint probability that along "v1" the codon did not change state, while along "v2" the codon changed from "AAT" to "AAC" (see text). We can then fix the ancestral node at "AAC," and perform the calculation again, and so on until we have the probability for all 61 possible codons. The likelihood of the node is the sum of the individual ancestral state probabilities

In this expression, v_i denotes the length of time between i and k, which is the branch length in the tree between i and k. The sum within the left set of parentheses denotes the probability of i changing to k ($P_{ki}(v_i)$), weighted by the likelihood of i (L_i), over all possible i ($\sum_i$) conditional on a certain k. The right set of parentheses contain the same quantity evaluated for the change between j and k. The product of these two parentheses for a specific k gives the joint probability of that k changing to any i given time v_i and to any j given time v_j. This process can essentially be repeated until the root of the tree is reached in order to obtain a likelihood for the whole tree [15].

The model implemented in CODEML generalizes this Markov process [16], that was originally developed for four nucleotide bases (and thus had four possible states), to a model that includes all 61 *sense* codons (codons that are not **stop** codons), and thus has 61 possible states. What is required, therefore, is some expression that provides us with the probability of codon i changing to codon j. To obtain this probability, first we must describe the transition rates between codon i and j ($i \neq j$):

$$q_{ij} = \begin{cases} 0, & \text{if codons } i \text{ and } j \text{ differ at more than one position} \\ \mu\pi_j, & \text{synonymous transversions} \\ \mu\kappa\pi_j, & \text{synonymous transitions} \\ \mu\omega\pi_j, & \text{non} - \text{synonymous transversions} \\ \mu\kappa\omega\pi_j, & \text{non} - \text{synonymous transitions} \end{cases}$$

As we can see, the rates are weighted according to whether they involve a transition or a transversion (κ, explained below in Subheading 4.2), and whether the change results in a synonymous or non-synonymous change (ω). The rate also depends on the prior probability of observing codon j (π_j). The individual q_{ij} can be laid out in a 61 × 61 rate matrix **Q**, that describes the transition rates between any codon i to any other codon j. The elements of the diagonal q_{ii}, which denote the rate of *not changing*, are determined by the constraint that the rows of **Q** sum to zero. This constraint is required because equilibrium is assumed between the rate of changing from i to j and from j to i. The parameter μ is a scaling factor that ensures that, for a single unit of time, we expect a single change to occur. It is determined such that:

$$-\sum_i \pi_i q_{ii} = \sum_i \pi_i \sum_{j \neq i} q_{ij} = 1$$

The matrix of transition probabilities over a time t can be modelled as:

$$\mathbf{P}(t) = e^{\mathbf{Q}t}$$

The individual elements of the matrix **P** give us the probability of change from codon i to j (P_{ij}), which is what we need to calculate the likelihood of a tree. Thus, for some codon site at two or more sequences (if more than two, they must be connected by a phylogenetic tree), we can ask what is the likelihood of the observed changes given some κ, ω, equilibrium probability for the codons (π_i), and the time separating each site from its ancestor (branch lengths). Formally, we are calculating $L(data|\kappa, \omega, \pi, t, \mathcal{g})$ where t represents the time separating the data along the tree with topology $\mathcal{g}$, and π the equilibrium probabilities for each codon. The values of these parameters that maximize this likelihood are said to be the most likely to describe the process that generated the data. To maximize the likelihood various optimization algorithms are used [12, 15]

The naive example (Subheading 3) is a special case of this model in which we assumed every change was equally likely; this means that our **Q** matrix has all elements equal to 1/61. This is achieved by setting $\kappa = \omega = \pi = t = 1$, $\mu = 1/61$, and $\mathcal{g}$ is unimportant because we only had two sequences. Now that we understand the basic model, we can examine how the different elements affect estimates, and how it has been modified to better accommodate biological data.

4.2 The Importance of the Transition/Transversion Ratio

The basic model (Subheading 4.1) is based on rates of change from one codon to another. To understand why transition/transversion bias might affect these rates we first need to define non-degenerate, two-fold degenerate, and four-fold degenerate sites [18]. Two-fold degenerate sites are those for which out of the three possible events that would change the nucleotide at that position, one results in a synonymous change. Four-fold degenerate sites are those for which all three possible events result in a synonymous change. Finally, non-degenerate sites are those for which any change is non-synonymous (Subheading 2).

It is believed that the reason for the often observed transition bias lies in the fact that in two-fold degenerate sites the synonymous change is (almost) always a transition. This can be verified by examining a table defining the relationship between codons and amino acids. First, most amino acids are coded by two or four different codons (Table 1). Furthermore, almost all codons that code for the same amino acid are different at only the third position. Finally, for those amino acids coded by only two codons, the codons are distinct by a single transition event. Thus, the majority of synonymous mutations are transitions, and would, by our expectations of how the evolutionary process works, be more frequent than non-synonymous transversions.

Thus, in the basic Markov model transitions are weighted by κ (the transition/transversion ratio). This has important consequences to estimates of ω. To illustrate this, I have run the HIV

Table 1
The universal genetic code

Codon	Amino acid	Codon	Amino acid	Codon	Amino acid	Codon	Amino acid
UUU	Phe	UCU	Ser	UAU	Tyr	UGU	Cys
UUC	Phe	UCC	Ser	UAC	Tyr	UGC	Cys
UUA	Leu	UCA	Ser	UAA	Stop	UGA	Stop
UUG	Leu	UCG	Ser	UAG	Stop	UGG	Trp
CUU	Leu	CCU	Pro	CAU	His	CGU	Arg
CUC	Leu	CCC	Pro	CAC	His	CGC	Arg
CUA	Leu	CCA	Pro	CAA	Gln	CGA	Arg
CUG	Leu	CCG	Pro	CAG	Gln	CGG	Arg
AUU	Ile	ACU	Thr	AAU	Asn	AGU	Ser
AUC	Ile	ACC	Thr	AAC	Asn	AGC	Ser
AUA	Ile	ACA	Thr	AAA	Lys	AGA	Arg
AUG	Met	ACG	Thr	AAG	Lys	AGG	Arg
GUU	Val	GCU	Ala	GAU	Asp	GGU	Gly
GUC	Val	GCC	Ala	GAC	Asp	GGC	Gly
GUA	Val	GCA	Ala	GAA	Glu	GGA	Gly
GUG	Val	GCG	Ala	GAG	Glu	GGG	Gly

example data-set provided with the PAML package (the full files run for this example can be found at: https://github.com/andersgs/measuringNaturalSelection. The full data-set includes 13 sequences spanning 273 bases of the envelope glycoprotein (env) gene, V3 region, of the HIV virus [19, 20]. The first thing to notice is that the sequences are from a coding region, an important assumption of the CODEML model, and that they start at the first position of a codon, and end at the third position of the last codon (273 bases equals 91 codons). Finally, none of the codons code for a **stop** codon (*see* **Note 1**).

To demonstrate the effect of κ on how CODEML counts synonymous and non-synonymous mutations, I have analyzed the data by fixing κ at 0.01, 0.10, 1.00, and 10.00. In Table 2, we can see the results for a single branch of the tree. We can see that as the bias towards transitions increases the expected number of non-synonymous changes (N) decreases (from 235.6 to 228.8), while the expected number of synonymous changes (S) increases (from 37.4 to 44.2). This makes sense, because the model implemented in CODEML weights transitions by κ; and as we saw above, all transitions are synonymous. Not only does κ affect the estimate of

Table 2
Estimates of synonymous and non-synonymous changes across different transition/transversion ratios for the same data-set

κ	t	N	S	dN/dS	dN	dS	N_dN	S_dS
0.01	0.035	235.6	37.4	1.1983	0.0121	0.0101	2.8	0.4
0.10	0.028	235.2	37.8	0.8785	0.0090	0.0103	2.1	0.4
1.00	0.024	232.9	40.1	0.8389	0.0077	0.0092	1.8	0.4
10.00	0.024	228.8	44.2	0.9942	0.0081	0.0082	1.9	0.4

Table 3
Nucleotide frequencies across the three codon positions and the whole data-set

	Position 1	Position 2	Position 3	All
A	0.49	0.36	0.54	0.46
C	0.12	0.20	0.12	0.15
G	0.24	0.18	0.06	0.16
T	0.15	0.26	0.28	0.23

number of non-synonymous sites in a data-set, but it also affects the estimate of the number of non-synonymous differences (N*dN). Thus, depending on the value of κ, we can arrive at very different conclusions about the effects of natural selection (*see* **Note 2** for strategies on picking a κ value for your analysis).

4.3 Codon Frequency Model

Another important parameter of the model is the equilibrium probability of each codon (π_i). In the naive implementation, we assumed that all codons were equally likely. However, sequences and genomes of organisms often have bias in their nucleotide content, and in their codon usage [21]. As illustrated in Table 3 and Fig. 2, nucleotide and codon frequencies are highly skewed in the HIV data-set.

`CODEML` implements a number of codon frequency models that help account for these biases. The simplest is the naive implementation: all codons have equal frequency (1/61). This model is sometimes referred to as FEqual or F0. A more realistic approach involves calculating the expected codon frequency based on the nucleotide frequencies at each codon position across the whole data-set. As an example, under this model, the frequency of the codon "TTT" for the HIV data-set would be $0.23^3 = 0.0121$, as the frequency of "T" in the data-set is 0.23, and there are 3 "T"s in this codon (Table 3). This approach has two implementations in

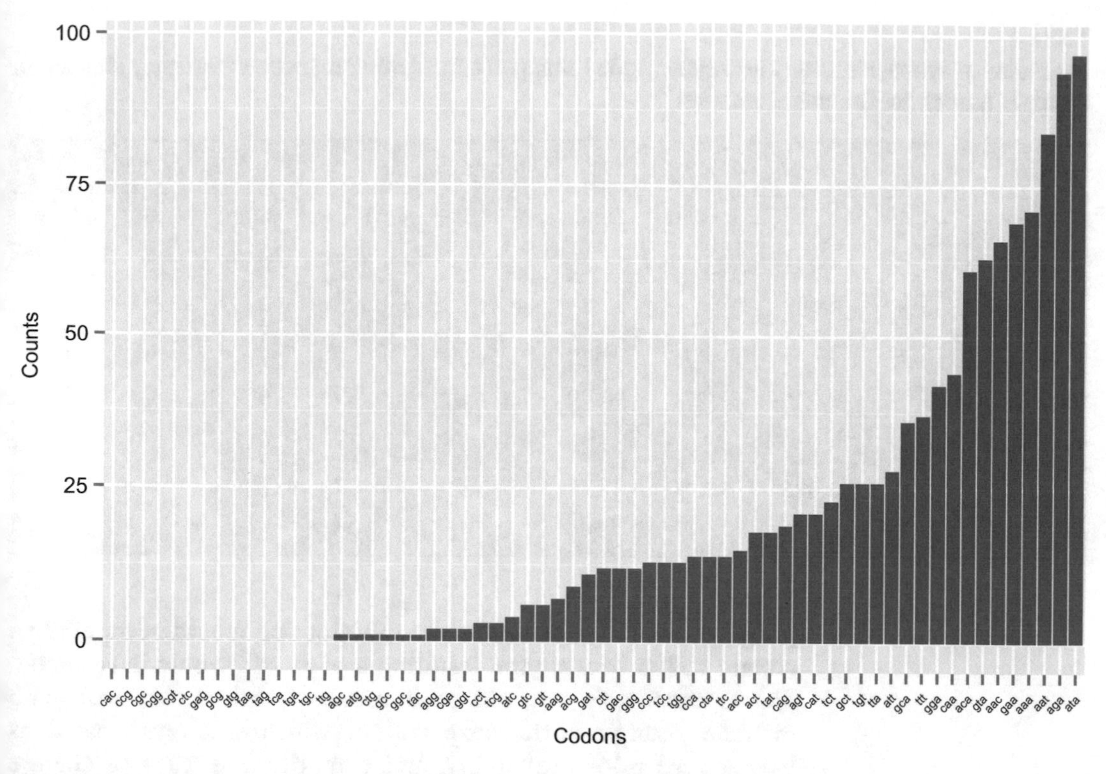

Fig. 2 Counts of observed codons across 13 sequences spanning 273 bases of the envelope glycoprotein (env) gene, V3 region, of the HIV virus

CODEML: (1) F1X4 and (2) F1X4MG. The "1X4" stands for the dimensions of the matrix used to calculate the codon frequencies: it has four columns (one for each nucleotide), and a single row (the frequency of each base across the data-set). The "MG" suffix differentiates between the model described by Muse and Gaut [11] and that described by Goldman and Yang [16].

The next level up in complexity estimates the expected codon frequency based on the nucleotide frequencies at each codon position. Under this model, the frequency of the "TTT" codon in our data-set would be $0.15 * 0.26 * 0.28 = 0.0109$, with 0.15, 0.26, and 0.28 being the respective frequencies of "T" at the first, second, and third codon positions in the data-set (Table 3). This approach also has two implementations: (1) F3X4 and (2) F3X4MG. The nomenclature has the same rationale as the previous one. The "3" now refers to the three rows needed to account for all codon positions.

It is also possible to simply estimate codon frequencies based solely on the frequencies of the codons in the data-set (called "Codon Table" in CODEML, and referred here as FObs, Table 4). This carries the obvious caveat that if the codon was not observed in the data-set, it will have probability of zero. The appropriateness of this assumption must be evaluated on a case-by-case basis. It is

Table 4
Estimates of synonymous and non-synonymous changes across different codon frequency models for a single branch for the same data-set

Model	κ	t	N	S	dN/dS	dN	dS	N_dN	S_dS
FEqual	2.59	0.026	194.50	78.50	2.0386	0.0102	0.0050	2.00	0.40
F1X4	2.61	0.024	211.30	61.70	1.2401	0.0083	0.0067	1.70	0.40
F1X4MG	2.83	0.025	204.30	68.70	1.6267	0.0093	0.0057	1.90	0.40
F3X4	2.47	0.024	231.10	41.90	0.9013	0.0078	0.0086	1.80	0.40
F3X4MG	2.70	0.026	218.20	54.80	1.2484	0.0089	0.0072	2.00	0.40
FObs	2.28	0.024	234.40	38.60	0.9109	0.0078	0.0086	1.80	0.30
FMutSel0	3.08	0.026	205.30	67.70	1.7279	0.0097	0.0056	2.00	0.40
FMutSel	3.09	0.030	220.10	52.90	1.2607	0.0106	0.0084	2.30	0.40

possible that the sequence under examination would never allow a certain codon because the resulting amino acid would have such a drastic effect that it would be immediately selected against. The extreme example is the **stop** codon, which is always treated as having zero prior probability, and thus the frequency of change from one codon to a **stop** codon is always assumed to be zero.

The last two models of codon frequency attempt to incorporate additional biological information. In particular, we know that codon-usage bias is widespread. Bias in codon-usage manifests itself when one or more codons that code for the same amino acid occur at much higher frequency in a data-set than the others. In the HIV data-set, we can see that Arginine (Arg), which is coded by six possible codons (Table 1), occurs 98 times in the 13 sequences. In 94 occurrences it is coded by the codon "AGA," twice it is coded by the codon "AGG," and twice it is coded by the codon "CGA." There are two alternative explanations for this phenomenon: (1) mutational bias and (2) selection for a particular synonymous codon.

Evidence suggests that mutational bias exists and is widespread [22]. We just need to examine transition/transversion ratios to convince ourselves of this fact. Selective pressure determining that a codon be preferentially used over another has less support. Yet, it is plausible, and it could happen at two levels. First, it could occur between different amino acids with similar biochemical properties. The second could occur between codons that code for the same amino acid. The classical situation is one in which speed of translation is important in order to meet some physiological demand [23, 24]. If the tRNAs (the molecules that match codons to amino acids) for the different codons of the same amino acid occur at different frequencies in the cell, then there would be pressure to preferentially use the tRNA that is most abundant, thus reducing the time to produce the protein.

To account for this possibility, two codon frequency models are available in CODEML, the "FMutSel0" and the "FMutSel" models [25]. These models elaborate on the previous F3X4 by adding an additional fitness parameter. The FMutSel0 specifies that synonymous codons have the same fitness value, while the FMutSel specifies that each codon has a distinct fitness value. Thus, the FMutSel0 represents a model in which there is only mutational bias among synonymous codons, and selection might occur among amino acids; while the FMutSel represents a model in which both mutational bias and selection are occurring across synonymous codons.

In Table 4, the expected number of synonymous and non-synonymous sites and the expected number of synonymous and non-synonymous differences were calculated for the HIV data-set using each codon frequency model available on CODEML. In this particular case, the change in codon frequency models does not affect the expected number of synonymous and non-synonymous differences (S_dS and N_dN, respectively), however, it strongly affects the expected number of synonymous and non-synonymous sites (S and N, respectively), and thus our estimate of dN/dS. In light of what we know about these sequences, the results are reasonable. We know, for instance, that there is some synonymous codon-usage bias (as outlined above in the case of Arginine, where a number of the synonymous codons are not recorded in the data-set). This translates into a lower number of expected synonymous sites when using the Fobs model (S = 38.60), which gives zero prior weight to the non-observed synonymous codons, than estimated under the FEqual model (S = 78.50), which states that even though they were not observed we still expect those synonymous codons to be equally probable.

Likelihood ratio tests (**LRT**, described in Subheading 5) can be used to pick the most appropriate codon model.

4.4 Amount of Divergence

The amount of divergence between sequences is also an important parameter affecting principally the expected number of synonymous and non-synonymous differences among sequences in a data-set [10]. Remember that in our model, the number of expected changes is dependent on t and the topology of the tree, $\mathcal{g}$. In general, we expect sequences with fewer differences between them to be closer in the tree than those with a greater number of differences. However, this is only true within certain bounds of divergence. This is because past a certain amount of divergence the number of differences between two sequences does not continue to increase linearly with the amount of time since the sequences shared a common ancestor. Instead, an asymptote is reached where the rate at which new mutations generate a similarity between two sequences is at equilibrium with the rate at which new mutations generate a difference. This is called mutation saturation [10].

To account for this bias, and ensure that the amount of divergence continues to increase linearly with time, CODEML employs a bias correction algorithm. The algorithm introduces a bias correction term that adds mutation events to the observed number of differences at a rate that depends on the number of observed differences. In CODEML the F84 model of nucleotide substitution is used in order to correct for possible multiple mutation events at the same site [26, 27]. This model advances the Jukes and Cantor model seen earlier by accounting for transition/transversion bias.

An additional assumption is that the number of changes is only a function of time. Thus, we assume that all mutations occur at the same rate. If we relax this assumption, we could have two lineages that share a relatively recent common ancestor but have more differences between them than other sequences that share a more distant common ancestor. In this case, it is possible that natural selection is affecting the rate at which mutations are fixed.

To illustrate how this might affect the counts of synonymous and non-synonymous differences, I have analyzed the 13 HIV sequence data-set under four separate phylogenetic hypotheses (Fig. 3). The first hypothesis assumes that mutation rates are the same across lineages, and thus divergence is a function of time. For the other three hypotheses, I have exchanged the position of sequence 1 with another sequence in the tree. This violates the assumption of a linear increase of number of changes between two sequences with time. The tree file now looks like this:

```
13 4
(1,2,((3,(4,5)),(6,(7,(8,((((9,11),10),12),13))))));
(4,2,((3,(1,5)),(6,(7,(8,((((9,11),10),12),13))))));
(9,2,((3,(4,5)),(6,(7,(8,((((1,11),10),12),13))))));
(12,2,((3,(4,5)),(6,(7,(8,((((9,11),10),1),13))))));
```

As predicted, estimates of the number of synonymous and non-synonymous differences for sequence 1 (S_dS and N_dN, respectively) vary considerably across the different phylogenetic hypotheses (Table 5). When sequence 1 is not at its most parsimonious location at the base of the tree given its divergence relative to the other sequences, the model must accommodate the extra divergence by assuming additional mutational events. In the Markov model, this is equivalent to assuming that *v*, the time between the sequences, is large, resulting in the characteristic long-branch normally associated with positive selection. Indeed, in phylogenetic hypotheses B-D, the estimate of dN/dS is larger than 1, and all have long-branches (*t* in Table 5). In the case that there is independent evidence to suggest that the sequence should be placed in that

Table 5
Estimates of synonymous and non-synonymous change across different phylogenetic hypotheses for the same branch and data-set

Trees	κ	t	N	S	dN/dS	dN	dS	N_dN	S_dS
A	2.28	0.024	234.40	38.60	0.9109	0.0078	0.0086	1.80	0.30
B	2.23	0.196	234.50	38.50	1.0084	0.0653	0.0648	15.30	2.50
C	2.61	0.293	233.40	39.60	1.1188	0.0992	0.0887	23.20	3.50
D	2.58	0.227	233.50	39.50	1.1106	0.0768	0.0691	3.50	2.70

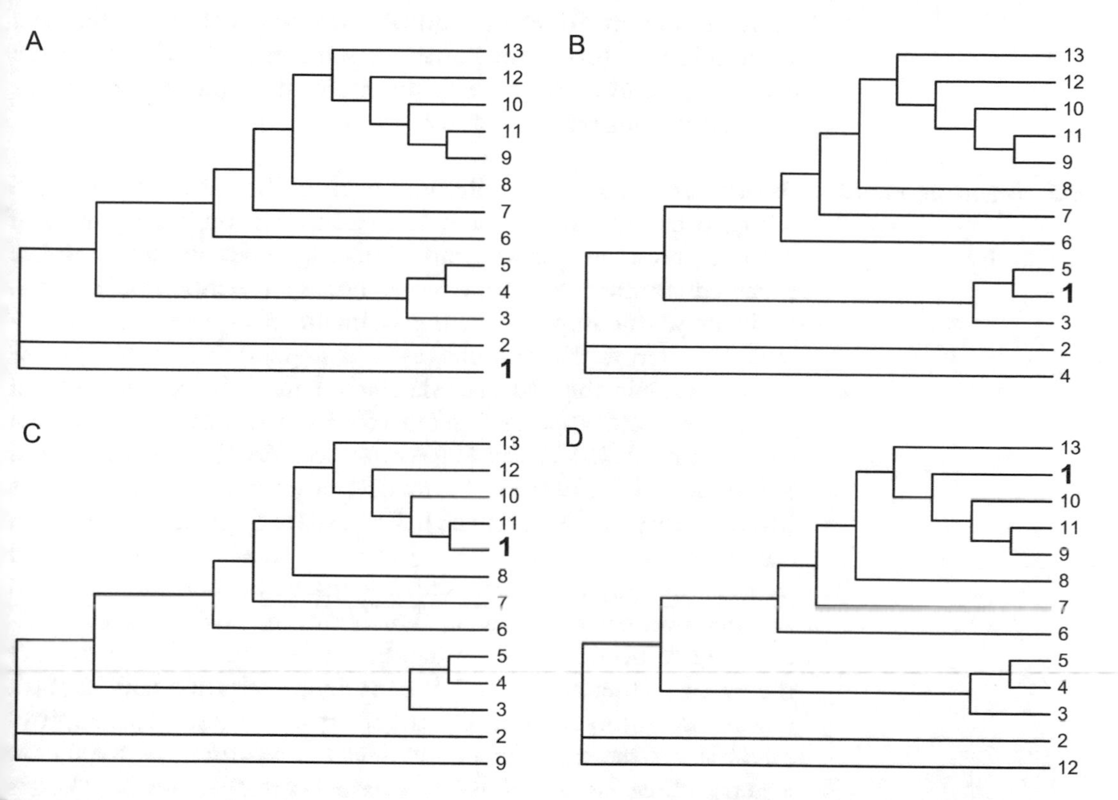

Fig. 3 Alternative phylogenetic hypotheses for 13 HIV sequences. (**A**) Tree assuming mutation rate is equal across all lineages and amount of divergence is only a function of time; the three remaining topologies shift the position of sequence 1 on the tree, thus forcing mutation rates to be different across lineages. (**B**) Sequence 1 (highlighted in bold) is exchanged with sequence 4; (**C**) Sequence 1 is exchanged with sequence 9; and (**D**) Sequence 1 is exchanged with sequence 12

position of the tree, this would constitute good evidence for positive natural selection on this particular lineage.

A much more common case can be demonstrated by taking only sequences 1, 9, 10, and 11 (as in the sub-clade of tree D from Fig. 3 that has sequence 1 as the basal sequence). In this case, sequence 1 is a highly diverged outgroup. The long divergence time means that the number of codons that have more than one nucleotide change increases. When this happens, there is large uncertainty about the single mutational steps taken to transition from one codon to the other, which gets compounded across a large number of sites, leading to generally untrustworthy results (Subheading 3.2). This can lead to either over- or underestimation of dN and dS [10]. When analyzing only sequences 1, 9, 10, and 11, dN/dS was estimated at 0.45, with S_dS and N_dN for sequence 1 being estimated at 7.1 and 21.6, respectively. This would lead to an erroneous conclusion of relatively strong purifying selection. Relative to the results in Table 5, it suggests that the excessive divergence between sequence 1 and sequences 9, 10, and 11 is causing an overestimation of S_dS, when they are analyzed without the context of the remaining sequences. This is a common occurrence, and illustrates why the choice of sequences to include in a CODEML analysis can matter.

4.5 Estimating dN/dS

As discussed above, κ, equilibrium probabilities of codons (π) and divergence time and topology all have an effect on the estimates of expected number of synonymous/non-synonymous sites and the expected number of synonymous/non-synonymous differences. Ultimately, this impacts on the estimate of ω, the evolutionary rate. Thus far, we have assumed that a single dN/dS ratio is sufficient to explain the observed data-set. This is the simplest model possible in CODEML, and is often referred to as the "M0" model (which is modelled by the Markov process described at the end of Subheading 4.1). Empirical evidence, however, suggests that rate heterogeneity is frequent [28]. For instance, purifying selection might be stronger at a section of the gene that is closely related to its function (say a binding site, for instance), and be less so in another portion of the gene. Furthermore, positive selection is expected to occur on only a handful of codons, with the bulk of the codons either neutral or under purifying selection [20, 29, 30].

To accommodate rate heterogeneity across sites, and improve our ability to detect distinct natural selection patterns across sites in a gene, three classes of models have been developed: (1) sites models, of which the "M0" is the most basic [20, 30]; (2) branch models [31, 32]; and (3) branch-site models [33–36]. As the names suggest, the models differ on where ω heterogeneity occurs in a gene: it is either among sites; or among lineages; or both.

These different models are accommodated within our framework by generating different **Q** matrices, one for each category of ω identified in the model. The appropriate **Q** matrix is then invoked depending on the site, the branch of the tree, or both. Essentially, this adds an additional variable to the model, y_{ij}, which specifies to what category of ω codon i from branch j belongs to.

The likelihood of the data at a site must now be additionally summed across all possible categories, each weighted by the proportion of that category in the data.

4.5.1 Sites Models

The sites models are some of the most developed and tested models available in CODEML [37]. These models attempt to estimate two basic sets of parameters: values of ω for each category of ω; and the proportion of sites that are evolving under each category. The number of ω categories is specified *a priori* for each model, and is usually between one and three. These cover the following situations: $\omega < 1$ (purifying selection); $\omega \sim 1$ (neutral); and $\omega > 1$ (positive selection). It is possible to model additional categories, but this is generally not advised. The reason is that increasing the number of parameters in the model will increase the bias in the estimates, with a concomitant decrease in the variance.

Among the site models, there are two separate classes of models. One set assumes that we can group sites into different categories, and each category has a single estimate of ω. The other assumes that the category of sites under purifying selection has ω values that follow a Beta distribution, with two parameters α and β. This allows greater complexity to be incorporated into the model (Fig. 4), but without adding too many new parameters.

To illustrate the different possibilities, I have run the models "M0," "M1a," "M2a," "M7," and "M8." The "M0" is the basic model used thus far (and presented in Subheading 4.1), in which

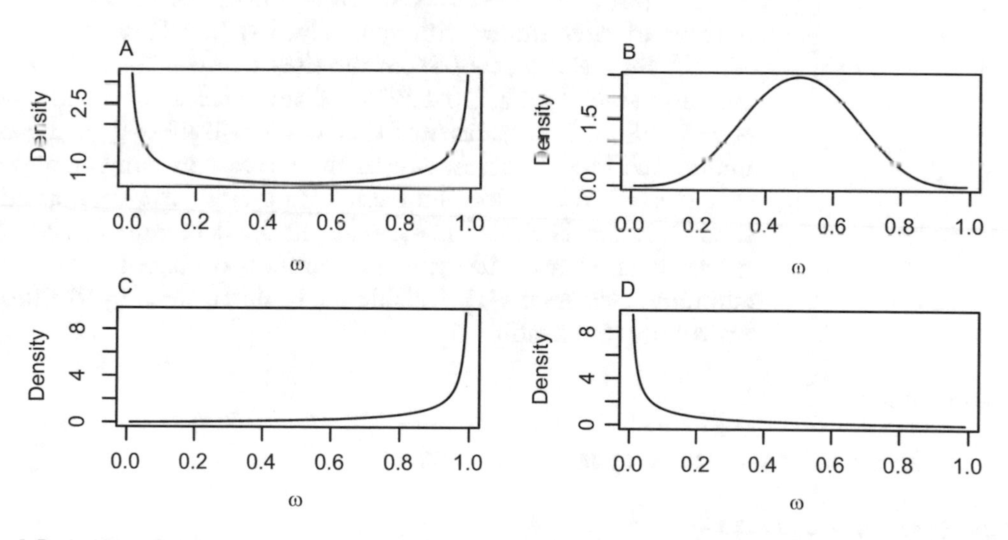

Fig. 4 Probability distribution of sites with different ω value under different Beta distribution hypotheses. (**A**) A scenario where there are many sites with strong purifying selection and many that are nearly neutral (Beta distribution parameters $\alpha = 0.5$; $\beta = 0.5$). (**B**) A scenario where most of the sites are under intermediate levels of purifying selection ($\alpha = 5$; $\beta = 5$). (**C**) A scenario where the majority of the sites are nearly neutral ($\alpha = 2$; $\beta = 0.2$). (**D**) A scenario where the majority of the sites are under strong purifying selection ($\alpha = 0.2$; $\beta = 2$)

there is a single ω. Model "M1a" has two ω categories ($\omega_0 < 1$; $\omega_1 = 1$); ω_1 is fixed at one, and is thus not a free parameter of the model. This model also estimates the proportion of sites evolving under ω_0, called p_0, with the proportion of sites under ω_1 being estimated as $p_1 = 1 - p_0$. Model "M2a" is the extension of the "M1a" model adding a third evolutionary rate (ω_2), which is constrained to be > 1. The latter two models are generally referred to as the "nearly neutral" and "positive selection" models.

Below, we can see a portion of the output file that relates to the results of models "M0," "M1a," and "M2a," respectively. The lines that start with *lnL* report the log-likelihood of the data given the model. So, for "M0," the log-likelihood of the data is -1133.89. The values in parenthesis (*ntime* and *np*) refer to the number of branch lengths that need to be estimated in the model and the total number of parameters in the model, respectively. Thus, *np-ntime* gives the total non-branch parameters that are estimated in the current model (usually, ω, proportion of sites under each ω categories minus 1, and κ). Because I allowed the program to estimate both κ and ω, the number of parameters for "M0" is $25 - 23 = 2$. The final value ($+0.000$) gives the log-likelihood difference of the model evaluated at different topologies. Therefore, if our tree file had *n* different topologies (as seen in Fig. 3), there would be *n* log-likelihood values for each evaluated model. The topology with the maximum likelihood would have a value of +0.000 in this position, and the others would have some negative value.

The results for "M1a" and "M2a" include estimates for proportion of sites under different values of ω. Thus, for model "M1a," for instance, 0.5056 of the sites are evolving with $\omega = 1$; and in model "M2a," 0.19059 of the sites are evolving with $\omega = 5.00508$. It is noticeable that the log-likelihood of the data under each model decreases with the increase in number of parameters used to describe the data. This is not surprising, as additional parameters allow the model to fit the data more closely. The question of whether the gain in goodness-of-fit obtained by the additional parameters is justifiable can be determined by **likelihood ratio tests** (Subheading 5).

```
Model 0: one-ratio
...
lnL(ntime: 23 np: 25): -1133.892429    +0.000000
...
kappa (ts/tv) = 2.27532
omega (dN/dS) = 0.91089
 branch        t       N      S  dN/dS     dN       dS  N*dN  S*dS
  14..1    0.024  234.4   38.6 0.9109 0.0078 0.0086   1.8   0.3
...
Model 1: NearlyNeutral (2 categories)
...
```

```
lnL(ntime: 23 np: 26): -1110.361397   +0.000000
...
kappa (ts/tv) = 2.28035
dN/dS (w) for site classes (K=2)
p:  0.49440 0.50560
w:  0.08871 1.00000
...
 branch      t      N      S   dN/dS     dN      dS N*dN   S*dS
  14..1    0.025  234.4   38.6  0.5495  0.0074  0.0134    1.7    0.5
Model 2: PositiveSelection (3 categories)
...
lnL(ntime: 23 np: 28): -1097.720296   +0.000000
...
kappa (ts/tv) = 2.69691
dN/dS (w) for site classes (K=3)
p:  0.30445 0.50496 0.19059
w:  0.01184 1.00000 5.00508
branch       t    N      S    dN/dS     dN      dS N*dN  S*dS
14..1    0.028  233.2  39.8   1.4625 0.0098 0.0067  2.3  0.3
```

The "M7" and "M8" models are similar to the "M1a" and "M2a," respectively, except that the "M7" assumes that all sites are evolving under ω values that are drawn from a Beta distribution (as in Fig. 4). The "M8" assumes that only a portion of the sites are evolving under ω drawn from a Beta distribution, with another portion having some $\omega > 1$.

The output for these models is similar to the above models, but has some important distinctions. First, in the model header there is some additional information contained in the parentheses. In the analyses performed here, we see that for "M7" there are 10 categories. These refer to the number of bins the interval between 0 and 1 was divided into in order to estimate the shape parameters of the Beta distribution. Each bin has an equal proportion of the sites (as seen in the line starting with "p:"), with the boundaries of the bins allowed to vary in order to accommodate this constraint (as seen in the line starting with "w:"). Thus, if we imagine the extreme case where 90 % of the sites were evolving at $\omega < 0.1$, then there would be nine bins included in the interval $(0,0.1)$, and a single bin in the interval $[0.1,1)$, leading to a Beta distribution with most of its density in values < 0.1 (e.g., Fig. 4, panel D). In the case of "M8," there are 11 categories, the 10 categories to estimate the Beta distribution, and a single category for sites evolving at $\omega > 1$.

The parameters of the Beta distribution (called p and q in the CODEML output) are not very different for the "M7" and "M8" models in this data-set ("M7": $p = 0.179$ and $q = 0.153$; "M8": $p = 0.138$ and $q = 0.102$), and are not very different between themselves. This suggests that, under these models, sites evolving at $\omega > 0$ and $\omega < 1$ are either under strong purifying selection or are nearly neutral (Fig. 4, panel A). Furthermore, under "M8," a proportion (0.194) of the sites are evolving under $\omega = 4.655$.

A guide to the models can be found in Yang et al. [37] and by reading the PAML manual.

```
Model 7: beta (10 categories)
...
lnL(ntime: 23 np: 26): -1110.888044   +0.000000
...
kappa (ts/tv) = 2.23791
Parameters in M7 (beta):
 p = 0.17955 q = 0.15309
dN/dS (w) for site classes (K=10)
p:  0.10000 0.10000 0.10000 0.10000 0.10000
    0.10000 0.10000 0.10000 0.10000 0.10000
w:  0.00000 0.00158 0.02675 0.15733 0.46705
    0.79547 0.95500 0.99486 0.99982 1.00000
...
 branch      t       N      S   dN/dS    dN      dS  N*dN  S*dS
 14..1    0.025  234.5   38.5  0.5398  0.0073  0.0135  1.7  0.5
...
Model 8: beta&w>1 (11 categories)
...
lnL(ntime: 23 np: 28): -1097.634408   +0.000000
...
kappa (ts/tv) = 2.68911
Parameters in M8 (beta&w>1):
 p0 =  0.80595  p =  0.13834 q =  0.10299
 (p1 =  0.19405) w =  4.65519
dN/dS (w) for site classes (K=11)
p:  0.08059 0.08059 0.08059 0.08059 0.08059
    0.08059 0.08059 0.08059 0.08059 0.08059
    0.19405
w:  0.00000 0.00045 0.01789 0.17735 0.61798
    0.92604 0.99319 0.99974 1.00000 1.00000
    4.65519
 branch     t       N      S   dN/dS     dN      dS    N*dN   S*dS
 14..1   0.028  233.2   39.8  1.3654  0.0097  0.0071   2.3   0.3
```

4.5.2 Branch Models

The branch models are similar to the site models, but instead of assuming that rate heterogeneity occurs among sites in the data-set, it assumes that it occurs among lineages [31, 32]. In Subheading 4.4 I examined the effect of moving a sequence to a different location of the tree (Fig. 3, panels B–D). Here, we will re-analyze the trees A and D using the branch-site model in order to test the hypothesis that ω is not constant across branches in the tree.

There are two possible ways of doing this: (1) assume that each branch of the tree has a unique ω; or (2) modify the tree file in order to group branches by two or more ω rate. The first approach is generally not recommended, as the number of parameters will grow with the number of sequences on the tree.

There are good instructions in the manual on how to modify a tree file in order to group branches by evolutionary rate. In short, modifying the tree file involves adding a `#<integer>` or `$<integer>` to the appropriate locations. For instance, any branch with `#1` would mean that the assigned branch would have ω_1; while a `\$1` to a basal branch would automatically assign `#1` to the whole clade. All branches not assigned a specific ω value are treated as evolving under ω_0. Below, I have reproduced the tree file I used for obtaining results presented here for the second strategy. As can be seen, there are two standard Newick formatted trees, each with 13 sequences (hence the "13 2" in the first row). In both trees, Sequence 1 has a `#1` next to it, while all others have no additional notations. This is equivalent to saying that I am interested in estimating two separate ω values, and that all sites in all sequences but Sequence 1 have their non-synonymous changes weighted by ω_0 in the Markov model (Subheading 4.1), and those of Sequence 1 are weighted by ω_1.

```
13 2
(1 #1,2,((3,(4,5)),(6,(7,(8,((((9,11),10),12),13))))));
(12,2,((3,(4,5)),(6,(7,(8,((((9,11),10),1 #1),13))))));
```

The tree for this data-set has 23 branches (Fig. 3). Thus, under strategy one, where each branch has a unique ω, 23 ω values are estimated:

```
lnL(ntime: 23 np: 47): -1116.995163    +0.000000
...
kappa (ts/tv) = 2.29522
w (dN/dS) for branches: 0.08998 0.57791 999.00000 999.00000 999.00000 999.00000
        999.00000 0.82760 0.59207 999.00000 999.00000 1.14794 0.48694 1.04106
        0.24161 999.00000 999.00000 0.08994 0.00010 0.40932 0.30594 0.67798
        2.28343
 branch    t      N      S     dN/dS     dN     dS    N*dN  S*dS
 14..1   0.027  234.3  38.7  0.0900    0.0037 0.0408  0.9   1.6
 14..2   0.076  234.3  38.7  0.5779    0.0229 0.0396  5.4   1.5
 14..15  0.160  234.3  38.7  999.0000  0.0621 0.0001 14.5   0.0
 15..16  0.022  234.3  38.7  999.0000  0.0087 0.0000  2.0   0.0
```

The results suggest high variation in ω values across branches. The value of **999.00** is given to branches along which there were no inferred or observed synonymous differences, as we can see for branches "14..15" and "15..16" in the excerpt above. In such cases, the dN/dS ratio is undetermined, as we are dividing by zero. In terms of how well the data are explained by the model, a cursory assessment suggests that this model is not much better than the "M0" model. The model of individual branch ω adds an additional 22 parameters relative to the "M0" model (a model with a single ω for all the branches), yet the data does not seem to fit the model much better than it did to "M0" (log-likelihood of "M0" = −1133. 89 vs − 1116. 99). This can be contrasted, for instance, to the fit of the data to the "M2a" model (log-likelihood of the "M2a" model = −1097. 72), which only added one extra parameter.

The alternative approach, one in which we wish to test whether a specific branch or clade has a different evolutionary rate from the rest, generally adds far fewer parameters. As we can see in the results for trees A and D in Fig. 3, by asking if dN/dS is different on the branch leading up to Sequence 1 to the dN/dS ratio of all other branches we add a single extra parameter to the model ($\omega_{\text{Sequence1}}$). From the excerpts below, we can see that the location of Sequence 1 on the tree, as seen before, matters in regard to the estimate of dN/dS. In Tree A, dN/dS for the branch leading to Sequence 1 is 0.088. In Tree D, the estimate is 999.00. Fig. 5B illustrates that the

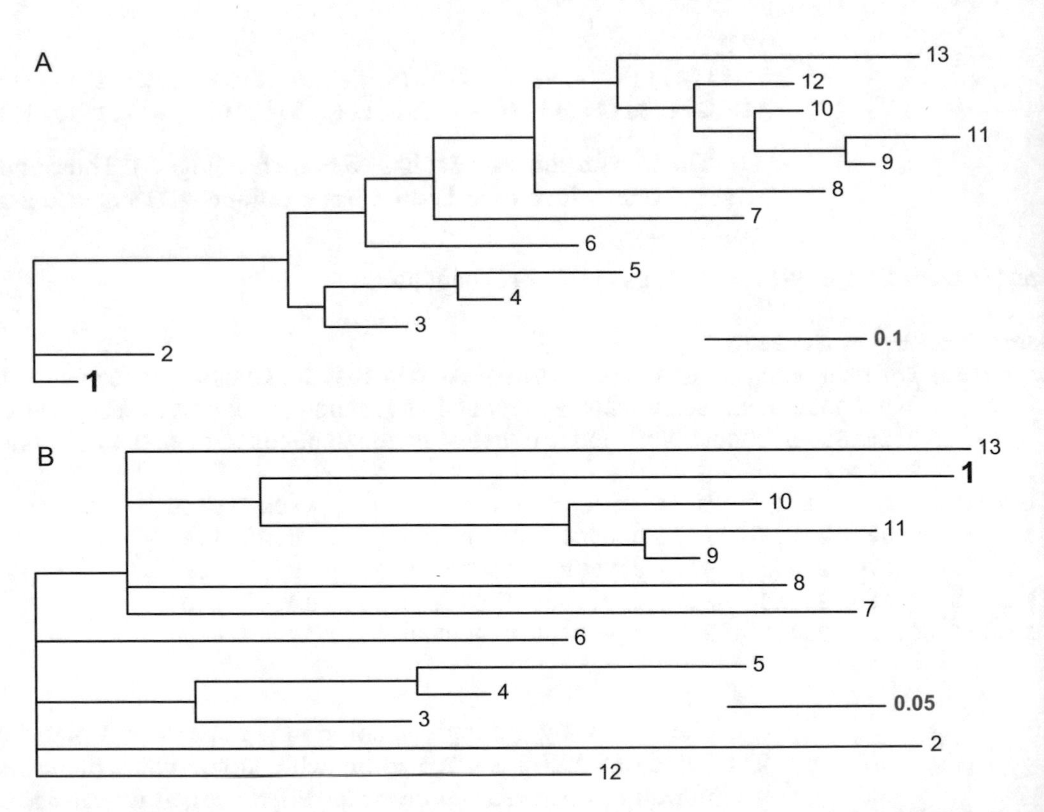

Fig. 5 Phylogenetic tree for the 13 HIV sequences plotted with branch lengths inferred using CODEML. Branch lengths indicated on the scales are the expected number of substitutions per codon. (**A**) The same as tree A in Fig. 3. (**B**) The same as tree D in Fig. 3

length of the branch leading to Sequence 1 is long, due to the number of differences it has relative to the sequences near which it has been placed.

```
Tree A
lnL(ntime: 23 np: 26): -1132.801466    +0.000000
kappa (ts/tv) = 2.27454
w (dN/dS) for branches: 0.98143 0.08790
 branch        t       N     S  dN/dS    dN         dS  N*dN  S*dS
  14..1    0.027  234.4  38.6 0.0879 0.0037 0.0420  0.9  1.6
  14..2    0.075  234.4  38.6 0.9814 0.0250 0.0255  5.9  1.0
  14..15   0.159  234.4  38.6 0.9814 0.0530 0.0540 12.4  2.1
  15..16   0.022  234.4  38.6 0.9814 0.0074 0.0076  1.7  0.3
Tree D
lnL(ntime: 23 np: 26): -1233.751712  -100.950246
kappa (ts/tv) = 2.57528
w (dN/dS) for branches: 0.99875 999.00000
dN & dS for each branch
 branch        t       N     S  dN/dS      dN       dS N*dN   S*dS
  14..12   0.176  233.5  39.5 0.9987 0.0588 0.0589 13.7    2.3
  14..2    0.282  233.5  39.5 0.9987 0.0938 0.0940 21.9    3.7
  14..15   0.000  233.5  39.5 0.9987 0.0000 0.0000  0.0    0.0
  15..16   0.051  233.5  39.5 0.9987 0.0169 0.0169  3.9    0.7
```

4.5.3 Branch-Site Models

The branch-site models combine the models above, and allow for both heterogeneity among sites and branches [33–36]. It achieves this in a hierarchical fashion, where some proportion of the sites are modelled under a "sites model" and the remainder are modelled under the "branch model." The number of ω values to be estimated will be determined by the number of "site" categories, and the number of "branch" categories. When the model is specified, the user defines the number of categories to be modelled, which is the number of "site" categories plus 1. The additional category will accommodate the different ω values specified in the tree file.

The available models increase in complexity, and are named "Clade Model A" through to "Clade Model D." In Model A there are four categories of sites: (1) sites under purifying selection ($0 < \omega_0 < 1$); (2) neutral sites ($\omega_1 = 1$); (3) sites that are under positive selection in one clade/branch ($\omega_2 > 1$), but under purifying selection in the rest of the tree ($0 < \omega_0 < 1$); and (4) sites that are under positive selection in one clade/branch ($\omega_2 > 1$), but are neutral in the rest of the tree ($\omega = 1$). In this model, ω_0 and ω_2 are estimated from the data, while ω_1 is fixed at 1. The model also estimates the proportion of sites in categories 1 and 2. The remaining sites are split between categories 3 and 4 in proportion to the sites in categories 1 and 2, respectively. Model B differs from Model A in that it allows ω_1 to be estimated from the data-set.

Model C specifies three categories of sites: (1) sites under purifying selection ($0<\omega_0<1$); (2) neutral sites ($\omega_1=1$); and (3) sites that have clade/branch specific rates. Thus, if there are two different branch rates specified in a tree, category 3 sites would include estimates for two ω values (ω_2 and ω_3). Finally, Model D generalizes Model C by allowing for 1 or 2 "site" categories, and for the ω values at these categories to be estimated freely from the data. Thus, a three category model D would have the same three categories outlined for Model C above, but ω_1 would be free to vary, and ω_0 would not be constrained to be $0<\omega_0<1$. The total number of ω values estimated with Models C and D depends on how many branch/clade specific values are specified in the tree.

The output from these models (shown below) is similar to those above. One distinguishing feature is the table reporting the ω values. As we can see in an output below for Model C, 37 % of the sites are evolving at category $\omega_0 = 0.024$. As sites evolving under this category do not distinguish between branch types, both branch types have the same value of ω for sites under this category. On the other hand, 18.5 % of the sites are evolving under category ω_2. Yet, these have ω values that are conditional on the branch type ($\omega_{2_0} = 4.89$ and $\omega_{2_1} = 10.63$). In the print-out from `CODEML` below, the columns indicate the ω categories (0, 1, and 2), and the rows below `proportion` indicate the ω values for the distinct branches. As we can see, ω_0 and ω_1 have the same values independently of the branch type, meanwhile ω_2 has values that vary across the different branch types.

```
Output from Clade Model C
site class          0         1         2
proportion      0.37093   0.44417   0.18489
branch type 0:  0.02399   1.00000   4.89783
branch type 1:  0.02399   1.00000  10.63394
```

Unlike the previous models, however, the optimization algorithm is much more likely to get stuck at local optima. It is thus highly recommended that the program be run multiple times, with different starting values of ω and κ. Unfortunately, the program does not offer an easy way of performing replicates (*see* **Note 3**). In Fig. 6, we can see how the log-likelihood of the data given the Clade Model D with three site categories ($k = 3$) using Tree A from Fig. 3 changes across different starting values of ω, and across different runs with the same starting value.

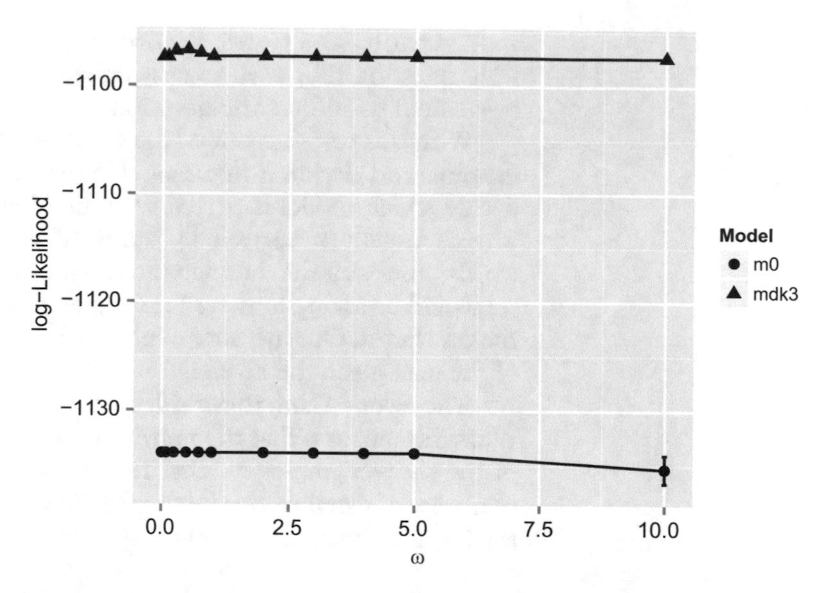

Fig. 6 Maximum log-likelihood values observed across different starting values of ω using the HIV data-set run under the M0 sites model and the branch-site model D. Starting values were 0.001, 0.01, 0.1, 0.25, 0.5, 0.75, 1.0, 2.0, 3.0, 4.0, 5.0, and 10.0. κ was held constant at 2.5. Error bars were obtained by running each model/starting value pair for 10 replicates. In this case, results for a particular starting value were highly consistent across replicates. This is not always the case. For the branch-site model D, a starting value $\omega = 0.25$ results in the algorithm finding a global maximum, whereas all other starting values converge on local optima

5 Model Selection

It should be clear by now that there is an almost endless possibility of models that can be tested with CODEML. The question then remains, which is best given a certain data-set? Model selection is easy to implement, and easier to get wrong. The authors of CODEML recommend an approach based on likelihood ratio tests (**LRT**) [31, 38, 39]. However, before we get to LRTs, it is important to note that model selection starts before any computation is carried out. The point of the model is to understand how evolution is shaping the sequence at hand. When approaching the problem, most of us will already have some prior assumptions about what is the most appropriate model (*see* **Note 4**).

For instance, if the hypothesis is that a certain site might be under positive selection, and there is little reason to suppose that this would vary wildly across lineages, then perhaps a "sites" model is sufficient to answer the question at hand. Thus, additional complexity should not be added to a model simply because it is possible. Additional complexity should be added if there is justification to do so. Therefore, the reader is encouraged to think carefully about what question they wish to answer, and which model(s) would be best suited to address the question.

With this warning out of the way, we can examine how we can measure and decide if one model is better than another. First, to decide which model is better, we must define what better means. This is usually measured in terms of goodness-of-fit. In other words, how well the data fits the model (in a Bayesian framework, we would be asking how well the model fits the data, but that is not the case here). Our measure of goodness-of-fit is the log-likelihood of the data given the model.

The second thing to consider is that the models must be nested. What this means is that the more complex model is a generalization of the simpler model; or that the complex model is obtained by removing constraints on the simpler model. If such is the case, then the log-likelihood ratio is given by:

$$\Delta = -2 * (lnL(simple) - lnL(complex))$$

An outcome of the "nested models" rule is that the model with more parameters should always fit the data better, and thus should have a higher log-likelihood value. Therefore, Δ will always be positive, and is expected to be distributed according to a χ^2 distribution, with degrees of freedom specified by the difference in the number of parameters between the two models (which, again, because these models are nested will always be positive and larger than 0). We can then ask what is the probability of observing a specific Δ by chance. If such a Δ is unlikely, then we might consider accepting the more complex model, as the information gain is greater than we would expect by chance. The threshold probability beyond which we are willing to accept the more complex model is arbitrary, and should be carefully considered before the analyses begin (or perhaps even before any data is collected). However, it is generally accepted to be 0.05.

The assumption that Δ follows a χ^2 distribution does not always hold [39]. The authors of CODEML have attempted to determine empirically whether this assumption holds for many cases [38, 39]. These are carefully outlined in the manual, and the numerous papers produced during the development of CODEML. Unless it has been specifically demonstrated for the model comparison you wish to undertake, it might not be prudent to expect this assumption to hold. In the CODEML manual the authors outline which models can be compared using this approach.

Finally, it is important to note that there is an expectation, not often explicitly stated, that the two models being compared differ only in their assumptions about ω. All other parameters are, for the most part, kept constant, as is the data. Thus, the topology should be the same, as should be the codon frequency model. If κ is fixed in the simpler model, it should also be fixed in the more complex model. Thus, the only thing effectively changing is how many categories of ω are being modelled. Alternatively, to test which topology is best, all other parameters should be kept constant.

To give a numeric example, let us consider models "M0" and "M2a," for which results are presented in Subheading 4.5.1. The "M0" model assumes a single category of ω for the whole data-set; while model "M2a" assumes three. However, only two of the categories are estimated from the data, with the third being constrained to 1. Model "M0" estimates a single parameter, while "M2a" estimates four (two ω values, and the proportions of two of the three classes—the proportion of the third class is known once we know the proportion of the other two). Thus, the degrees of freedom of our χ^2 distribution will be $4 - 1 = 3$. The value of Δ is calculated below:

$$\begin{aligned} lnL(M0) &= -1133.97 \\ lnL(M2a) &= -1097.72 \\ \Delta &= -2*(-1133.97 - (-1097.72)) \\ &= 72.34 \end{aligned}$$

The probability of observing $\Delta > 72.34$ under a χ_3^2 is $\sim 10^{-15}$ (*see* **Note 5** for how to calculate the *p*-value). Given a threshold of 0.05, it seems highly unlikely that we should observe such a Δ, and model "M2a" is providing a significantly better fit to the data than model "M0."

If it is necessary to compare models that are not nested, then Akaike Information Criterion (AIC) could be used. Information on how to apply the AIC and how to calculate it can be found in Felsenstein [40].

6 Assumptions to Keep in Mind

Measuring natural selection can be complicated. The models implemented in `CODEML` allow us to investigate the possibility of natural selection under certain conditions. Under these conditions, they can be powerful (although, read Nei [41]). However, they constrain us in the types of questions we can ask. For instance, they preclude us from investigating natural selection outside of coding regions, and only allow for natural selection to be measured across lineages that have diverged generally to the point where they are distinct (albeit close) species—viruses, like HIV, are an exception because of their large effective population sizes and high mutation rate.

The constraint of needing data across multiple species, or sequences that are reasonably diverged, imposes one of the most fundamental assumptions of the method. We must necessarily assume that they have the same effective population sizes (or independently derive estimates of effective population size) in order to be able to conclude that actual selective pressures are different among lineages [23]. This is because the amount of selection experienced by a lineage will be the product of $N_e s$, where N_e is the effective population size and s is the selection coefficient.

The CODEML models also assume that codon substitutions that instantaneously involve more than a single substitution are not possible. Yet, recent work suggests that, at least in some cases, allowing for two or three simultaneous substitutions can lead to a better fit of the data to the models [42, 43]. Unfortunately, to my knowledge, software that implement these extensions to the basic Markov model are yet to be developed.

7 Final Remarks

The search for methods to identify the footprints of natural selection along the genome of organisms is still ongoing. For years, the focus has been on the coding sequence of genes thought *a priori* to be under natural selection. Yet, from many decades of experiments on identifying the genetic origins of particular polymorphisms, we know that this is not the best approach [5]. The reason is quite simple, our knowledge of the genome is incomplete, and the same phenotype can map to a number of distinct polymorphisms in the genome. Furthermore, it is not even known whether coding sequences are indeed the most frequent targets for natural selection [44]. New sequencing technologies have opened up the possibility of searching through whole genomes. These data have opened up new research opportunities, in particular in regard to understanding where in the genome natural selection acts, and how does evolutionary pressure on a phenotype translate into changes in the genome.

In this chapter, I have reviewed the implementation of the model used by CODEML to estimate the expected number of synonymous/non-synonymous substitutions and synonymous/non-synonymous differences. This is one method that can be used to identify signatures of natural selection in a small portion of the genome. Because of the extensive work that has gone into developing these models, the review is necessarily far from comprehensive. There is much detail that is left out, and I did not review the two methods used to estimate which sites are under positive selection. Yet, I hope that this chapter will still be useful to the reader, assisting the reader in developing a deeper understanding of what CODEML is capable, how it operates, what sort of answers one can obtain, and perhaps, in developing strategies to further explore CODEML with their own data.

8 Notes

1. CODEML does not allow for *stop* codons in its data. This is often a source of frustration among PAML users, as the program will often end abruptly with no warning or error message if there are *stop* codons in the input sequences. If the program stops while reading the input sequences, it is quite possible this is the issue. One can manually check sequences the ExPASy translatetool. One should also ensure that CODEML is set to read the codons using the appropriate translation table (e.g., universal code).
2. There are different strategies employed to identify the best κ value to use in an analysis. Some decide to use κ values obtained from the literature, others decide to estimate κ from the data. If estimating κ from the data, it is often the case that κ is first estimated by fixing $\omega = 1$. Then, subsequently, setting κ to the value estimated, and then estimating ω. Ideally, one should perform a sensitivity analysis, where changes in the estimates of ω are explored along a range of sensible κ values.
3. In order to automate the process of replication, I have written a bash script that creates the necessary control files and re-runs the models *n* number of times (which can be downloaded at: https://github.com/andersgs/measuringNaturalSelection).
4. When performing model selection with CODEML one of the easiest traps to fall into is attempting to analyze all possible models. This strategy can easily result in any combination of models being only slightly different, with no significant statistical differences among models, but with large biological differences. Thinking carefully about which set of models to examine before starting the analysis can help avoid this common pitfall. For instance, if one is interested in whether there is evidence for differences in selective pressure across multiple lineages, then perhaps the suite of branch models might be sufficient. Ultimately, there are no simple heuristics for selecting the "correct" model. One's prior expectations along with one's question dictate which models are most appropriate. Finally, if one is found in such a situation, picking the simplest model is usually the recommended approach, however, it is not necessarily the best approach. One should use what information they have at their disposal to argue for a more complex model if the biology is such that the simpler model does not seem reasonable.
5. The *p*-value of LRT can be determined by using standard statistical tables. It can also be estimated in R by using the standard function pchisq(). In the example given in the text ($\Delta = 72.34$, and degrees of freedom = 3), the *p*-value can be obtained with the following command:
   pchisq(q = 72.34, df = 3, lower = F)

Acknowledgements

I would like to thank Alexandra Pavlova for comments and suggestions on an earlier draft of the chapter. I would like to express my deepest gratitude to my supervisor Rohan Clarke, who has given me the freedom and encouragement to explore evolution, adaptation, and bioinformatics in a whole new light, even though he would much rather I went bird-watching. I am also grateful to Paul Sunnucks, whom I had as an idol while still a bright-eyed, young, and naive biology student, and who turned out to be all that I expected and more. Finally, I would also like to thank Jonathan Keith for the opportunity, and for showing me the path to Bayesian theory in evolutionary work.

References

1. Darwin C, Wallace A (1858) On the tendency of species to form varieties; and on the perpetuation of varieties and species by natural means of selection. J Proc Linn Soc 3:45–62
2. Endler JA (1986) Natural selection in the wild. Princeton University Press, Princeton
3. Grant PR, Grant BR (2006) Evolution of character displacement in Darwin's finches. Science 313:224–226
4. Luikart G, England PR, Tallmon DA et al (2003) The power and promise of population genomics: from genotyping to genome typing. Nat Rev Genet 4:981–994
5. Beutler B, Jiang Z, Georgel P et al (2006) Genetic analysis of host resistance: toll-like receptor signaling and immunity at large. Ann Rev Immunol 24:353–389
6. Kimura M (1968) Genetic variability maintained in a finite population due to mutational production of neutral and nearly neutral isoalleles. Genet Res 11:247–269
7. King JL, Jukes TH (1969) Non-Darwinian evolution. Science 164:788–798
8. Yang Z (2007) PAML 4: phylogenetic analysis by maximum likelihood. Mol Biol Evol 24:1586–1591
9. Yang Z (1997) PAML: a program package for phylogenetic analysis by maximum likelihood. Bioinformatics 13:555–556
10. Nei M, Gojobori T (1986) Simple methods for estimating the numbers of synonymous and nonsynonymous nucleotide substitutions. Mol Biol Evol 3:418–426
11. Muse SV, Gaut BS (1994) A likelihood approach for comparing synonymous and nonsynonymous nucleotide substitution rates, with application to the chloroplast genome. Mol Biol Evol 11:715–724
12. Yang Z, Nielsen R (2000) Estimating synonymous and nonsynonymous substitution rates under realistic evolutionary models. Mol Biol Evol 17:32–43
13. Jukes TH, Cantor CR (1969) Evolution of protein molecules. In: Munro HN (ed) Mammalian protein metabolism. Academic Press, New York, pp 21–123
14. Yang Z, Yoder AD (1999) Estimation of the transition/transversion rate bias and species sampling. J Mol Evol 48:274–283
15. Felsenstein J (1981) Evolutionary trees from DNA sequences: a maximum likelihood approach. J Mol Evol 17:368–376
16. Goldman N, Yang Z (1994) A codon-based model of nucleotide substitution for protein-coding DNA sequences. Mol Biol Evol 11:725–736
17. Bielawski JP, Yang Z (2005) Maximum likelihood methods for detecting adaptive protein evolution. In: Nielsen R (ed) Statistical methods in molecular evolution. Springer, New York, pp 103–124
18. Li WH, Wu CI, Luo CC (1985) A new method for estimating synonymous and nonsynonymous rates of nucleotide substitution considering the relative likelihood of nucleotide and codon changes. Mol Biol Evol 2:150–174
19. Leitner T, Kumar S, Albert J (1997) Tempo and mode of nucleotide substitutions in gag and env gene fragments in human immunodeficiency virus type 1 populations with a known transmission history. J Virol 71:4761–4770
20. Yang Z, Nielsen R, Goldman N et al (2000) Codon-substitution models for heterogeneous

selection pressure at amino acid sites. Genetics 155:431–449

21. Grantham R, Gautier C, Gouy M et al (1980) Codon catalog usage and the genome hypothesis. Nucleic Acids Res 8:r49–r62
22. Duret L (2002) Evolution of synonymous codon usage in metazoans. Curr Opin Genet Dev 12:640–649
23. Akashi H (1995) Inferring weak selection from patterns of polymorphism and divergence at "silent" sites in Drosophila DNA. Genetics 139:1067–1076
24. Sharp PM, Averof M, Lloyd AT et al (1995) DNA sequence evolution: the sounds of silence. Philos T Roy Soc B 349:241–247
25. Yang Z, Nielsen R (2008) Mutation-selection models of codon substitution and their use to estimate selective strengths on codon usage. Mol Biol Evol 25:568–579
26. Thorne JL, Kishino H, Felsenstein J (1992) Inching toward reality: an improved likelihood model of sequence evolution. J Mol Evol 34:3–16
27. Yang Z (1994) Maximum likelihood phylogenetic estimation from DNA sequences with variable rates over sites: approximate methods. J Mol Evol 39:306–314
28. Yang Z, Swanson WJ (2002) Codon-substitution models to detect adaptive evolution that account for heterogeneous selective pressures among site classes. Mol Biol Evol 19:49–57
29. Nielsen R (1997) The ratio of replacement to silent divergence and tests of neutrality. J Evol Biol 10:217–231
30. Nielsen R, Yang Z (1998) Likelihood models for detecting positively selected amino acid sites and applications to the HIV-1 envelope gene. Genetics 148:929–936
31. Yang Z (1998) Likelihood ratio tests for detecting positive selection and application to primate lysozyme evolution. Mol Biol Evol 15:568–573
32. Yang Z, Nielsen R (1998) Synonymous and nonsynonymous rate variation in nuclear genes of mammals. J Mol Evol 46:409–418
33. Yang Z, Nielsen R (2002) Codon-substitution models for detecting molecular adaptation at individual sites along specific lineages. Mol Biol Evol 19:908–917
34. Bielawski JP, Yang ZH (2004) A maximum likelihood method for detecting functional divergence at individual codon sites, with application to gene family evolution. J Mol Evol 59:121–132
35. Yang Z, Wong WSW, Nielsen R (2005) Bayes empirical bayes inference of amino acid sites under positive selection. Mol Biol Evol 22:1107–1118
36. Zhang J, Nielsen R, Yang Z (2005) Evaluation of an improved branch-site likelihood method for detecting positive selection at the molecular level. Mol Biol Evol 22:2472–2479
37. Yang Z, Bielawski J (2000) Statistical methods for detecting molecular adaptation. Trends Ecol Evol 15:496–503
38. Whelan S, Goldman N (1999) Distribution of statistics used for comparison of models of sequence evolution in phylogenetics. Mol Biol Evol 16:1292–1299
39. Anisimova M, Bielawski JP, Yang Z (2001) Accuracy and power of the likelihood ratio test in detecting adaptive molecular evolution. Mol Biol Evol 18:1585–1592
40. Felsenstein J (2004) Inferring phylogenies. Sinauer Associates, Inc., Sunderland, MA
41. Nei MM, Suzuki YY, Nozawa MM (2010) The neutral theory of molecular evolution in the genomic era. Ann Rev Genom Hum G 11:265–289
42. Whelan S, Goldman N (2004) Estimating the frequency of events that cause multiple-nucleotide changes. Genetics 167:2027–2043
43. Kosiol C, Holmes I, Goldman N (2007) An empirical codon model for protein sequence evolution. Mol Biol Evol 24:1464–1479
44. Harrisson KA, Pavlova A, Telonis-Scott M et al (2014) Using genomics to characterize evolutionary potential for conservation of wild populations. Evol Appl. doi:10.1111/eva.12149

Chapter 14

Inferring Trees

Simon Whelan and David A. Morrison

Abstract

Molecular evolution can reveal the relationship between sets of homologous sequences and the patterns of change that occur during their evolution. An important aspect of these studies is the inference of a phylogenetic tree, which explicitly describes evolutionary relationships between homologous sequences. This chapter provides an introduction to evolutionary trees and how to infer them from sequence data using some commonly used inferential methodology. It focuses on statistical methods for inferring trees and how to assess the confidence one should have in any resulting tree, with a particular emphasis on the underlying assumptions of the methods and how they might affect the tree estimate. There is also some discussion of the underlying algorithms used to perform tree search and recommendations regarding the performance of different algorithms. Finally, there are a few practical guidelines, including how to combine multiple software packages to improve inference, and a comparison between Bayesian and Maximum likelihood phylogenetics.

Key words Phylogenetic inference, Evolutionary trees, Maximum likelihood, Parsimony, Distance methods, Review

1 Introduction

Phylogenetics and comparative genomics use multiple alignments of homologous sequences—such as those from a gene or another locus—to study how the genetic material changes over evolutionary time and to draw biologically interesting inferences. The primary aim of many studies, and an important by-product of others, is finding the phylogenetic tree that best describes the evolutionary relationship of the sequences—that is, a tree representing the phylogenetic history of the gene or locus.

Trees have proved useful in many areas of molecular biology. In studies of pathogens, trees have offered a wealth of insights into the interaction between viruses and their hosts during evolution. Preeminent among these have been studies of HIV where, for example, trees were crucial in demonstrating that HIV was the result of at least two different zoonoses from chimpanzees [1]. Trees have also provided valuable insights in molecular and physiological studies,

Jonathan M. Keith (ed.), *Bioinformatics: Volume I: Data, Sequence Analysis, and Evolution*, Methods in Molecular Biology, vol. 1525, DOI 10.1007/978-1-4939-6622-6_14, © Springer Science+Business Media New York 2017

including how the protein repertoire of different organisms has evolved [2–5], genome evolution [6, 7], and the development of taxonomic classifications and species concepts [8, 9]. Accurate reconstruction of evolutionary relationships is also important in studies where the primary aim is not necessarily tree inference, such as investigating the selective pressures acting on proteins (*see* Chapter 13) or the identification of conserved elements in genomes [10]. Failure to correctly account for the historical relationship of sequences can lead to inaccurate hypothesis testing and impaired biological conclusions [11].

Despite the importance of phylogenetic trees, obtaining an accurate estimate of one can seem complicated. There are a large number of different methods available, which are implemented in a vast number of software packages, and often each has unique advantages and disadvantages. An overview of many of these packages is provided at the URL: http://evolution.genetics.washington.edu/phylip/software.html. This chapter aims to provide an introductory guide to help users understand the key assumptions that must be made when inferring a phylogenetic tree, which will allow users to make informed decisions about the phylogenetic software they use.

2 Before Tree Inference: Theoretical Background

In order to infer trees we need to make assumptions about our sequence data and how it changes over evolutionary time. These assumptions affect how we estimate our tree and the accuracy we expect from the result. Some important concepts and key words are highlighted in Box 1.

Box 1: Key Words and Phrases When Inferring Trees:

Homology: Characters in a sequence are homologous when they have been inherited from a common ancestor, although they may have been modified by various evolutionary processes (e.g., substitution or duplication). Multiple sequence alignments assume that all characters in a column are homologous.

Likelihood function: Statistical methods use the likelihood function, $L(\mathbf{X}|\theta)$, to express how likely a set of observed data, $\mathbf{X}$, are to have occurred given a model, θ. When estimating trees, nearly all programs assume that $\mathbf{X}$ represents only the observable characters in the multiple sequence alignment—the nucleotides and amino acids, but not the gaps—given a substitution model.

Multiple sequence alignment: A tabular representation of the sequence data used as the input for tree building. The rows represent the sequences and the columns represent aligned homologous characters. *See* Chapter 8 for more details.

(continued

Box 1 (continued)

Phylogenetic tree is a representation of the evolutionary history of a group of sequences. Each branch in the tree is a split between groups of sequences attributable to their shared ancestry, whereas each node in the tree represents the common ancestor, where sequences diverged from each other through duplication or speciation. The branch lengths of trees estimated by statistical methods are usually in units of expected number of substitutions per sequence site.

Substitution is a sequence mutation that has propagated and become fixed within a population. This fixation may arise despite purifying selection, through random change (genetic drift) or through positive selection and adaptation.

Substitution models are used by statistical methods to describe how sequences change over time. They are parameterized in terms of the relative rates of substitution between characters, such as nucleotides or amino acids, and other biologically important factors, such as rate variation between sites. *See* Chapters 13 and 15 and other sources for more information.

Tree estimation is when methods of tree inference attempt to estimate or select a phylogenetic tree topology according to some scoring criterion.

Tree search: Many of the most powerful tree estimation methods produce scores for each possible tree topology, but need to perform heuristic (approximate) tree search to find a putative highest scoring tree.

Tree space refers to the graph expressing the set of all possible tree topologies, where each node is a possible tree topology and the edges represent specific operations available during tree search to move through tree space.

Tree topology often refers to the phylogenetic tree structure (branching order) without branch lengths.

2.1 Assumptions About the Sequence Data

Nearly all methods of tree inference take as input a multiple sequence alignment, often referred to as the data matrix (*see* **Note 1**). Here, this matrix shall be a set of homologous nucleotide sequences, meaning that they have all arisen by descent from the same locus in their common ancestor [4, 12–14]. The columns of this matrix represent the homology relationships between the nucleotide characters in the sequence, so that if an A from one sequence is in the same column as a G from another sequence then it means they must have evolved from the same common ancestral nucleotide through a series of substitutions. Indels—a contraction of 'insertions and deletions' and often referred to as gaps—describe one or more characters that are missing from a sequence with a '-' character. Indels arise either due to a deletion (where it has been lost in a lineage leading to that sequence) or due to an insertion (where the ancestor for other sequences in the column acquired these extra characters for this sequence). There may also be ambiguity characters, such as 'W' (weak), representing either A or T, or missing data characters, such as 'X' or 'N.'

When inferring trees it is important to realize that most of the methods described here use only substitutions to infer trees, and they completely ignore missing data and indel characters (i.e., gaps are treated as missing) [15]. They often also treat ambiguity characters as missing data. A small number of methods do attempt to incorporate information about insertions and deletions when inferring trees, but these are beyond the scope of an introductory text [16–18]. The accuracy of the alignment is also known to be very important for tree inference [14, 19–21]. Chapter 8 discusses the multiple sequence alignment problem in detail, so it is sufficient to say here that the approach to multiple sequence alignment can have a substantial effect on downstream analyses, including the tree estimate and the confidence one has in that tree estimate.

2.2 The Tree Assumption

All of the methods discussed here take the data matrix and use it to estimate a tree, but aiming to infer a tree requires some assumptions about the evolutionary process. The most important set of assumptions are related to the sequences evolving in a tree-like manner (*see* **Note 2**), which involves assuming that all sequences share a common ancestor and that sequences evolving along all of the branches in the tree evolve independently. Violations of the former assumption occur when unrelated regions are included in the data. This occurs, for example, when only subsets of protein domains are shared between sequences, or when data have entered the tree from other sources such as sequence contamination, gene flow (e.g., hybridization, lateral gene transfer), or mobile genetic elements, such as transposons. Violations of the second assumption occur when information in one part of a tree affects sequences in another part. This occurs, for example, in gene families under gene conversion or gene flow.

Before assuming a bifurcating tree for phylogenetic analyses one should try to ensure these implicit assumptions are not violated. This is done by first exploring the data without assuming a tree structure [22]. Possible tree-independent approaches include spectral analysis [23] and splits graphs [24], the latter now being a very popular choice. When the first set of assumptions is not met one should consider whether the evolution of the sequences could be better represented by a network rather than a tree, as this incorporates reticulations representing gene flow [25, 26].

The second set of assumptions relates to assigning directionality to the evolutionary process. When we have directional

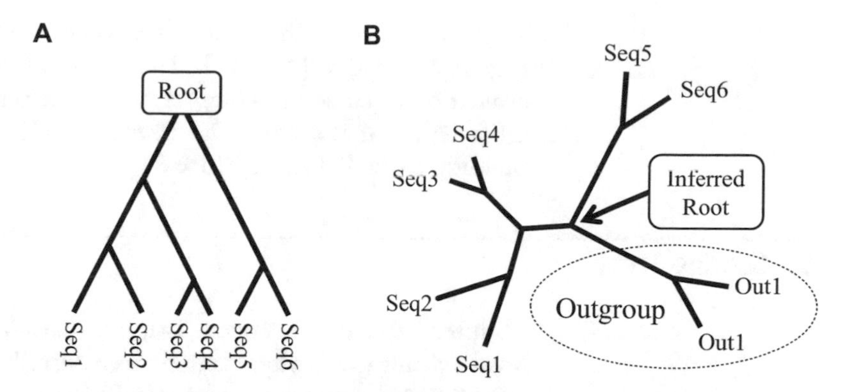

Fig. 1 Two common forms of bifurcating tree are used in phylogenetics. (**a**) Rooted trees make explicit assumptions about the most recent common ancestor of sequences and can imply directionality of the evolutionary process. (**b**) Unrooted trees assume a time-reversible evolutionary process. The root of sequences 1–6 can be inferred by adding an outgroup

information we can infer rooted trees, which explicitly mark the node representing the most recent common ancestor (Fig. 1a), whereas if we do not have directional information then we cannot place a root, and so estimate unrooted trees (Fig. 1b). Nearly all commonly used substitution models (Box 1) assume that evolution appears the same going forwards as backwards, meaning the models cannot be used to estimate rooted trees, so the output of most tree inference programs is an unrooted tree. This must then be rooted, because only a rooted tree can represent a phylogeny. For the majority of applications, the root of a tree can be inferred *post hoc* through the use of an outgroup (Fig. 1b).

2.3 Model Assumptions

Details of the nucleotide substitution models used when inferring trees are provided in Chapters 13 and 15, but here we discuss how some of the common assumptions might affect tree inference. Nearly all of the models that are widely used for the evolution of nucleotides make three key assumptions [15]: (1) reversibility, which means that observations about the evolutionary process seem to be the same going in any direction across the tree; (2) stationarity, which means that the frequencies with which we see nucleotides (or amino acids) are the same throughout the tree; and (3) homogeneity, which means that we expect, on average, the same substitution process to be acting throughout the tree. The methods of tree inference described later are thought to work relatively well when these assumptions approximately hold [11]. When these assumptions are seriously violated it is often referred to as model misspecification, which can have serious effects on the quality of the trees inferred, by causing phylogenetic artifacts. Two widely known artifacts are long branch attraction, where long and often deep branches in a tree incorrectly group together [22, 27, 28], and compositional artifacts, where sequences with similar %GC content are incorrectly grouped together in a tree [19, 29]. The methods of phylogenetic model evaluation discussed in Chapter 15 can help to avoid some of these artifacts, but it is always wise to examine these assumptions before building a tree ([30]; *see* **Note 3**).

3 Scoring Trees

To infer a tree one needs to assign a scoring function that assesses how well any given phylogenetic tree describes the variation among the observed sequences. The last 40 years have led to the proposal of many different scoring functions, summarized briefly as statistical, parsimony, and distance based. There has been considerable debate in the literature over which methodology is the most appropriate for inferring trees. A brief discussion of each method is provided later, although a full discussion is beyond this chapter's scope (*see* [11, 15] for an introduction).

3.1 Scores

Fortunately, there is now a broad consensus that statistical methods provide the most reliable results, since they use the information stored in the sequence data most efficiently and they may be relatively robust to modeling errors. All statistical methods use a likelihood function to assess how likely the alignment of sequences is to occur, conditional on the substitution model used and the phylogenetic tree (*see* Box 1). Statistical methods come in two flavors: maximum likelihood (ML), which searches for the tree that maximizes the likelihood function, and Bayesian inference, which samples trees in proportion to the likelihood function and prior expectations. The primary strength behind statistical methods is that they are based on established and reliable mathematical and statistical methodology that has been applied to many areas of research, from classical population genetics to modeling world economies [31]. By using substitution models, statistical methods allow us to capture and correct for the important evolutionary forces known to affect sequence evolution, which in turn leads to more accurate estimates of evolutionary trees. Statistical methods are also statistically consistent: under an accurate evolutionary model they tend to converge to the 'true tree' as progressively longer sequences are used [32, 33]. This, and other associated properties, enables statistical methodology to produce high-quality phylogenetic estimates with a minimum of bias under a wide range of conditions. Statistical methods are computationally intensive, which has been an issue in the past, but modern programs and computing power mean that we can now use these statistical methods even on thousands of sequences [34, 35].

Parsimony counts the minimum number of changes required on a tree to describe the observed variation in the sequence data. Parsimony is intuitive to understand because it reconstructs the precise history of changes on a tree in a deterministic manner, including assigning observations to the internal nodes of the tree. This approach is in contrast to statistical methods that explicitly account for uncertainty in this history through probability models of sequence change. As a consequence, parsimony is computationally fast relative to statistical methods. It is often criticized for being statistically inconsistent: increasing sequence length in certain conditions can lead to greater confidence in the wrong tree. Parsimony does not include an explicit evolutionary model of substitutions (although *see* [36]). In some quarters, this is seen as an advantage because of a belief that evolution cannot be modeled (e.g., [37]). Others view this as a disadvantage because it does not account for widely acknowledged variation in the evolutionary process. In practice—and to some extent in theory—parsimony can be viewed as a heuristic that approximates the results of the statistical methods when the branch lengths on a tree are short [36].

Distance-based methods infer trees from pairwise distance matrices, which contain estimates for the evolutionary distance between each pair of sequences in an alignment (e.g., the estimated number of nucleotide substitutions). These distances are then taken as the input for an agglomerative hierarchical clustering algorithm that constructs a tree. The best known of these algorithms is Neighbor-Joining (NJ), which starts with an unresolved star tree and aggregates groups of sequences together according to a simple scoring scheme [38]. The great advantage of distance-based methods is their speed: a phylogeny can be produced from thousands of sequences within minutes, with the limiting step being the complexity of the statistical method used to infer the pairwise distances. This speed advantage has historically made distance-based methods the most widely used tree-building approach.

There are three main disadvantages of distance methods when compared to statistical methods. First, by only considering pairs of sequences, distances dramatically reduce the information compared to the full sequence alignment. The next two disadvantages are the result of this information reduction. Statistical methods use the full information in the sequences, which effectively breaks up distances with information from all of the other sequences. This means that all sequences contribute towards accurately estimating branch lengths, and this also reduces their variance. Finally, the information reduction incurred when using pairs of sequences restricts the use of sophisticated substitution models, which potentially makes their estimates more prone to modeling error [39].

In practice, distance methods can provide reasonable heuristic approximations of the results from more sophisticated statistical methods. Their speed makes them an excellent choice for exploratory data analysis, and they are the only feasible method for the analysis of large protein families with 10,000s to 100,000s of members. However, for most applications we strongly recommend the use of statistical methods for inferring phylogenetic trees.

3.2 Why Estimating Trees Is Difficult

The effective and accurate estimation of phylogenetic trees remains difficult, despite their wide-ranging importance in experimental and computational studies. Statistical and parsimony methods need to identify the highest scoring (optimal) tree, which can only be done by searching through the entirety of tree space [11]. This approach of exhaustive tree search is infeasible because the size of tree space increases rapidly with the number of sequences. For 50 sequences, there are approximately 10^{76} possible trees, a number comparable to the estimated number of atoms in the observable universe. This necessitates heuristic tree search methods for searching tree space that speed up computation, often at the expense of accuracy. The phylogenetic tree estimation problem is statistically unusual, and there are few well-studied examples from other research disciplines to draw on for heuristics [15].

Consequently, there has been a lot of active research into methodology for finding the optimal tree using a variety of novel heuristic algorithms. Many of these approaches take a 'hill-climbing' optimization approach and progressively look to improve the tree estimate by iteratively examining the score of nearby trees, then making the highest scoring the new best estimate, and stopping when no further improvements can be found. The nature of these heuristics means that there is often no way of deciding whether the newly discovered optimum is the globally best tree or whether it is one of many other local optima in tree space (although *see* [40], for a description of a Branch and Bound algorithm).

By acknowledging this problem and applying phylogenetic software to its full potential, it is possible to produce good estimates of trees that exhibit many characteristics of the sequences' evolutionary history [41].

4 Inferring Trees Using Maximum Likelihood

The majority of software for inferring trees results in a single (point) estimate of the tree that may best describe the evolutionary relationships in the data (*see* **Note 4**). This tree is ultimately what most researchers are interested in and what is usually included in any published work. In order to obtain the best possible estimate, it is valuable to understand how phylogenetic software functions and the strengths and weaknesses of different approaches. Phylogenetic tree heuristics can be summarized by the following four-step approach [42]:

1. Propose a tree.
2. Refine the tree using a traversal scheme until no further improvement found.

3. Check a stopping criterion.
4. Resample from tree space and go to (2).

Not all four steps are employed by all phylogenetic software. Many distance methods, for example, use only step (1) as a heuristic, while many others stop after refining a tree estimate.

4.1 Proposing an Initial Tree

There is no substitute for a good starting topology when inferring trees. An initial tree can use distance-based clustering methods, chosen for computational speed, most commonly Neighbor-Joining [38]. Occasionally, more sophisticated approaches such as quartet puzzling [43] are used, which can be highly effective for smaller datasets, but may not scale so well for large numbers of sequences. An alternative, widely used approach is to use sequence-based clustering algorithms [11, 40]. These use something akin to full statistical or parsimony approaches. A popular choice is Stepwise addition (Fig. 2a), which starts with a tree of three sequences and randomly adds the remaining sequences to the tree in the location that maximizes the scoring criterion. Since the proposed

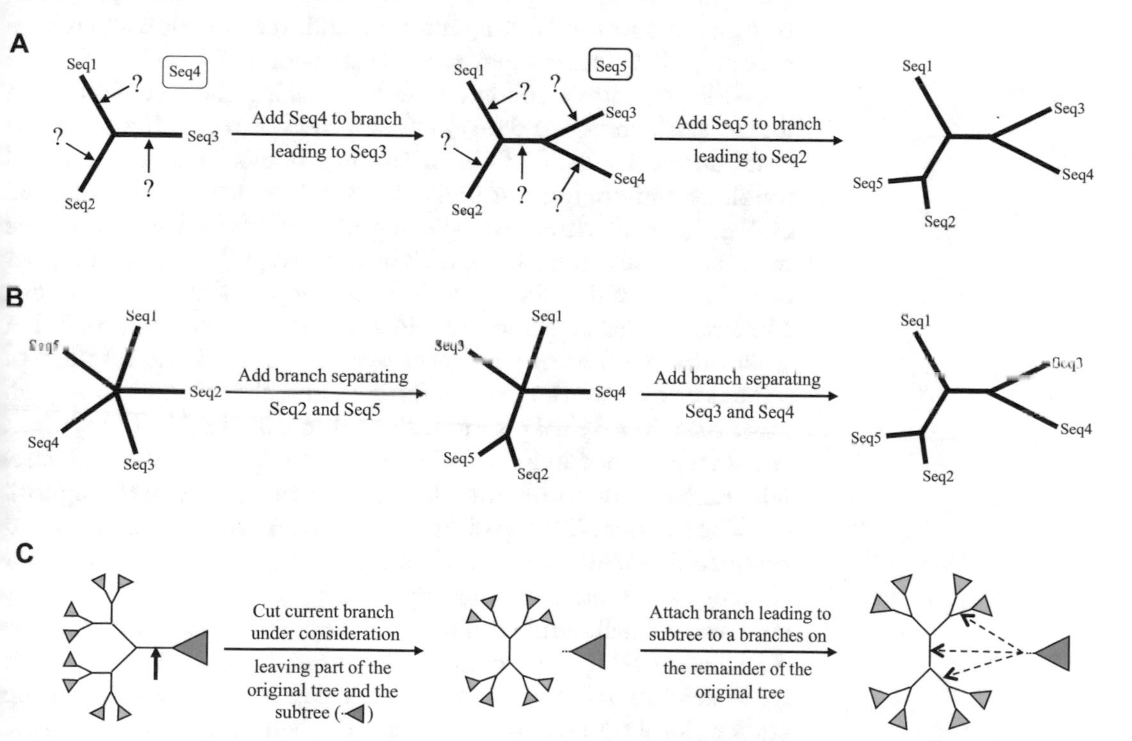

Fig. 2 Common algorithms used in tree search. The first two rows show common sequence-based clustering algorithms that are frequently used to propose trees: (**a**) stepwise addition progressively adds sequences to a tree at a location that maximizes the score function; and (**b**) star decomposition starts with a topology with no defined internal branches and serially adds branches that maximize the score function. The final row, (**c**), shows the *s*ubtree-*p*runing-and-*r*egrafting (SPR) tree rearrangement algorithm. *Dotted arrows* demonstrate some potential regrafting points

topology can be affected by the addition order, a random order is often used as a resampling step. Alternatively, Star-decomposition (Fig. 2b) starts with all of the sequences related by a star phylogeny: a tree with no defined internal branches. This algorithm is a close relative to Neighbor-Joining and progressively adds branches to this tree by resolving multifurcations (undefined regions of a tree) that increase the scoring criteria by the largest amount.

4.2 Refining the Tree Estimate

Refining the tree estimate is the heuristic optimization step, which uses an iterative procedure similar to hill climbing that stops when no further improvement can be found. For each iteration a traversal scheme is used to move around tree space and propose a set of candidate trees from the current tree estimate. Each tree is assessed using the score function and one (or more) trees are chosen as the starting point for the next round of iteration. The set of candidate trees are proposed by traversal schemes that make small rearrangements to the current tree, usually examining each internal branch of a tree in turn and they vary in the way they propose trees from it. The most effective method is probably subtree-pruning-and-regrafting (SPR) [44], with nearest neighbor interchange (NNI) being a more restricted simplification, and tree bisection and reconnection (TBR) being a generalized extension [45].

SPR generates candidate trees by breaking the internal branch under consideration and proposes new trees by regrafting the resultant subtree to each of the remaining branches of the original topology and computing their likelihood, which is demonstrated in Fig. 2c with three example regraftings (dotted arrows). The number of trees proposed by SPR increases rapidly with the number of sequences and makes pure SPR impractical for larger datasets. Modern tree search programs add a further restriction to SPR that makes the number of candidate trees linear with the number of sequences by bounding the number of branches that a subtree can move from its original position. The subtree in Fig. 2c, for example, could be bounded in its movement to a maximum of two branches (all branches not represented by a triangular subtree in the figure).

The simpler NNI algorithm merely breaks an internal branch to produce four subtrees, which can be arranged to form two new topologies per branch. This approach is very fast because it only considers a small number of trees per step, but it also represents the weakness of NNI. The small number of trees means NNI can only make small steps in tree space, which makes it more liable to getting stuck at local optima during tree search than other more expansive schemes [46]. On the other hand, TBR is a more generalized form of SPR and allows any branch of the pruned subtree to be regrafted on to the original tree and not just the one cut. In practice, TBR does not seem much more effective than SPR and the large number of trees that must be examined at each step in the iteration makes TBR computationally impractical for large numbers of sequences.

4.3 Stopping Criteria

Many phylogenetic software packages do not resample tree space and stop after a single round of refinement. When resampling is used, a stopping rule is required. These are usually arbitrary, allowing only a prespecified number of resamples or refinements. An alternative is to base the stopping rule on how frequently improvements in the overall optimal tree are observed [43].

4.4 Resampling from Tree Space

Sampling from one place in tree space and refining will lead to a local likelihood optimum, which may or may not be the globally optimal tree. Resampling expands the area of tree space searched by the heuristic and may uncover better optima. Each optimum has an area of tree space associated with it, and the majority of phylogenetic problems have potentially large numbers of local optima. Resampling is achieved by starting the refinement procedure from another point in tree space. Two of the many possible resampling schemes are discussed here: stepwise addition and the ratchet.

Stepwise addition with random sequence ordering is a viable resampling strategy, because adding sequences in a different order tends to produce relatively good starting trees [34, 40]. An alternative approach is to try and use information from the current best tree estimate as a means of finding other optima. The ratchet procedure is one such approach and obtains a new starting tree by reweighting the characters and then refining the current tree [41]. All resampling strategies can also be helped by tabu search, where regions of tree space close to currently identified optima are not allowed as new starting trees [42, 47].

4.5 Other Approaches to Point Estimation

Among other popular approaches to phylogenetic tree estimation is the use of genetic algorithms, as in the GARLI program [48]. Genetic algorithms are a general approach for numerical optimization that use an analogy with evolutionary biology, allowing a population of potential trees to adapt by improving their fitness (score function) according to a refinement scheme defined by the software designer. As the algorithm progresses, a proxy for natural selection weeds out trees with lower fitness, and so higher fitness trees tend to dominate the population. This method thus proposes an initial set of starting trees (*see* Subheading 4.1), each of which is refined (*see* Subheading 4.2), and tree space is explored by examining and keeping suboptimal trees, until the stopping criterion is reached (Subheading 4.3). The construction of genetic algorithms is very much an art, and highly dependent on the designer's ability to construct a coherent and effective fitness scale, and the application of quasi-natural selection. Some approaches for estimating trees using genetic algorithms have been noticeably successful [49, 50].

4.6 Choosing a Model and Partitioning Data

In addition to the tree model (containing the topology and branch-length estimates), statistical methods require a substitution model that captures the major factors affecting sequence evolution in the observed sequence data. Many models exist, with variable numbers of parameters, associated with different assumptions about the evolutionary forces acting on sequences during their evolution. By far the most popular approach to choosing a model is to use a formal model selection procedure. Here, an information-theoretic approach is taken that attempts to measure the fit of each model to the observed data, using distances derived from Akaike's Information Criterion (AIC) or the Bayesian Information Criterion (BIC), which adjust the likelihoods for the number of parameters in the model. The procedure for model selection is now fully automated, with programs such as jModelTest [51] and ProtTest [52] capable of selecting the best-fit nucleotide and amino acid model, respectively.

For protein coding sequences, this approach to model selection requires users to decide in what form they should analyze their data, for example, as nucleotides or amino acids. This arbitrary decision can have a substantial effect on tree inference and has been addressed using a more general form of model selection that assesses both the model and the data type and is available in the ModelOMatic program [53]. Similar approaches to model selection can be taken for RNA sequences as well using PHASE [54].

Performance-based model selection provides an alternative approach and attempts to select models based both on their model fit and how likely they are to result in inferential errors. Although not widely used at present, performance-based model selection has the potential to offer a valuable alternative to information-theoretic approaches.

A second important factor to consider when selecting a model for likelihood-based tree inference is possible nonindependence of evolutionary pressures acting along the sequences. There are many biological reasons for variation in these forces and they affect the global applicability of the models along the sequences. One common solution to this potential problem is to use different models for different parts of the sequence, which is called partitioning [55]. In principle the different partitions should have greater inter- than intrapartition variability in substitution rates. The usual procedure is to apply the above tests to different data subsets defined by biological motivation (e.g., different genes, different codon positions, paired versus unpaired rRNA), which can be automated by PartitionFinder [56]. Programs such as GARLI, MrBayes, and RAxML can also accommodate partitioning.

An alternative approach to dealing with sequence heterogeneity is through the use of mixture models. Here, the likelihood of each character is calculated under more than one model, and these likelihoods are then combined. Such models have been developed for nucleotide [57] and amino-acid [58] sequences, but this is otherwise a very underexplored part of phylogenetic analysis. Some programs, such as PhyML, have been developed to include relatively sophisticated mixture models based on (e.g.) protein structure [59].

5 How Good Is My Tree and How to Test Tree-Based Hypotheses

The putative ML tree identified by a tree search program represents solely a point estimate, and it is of limited use unless one knows how much confidence one should place in that estimate. Typically the confidence we place in phylogenetic trees is measured in one of two ways, either through measures of support for specific branches on the tree or through the construction of confidence sets by comparing groups of trees. These two approaches are complementary and are regularly used to address different types of questions.

5.1 Measures of General Branch Support and Bootstrapping

The probability of the point estimate of a tree being correct is vanishingly small. Therefore, it is usual to consider the combined branch support values as a general measure of confidence in a tree. Each branch of a rooted tree represents a subtree or clade, which is all of the descendants of a single common ancestor. The support provided by the data for clades, and therefore branches, is of primary importance.

There are many proposed ways to quantify this support, which can be grouped into: (1) analytical procedures, such as interior-branch tests, likelihood-ratio tests, clade significance, and the incongruence length difference test; (2) statistical procedures, such as the nonparametric bootstrap, posterior probabilities, the jackknife, topology-dependent permutation, and clade credibility; and (3) nonstatistical procedures, such as the decay index, clade stability, data decisiveness, spectral signals, splits graphs, and cloudograms. From the third group, splits graphs are the most popular approach [24], while cloudograms provide an alternative visualization [60] (*see* **Note 2**).

From among the statistical procedures, the two most commonly used measures are bootstrap proportions and posterior probabilities (Subheading 6). The nonparametric bootstrap uses a resampling scheme to assess confidence in a tree on a branch-by-branch basis [61]. Tree estimates are obtained for a large number of simulated datasets, obtained by resampling the original data. That is, sampling with replacement is used to produce a simulated dataset by repeatedly drawing samples from the original data to make a new dataset of suitable length. Bootstrap values are placed on branches of the original point estimate of the tree, representing the frequency that implied bipartitions are observed in the trees obtained from the simulated data. This procedure assumes that if evolution had produced multiple copies of the original data, the average frequency of each sequence pattern would be exactly that observed in the original data; this will be approximately true if the original data form a representative sample.

This approach is useful for examining the evidence for particular subtrees, but becomes difficult to interpret for a whole tree because it hides information about how frequently alternative trees are estimated. This can be addressed by describing the bootstrap probability of different trees. In Fig. 3, the two bootstrap values are expanded to five bootstrap probabilities and, using hypothesis testing, three of the five can be rejected. These forms of simple bootstrapping are demonstrably biased and often place too much confidence in a small number of trees [62, 63], but due to their simplicity they remain practical and useful tools for exploring confidence in a tree estimate.

The primary practical limitation of bootstrapping methods is that they are computationally very intensive, although there are many texts detailing computational approximations to make them computationally more efficient (Subheading 5.3).

5.2 Confidence Sets of Trees: The SH and AU Test

When one has a set of competing hypotheses, each represented by a specific tree topology, one often wishes to test which of those hypotheses (topologies) are supported and which can be rejected. Often, this question is associated with the placement of an outgroup that roots a particular clade on the tree, for instance, when one is studying the origins of the organelles [64, 65] or testing molecular evidence for macroevolutionary transitions, such as those in cetaceans [66].

There are two popular tests that are suitable for addressing this type of question: the Shimodaira–Hasegawa (SH-)test [67] and the Approximate-Unbiased (AU-)test [68]. These tests are variants of the bootstrapping procedure discussed earlier. Instead of looking at how frequently particular branches are recovered, they examine the differences in likelihood between the ML tree for the bootstrapped data and the other hypotheses in the resampled data. Both tests use these differences in likelihood to form a confidence set of trees, which can reject those hypotheses that fall below the critical value.

Both tests control the level of type I (false positive) error successfully. In other words, the confidence interval is conservative and does not place unwarranted confidence in a small number of trees. The AU test is constructed in a subtly different manner to the SH test, which removes a potential bias and can increase statistical power in some cases. This difference may allow the AU test to reject more trees than the SH test and produce tighter confidence intervals (demonstrated in Fig. 3).

Readers may also come across the Kishino–Hasegawa (KH-) test [69] for comparing tree hypotheses. The KH test is only applicable when all of the potential hypotheses (trees) can be

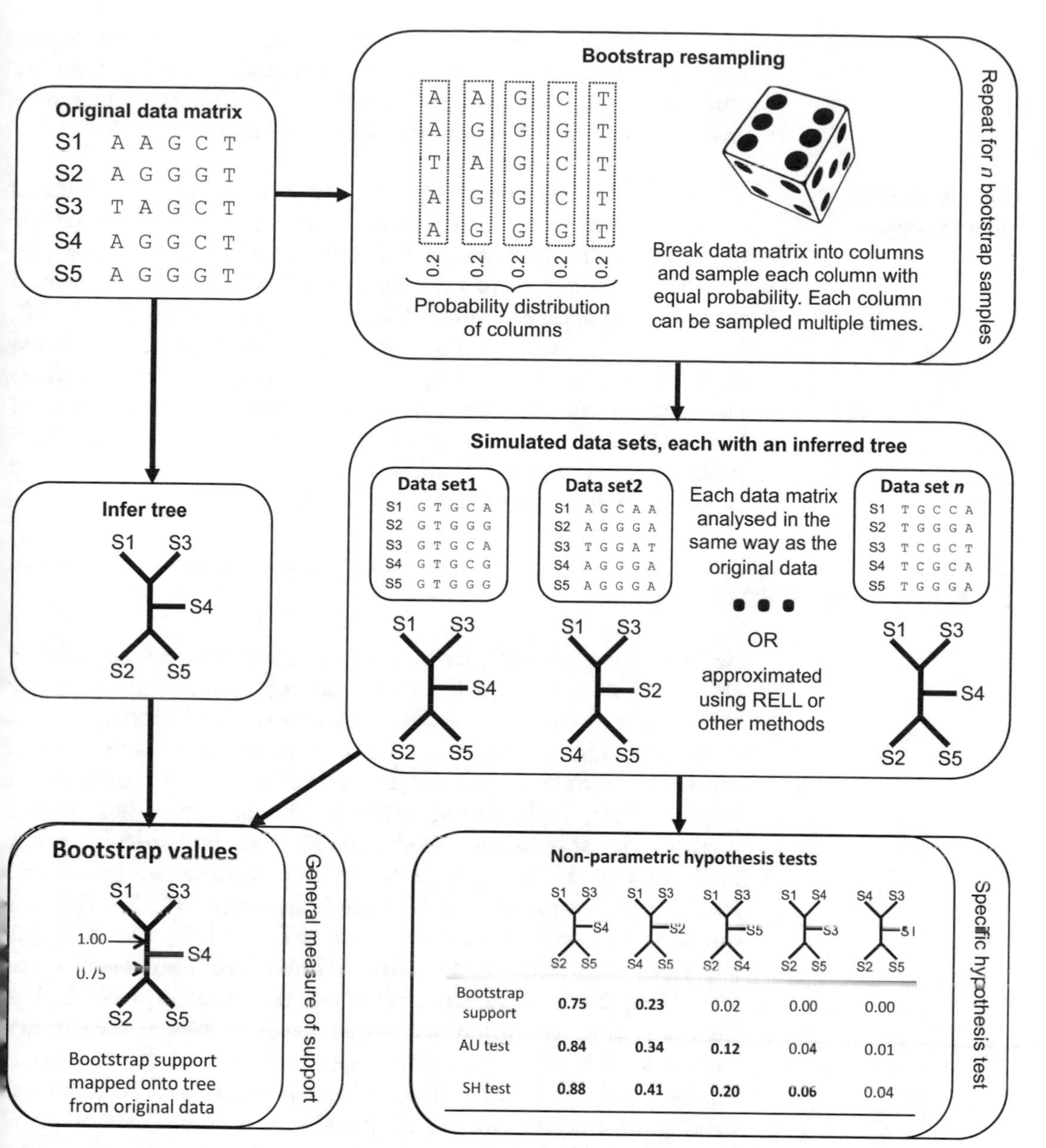

Fig. 3 Two common ways to assess confidence in tree estimate are to use bootstrap resampling (*1*) to provide a general measure of confidence in a tree or (*2*) for hypothesis testing. These approaches resample with replacement columns from the original data matrix, to generate simulated datasets with properties that reflect the original data. Full bootstrapping approaches analyze these data in the manner of the original dataset to produce a tree estimate for each bootstrap replicate, although there are also heuristic approaches to obtain these estimates. The results from the simulation can be summarized through the tree estimates (for bootstrap values or support) or the difference in likelihood between the ML tree and specific topologies (AU- and SH-test)

presented prior to examining the data. Failure to meet this condition means the KH test can falsely reject hypotheses and give undue confidence in a small number of hypotheses. In nearly all circumstances it is preferable to use the SH or AU tests.

5.3 Other Tests of Tree Topologies

The methods described earlier are very general and used by many different tree search programs. There are, however, several program-specific approaches to assessing the confidence in a tree worth mentioning. Both RAxML and IQPNNI have implemented different fast approximations to the standard bootstrap [34, 70, 71]. These fast bootstrapping approaches use information gained during tree search to speed up the assessment of bootstrapped trees and thus reduce computational costs. An alternative approach, used in PhyML, is the approximate likelihood ratio test (aLRT), which performs a per branch statistical test to produce statistics that may be comparable to bootstrap proportions [72].

6 Bayesian Inference of Trees

Bayesian inference of phylogenetic trees is now an established alternative to maximum likelihood inference, and simultaneously estimates the tree and the parameters in the evolutionary model while providing a measure of confidence in those estimates (*see* **Note 5**). The following provides a limited introduction to Bayesian phylogenetics, highlighting some of the principles behind the methods, its advantages, disadvantages, and similarities to other methodology. More comprehensive introductions and guides to Bayesian phylogenetics can be found elsewhere [15, 73–76] and online in the documentation for the BEAST [77] and MrBayes software [78]. There are many theoretical differences in how Bayesian inference and maximum likelihood treat data and models, but the major practical difference is that Bayesian inference examines the probability of a tree based on the product of the likelihood function and a factor describing prior expectations about that tree. More precisely, the prior is a probability distribution over all parameters, including the tree and the model, describing how frequently one would expect values to be observed before evaluating evidence from the data. Bayesian inference uses information in the data, expressed through the likelihood function, to update the prior and estimate the posterior distribution of the parameters of interest. The outcome of a Bayesian phylogenetic tree analysis produces a posterior distribution over all parameters, which needs to be processed to retrieve posterior distributions or posterior probabilities of the parameters or trees of interest.

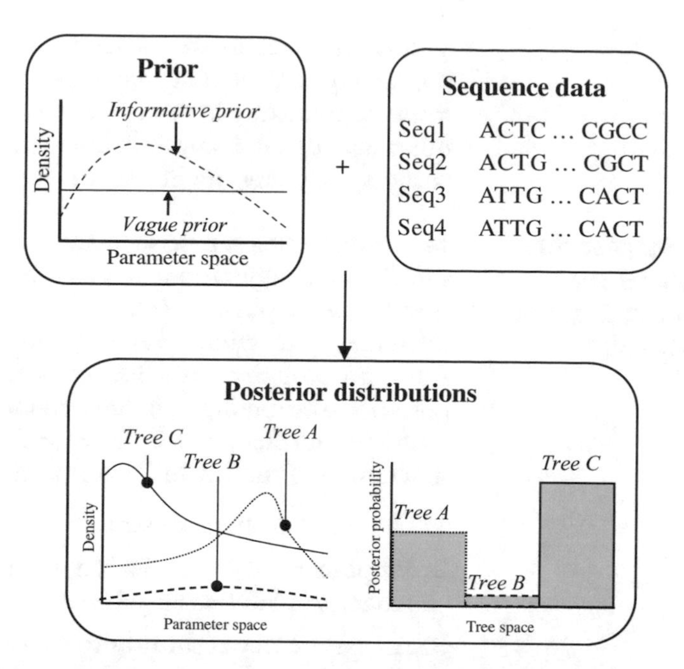

Fig. 4 A schematic showing Bayesian tree inference. The prior (*left*) contains the information or beliefs one has about the parameters contained in the tree and the evolutionary model before seeing the data. During Bayesian inference, this is combined with the information about the tree and model parameter values held in the original data (*right*) to produce the posterior distribution. These may be summarized to provide estimates of the parameter values (*bottom left*) and the trees (*bottom right*)

6.1 Bayesian Estimation of Trees

To obtain the posterior probability of trees, those parameters that are not of direct interest to the analysis need to be 'integrated out' of the posterior distribution. These extra parameters are sometimes referred to as 'nuisance parameters,' and include components of the evolutionary model and branch lengths, that is we want the tree topologies averaged across all parameter values.

A common summary measure in Bayesian phylogenetics is the maximum a posteriori probability (MAP) tree, which is the tree with the highest posterior probability in tree space. The integration required to obtain the MAP tree is represented in the transition from left to right in the posterior distribution section of Fig. 4, where the area under the curve for each tree on the left equates to the posterior probability for each tree on the right. The confidence in the MAP tree can naturally be estimated from the posterior distribution and requires no additional computation. In Bayesian parlance, this is achieved by constructing a credibility interval, which can roughly be considered similar to the confidence interval of classical statistics and is constructed by adding trees to the credible set in order of decreasing probability. For example, the credibility interval for the data in Fig. 4 would be constructed by adding the trees in the order C, A, and B. The small posterior probability contained in tree B means that it is likely to be rejected from the credibility interval. Readers should be aware that there are other equally valid ways of summarizing the results of a Bayesian analysis, including the majority rule consensus tree [79].

6.2 Sampling the Posterior Using Markov Chain Monte Carlo (MCMC)

In phylogenetics, a precise analysis of the posterior distribution is usually not computationally possible because it requires a summation across all possible tree topologies. MCMC rescues Bayesian inference by forming a series (chain) of pseudo-random samples from the posterior distribution as an approximation to the true posterior distribution. Understanding how this sampling works is useful to further explain Bayesian tree inference. A simplified description of the MCMC algorithm for tree estimation is:

1. Get initial estimate of tree.
2. Propose a new tree (often by methods similar to traversal schemes in Subheading 3.2).
3. Accept the tree according to a specified probability function.
4. Go to (2).

For a more complete version of the MCMC algorithm, *see* Larget and Simon [79] and Yang [15]. The options for obtaining an initial tree estimate, (1), are the same as for point estimation (e.g., maximum likelihood or maximum parsimony), although a random starting place can also be a good choice, because starting different MCMC chains from very different places can be useful for assessing their convergence (*see later*). The tree proposal mechanism in (2) needs to satisfy at least three criteria: (a) the proposal process is random, (b) every tree is connected to every other tree, and (c) the sampling chain does not periodically repeat itself. The necessity for (a) and (b) allows the chain potential access to all points in tree space, which in principle allows complete sampling if the chain is allowed to run for long enough. The final point, (c), is a technical requirement that ensures the chain does not repeatedly visit the areas of tree space in the same order (it is aperiodic). The ability of MCMC to effectively sample tree space is highly dependent on the proposal scheme, and the most popular schemes are similar to those used in point estimation (*see earlier*).

The probability of a new tree being accepted, (3), is the function that enables the MCMC algorithm to correctly sample from the posterior distribution. The probability of acceptance depends on the difference in likelihood scores of the current and new tree their chances of occurring under the prior, and an additional correction factor that depends on the sampling approach. A good sampling scheme coupled with this acceptance probability enables the chain to frequently accept trees with a higher posterior probability but to also occasionally accept mildly poorer trees. Trees with very low posterior probability are rarely visited. The overall result is that the amount of time a chain spends in regions of tree space is directly proportional to the posterior distribution. This approach allows the posterior probability of trees to be easily calculated from the frequency of time that the chain spends visiting different topologies.

6.3 How Long to Run a Markov Chain Monte Carlo

The number of samples required for MCMC to successfully sample the posterior distribution is dependent on two factors: convergence and mixing (*see* **Note 6**). A chain is said to have converged when it begins to accurately sample from the posterior distribution, and the period before this happens is called burn in. The mixing of a chain is a measure of how similar one sample from the MCMC is to the next, with good mixing allowing relatively close samples to provide independent draws from the posterior distribution. Mixing is important because it controls both how quickly a chain converges and its ability to sample effectively from the posterior distribution afterward. When a chain mixes well, all trees can be quickly reached from all other trees and MCMC is a highly effective method. When mixing is poor the chain's ability to sample effectively from the posterior is compromised. It is common for samples to be recorded only periodically in the chain, say every 100 or 1000 iterations, to ensure samples are as close to independent draws from the posterior as possible.

It is notoriously difficult to confirm that the chain has converged and is successfully mixing, but there are diagnostic tools available to help. A powerful way to examine these conditions is to run multiple chains and compare them. If a majority of chains starting from substantially different points in tree space concentrate their sampling in the same region it is indicative that the chains have converged. Evidence for successful mixing can be found by comparing samples between converged chains. When samples are clearly different, it is strong evidence that the chain is not mixing well. These comparative approaches can go awry; for example, when a small number of good tree topologies with large centers of attraction are separated by long and deep troughs in the surface of posterior probabilities. If by chance all the chains start in the same centre of attraction, they can misleadingly appear to have converged and mixed well, even when they have poorly sampled the posterior. This behavior has been induced, for example, for small trees under artificial misspecifications of the evolutionary model, although the general prevalence of this problem is currently unknown.

An alternative diagnostic is to examine a plot of the likelihood and/or model parameter values, such as rate variation parameters and sums of branches in the tree, against sample number. Before convergence these values may tend to show discernable patterns of change. The likelihood function, for example, may appear on average to steadily increase while the chain moves to progressively better areas of tree space. When the chain converges these values may appear to have quite large random fluctuations with no apparent trend. Fast fluctuation accompanied by quite large differences in likelihood, for example, would be indicative of successful mixing. This character alone is a weak indicator of convergence, because chains commonly fluctuate before they find better regions of tree space. New sampling procedures, such as Metropolis Coupled MCMC (MC^3), are being introduced that can address more difficult sampling and mixing problems, and are likely to feature more frequently in phylogenetic inference (*see* **Notes 5** and **6**).

6.4 The Specification of Priors

All Bayesian phylogenetic analyses require the specification of prior distributions for all of the parameters in the model, including the substitution model and the tree and its branch lengths. There are two broad approaches to specifying priors that divide Bayesian inference into two groups [80]. The first approach is Subjective Bayes, where the informative prior represents a researcher's prior belief in the question at hand. Bayesian inference adjusts this prior belief to obtain a posterior that reveals how much the researcher's opinion should be changed by the observed data. At a superficial level Subjective Bayes is exactly how one may wish to conduct scientific research, but it is difficult to reconcile with the more general approach to scientific method. The prior beliefs for particular parameters or models vary between researchers, making it difficult for researchers to agree on the accuracy, or even relevance, of the posterior. The rarity with which research fields can agree on a prior makes publishing using Subjective Bayes almost impossible.

The second approach is Objective Bayes, where a researcher attempts to express ignorance of the problem at hand in their priors, and then use the posterior distribution as a way of assessing how much information is in their data. This approach is closely linked to the idea of uninformative or flat priors, which define all outcomes as equally likely. In tree inference this can be broadly interpreted as each tree being equally likely, which is philosophically similar to how other methods, such as likelihood and parsimony, treat tree estimation. The problem with Objective Bayes is that there is no such thing as 'uninformative priors,' since even relatively innocent-looking flat priors actually make strong statements about the data, and this can have a major impact on the posterior distribution. A simple example of this problem can be demonstrated by trying to specify a prior distribution on a square with edge length l. One can assign a flat prior to l so that all edge lengths are equally likely. An alternative, and equally valid approach, is to assign a flat prior on the area of the square, l^2, so that all areas are equally likely. Both of these approaches produce different and incompatible prior distributions, since a uniform distribution over l can never be the same as a uniform distribution over l^2.

The unacceptability of Subjective Bayes and the impossibility of Objective Bayes leave Bayesian inference in a quandary. Methodology research has focused on producing vague or disperse priors that have a minimal impact on inference (e.g. [81, 82]), but the literature of Bayesian phylogenetics (along with other fields using Bayesian inference) is littered with cases where apparently innocent priors have had a strong and negative effect on Bayesian inference [83–85]. Users of Bayesian phylogenetics should ensure that they use up-to-date versions of programs, keep abreast of developments in the area, and employ due care and diligence, just as with any other tree estimation method.

7 Strengths and Weaknesses of Statistical Methods

Bayesian and ML approaches to statistical tree inference are highly complementary, sharing the same likelihood function and evolutionary models. Although both are rigorous forms of statistical inference, there are differences in their underlying philosophy and how they treat data. For example, ML estimates the tree and parameters under the assumption that they come from a single fixed set of values, whereas Bayesian inference treats the tree and parameters as random variables in the model. Rather than dwell on the philosophical and technical differences between them, we choose to present some of their relative advantages and disadvantages for practical data analysis.

The programs used for ML tree inference of trees have been under development for many years and have benefited from high-performance computing approaches, meaning that they can be used to estimate trees from relatively large datasets consisting of thousands of sequences. The newer fast bootstrapping approaches, aLRT, and SH and AU test are all (relatively) computationally fast after the initial ML tree(s) is estimated, allowing quick assessment of the tree and the testing of phylogenetic hypotheses. In contrast, Bayesian inference tends to be somewhat slower, since the MCMC algorithm needs to burn-in and then adequately sample from the posterior distribution, although recent computational developments promise faster run-times for Bayesian inference. After the MCMC is complete, however, one has all the information needed to assess confidence in the tree and the parameter estimates, whereas ML estimates require additional calculations for bootstrap estimates of confidence, for example.

Although Bayesian inference and ML both use the same likelihood function, they differ in their ability to deal with very complex models. ML relies on 'hill-climbing' to find the optimal combination of parameter estimates for a given model (or models, if partitioning is used), which can get stuck when there are many different optima (hills) or when the landscape is mostly flat. This allows ML to use only moderately complex models that capture many biologically important factors during evolution. On the other hand, the MCMC algorithm means that Bayesian inference can sample from very difficult landscapes, allowing inference under extremely complex models, providing that the chain mixes adequately. Researchers have taken advantage of MCMC to create very sophisticated, and potentially more realistic, models that attempt to capture accurate phylogenetic signal in deep phylogeny [86, 87], protein structure [4, 88], and the integration of phylogeny with life history traits [89]. The ability of Bayesian inference to cope with very complex models, while also incorporating uncertainty, has also made it a popular choice when attempting to date species divergences based on calibrations from fossils [77, 90]. These models are often based on a relaxed version of the molecular clock [91], which allows the rate of molecular evolution to change over time, and integrate over uncertainty in the placement of fossil data by representing them by a range of ages rather than just point estimates. It is also worth noting that ML programs sometimes do not agree on the choice of optimal tree, because they estimate the likelihoods slightly differently. There are usually a very large number of near-optimal trees, and thus even minor difference in the estimates can create discrepancies. This problem should not occur with Bayesian analyses providing there are enough samples from the MCMC to differentiate between trees.

The thorniest difference between the statistical methods is that of branch support, usually measured by bootstrap proportions in ML and posterior probabilities in Bayesian inference. In principle, posterior probabilities provide a clear and easy to interpret measure of branch support, but there is widespread concern that they offer artificially high support when compared to bootstrap proportions. Some of this overconfidence may be attributable to problems with the prior distribution over tree topologies needed by Bayesian inference, but even after careful correction the posterior probabilities still often appear higher than their bootstrap proportion counterparts. Bootstrap proportions, on the other hand, are based on a simplistic model that in some circumstances underestimates the branch support in a variable manner that depends on both the number of sequences and sequence length.

Both approaches share the limitations of their assumptions about models and data, so that there is no guarantee about what will happen when things go wrong. Furthermore, conflicting minority signals in the data are ignored by virtue of the fact that a tree is produced as the final estimate of the phylogeny. Biological processes that follow a non-tree-like structure, such as hybridization and gene flow, can create such signals, which may be important.

Finally, phylogenomics [92], the idea of applying genomic data to phylogenetic studies, offers the enticing possibility of a fast and cost-effective means of generating large amounts of multilocus sequence data. However, phylogenetic analyses might need to change in response to these increasing dataset sizes. For example, tree search strategies may need to be revised, and incongruence between trees constructed from different genes will be a major issue. Tree inference has an exciting future ahead of it.

8 Notes

1. *Examining prior assumptions*: Tree building depends critically on the validity of the assumptions made by each method. Common to all methods is the idea that the multiple sequence alignment used as input is correct. This cannot be known, but it requires careful attention to ensure that each column of the matrix does contain homologous characters. There are no formal tests by which this can be assessed, although there are innumerable heuristic methods for filtering [93, 94] that have mixed performance when applied to both simulated and real data [95]. For a more extensive list of alignment filtering strategies, *see* URL: http://phylonetworks.blogspot.se/2014/10/uncertainty-in-multiple-sequence.html.
2. *Examining the assumption of a tree*: Tree-building methods assume that the evolutionary history has been tree like, rather than being a reticulating network, and that there is some phylogenetic signal in the data in the first place. If the predominant pattern of history has been from parent to offspring then the data will be tree like, but gene flow will create reticulations. Lack of signal produces a star-like tree rather than one with dichotomous branching. These possibilities are best assessed with a priori exploratory data analysis, to assess the predominant patterns in the data in a manner independent of a tree structure.

   Tree-independent approaches take the usual sequence alignment as input, and output visualizations that indicate the nature of all of the predominant signals in the data, irrespective of whether they fit a single tree structure. These include spectral analysis, carried out using the SpectroNet program [96], which measures the signal for all possible splits supported by the data. Alternatively, splits graphs, produced using the SplitsTree program [97], directly visualize the supported splits, and comparison between the tree topology and the degree of reticulation in the splits graph provides a very effective assessment of how tree like are the data, as well-supported branches will have little or no reticulation. Cloudograms provide an alternative visualization [21], which superimposes the set of all optimal or nearly optimal trees arising from an analysis. Both splits graphs and cloudograms provide fast and efficient methods, but suffer the weakness that they provide no distinction between sampling error, model misspecification, and real violations of the tree hypothesis.

3. *Choosing a model*: Model choice is key to the successful use of statistical methods for building trees. The model selected must adequately describe the data variation both within and among the sequences. In general, models that more realistically describe evolution are thought to be more likely to produce accurate estimates of the true tree. DNA models describe the relative frequency of the different bases and the bias toward transition mutation and can be assessed using the jModelTest program [51]. RNA models take into account the stem pairing that typically involves more than half of the sequence length and can be assessed using the PHASE program [54]. The replacement rates between amino acids in protein models are usually based on empirical estimates and can be assessed using the ProtTest program [52]. Models describing the evolution of codons can also be used for phylogeny estimation [98]; and these can be compared to the other models using the ModelOMatic program [53]. *See* Chapter 15 for more details about phylogenetic model evaluation.
4. *Maximum likelihood inference*: There are a number of commonly used programs for ML estimation of trees, which differ considerably in their heuristic approach, with each focused on different user goals. PAUP* [99] is oldest of these, and is thus the slowest. It proposes a starting tree using stepwise addition, refines the tree using TBR branch swapping, and resamples tree space using randomized sequence orders. It has numerous user-programmable options, and can be programmed to use the ratchet for resampling, which speeds up the search enormously. PhyML proposes a starting tree using stepwise addition, refines the tree using a combination of NNI and SPR branch swapping [100], but does not do resampling. RAxML has been optimized for large numbers of sequences and considerable sequence length [34], trying to save both computer memory and time, and it thus represents the current state of the art. It proposes a starting tree using parsimony-based stepwise addition, refines the tree using a modified version of SPR branch swapping, and resamples tree space using randomized sequence orders. It also has a fast bootstrap option [70]. There are a number of models available, some of them based on approximations, and an increasing number of user options. It is available via a web server for high-speed computing. GARLI uses a genetic algorithm to explore tree space stochastically [48], thus differing notably from the other choices (which act deterministically). It focuses on flexibility of model choice (including partitions) and rigor of parameter estimation, while still allowing large datasets.

Multiple runs of the program are needed to avoid local optima. It is also available via a web server for high-speed computing.

5. *Bayesian inference*: There are also a number of commonly used programs for Bayesian estimation of trees, with each spotlighting different user goals. MrBayes is a generalist program [78], focusing on ease of use to produce a single tree estimate. It provides a range of models, including RNA and codon models, each with user-selectable defaults. BEAST is focused on parameter estimation of its wide range of specialist models [77], using MCMC to integrate across possible trees, rather than focusing on the trees themselves. These parameters include dating of nodes, estimation of population sizes and their changes through time, and quantification of selection pressures. This program requires considerable user input to set options, as there are few default values, and it uses its own XML input format. Options are available to use multiple cores and processors, and also graphics cards for BEAST and MrBayes. PhyloBayes is mainly distinct in employing the CAT infinite-mixture model, which accounts for site-specific amino acid or nucleotide preferences, and is thus suitable for large multigene datasets [101].
6. *MCMC convergence and mixing*: The success of Bayesian analysis depends on convergence and adequate mixing of the chain(s). Several programs have been developed to assess this by examining the parameter posteriors stored in the output files from the programs listed earlier. These diagnostic programs include TRACER, which produces trace plots of the likelihood score, plus summary statistics and frequency distributions of the posteriors of all parameters, for each chain as well as summed across all chains [77]. Another program is AWTY (Are We There Yet), which also estimates convergence rates of posterior split probabilities and branch lengths [102].

Acknowledgments

SW is funded by Uppsala University. DAM is funded by Akademikernas A-kassa and Trygghetsstiftelsen.

References

1. Hahn BH et al (2000) AIDS—AIDS as a zoonosis: scientific and public health implications. Science 287:607–614
2. Pellegrini M et al (1999) Assigning protein functions by comparative genome analysis: protein phylogenetic profiles. Proc Natl Acad Sci U S A 96:4285–4288
3. Ames RM et al (2012) Determining the evolutionary history of gene families. Bioinformatics 28:48–55
4. Liberles DA et al (2012) The interface of protein structure, protein biophysics, and molecular evolution. Protein Sci 21:769–785

5. Hahn MW, Han MV, Han S-G (2007) Gene family evolution across 12 Drosophila genomes. PLoS Genet 3:e197
6. Mouse Genome Sequencing Consortium (2002) Initial sequencing and comparative analysis of the mouse genome. Nature 420:520–562
7. Lynch M, Walsh B (2007) The origins of genome architecture. Sinauer Associates, Sunderland, MA
8. Gogarten JP, Doolittle WF, Lawrence JG (2002) Prokaryotic evolution in light of gene transfer. Mol Biol Evol 19:2226–2238
9. Yang Z, Rannala B (2010) Bayesian species delimitation using multilocus sequence data. Proc Natl Acad Sci U S A 107:9264–9269
10. Siepel A et al (2005) Evolutionarily conserved elements in vertebrate, insect, worm, and yeast genomes. Genome Res 15:1034–1050
11. Felsenstein J (2003) Inferring Phylogenies. Sinauer Associates, Sunderland, MA
12. Löytynoja A, Goldman N (2008) Phylogeny-aware gap placement prevents errors in sequence alignment and evolutionary analysis. Science 320:1632–1635
13. Anisimova M, Cannarozzi G, Liberles DA (2010) Finding the balance between the mathematical and biological optima in multiple sequence alignment. Trends Evol Biol 2:e7
14. Löytynoja A (2012) Alignment methods: strategies, challenges, benchmarking, and comparative overview. In: Evolutionary genomics. Springer, New York, pp 203–235.
15. Yang Z (2006) Computational molecular evolution. Oxford University Press, Oxford
16. Redelings B, Suchard M (2005) Joint Bayesian estimation of alignment and phylogeny. Syst Biol 54:401–418
17. Thorne JL, Kishino H, Felsenstein J (1991) An evolutionary model for maximum likelihood alignment of DNA sequences. J Mol Evol 33:114–124
18. McGuire G, Denham MC, Balding DJ (2001) Models of sequence evolution for DNA sequences containing gaps. Mol Biol Evol 18:481–490
19. Morrison DA, Ellis JT (1997) Effects of nucleotide sequence alignment on phylogeny estimation: a case study of 18S rDNAs of Apicomplexa. Mol Biol Evol 14:428–441
20. Wong K, Suchard M, Huelsenbeck J (2008) Alignment uncertainty and genomic analysis. Science 319:473–476
21. Blackburne BP, Whelan S (2013) Class of multiple sequence alignment algorithm affects genomic analysis. Mol Biol Evol 30:642–653
22. Wägele JW, Mayer C (2007) Visualizing differences in phylogenetic information content of alignments and distinction of three classes of long-branch effects. BMC Evol Biol 7:147
23. Hendy MD, Penny D (1993) Spectral analysis of phylogenetic data. J Classif 10:5–24
24. Morrison DA (2010) Using data-display networks for exploratory data analysis in phylogenetic studies. Mol Biol Evol 27:1044–1057
25. Huson DH, Bryant D (2006) Application of phylogenetic networks in evolutionary studies. Mol Biol Evol 23:254–267
26. Morrison DA (2011) Introduction to phylogenetic networks. RJR Productions, Uppsala, Sweden
27. Philippe H, Germot A (2000) Phylogeny of eukaryotes based on ribosomal RNA: long-branch attraction and models of sequence evolution. Mol Biol Evol 17:830–834
28. Inagaki Y et al (2004) Covarion shifts cause a long-branch attraction artifact that unites microsporidia and archaebacteria in EF-1α phylogenies. Mol Biol Evol 21:1340–1349
29. Viklund J, Ettema TJ, Andersson SG (2011) Independent genome reduction and phylogenetic reclassification of the oceanic SAR11 clade. Mol Biol Evol 29:599–615
30. Morrison DA (2006) Phylogenetic analyses of parasites in the new millennium. Adv Parasitol 63:1–124
31. Edwards AWF (1972) Likelihood: an account of the statistical concept of likelihood and its application to scientific inference. Cambridge University Press, New York
32. Chang JT (1996) Full reconstruction of Markov models on evolutionary trees: identifiability and consistency. Math Biosci 137:51–73
33. Rogers JS (1997) On the consistency of maximum likelihood estimation of phylogenetic trees from nucleotide sequences. Syst Biol 46:354–357
34. Stamatakis A (2014) RAxML version 8: a tool for phylogenetic analysis and post-analysis of large phylogenies. Bioinformatics 30:1312–1313
35. Izquierdo-Carrasco F, Smith SA, Stamatakis A (2011) Algorithms, data structures, and numerics for likelihood-based phylogenetic inference of huge trees. BMC Bioinformatics 12:470
36. Steel M, Penny D (2000) Parsimony, likelihood, and the role of models in molecular phylogenetics. Mol Biol Evol 17:839–850

37. Siddall ME, Kluge AG (1997) Probabilism and phylogenetic inference. Cladistics 13:313–336
38. Saitou N, Nei M (1987) The neighbor-joining method—a new method for reconstructing phylogenetic trees. Mol Biol Evol 4:406–425
39. Allman ES, Rhodes JA (2006) The identifiability of tree topology for phylogenetic models, including covarion and mixture models. J Comput Biol 13:1101–1113
40. Swofford DL et al (1996) Phylogenetic inference. In: Hillis DM, Moritz C, Mable BK (eds) Molecular systematics. Sinauer Associates, Sunderland, MA, pp 407–514
41. Morrison DA (2007) Increasing the efficiency of searches for the maximum likelihood tree in a phylogenetic analysis of up to 150 nucleotide sequences. Syst Biol 56:988–1010
42. Whelan S (2007) New approaches to phylogenetic tree search and their application to large numbers of protein alignments. Syst Biol 56:727–740
43. Vinh LS, von Haeseler A (2004) IQPNNI: moving fast through tree space and stopping in time. Mol Biol Evol 21:1565–1571
44. Money D, Whelan S (2012) Characterizing the phylogenetic tree-search problem. Syst Biol 61:228–239
45. Bryant D (2004) The splits in the neighborhood of a tree. Ann Combin 8:1–11
46. Whelan S, Money D (2010) The prevalence of multifurcations in tree-space and their implications for tree-search. Mol Biol Evol 27:2674–2677
47. Lin Y-M, Fang S-C, Thorne JL (2007) A tabu search algorithm for maximum parsimony phylogeny inference. Eur J Oper Res 176:1908–1917
48. Zwickl D (2006) Genetic algorithm approaches for the phylogenetic analysis of large biological sequence datasets under the maximum likelihood criterion. Ph.D. thesis, University of Texas, USA
49. Lewis PO (1998) A genetic algorithm for maximum-likelihood phylogeny inference using nucleotide sequence data. Mol Biol Evol 15:277–283
50. Lemmon AR, Milinkovitch MC (2002) The metapopulation genetic algorithm: an efficient solution for the problem of large phylogeny estimation. Proc Natl Acad Sci U S A 99:10516–10521
51. Darriba D et al (2012) jModelTest 2: more models, new heuristics and parallel computing. Nat Methods 9:772
52. Darriba D et al (2011) ProtTest 3: fast selection of best-fit models of protein evolution. Bioinformatics 27:1164–1165
53. Whelan S et al (2015) ModelOMatic: fast and automated model selection between RY, nucleotide, amino acid, and codon substitution models. Syst Biol 64:42–55
54. Allen JE, Whelan S (2014) Assessing the state of substitution models describing noncoding RNA evolution. Genome Biol Evol 6:65–75
55. Blair C, Murphy RW (2011) Recent trends in molecular phylogenetic analysis: where to next? J Hered 102:130–138
56. Lanfear R et al (2012) PartitionFinder: combined selection of partitioning schemes and substitution models for phylogenetic analyses. Mol Biol Evol 29:1695–1701
57. Pagel M, Meade A (2004) A phylogenetic mixture model for detecting pattern-heterogeneity in gene sequence or character-state data. Syst Biol 53:571–581
58. Le SQ, Lartillot N, Gascuel O (2008) Phylogenetic mixture models for proteins. Philos Trans R Soc B Biol Sci 363:3965–3976
59. Le SQ, Gascuel O (2010) Accounting for solvent accessibility and secondary structure in protein phylogenetics is clearly beneficial. Syst Biol 59:277–287
60. Bouckaert RR (2010) DensiTree: making sense of sets of phylogenetic trees. Bioinformatics 26:1372–1373
61. Felsenstein J (1985) Confidence limits on phylogenies: an approach using the bootstrap. Evolution 39:783–791
62. Hillis DM, Bull JJ (1993) An empirical test of bootstrapping as a method for assessing confidence in phylogenetic analysis. Syst Biol 42:182–192
63. Efron B, Halloran E, Holmes S (1996) Bootstrap confidence levels for phylogenetic trees. Proc Natl Acad Sci U S A 93:13429
64. Embley TM, Martin W (2006) Eukaryotic evolution, changes and challenges. Nature 440:623–630
65. Fitzpatrick DA, Creevey CJ, McInerney JO (2006) Genome phylogenies indicate a meaningful α-proteobacterial phylogeny and support a grouping of the mitochondria with the Rickettsiales. Mol Biol Evol 23:74–85
66. McGowen MR, Gatesy J, Wildman DE (2014) Molecular evolution tracks macroevolutionary transitions in Cetacea. Trends Ecol Evol 29:336–346
67. Shimodaira H, Hasegawa M (1999) Multiple comparisons of log-likelihoods with

applications to phylogenetic inference. Mol Biol Evol 16:1114–1116

68. Shimodaira H (2002) An approximately unbiased test of phylogenetic tree selection. Syst Biol 51:492–508
69. Kishino H, Hasegawa M (1989) Evaluation of the maximum likelihood estimate of the evolutionary tree topologies from DNA sequence data, and the branching order in Hominoidea. J Mol Evol 29:170–179
70. Stamatakis A, Hoover P, Rougemont J (2008) A rapid bootstrap algorithm for the RAxML web servers. Syst Biol 57:758–771
71. Minh BQ, Nguyen MAT, von Haeseler A (2013) Ultrafast approximation for phylogenetic bootstrap. Mol Biol Evol 30:1188–1195. doi:10.1093/molbev/mst024
72. Anisimova M, Gascuel O (2006) Approximate likelihood-ratio test for branches: a fast, accurate, and powerful alternative. Syst Biol 55:539–552
73. Huelsenbeck JP et al (2002) Potential applications and pitfalls of Bayesian inference of phylogeny. Syst Biol 51:673–688
74. Holder M, Lewis PO (2003) Phylogeny estimation: traditional and Bayesian approaches. Nat Rev Genet 4:275–284
75. Ronquist F, Deans AR (2010) Bayesian phylogenetics and its influence on insect systematics. Annu Rev Entomol 55:189–206
76. Yang Z, Rannala B (2012) Molecular phylogenetics: principles and practice. Nat Rev Genet 13:303–314
77. Drummond AJ et al (2012) Bayesian phylogenetics with BEAUti and the BEAST 1.7. Mol Biol Evol 29:1969–1973
78. Ronquist F et al (2012) MrBayes 3.2: efficient Bayesian phylogenetic inference and model choice across a large model space. Syst Biol 61:539–542
79. Larget B, Simon DL (1999) Markov chain Monte Carlo algorithms for the Bayesian analysis of phylogenetic trees. Mol Biol Evol 16:750–759
80. Alfaro ME, Holder MT (2006) The posterior and the prior in Bayesian phylogenetics. Annu Rev Ecol Evol Syst 37:19–42
81. Zhang C, Rannala B, Yang Z (2012) Robustness of compound Dirichlet priors for Bayesian inference of branch lengths. Syst Biol 61:779–784
82. Bergsten J, Nilsson AN, Ronquist F (2013) Bayesian tests of topology hypotheses with an example from diving beetles. Syst Biol 62:660–673
83. Rannala B, Zhu T, Yang Z (2012) Tail paradox, partial identifiability, and influential priors in Bayesian branch length inference. Mol Biol Evol 29:325–335
84. Lewis PO, Holder MT, Holsinger KE (2005) Polytomies and Bayesian phylogenetic inference. Syst Biol 54:241–253
85. Yang ZH (2007) Fair-balance paradox, star-tree paradox, and Bayesian phylogenetics. Mol Biol Evol 24:1639–1655
86. Lartillot N, Philippe H (2004) A Bayesian mixture model for across-site heterogeneities in the amino-acid replacement process. Mol Biol Evol 21:1095–1109
87. Lartillot N, Brinkmann H, Philippe H (2007) Suppression of long-branch attraction artefacts in the animal phylogeny using a site-heterogeneous model. BMC Evol Biol 7:S4
88. Robinson D et al (2003) Protein evolution with dependence among codons due to tertiary structure. Mol Biol Evol 20:1692–1704
89. Lartillot N, Poujol R (2011) A phylogenetic model for investigating correlated evolution of substitution rates and continuous phenotypic characters. Mol Biol Evol 28:729–744
90. Lukoschek V, Keogh JS, Avise JC (2012) Evaluating fossil calibrations for dating phylogenies in light of rates of molecular evolution: a comparison of three approaches. Syst Biol 61:22–43
91. Baele G et al (2012) Improving the accuracy of demographic and molecular clock model comparison while accommodating phylogenetic uncertainty. Mol Biol Evol 29:2157–2167
92. Delsuc F, Brinkmann H, Philippe H (2005) Phylogenomics and the reconstruction of the tree of life. Nat Rev Genet 6:361–375
93. Landan G, Graur D (2007) Heads or tails: a simple reliability check for multiple sequence alignments. Mol Biol Evol 24:1380–1383
94. Penn O et al (2010) An alignment confidence score capturing robustness to guide tree uncertainty. Mol Biol Evol 27:1759–1767
95. Jordan G, Goldman N (2012) The effects of alignment error and alignment filtering on the sitewise detection of positive selection. Mol Biol Evol 29:1125–1139
96. Huber KT et al (2002) Spectronet: a package for computing spectra and median networks. Appl Bioinformatics 1:2041–2059
97. Huson DH (1998) SplitsTree: analyzing and visualizing evolutionary data. Bioinformatics 14:68–73
98. Gil M et al (2013) CodonPhyML: fast maximum likelihood phylogeny estimation under

codon substitution models. Mol Biol Evol 30:1270–1280

99. Swofford DL (2002) Phylogenetic analysis using parsimony (*and other methods). Sinauer Associates, Sunderland, MA

100. Guindon S et al (2010) New algorithms and methods to estimate maximum-likelihood phylogenies: assessing the performance of PhyML 3.0. Syst Biol 59:307–321

101. Lartillot N, Lepage T, Blanquart S (2009) PhyloBayes 3: a Bayesian software package for phylogenetic reconstruction and molecular dating. Bioinformatics 25:2286–2288

102. Nylander JA et al (2008) AWTY (are we there yet?): a system for graphical exploration of MCMC convergence in Bayesian phylogenetics. Bioinformatics 24:581–583

Chapter 15

Identifying Optimal Models of Evolution

Lars S. Jermiin, Vivek Jayaswal, Faisal M. Ababneh, and John Robinson

Abstract

Most phylogenetic methods are model-based and depend on models of evolution designed to approximate the evolutionary processes. Several methods have been developed to identify suitable models of evolution for phylogenetic analysis of alignments of nucleotide or amino acid sequences and some of these methods are now firmly embedded in the phylogenetic protocol. However, in a disturbingly large number of cases, it appears that these models were used without acknowledgement of their inherent shortcomings. In this chapter, we discuss the problem of model selection and show how some of the inherent shortcomings may be identified and overcome.

Key words Evolutionary processes, Phylogenetic assumptions, Stationary conditions, Reversible conditions, Homogeneous conditions, Rate-heterogeneity across sites, Markov models, Model selection, Model evaluation

1 Introduction

Molecular phylogenetics is a fascinating aspect of bioinformatics with an increasing impact on the Life Sciences. It allows us to infer historical relationships among species [1], genomes [2], and genes [3], and provides us with a framework for classifying organisms [4] and genes [5], and for studying coevolution of traits [6]. Phylogenetic trees are the intended products of many studies and the inputs of others. Charleston and Robertson [7], for example, compared a phylogeny of pathogens to that of their hosts and found that the evolution of the pathogens had involved co-divergence and host switching. Jermann et al. [8], on the other hand, used a phylogeny of artiodactyls to manufacture enzymes that may have been expressed in the cells of their 8–50-million-year-old common ancestors. In some cases, phylogenetic trees are inferred from alignments of a single gene or gene product when, for example, the objective is to infer its function [9]; in other cases, they are inferred from concatenations of such alignments when, for example, the objective is to infer the species tree [10]. In the majority of cases,

Jonathan M. Keith (ed.), *Bioinformatics: Volume I: Data, Sequence Analysis, and Evolution*, Methods in Molecular Biology, vol. 1525, DOI 10.1007/978-1-4939-6622-6_15,

including those cited above, the phylogeny is unknown, so it is useful to know how to infer it.

Phylogenetic inference is often considered a challenge, and many people still shy away from approaching scientific questions from an evolutionary perspective because they consider the phylogenetic approach too hard. Admittedly, the phylogenetic methods are underpinned by mathematics, statistics, and computer science, so a good basic understanding of these sciences goes a long way towards establishing a sound theoretical and practical basis for phylogenetic research. However, it need not be that difficult because flexible and user-friendly phylogenetic programs are available for most computer platforms. Instead, the challenges lie in: (a) choosing appropriate phylogenetic data for the question in mind, (b) choosing a phylogenetic method to analyze these data, and (c) determining the extent to which the phylogenetic results are reliable.

Most molecular phylogenetic methods rely on substitution models that are designed to approximate the evolutionary process of change from one nucleotide (or amino acid) to another. The models are usually selected by the operator, increasingly often with the help of model-selection methods, which have been designed for nucleotides and amino acids [11, 12], and it is even possible to accommodate cases where different sets of sites have evolved under different conditions [13, 14].

The substitution models considered by these model-selection methods implicitly assume that the sequences have evolved under stationary, reversible, and homogeneous conditions (defined below). However, based on a growing body of data (*see* citations 2, 3, 17–40 in Ref. 15), it is now clear that homologous sequences of nucleotides or amino acids, more often than not, have evolved under more complex conditions, implying that it would be: (a) unwise to use popular implementations of model-selection methods (because they only consider stationary, reversible, and homogeneous models of evolution); and (b) wise to use phylogenetic methods that consider more general Markov models of molecular evolution [16–43].

The choice of substitution model clearly is an important one for phylogenetic studies, but many investigators are still either: (a) unaware that the model chosen by them may not be appropriate for analysis of their data, (b) unaware that using an inappropriate model may lead to errors in phylogenetic estimates, or (c) unsure about how to select an appropriate model. Realizing that a clear explanation of the problem and its potential solutions was not available in the literature, we [44] published a book chapter that aimed to address this issue. Here, we update this explanation in the light of new research bearing on the issue.

The approach taken in this chapter is based on the idea that the *evolutionary pattern* and the *evolutionary process* are two sides of the same coin: the former is the phylogeny, a rooted binary tree that depicts the time and order of different divergence events, and the latter is the process by which mutations in DNA accumulate over time along diverging lineages. It makes no sense to consider the pattern and the process separately, even though only one of the two might be of interest, because the estimate of evolutionary pattern depends on the evolutionary process, and vice versa. Underpinning this chapter is also a hope to raise an awareness of what the term *rate of molecular evolution* means: it is not just a single variable, as commonly portrayed, but rather, in mathematical terms, a matrix of variables. Finally, although many types of mutations are known to affect DNA and protein, the focus of this chapter is on point mutations in DNA. Our reasons for limiting the focus to these changes is that phylogenetic studies frequently rely on the products of point mutations as the main source of phylogenetic information, and that the substitution models used in phylogenetic methods usually focus on those types of changes (much of what is written below applies equally well, albeit with some modifications, to sequences of amino acids).

In the following sections, we first describe the phylogenetic assumptions, including some of the relevant aspects of the Markov models commonly used in phylogenetic studies. We then discuss some of the terminology used to characterize phylogenetic data and describe several methods that can be used for identifying the optimal Markov model. We also discuss why it is necessary to use data-surveying methods before and after phylogenetic analysis. Using such methods prior to phylogenetic analyses is becoming increasingly popular, but it is still rare to see phylogenetic results being properly evaluated using the parametric bootstrap.

2 Underlying Principles

The evolutionary processes that result in the accumulation of substitutions in nucleotide sequences are most conveniently described in statistical terms. From a biologist's point of view, the descriptions may appear both complex and removed from what is known about the biochemistry of DNA (e.g., the order of the nucleotides in DNA is usually important, but is generally ignored in phylogenetic studies). However, research using the parametric bootstrap has revealed that the differences found in alignments of real sequence data can be modeled remarkably well using statistical descriptions of the evolutionary processes [34, 37, 38, 42, 45–47], so there is reason to be confident about the use of a statistical approach to describe the evolutionary processes.

From a statistical point of view, it is convenient to assume that the evolutionary processes operating over an edge in a phylogeny (edges are sometimes called branches, but we recommend that the term be avoided as sometimes it is used to refer to a set of edges and other times it is used to refer to a single edge: [48]) can be modeled as a Markovian process (i.e., a process where the conditional probability of change at a site in a sequence depends only on the current state and is independent of previous states). Markovian processes are most easily described in terms of Markov models, which, in molecular phylogenetics, are matrices that describe the conditional probability of change from one nucleotide to another. In molecular phylogenetics, it is usually assumed that these Markov models (or substitution models) are good approximations of the evolutionary processes, so it is preferable to know and understand the assumptions of these Markov models if selecting such models is on the agenda.

In the next three subsections, we describe some of the phylogenetic assumptions and then outline how the evolutionary process can be modeled for DNA; this description is based mainly on papers by Tavaré [49] and Ababneh et al. [50]—for an alternative description, please *see* Bryant et al. [51].

2.1 The Phylogenetic Assumptions

In the context of the *evolutionary pattern*, it is generally assumed that the sequences have evolved along a bifurcating tree, where each edge in this tree represents the period of time over which point mutations have accumulated and each bifurcation represents a speciation event. Sequences that evolve in this manner are considered useful for studies of many aspects of evolution. A violation of this assumption occurs when gene duplication, recombination between homologous chromosomes and/or lateral gene transfer between genomes has occurred. In phylogenetic trees, gene duplications resemble speciation events and might be interpreted as such unless all descendant copies of each gene duplication are accounted for, which is neither always possible nor always the case. One solution to this problem would be to carry out a probabilistic orthology analysis [52]. Recombination between homologous chromosomes is most easily detected in sequences with a recent common origin, and the phylogenetically confounding effect of it diminishes with the age of the recombination event (due to the subsequent accumulation of point mutations). Lateral gene transfer is more difficult to detect, but it is thought to affect studies of phylogenetic data with recent as well as ancient origins. Methods to detect recombination [53–57] and lateral gene transfer [58–64] are available, but it is beyond the scope of this chapter to review them. In the following, we simply assume that the sequences evolved on a bifurcating tree, without gene duplication, recombination and lateral gene transfer.

In the context of the *evolutionary process*, it is generally assumed that the sites in a gene have evolved independently under the same Markovian conditions (and the sites then are said to be *independent and identically distributed*). The advantage of this is that only one

Markov model is needed to approximate the evolutionary process. The simplest variation of this assumption is that some of the sites are *invariable* (i.e., unable to change) and the other sites have evolved independently, albeit under the same Markovian conditions [34, 65, 66]. Other variations are that the variable sites: (a) have evolved independently under different Markovian conditions [13, 67–69], (b) have evolved in a correlated manner [70–81], or (c) have alternated between being variable and invariable [82–91].

It is generally also assumed that the evolutionary process at each site is *stationary*, *reversible*, and *homogeneous*. Here *stationarity* implies that the marginal probability of the nucleotides remains the same, *reversibility* implies that the probability of sampling nucleotide i from the stationary distribution and going to nucleotide j is the same as the probability of sampling nucleotide j from the stationary distribution and going to nucleotide i, and *homogeneity* implies constant rates of change over an edge (for details, *see* Refs. [32, 50, 51]). An evolutionary process may be *locally homogeneous*, in which case the homogeneity pertains to a single edge in the tree, or it may be *globally homogeneous*, in which case the homogeneity pertains to all edges in the tree (in both cases, the processes are said to be *time-homogeneous*). However, the relationship between these three conditions is complex, with six possible combinations (Table 1) for each edge in a tree (by definition, a reversible process is also a stationary process: [92]). The advantages of assuming a stationary, reversible, and homogeneous process over the whole tree are that we only need to use one Markov model to approximate the evolutionary processes, and that we then can ignore the direction of evolution over each edge. In practice, this entails (a) a reduction in the number of parameters that need to be estimated during a phylogenetic analysis, and (b) that there is no need to specify the position of the ancestral sequence.

Table 1
The spectrum of conditions relating to the phylogenetic assumptions

Scenario	Stationarity	Reversibility	Homogeneity	Comment on each scenario
1	+	+	+	Possible
2	–	+	+	Impossible (by definition)
3	+	–	+	Possible
4	–	–	+	Possible
5	+	+	–	Possible
6	–	+	–	Impossible (by definition)
7	+	–	–	Possible
8	–	–	–	Possible

NOTE: '+' implies the condition is met; '–' implies the condition is not met

Because most model-based phylogenetic methods assume that the sequences evolved under stationary, reversible, and homogeneous conditions (Scenario 1) and features in the data often suggest that the other scenarios better describe the conditions under which the data may have evolved (e.g., compositional heterogeneity across the sequences in an alignment), it is wise to question whether the data have evolved under more complex conditions than those included in Scenario 1 (for a different point of view, *see* page 125 of Ref. 93)—we will return to this question in the next section.

When the evolutionary processes are globally stationary, reversible, and homogeneous, we can use time-reversible Markov models to approximate the evolutionary processes over all edges in an unrooted phylogeny. The Markov models available for these restricted conditions range from the one-parameter model [94] to the general time-reversible model [95]. When these conditions are not met by the data, alternative methods are available [16–42].

Given that the functions of many proteins and RNA molecules are maintained by natural selection, there is reason to assume that the evolutionary process at different sites is more heterogeneous than described above. For example, it is likely that some sites have evolved under time-reversible conditions and that other sites have evolved under more general conditions, leaving a complex signal in the alignment that we might not be able to detect using the currently available phylogenetic methods, so we recommend that phylogenetic results be assessed using the parametric bootstrap.

2.2 Modeling the Evolutionary Process at a Site in a Sequence

We now describe, in statistical terms, an evolutionary process operating at a site in a sequence of nucleotides. Allow the site to be in one of four possible states (i.e., A, C, G, and T—for the sake of convenience, indexed as 1, 2, 3, and 4). Over time, the site may change from state i to state j, where $i, j = 1, \ldots, 4$. Let X be a Markov process that approximates the underlying evolutionary process, and let it take the values 1, 2, 3, or 4 at any point in continuous time, t. The process $X(t)$ may then be described by the transition function

$$p_{ij}(t) = Pr[X(t) = j | X(0) = i] \quad (1)$$

where $p_{ij}(t)$ is the probability that the nucleotide is j at time $t > 0$, given that it was i at time $t = 0$. Assuming a time-homogeneous Markov process (i.e., the process is constant over time), let r_{ij} be the instantaneous rate of change from nucleotide i to nucleotide j, and let $\mathbf{R}$ be the 4×4 matrix with these rates of change. Then, representing $p_{ij}(t)$ in matrix notation as $\mathbf{P}(t)$, we can write Eq. (1) as

$$\mathbf{P}(t) = \mathbf{I} + \mathbf{R}t + \frac{(\mathbf{R}t)^2}{2!} + \frac{(\mathbf{R}t)^3}{3!} + \cdots$$
$$= \sum_{k=0}^{\infty} \frac{(\mathbf{R}t)^k}{k!} \quad (2)$$
$$= e^{\mathbf{R}t}$$

where **R** is a time-independent rate matrix satisfying three conditions:

1. $r_{ij} > 0$ for $i \neq j$;
2. $r_{ii} = -\sum_{j \neq i} r_{ij}$, implying that $\mathbf{R}1 = 0$, where $1^T = (1, 1, 1, 1)$ and $0^T = (0, 0, 0, 0)$;
3. $\pi^T \mathbf{R} = 0^T$, where $\pi^T = (\pi_1, \pi_2, \pi_3, \pi_4)$ is the stationary distribution of **R**, $0 < \pi_j < 1$, and $1 = \sum \pi_j$.

The second of these conditions is required to ensure that $\mathbf{P}(t)$ is a valid transition matrix for $t > 0$.

Let f_{0j} be the frequency of the j-th nucleotide in the ancestral sequence. Then the process is

1. *stationary*, if $Pr(X(t) = j) = f_{0j} = \pi_j$, for $j = 1, \ldots, 4$, and
2. *reversible*, if the balance equation $\pi_i r_{ij} = \pi_j r_{ji}$ is met for $1 \leq i, j \leq 4$.

R is a key component in the context of modeling the accumulation of point mutations at sites in a nucleotide sequence. Each element of **R** has a role to play when modeling the state a site will be in at time t, so it is useful to understand the implications of changing **R**. For this reason, it is unwise to consider the rate of evolution as a single variable when, in fact, it is better represented by a matrix of variables.

2.3 Modeling the Evolutionary Processes at a Site in two Sequences

Consider a site in a pair of nucleotide sequences that have diverged from their common ancestor by two independent Markov processes. Let X and Y denote the Markov processes operating at the site, one for each edge, and let $\mathbf{P}^X(t)$ and $\mathbf{P}^Y(t)$ be the transition functions that describe the Markov processes $X(t)$ and $Y(t)$. The joint probability that the sequences contain nucleotide i and j, respectively, is then given by

$$f_{ij}(t) = Pr\left[X(t) = i, Y(t) = j \middle| X(0) = Y(0)\right], \quad (3)$$

where $i, j = 1, \ldots, 4$. Because $X(t)$ and $Y(t)$ are independent Markov processes, the joint probability at time t can be expressed, in matrix notation, as

$$\mathbf{F}(t) = \mathbf{P}^X(t)^T\mathbf{F}(0)\mathbf{P}^\Upsilon(t), \quad (4)$$

where $\mathbf{F}(0) = diag(f_{01}, f_{02}, f_{03}, f_{04}), f_{0k} = \Pr[X(0) = \Upsilon(0) = k]$ and $k = 1, \ldots, 4$. Equation (4) can be extended to n sequences, as described in Ababneh et al. [50]. The joint probability function has useful properties that will be relied upon in the next section.

3 Choosing a Substitution Model

Before describing the methods available to select models of evolution for analysis of a given data set, it is necessary to consider the terminology used to describe some of the properties of sequence data.

3.1 Bias

The term bias has variously been used to describe (a) a systematic distortion of a statistical result due to a factor not allowed for in its derivation, (b) a nonuniform distribution of the frequencies of nucleotides, codons or amino acids, and (c) compositional heterogeneity among homologous sequences. In some instances, there is little doubt about the meaning but in others, the authors have inadvertently provided grounds for confusion. Because of this, we recommend that the term bias be reserved for statistical purposes, and that four other terms be used to describe the observed nucleotide content:

1. The nucleotide content of a sequence is *uniform* if the nucleotide frequencies are identical; otherwise, it is *nonuniform*;
2. The nucleotide content of two sequences is *compositionally homogeneous* if they have the same nucleotide content; otherwise, it is *compositionally heterogeneous*.

The advantages of adopting this terminology are: (a) that we can discuss model selection without the ambiguity that we otherwise might have had to deal with, and that (b) we are not forced to state what the unbiased condition is—it might not be a uniform nucleotide content, as implied in many instances. The five terms are also applicable to codons and amino acids without loss of clarity.

3.2 Signal

When discussing the complexity of an alignment of nucleotides, it is frequently useful to consider variation in an alignment in terms of the sources that led to its complexity. One such source is the order and time of divergence events, which leaves a signal in the alignment—it is this *historical signal* [23, 96–99] that is the target of most phylogenetic studies. The historical signal is detectable because the sequences have a tendency to (a) accumulate point mutations over time, and (b) diverge from one another at different points in time. Other sources of complexity, sometimes called *non-historical signals* [97], include: the *rate signal*, which may arise when the sites and/or lineages evolve at nonhomogeneous rates; the *compositional signal*, which may arise when the sites and/or lineages evolve under different stationary conditions; and the *covarion signal*, which may emerge when the sites evolve non-independently (e.g., switching between being variable and invariable along different edges).

Occasionally, the term *phylogenetic signal* is used, either synonymously with the historical signal or to represent the signals that the phylogenetic methods use during inference of a phylogeny. Due to its ambiguity and the fact that most popular phylogenetic methods are unable to distinguish the historical and non-historical signals [100], we recommend the term phylogenetic signal be used with caution.

Separating the different signals is difficult because their manifestations are similar, so an inspection of the inferred phylogeny is unlikely to offer the best solution to this problem. Recent simulation studies of nucleotide sequences generated under stationary, reversible, and homogeneous conditions as well as under more complex conditions have revealed a complex relationship between the historical and non-historical signals [100]. The results show that the historical signal decays over time whereas the other signals may increase over time, depending on the nature of the evolutionary processes operating over time. The results also show that the relative magnitude of the signals determines whether the phylogenetic methods are likely to infer the correct tree, and that it is possible to infer the correct tree even though the historical signal has been lost. Therefore, there are reasons to be cautious when studying ancient evolutionary events: what might be a well-supported phylogeny may in fact be a tree representing largely the non-historical signals.

3.3 Testing the Stationary, Reversible, and Homogeneous Condition

The composition of nucleotides in an alignment may vary across sequences and/or across sites. In either case, it would be unwise to assume that the evolutionary processes can be modeled accurately using a single time-reversible Markov model.

A solution to this problem is to determine whether there is compositional heterogeneity in the alignment of nucleotides and, if so, what type of compositional heterogeneity it is. If there is compositional heterogeneity across the sites but not across the sequences, then it is possible to model the evolutionary processes using a set of time-reversible Markov models applied to different sets of sites in the alignment; several methods facilitate such an analysis (e.g., [13, 14, 101–104]). On the other hand, if there is compositional heterogeneity across the sequences, then none of these strategies are appropriate because lineage-specific evolutionary processes must have been different.

The methods to detect compositional heterogeneity across sequences can be partitioned into six groups (reviewed in Refs. [15, 105]), with those belonging to the first group using graphs or tables to visualize the nucleotide composition of each sequence, and those belonging to the other groups returning test statistics, which can be evaluated by comparing them to the null distributions of these test statistics. However, many of the methods belonging to these other groups of methods are either statistically invalid, of limited use for surveys of species-rich data sets, or not yet accommodated by the wider scientific community. Given this backdrop and the knowledge that most popular phylogenetic methods assume that the sequences have evolved under globally stationary, reversible, and homogeneous conditions [100, 105], two papers were published in 2006, describing new statistical tests and visualization methods to assess whether homologous sequences have evolved under the same conditions [106, 107]. These methods allow us to get a clearer understanding of what might underpin observed variation in alignments of homologous nucleotides. We will now describe these methods.

3.3.1 Matched-pairs Tests of Homogeneity

Suppose we have n homologous sequences of m independent and identically distributed sites taking values in l categories (e.g., an alignment of nucleotides ($l = 4$) with $n = 8$ sequences and $m = 500$ sites). Such data can be summarized in an n-dimensional divergence matrix, $\mathbf{D}$, which has l^n categories and is the observed equivalent of $m\mathbf{F}(t)$. The null hypotheses of interest concern the symmetry, marginal symmetry, and internal symmetry of $\mathbf{D}$. In this regard, the matched-pairs tests of symmetry, marginal symmetry, and internal symmetry can be used to assess the fit between the data, $\mathbf{D}$, and the time-reversible Markov models commonly used to approximate the evolutionary processes. The rationale behind using these tests in this context has long been known [49, 68, 108–110], but, surprisingly, they are not yet widely used.

The matched-pairs tests of homogeneity can be divided into two groups depending on the number of sequences. In the simplest case, where only two sequences are considered, the matched-pairs tests of homogeneity can be used to test for symmetry [111], marginal symmetry [112], and internal symmetry [106] of a two-dimensional divergence matrix derived from the alignment. When more than two sequences are considered, an additional two tests (of marginal symmetry) are available [106, 109]. Ababneh et al. [106] reviewed these tests and put them in the context of other similar tests, so rather than describing them again, we will show how they might be used to determine whether homologous sequences have evolved under stationary, reversible, and homogeneous conditions.

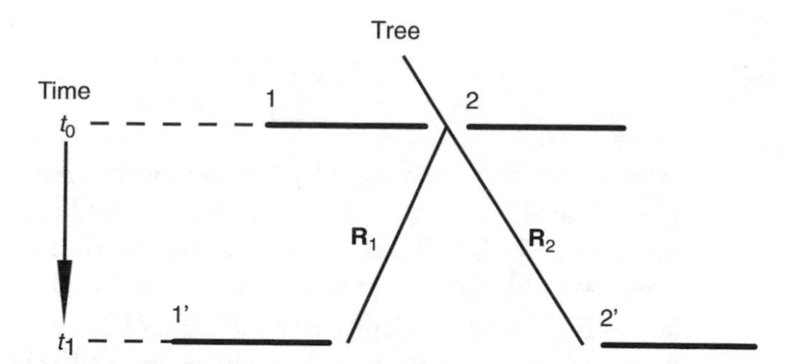

Fig. 1 A rooted 2-tipped tree with two diverging nucleotide sequences (i.e., fat horizontal lines) observed at time t_0 (i.e., sequences 1 and 2) and at time t_1 (i.e., sequences 1′ and 2′). The evolutionary processes operating along the terminal edges are labeled $\mathbf{R}_1$ and $\mathbf{R}_2$

3.3.2 Matched-pairs Tests of Homogeneity with Two Sequences

Consider a pair of homologous nucleotide sequences that have diverged from their common ancestor by independent Markov processes (Fig. 1), and let the sites within each sequence be independent and identically distributed. We then might want to know under what conditions the sequences have evolved. To address this challenge, we can test the symmetry, marginal symmetry, and internal symmetry of $\mathbf{D}$.

At time t_0, the divergence matrix might looking like this:

$$\mathbf{D}(t_0) = \begin{bmatrix} 248 & 0 & 0 & 0 \\ 0 & 251 & 0 & 0 \\ 0 & 0 & 252 & 0 \\ 0 & 0 & 0 & 249 \end{bmatrix} \tag{5}$$

(because the two sequences are identical) while at time t_1, the divergence matrix might look like this:

$$\mathbf{D}(t_1) = \begin{bmatrix} 191 & 71 & 68 & 57 \\ 14 & 142 & 22 & 33 \\ 16 & 12 & 144 & 29 \\ 26 & 19 & 18 & 138 \end{bmatrix}. \tag{6}$$

Each element of $\mathbf{D}$ (i.e., d_{ij}) represents the number of sites where the descendant sequences have nucleotides i and j, respectively. It is easy to see that the two sequences have different nucleotide frequencies and that the matrix is asymmetrical (i.e., $d_{ij} \neq d_{ji}$ for $i \neq j$), so it is tempting to conclude that they have evolved under different conditions. However, doing so would be unwise before testing whether the two sequences have evolved under the same conditions. This can be done using the matched-pairs tests of symmetry, marginal symmetry, and internal symmetry.

In order to use the *matched-pairs test of symmetry* [111], we enter the elements of $\mathbf{D}(t_1)$ into the following equation:

$$S_S^2 = \sum_{i<j} \frac{\left(d_{ij} - d_{ji}\right)^2}{d_{ij} + d_{ji}}, \tag{7}$$

where the test statistic (S_S^2) is asymptotically distributed as a χ^2-variate on $v = l(l-1)/2$ degrees of freedom. In the present case, $S_S^2 = 91.2772$ and $v = 6$. Assuming that the two sequences have evolved under the same conditions (which would be true if $\mathbf{R}_1 = \mathbf{R}_2$), the probability of $S_S^2 \geq 91.2772$ is 1.6×10^{-17}. Therefore, we conclude that the sequences are unlikely to have evolved under the same conditions.

In order to use the *matched-pairs test of marginal symmetry* [112], we first obtain a vector of marginal differences ($\mathbf{u}$) and its variance-covariance matrix ($\mathbf{V}$). Here, $\mathbf{u}^T = (d_{1\bullet} - d_{\bullet 1}, d_{2\bullet} - d_{\bullet 2}, d_{3\bullet} - d_{\bullet 3})$ and $\mathbf{V}$ is a 3×3 matrix with elements given as follows:

$$v_{ij} = \begin{cases} d_{i\bullet} + d_{\bullet i} - 2d_{ii}, & i = j \\ -\left(d_{ij} + d_{ji}\right), & i \neq j \end{cases} \tag{8}$$

where $d_{1\bullet}$ is the sum of the first row of $\mathbf{D}(t_1)$, $d_{\bullet 1}$ is the sum of the first column of $\mathbf{D}(t_1)$, and so forth. Given $\mathbf{u}$ and $\mathbf{V}$, we then compute

$$S_M^2 = \mathbf{u}^T \mathbf{V}^{-1} \mathbf{u}, \tag{9}$$

where the test statistic (S_M^2) is asymptotically distributed as a χ^2-variate on $v = l - 1$ degrees of freedom. In the present case, $\mathbf{u}^T = (140,\ 33,\ 51)$ and

$$\mathbf{V} = \begin{bmatrix} 252 & -85 & -84 \\ -85 & 171 & -34 \\ -84 & -34 & 165 \end{bmatrix}, \tag{10}$$

so $S_M^2 = 79.2317$ and $v = 3$. Assuming that the two sequences have evolved under the same stationary conditions (which would be true if the stationary distributions of $\mathbf{R}_1$ and $\mathbf{R}_2$ were identical), the probability of $S_M^2 \geq 79.2317$ is 4.5×10^{-17}. Therefore, we conclude that the sequences are unlikely to have evolved under stationary conditions; in other words, they are likely to have evolved under nonstationary and nonhomogeneous conditions.

In order to use the *matched-pairs test of internal symmetry* [106], we rely on the fact that the test statistic $S_I^2 = S_S^2 - S_M^2$ is asymptotically distributed as a χ^2-variate on $v = (l-1)(l-2)/2$ degrees of freedom. In the present case, $S_I^2 = 12.0455$ and $v = 3$. Assuming evolution under the same homogeneous conditions, the probability of $S_I^2 \geq 12.0455$ is 7.2×10^{-3}. Therefore, we conclude that the sequences are unlikely to have evolved under globally homogeneous conditions.

Let τ be a user-defined threshold that can be used to determine whether a test is significant or not (historically, $\tau = 0.05$). Then, in general:

1. If $Pr(S_S^2|v) < \tau$, there is evidence that evolution has not occurred under stationary and globally homogeneous conditions, implying inconsistency with Scenario 1 in Table 1;
2. If $Pr(S_M^2|v) < \tau$, there is evidence that evolution has not occurred under stationary conditions, implying inconsistency with Scenarios 1, 3, 5 and 7 in Table 1;
3. If $Pr(S_I^2|v) < \tau$, there is evidence that evolution has not occurred under globally homogeneous conditions, implying inconsistency with Scenarios 1, 3 and 4 of Table 1.

It is obvious that the matched-pairs tests of homogeneity are unable to identify all the conditions under which pairs of sequences might have evolved, but the tests are still very useful because they are able to alert users to the fact that the sequences may have diverged under different conditions. When such a result is obtained, it is often useful to analyze the sequences using methods developed by Dutheil et al. [39] or Jayaswal et al. [42].

3.3.3 Matched-pairs Tests of Homogeneity with More than Two Sequences

We now turn to cases where the alignment contains more than two sequences. Suppose we have four nucleotide sequences that have evolved independently on a bifurcating tree with three bifurcations. The alignment might look like that in Fig. 2. Visual inspection of this alignment reveals how difficult it is to determine whether the sequences have evolved under stationary, reversible, and homogeneous conditions, emphasizing the need for statistical tests to address this issue. In the following, we demonstrate how such tests may be used.

Initially, we use the ***overall matched-pairs test of marginal symmetry*** [106], which returns a test statistic (T_M^2) that is asymptotically distributed as a χ^2-variate on $v = (n-1)(l-1)$ degrees of freedom. For the data in Fig. 2, $T_M^2 = 56.93$ and $v = 9$. Assuming evolution under stationary conditions, the probability of $T_M^2 \geq 56.93$ is ~5.22×10^{-09}, implying that these data are unlikely to have evolved under stationary conditions. This result fits well with what is known about the processes that generated the data (Fig. 2).

Notwithstanding this result, it is possible that some of the sequences have evolved under stationary, reversible, and homogeneous conditions. To test whether this is the case, the matched-pairs tests of symmetry, marginal symmetry, and internal symmetry for pairs of sequences may be used. Table 2 shows the corresponding three sets of six probabilities. After using the sequential Bonferroni correction to counteract the problem of

```
Seq1    TTTCTGTAGACTACAGCCGAACTGATACAATACAAGCACAAACAATTCACCGCGTCGCGCACAGT
Seq2    CGTCTGGGATCTTTTGCCGGGCTGGGTCGCTACACGAACGCAGAGTTCTACTCCGGTCGCACTTG
Seq3    CTACAGTTAAGTTCTGCAGAGCTGCTTGACTATACGATCAACGAATACAAGACGGGGCGCACAGG
Seq4    CTTCGGTATAGTTCTGCCGAGCTGGTTCGCTACATGATCAATGATTACGACCCTGGGCCCTCTGG

CGTCAAAGCGGCATTCCATAAAAGTTCATCCATACCCCGAGGTAACCTCACGTCGTCACGGGCTGACGTAATCAC
CGGATGAGTTGGTTACGGAGAGTGCGGGTCTTTTCCCAAAGTTCATTTCCCGTCGTTTCGGCCTGTTGTAATCAT
CATATAAGTGGGATTCCGTAAGATCATGTCTCTACCCAAAGGGTACATGTTGTCTTCACGGCCAAACCTAATCAC
CGTATGAGTGGGATGGTGTCAAATTTCTTCTTGACCGGCAGGTCACCTCTTGTCCTGAGGGCCGGGCGGCAGCAG

GAAAGCACCGCCCGACCGGTCAAGCCTCAGAAGGGTCGAACACGGACTCAGTCTCAAGTGCTCCTCCACAAACGT
GTGTGCTCCGCCCCATCGGTGAAGCCCCGCTAGCGTATTACTCGGAATGTGTATCTAGTGCCAATTCATATACGT
GGGTCACCTGCCCAACAGTTGAAGGCGCGCCAGGCCGGCCCACGCATACAGACTCCAGAGCAACTCCATCAACGT
GTCTGTGCTGTTTTGCCTGTAATGCCTCGTCAGGCGGGAGCACGGTTTTAGTATCCTTGCCTACTCTATTATTCT

CATACTTAGTTCACCATCCCCGAGCCTATTTCCCTTAAAATGCGGTAACCCGGCCAGGGAGGAGAGAAAGAGTGG
ATAAGTTAGTTTATAATCTCCTCGCCTATTTCCTTAGAAATAGTATTATCGATCTTTGACGGAGTGAACTATGCG
CAAACTTACCTCAAAGTCTCCGCGGCTAGTCCCATTGAAATACGATTATCTCACCTTGCAAGAGTGAAAAAATCG
GAAATTTAGCTGATAATCTCTTCAGCTAATTCTTTAGAAATAGGCTTATCGTCCCGGGTTGGTGCGAAACATCCG
```

Fig. 2 An alignment of nucleotides—the data were generated using Hetero [167] with default settings invoked. However, the marginal distributions of the evolutionary processes operating along the four terminal edges were nonuniform ($\pi_1^T = (0.4, 0.3, 0.2, 0.1)$ for Seq1 and Seq3, and $\pi_2^T = (0.1, 0.2, 0.3, 0.4)$ for Seq2 and Seq4), implying that the sequences evolution under nonstationary as well as nonreversible conditions

multiple comparisons [113], the probabilities may be interpreted as outlined above.

Two of the six matched-pairs tests of symmetry returned high probabilities (i.e., Seq1 vs. Seq3 and Seq2 vs. Seq4), implying that the sequences in these pairs are consistent with the assumption of evolution under stationary and globally homogeneous conditions. The other comparisons led to low probabilities, implying that the sequences in these pairs are unlikely to have evolved under these conditions. This result fits well with what is known about the processes that generated the data (Fig. 2).

The matched-pairs test of symmetry identified which pairs of sequences had evolved under different conditions but did not reveal what underpins the data's complexity. To obtain this information, we use the matched-pairs tests of marginal symmetry and internal symmetry. The probabilities obtained from the first of these tests are similar to those obtained using the matched-pairs test of symmetry while those obtained from the second of these tests are high for all pairs of sequences (Table 2). Therefore, the results are consistent with the notion that the data, as a whole, did not evolve under stationary, reversible, and homogeneous conditions.

Given this conclusion, it would be unwise to infer a phylogeny from this data set using phylogenetic programs that assume stationary, reversible, and homogeneous conditions.

Table 2
Results returned from the matched-pairs tests of symmetry, marginal symmetry, and internal symmetry for the sequences in Fig. 2

	Seq1	Seq2	Seq3
Symmetry			
Seq2	**0.0000256**		
Seq3	0.8064433	**0.0001914**	
Seq4	**0.0000030**	0.6146219	**0.0000139**
Marginal symmetry			
Seq2	**0.0000013**		
Seq3	0.6829269	**0.0000298**	
Seq4	**0.0000004**	0.9091629	**0.0000016**
Internal symmetry			
Seq2	0.8500044		
Seq3	0.6772105	0.4373846	
Seq4	0.3478985	0.2706100	0.4477230

NOTE: Each number corresponds to the probability of obtaining the observed test statistic by chance under the assumptions of symmetry, marginal symmetry, and internal symmetry. To counteract the problem of multiple comparisons and control the family-wise error rate, we recommend using the sequential Bonferroni correction [113]. Numbers in bold correspond to cases where the null hypothesis was rejected using a 5 % significance level (after application of the sequential Bonferroni correction)

Interestingly, the result from the matched-pairs test of symmetry (Table 2) revealed that some of the sequences are consistent with the assumption of evolution under stationary, reversible, and homogenous conditions. Such a result might actually be useful if, for example, a subset of the sequences were enough to answer the scientific question that the full set of sequences had been assembled for; in other words, the matched-pairs tests of homogeneity might be used to identify subsets of sequences that are consistent with evolution under stationary, reversible, and homogeneous conditions.

3.3.4 Analysis of Species-rich Alignments

When an alignment contains a large number of sequences, it becomes impractical to survey tables of probabilities (like those in Table 2). To address this problem, alternative strategies are available.

One such strategy [15, 107] rests on the idea that a de Finetti plot [114] can be extended to a tetrahedral plot with similar properties (i.e., each observation comprises four variables, a, b, c and d, where $a + b + c + d = 1$ and $0 \leq a, b, c, d \leq 1$). Each axis in the plot starts at the center of one of its four surfaces, at value 0, and

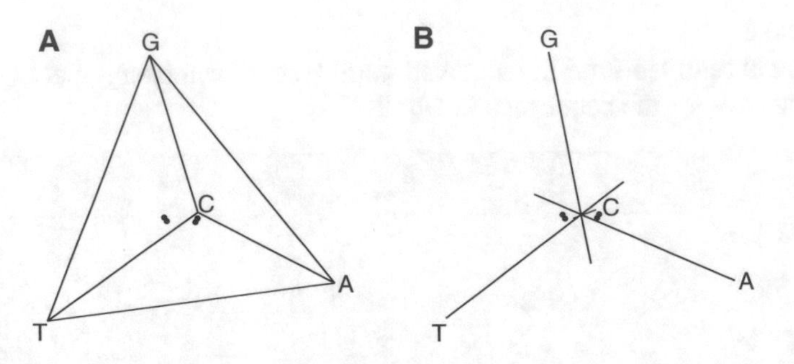

Fig. 3 Tetrahedral plot, with the borders (**a**) or axes (b) displayed. The four dots represent the four sequences in Fig. 2

stops at the opposite corner, at value 1. The nucleotide content of a nucleotide sequence is simply the set of shortest distances from its point within the tetrahedron to the surfaces of the tetrahedron (Fig. 3). Visual assessment of the spread of points shows the extent of compositional heterogeneity as well as the compositional trends that might exist in the data. Rotation of the tetrahedron permits inspection of the scatter of points from all angles, thus enhancing the chance of detecting compositional trends and/or sequences that may be outliers. Having assessed the distribution of points visually, it is useful to also do the matched-pairs test of symmetry to find out whether the sequences are consistent with the assumption of evolution under stationary, reversible, and homogeneous conditions.

The approach that uses a tetrahedral plot lends itself particularly well to studies of species-rich alignments of nucleotides. In addition, it can be combined with—or even replaced by—other strategies. To illustrate this, we analyzed an alignment of mitochondrial protein-coding nucleotide sequences from 53 species of animals. The alignment was first used by Bourlat et al. [115]. In the present study, we focused on the same 7542 sites that formed the basis for the phylogeny inferred by these authors. In their analysis, stationary, reversible, and homogeneous conditions were assumed.

First, we used SeqVis [107] and obtained three tetrahedral plots (Fig. 4), one for each codon position. The three plots reveal that the nucleotide composition varies extensively among the sequences, with the alignment of third codon position being the most heterogeneous set of sites, followed by that of first codon position and then that of second codon position. Given these plots, it might be concluded that it would be unwise to assume evolution under stationary, reversible, and homogeneous conditions.

To substantiate this preliminary conclusion and to shed more light on what might have led to the complexity of these data, we then analyzed these data using the matched-pairs tests of homogeneity. Because the alignment of second codon position was the

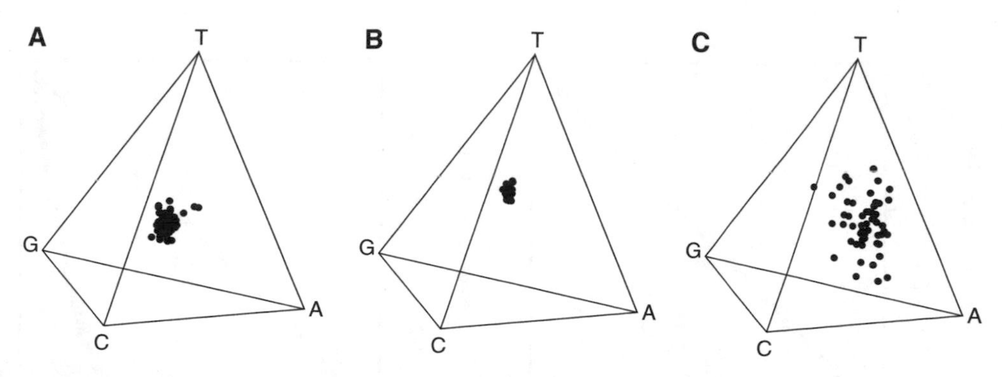

Fig. 4 Tetrahedral plots, based on first codon sites (**a**), second codon sites (**b**), and third codon sites (**c**) from an alignment of 2514 codons (extracted from a longer alignment of mitochondrial protein-coding genes from 53 animal species). This data set was originally analyzed by Bourlat et al. [115], who reported evidence of compositional heterogeneity

most compositionally homogeneous set of sites (Fig. 4b), we limited our analysis to this subset of sites. First, we used the overall matched-pairs test of marginal symmetry [106] and got $T_M^2 = 547.378$ and $\upsilon = 156$, implying that the sequences are highly unlikely to have evolved under stationary, reversible, and homogeneous conditions. This is in agreement with our earlier conclusion, which was based on a visual inspection of the tetrahedral plot.

To get a better understanding of under what conditions the data might have evolved, we used the matched-pairs tests of symmetry [111], marginal symmetry [112], and internal symmetry [106]. Assuming evolution under stationary, reversible, and homogeneous conditions, the distributions of probabilities returned from these three tests should be uniform, with 5 % of the probabilities taking values ≤ 0.05 (as in Fig. 5a).

The distribution of observed probabilities obtained with the matched-pairs test of symmetry is J-shaped and highly nonuniform (Fig. 5b). With more than 75 % of the observed probabilities below 0.05 (or 45 % after using the sequential Bonferroni correction), the distribution of the dots differs widely from the expected distribution (Fig. 5a), implying that a large proportion of the sequences have not evolved under stationary and globally homogeneous conditions.

The distribution of observed probabilities obtained with the matched-pairs tests of marginal symmetry is also J-shaped and highly nonuniform (Fig. 5c). With more than 78 % of the observed probabilities below 0.05 (or 48 % after using the sequential Bonferroni correction), the distribution of the dots differs widely from the expected distribution (Fig. 5a), implying that a large proportion of the sequences have not evolved under stationary conditions.

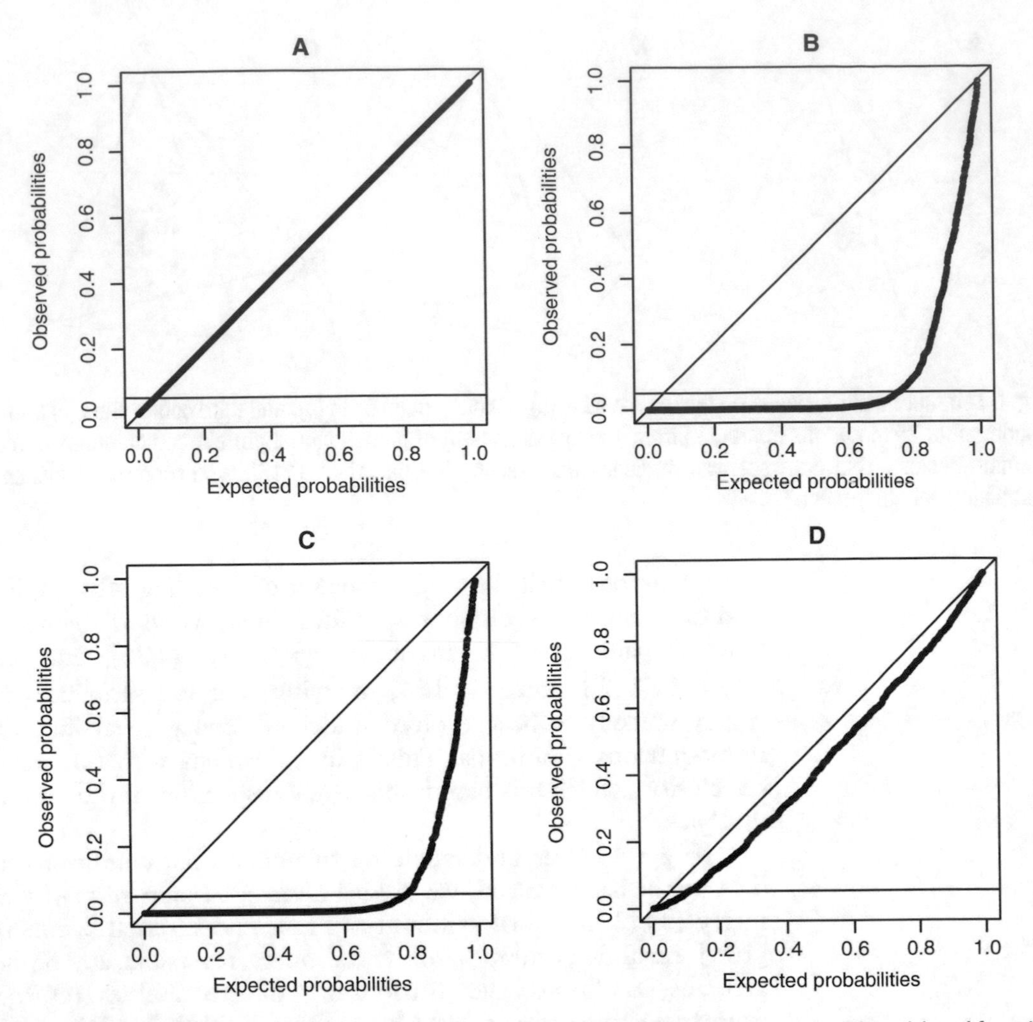

Fig. 5 PP plots from data generated under stationary, reversible, and homogeneous conditions (**a**) and from the alignment of 2514 second codon sites (for details, *see* Fig. 4). Results from the matched-pairs tests of symmetry, marginal symmetry, and internal symmetry (i.e., 1378 probabilities from each test) are presented in panels (**b**), (**c**), and (**d**), respectively. For each test, we first ordered the observed probabilities in a descending order before plotting them against a uniform distribution of probabilities. The horizontal line in each panel equals 0.05

The distribution of observed probabilities obtained with the matched-pairs tests of internal symmetry is slightly J-shaped and close to being uniform (Fig. 5d). With over 10 % of the observed probabilities below 0.05 (but 0 % after using the sequential Bonferroni correction), the distribution of the dots is in agreement with the expected distribution (Fig. 5a), implying that the abovementioned cases of evolution under nonhomogeneous conditions most likely are due to evolution under nonstationary conditions.

In summary, it is clear that a large proportion of the sequences in the alignment of second codon sites must have evolved under nonstationary, nonreversible, and nonhomogeneous conditions. Likewise, it is clear that some sequences are consistent with evolution under stationary, reversible, and homogeneous conditions (e.g., those associated with observed probabilities larger than 0.05 in Fig. 5c).

As stated above, it may be useful to identify the subset(s) of sequences that are consistent with evolution under stationary, reversible, and homogeneous conditions. To address this issue, we can use a heat map displaying the probabilities returned from the matched-pairs tests of homogeneity for pairs of sequences.

In this case, a heat map (Fig. 6) was obtained from the observed probabilities first presented in Fig. 5c. The distribution of the white pixels in this heat map indicates the presence of subsets of sequences that can be assumed to have evolved under stationary conditions. One such subset includes *Haliotis rubra*, *Aplysia californica*, *Pupa strigosa*, and *Albinaria coeerulea*, and another one includes *Protopterus dollei*, *Lampetra fluviatilis*, *Lepidosiren paradoxa*, *Raja porosa*, *Danio rerio*, *Cheliona mydas*, *Petromyzon marinus*, *Homo sapiens*, and *Boa constrictor*. However, the presence of black pixels in the heat map indicates that sequences in these two subsets have not evolved under the same condition.

Often it is not immediately clear how many good subsets of sequences a heat map contains. One way to tackle this issue is to rearrange the order of the rows and columns in the heat map in a synchronized manner (e.g., if we swap rows 16 and 34, then we also must swap columns 16 and 34). Using this approach may result in another heat map, which might reveal different subsets of sequences that are consistent with the assumption of evolution under stationary, reversible, and homogeneous conditions.

Figure 7 shows the result of such a row-and-column permutation done on the heat map in Fig. 6. The new heat map is similar to that in Fig. 6, but it also reveals a new group of sequences and that other sequences can be added to the previously identified groups. These groups of sequences appear to be consistent with evolution under stationary conditions and could be treated as such. However, given the abovementioned shortcomings of the matched-pairs tests of homogeneity, it is still highly recommended that the sequences within each group be analyzed phylogenetically using mixture-model methods like those developed by Dutheil et al. [39] and Jayaswal et al. [42]. The real strength of these heat maps is when they are used in conjunction with models of evolution with different rate matrices on different edges, as in Jayaswal et al. [42], where comparisons of results from the heat maps and the arrangements of rate matrices can show agreement.

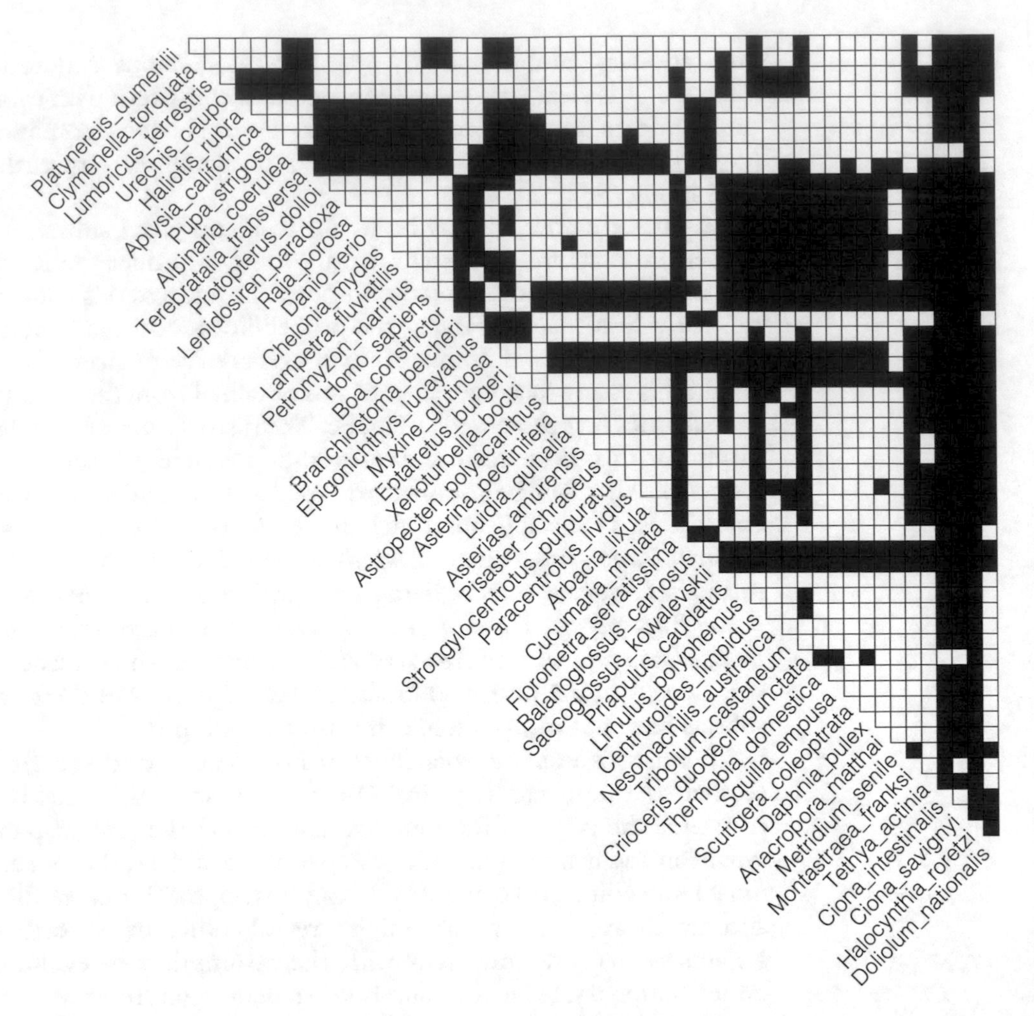

Fig. 6 Heat map with color-coded probabilities obtained using the matched-pairs test of marginal symmetry. Each cell in the heat map corresponds to an observed probability. If $Pr\left(S_M^2|v\right) < \tau'$, where τ' is the threshold determined using the sequential Bonferroni correction (to counteract the effect of multiple comparisons), then the corresponding cell is black; otherwise, it is white. The names of the sequences are shown along the diagonal (for further details about the data, *see* Fig. 4)

3.3.5 Phylogenetic Analysis of Sequences That Have Evolved Under Complex Conditions

So far, we have focused on determining whether homologous sequences are consistent with evolution under the conditions assumed by most popular molecular phylogenetic methods and, if this is not the case, whether subsets of the sequences are consistent with evolution under stationary, reversible, and homogeneous conditions. However, as our survey of the data from Bourlat et al. [115] shows, there are cases where the assumptions of popular molecular phylogenetic methods are violated extensively by the data. Here, we discuss some of the analytical options available.

One option is to assume that the non-historical signals found in the data are not big enough to affect the phylogenetic estimate. However, making this assumption entails taking a risk. For

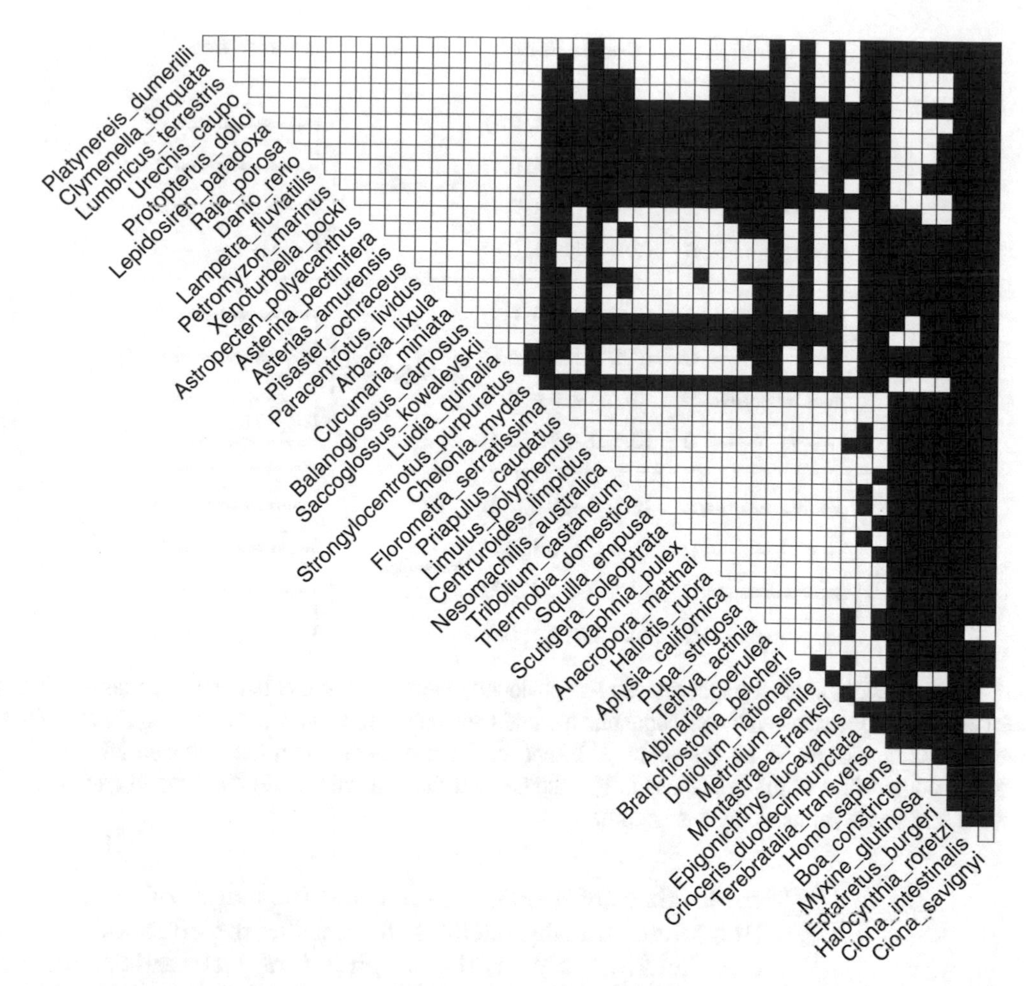

Fig. 7 Heat map with color-coded probabilities obtained using the matched-pairs test of marginal symmetry. The heat map was obtained by doing synchronized row-and-column permutations of the heat map in Fig. 6

example, if we were to infer the phylogeny of four compositionally heterogeneous sequences and the true tree were like that on the left in Fig. 8a, then it is likely that the tree inferred assuming a stationary, reversible, and homogeneous model of evolution would be similar to the true tree. This is because the historical and compositional signals concur. On the other hand, if the true tree were like those on the left in Fig. 8b and c, then the two signals would differ and the inferred tree might not look like the true tree. This is a likely outcome if the compositional signal were stronger than the historical signal [105]. Importantly, whether or not the true tree is inferred depends on several factors: the level of compositional heterogeneity across the sequences (the bigger the differences, the more likely a phylogenetic error is to occur: [105]); the length of the internal edges in the true tree, relative to the length of the whole tree (the shorter the internal edges in the true tree are, the

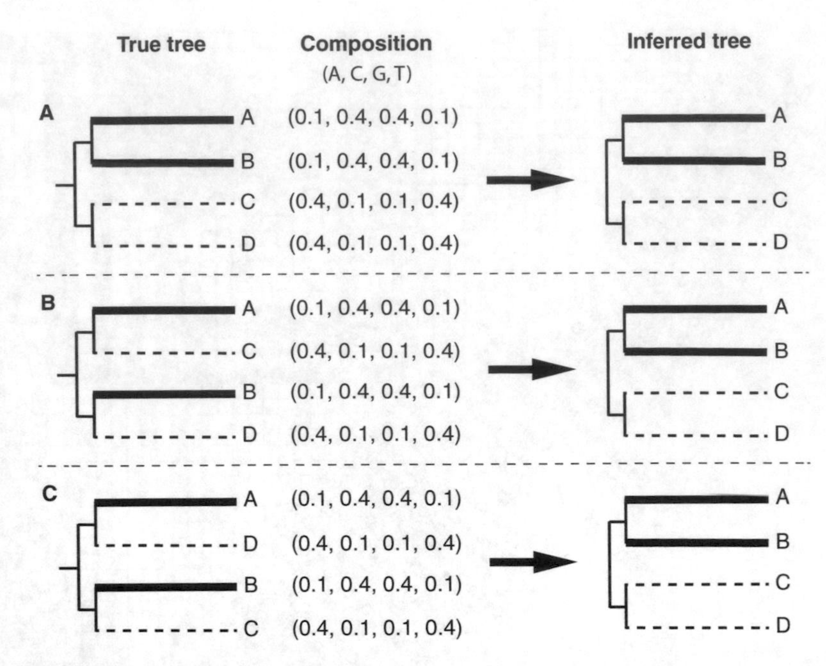

Fig. 8 Diagram showing what might happen if a phylogeny were to be estimated from four compositionally heterogeneous sequences and the phylogenetic methods assumed evolution under stationary, reversible, and homogeneous conditions. Three scenarios (**a**, **b**, and **c**) are considered, with the true tree on the left, the nucleotide composition of the sequences in the middle, and the inferred tree on the right. In each case, the ancestral nucleotide composition was uniform

harder they are to infer: [105]); and the length of the sequences (the longer the alignment is, the smaller the effect of stochastic error and the bigger the likelihood of systematic error [due to model misspecification]) (Wong and Jermiin, in prep.). In conclusion, this is not a recommended option.

Another option is to compare the sequences using phylogenetic methods that are able to accommodate more complex processes of molecular evolution. Broadly speaking, these phylogenetic methods can be divided into distance-based [19–22, 25, 26, 29, 31], parsimony-based [18, 23], likelihood-based [16, 17, 24, 27, 28, 30, 32, 34, 36–42] and Bayesian [30, 33, 35] methods. It is beyond the scope of this chapter to review these phylogenetic methods, but it is still worth highlighting some of their features, strengths and weaknesses.

The distance-based phylogenetic methods are designed to calculate accurate estimates of the evolutionary distances between pairs of sequences, which, in turn, may be used to infer a tree using a clustering algorithm (e.g., [116–119]). The methods are fast and, therefore, attractive. However, they are often of limited value because only one model of evolution is applied across the sites (for an exception, *see* Ref. [31]). Furthermore, it is often impossible to compare alternative hypotheses, especially if different models of evolution and/or different phylogenetic trees are considered.

The parsimony-based phylogenetic methods [18, 23] are designed to calculate estimates of support for different trees inferred by maximum parsimony, given a set of compositionally heterogeneous sequences. One advantage of these methods is that they allow users to bypass problems associated with randomization tests (i.e., rejection of the correct evolutionary hypothesis and acceptance of an incorrect hypothesis). Their shortcomings are that they focus on parsimony-informative sites, which depend on the sequences included in the alignment, and they do not consider the evolutionary processes that underpin the observed sequence variation.

The likelihood-based and Bayesian phylogenetic methods are more versatile than the other two types of phylogenetic methods, but they are also more challenging to use because a detailed understanding of the assumptions of the models of evolution is required. This is because they are designed to model evolutionary processes, which may be heterogeneous across lineages and/or across sites. The existing methods can be divided into three classes, depending on the way the evolutionary process is modeled.

One of these classes of methods models the evolutionary process over an edge in a tree using a matrix of joint probabilities (**Q**). Therefore, if the tree has n leaves, then the model of evolution comprises $2n - 3$ **Q** matrices (one for each edge in the tree). Each of these matrices represents the joint probability distribution of the nucleotides at the two ends of the corresponding edge. The most general of this class of methods is the Barry and Hartigan (BH) [16] model, which was published in 1987. It only assumes a Markovian process over each edge in the tree, and independent and identically distributed sites. A computational solution to the BH model appeared in 2005 [32], and 2 years later it was improved by allowing for invariable sites (BH+I) [34]. In 2011, two stationary BH models (SBH and SBH+I) appeared along with two reversible BH models (RBH and RBH+I) [37]. Although these six models of evolution are the most general ones available, they are also the most parameter-rich ones because each edge is assigned it own **Q** matrix (for details, *see* Ref. [38]). One obvious concern is that these models may be too parameter rich; another one is the amount of time it takes to estimate the most likely set of parameter values for a tree. The original studies considered alignments with less than eight sequences, but data sets with more sequences might still be analyzed using these phylogenetic methods, provided that the alignments are long enough and the number of trees considered is limited to a relevant subset of tree space.

Another class of methods models the evolutionary process over an edge in a tree using a matrix of instantaneous rates (**R**).

However, in this case it is the elements of $\mathbf{P} = e^{\mathbf{R}t}$ that are optimized, as in Eq. (2). Because t is an element in this equation, this class of methods assumes time-homogeneous processes between an ancestral node and its descendant nodes or leaves. This is a subtle but important feature that distinguishes the models of evolution that build on joint probability matrices from those that build on instantaneous rate matrices (for detail, *see* Ref. [37]). In practice, the process over an edge is often approximated by the general time-reversible (GTR) model [95], or constrained versions of this model, but with different stationary distributions for the edge-specific rate matrices [24, 27, 28, 30, 36, 39–41]. Like the previous class of methods, this one might suffer from over-parameterization (because each edge has its own **R** matrix). Unlike the previous class of methods, this one needs a rooted tree as input because the estimation of likelihood must be done from the root of the tree—obviously, in some cases, this might pose a problem because the location of the root is usually unknown.

The third class of methods is similar to the second one but the change from one **R** matrix to another can also occur on the edges [33, 35]. This has the advantage that it no longer is necessary for a change in the model of evolution to co-occur with a bifurcation in the tree. This ability to decouple the changes of model of evolution and the bifurcations in the tree may reduce the chance of over-parameterization. Developed within a Bayesian framework, the methods allow users to compare alternative scenarios involving different models of evolution and different phylogenetic trees.

Because many of the abovementioned phylogenetic methods assign a unique matrix of joint probabilities or instantaneous rates to each edge in the tree, it is very likely that the model of evolution over a tree might be more complex than is necessary. Foster [30] raised this issue during his analysis of five rDNA genes and discovered that two models of evolution were sufficient; the same conclusion was reached in study using another phylogenetic method [34]. To solve the problem of over-parameterization, Jayaswal et al. [38] developed an algorithm to reduce model complexity. In practice, it begins with a complex model of evolution (i.e., one where each edge is assigned a unique rate matrix: Fig. 9a). The algorithm then: finds the pair of edges whose rate matrices are most similar and assigns the same rate matrix to both edges, and the fit between tree, model and data is assessed. If the fit is better, then the process continues; if the fit cannot be improved, then the optimal model of evolution (Fig. 9b) is reported.

The advantage of this algorithm is obvious, but it has since become clear that the parameters of complex models of evolution might be unidentifiable [120], especially if the data includes a large number of sequences, and the development of an alternative approach became necessary. Two similar solutions appeared in

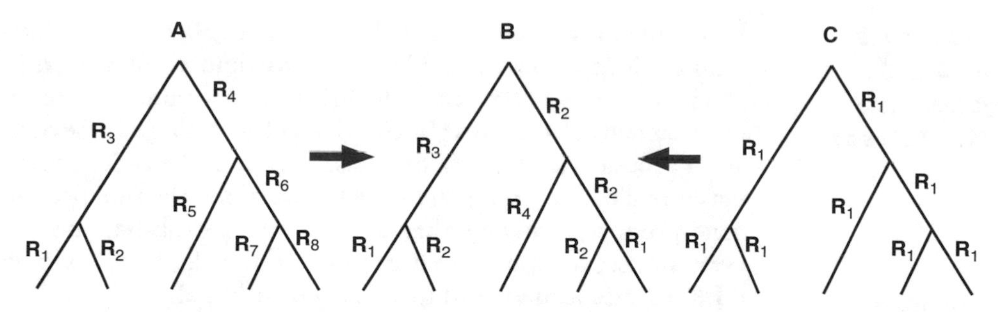

Fig. 9 Diagram illustrating two algorithms used to identify optimal models of evolution for sequence data. Starting with a unique model of evolution assigned to each edge in the tree (**a**), the CORE algorithm [38] will attempt to reduce complexity of the model until the best model of evolution is identified (**b**). Beginning with the same model of evolution assigned to all edges in the tree (**c**), the bottom-up algorithm [42] increases the complexity of the model until the best model of evolution is identified (**b**)

2012 [39] and 2014 [42], with the first one relying on substitution mapping [121] and the second one not doing so. In both cases, the search for the best model of evolution begins with the simplest model of evolution (i.e., one where the same rate matrix is assigned to all edges in the tree: Fig. 9c). By assigning different rate matrices to different edges, the complexity of the model of evolution can be increased. A comparison of the fit between tree, model and data can then be used to evaluate whether the increased complexity of any of the new models is a better explanation of the data than the older less-complex model of evolution. This process of increasing model complexity can be continued until the fit between tree, model and data cannot be improved further.

The results obtained using these three classes of phylogenetic methods are striking because they have revealed a more complex evolution for much of the sequence data that has been analyzed thus far. Not only is it now clear that phylogenetic methods that assume globally stationary, reversible, and homogeneous Markovian processes in many cases are inappropriate for the data and may produce biased phylogenetic results [100, 105]; it is also clear that there is a need for accurate and fast phylogenetic methods that simultaneously can search tree space while fitting the data optimally to each of the tree topologies considered. However, doing so is a big combinatorial problem that will require a smart heuristic solution because there are $\prod_{i=3}^{n}(2i-3)$ rooted binary trees, each with $2n-2$ edges, and a Bell number of distinct rate-matrix arrangements over the edges for each of these trees [38]. Most of the phylogenetic methods discussed above assume that the tree is known and that information about the models of evolution is the only unknown component, so better methods are clearly needed.

3.4 Testing the Assumption of Independent and Identical Processes

It is commonly assumed that the sites in an alignment of nucleotides are independent and identically distributed, or at least independent. The latter case includes a scenario where rate-heterogeneity across sites (RHAS) is modeled using a Γ distribution and a proportion of invariable sites. However, the order and number of nucleotides in a gene usually determine the function of the gene product, implying that it would be unrealistic, and maybe even unwise, to assume that the sites in the alignment of nucleotides are independent and identically distributed.

To determine whether sites in an alignment of nucleotides are independent and identically distributed, it is necessary to compare this simple model to the more complex models that describe the interrelationship among sites in the alignment of nucleotides. Descriptions of more complex models may depend on prior knowledge of the genes and gene products, and comparisons may require using likelihood-ratio tests, sometimes together with permutation tests or a parametric bootstrap.

The likelihood-ratio test provides a statistically sound framework for comparing alternative evolutionary hypotheses [122]. In statistical terms, the likelihood-ratio test statistic, Δ, is defined as

$$\Delta = \frac{max(L(\text{data} \mid \text{H}_0))}{max(L(\text{data} \mid \text{H}_1))} \qquad (11)$$

where the likelihood, L, of the data, given the null hypothesis (H_0), and the likelihood of the data, given the alternative hypothesis (H_1), both are maximized with respect to the parameters. If $\Delta > 1$, then the data favor H_0; otherwise, if $\Delta < 1$, the data favor the alternative hypothesis. When the hypotheses are nested and H_0 is a special case of H_1, then Δ is <1 and $-2\log(\Delta)$ is asymptotically distributed under H_0 as a χ^2-variate with υ degrees of freedom (υ is the extra number of parameters in H_1)—for a discussion of the likelihood-ratio test, *see* Whelan and Goldman [123] and Goldman and Whelan [124]. When the hypotheses are not nested, it is necessary to use the parametric bootstrap [122, 125], in which case pseudo-data must be generated under H_0. In some cases, permutation tests may be used instead [126].

The evolution of protein-coding genes and RNA-coding genes is likely to differ due to the structural and functional constraints of the gene products, so to determine whether sites in an alignment of such genes evolved under independent and identical conditions, it is useful to draw on knowledge of the structure and function of these genes and their gene products. For example, while a protein-coding gene may be regarded as a sequence of independently evolving sites (Fig. 10a), it might be more appropriate to consider it as a sequence of independently evolving codons (Fig. 10b) or a sequence of independently evolving codon positions, with sites in the same codon position evolving under identical and independent

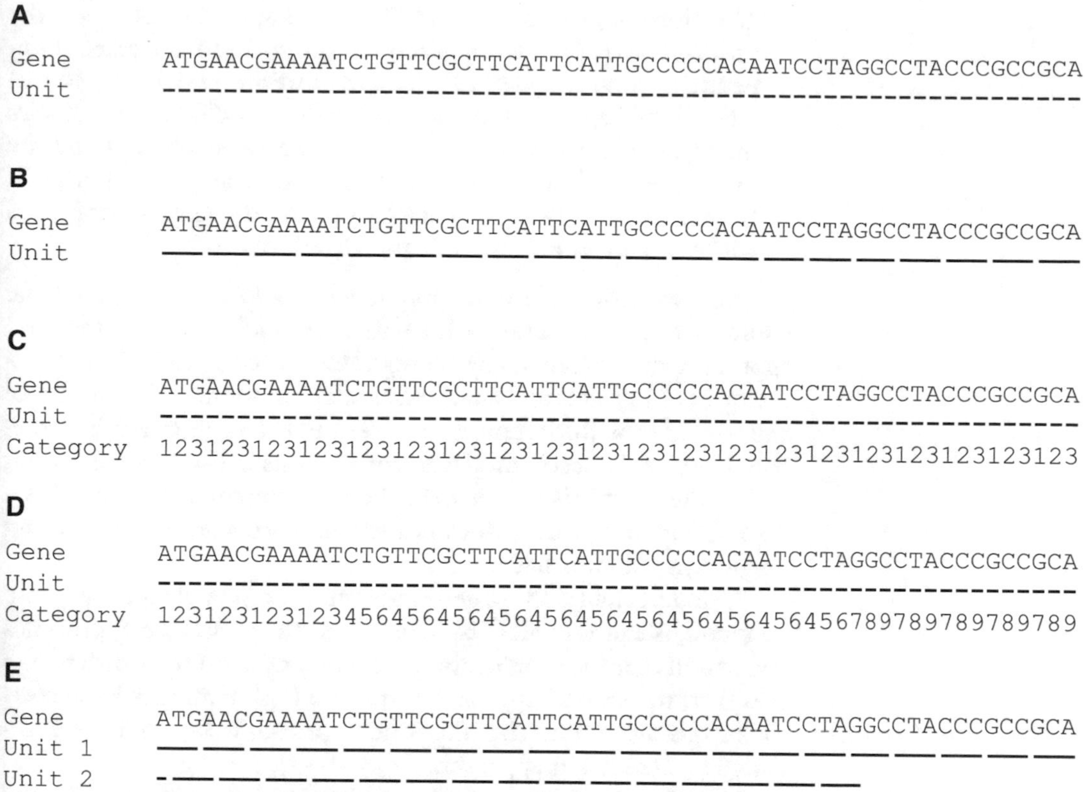

```
A
Gene      ATGAACGAAAATCTGTTCGCTTCATTCATTGCCCCCACAATCCTAGGCCTACCCGCCGCA
Unit      ------------------------------------------------------------

B
Gene      ATGAACGAAAATCTGTTCGCTTCATTCATTGCCCCCACAATCCTAGGCCTACCCGCCGCA
Unit      --- --- --- --- --- --- --- --- --- --- --- --- --- --- --- --- --- --- --- ---

C
Gene      ATGAACGAAAATCTGTTCGCTTCATTCATTGCCCCCACAATCCTAGGCCTACCCGCCGCA
Unit      ------------------------------------------------------------
Category  123123123123123123123123123123123123123123123123123123123123

D
Gene      ATGAACGAAAATCTGTTCGCTTCATTCATTGCCCCCACAATCCTAGGCCTACCCGCCGCA
Unit      ------------------------------------------------------------
Category  123123123123456456456456456456456456456456456456789789789789

E
Gene      ATGAACGAAAATCTGTTCGCTTCATTCATTGCCCCCACAATCCTAGGCCTACCCGCCGCA
Unit 1    --- --- --- --- --- --- --- --- --- --- --- --- --- --- --- --- --- --- --- ---
Unit 2    - --- --- --- --- --- --- --- --- --- --- --- --- --- --- --- ---
```

Fig. 10 Models used to describe the relationship among sites in a protein-coding gene. The protein-coding gene may be regarded as a sequence of independently evolving units, where each unit is a (**a**) site, (**b**) codon, or (**c**) site assigned its own model of evolution, given its position within a codon. More complex models include those that consider (**d**) information about the gene product's structure and function (here, categories 1, 2, and 3 correspond to models assigned to sites within the codons that encode amino acids in one structural domain, categories 4, 5, and 6 correspond to models assigned to sites in codons that encode amino acids in another structural domain, and so forth), and (**e**) overlapping reading frames (here, unit 1 corresponds to one reading frame of one gene whereas unit 2 corresponds to that of the other gene)

conditions (Fig. 10c). There are advantages and disadvantages of using each of these approaches:

1. The first approach (Fig. 10a) is fast because the number of parameters required to approximate the evolutionary processes is small (because **R** is a 4×4 matrix), and it is catered for by a large number of substitution models. However, the approach fails to consider that the evolution of neighboring sites might be correlated, which is highly likely for codon sites.
2. The second approach (Fig. 10b) is slow because the number of parameters needed to model the evolutionary process is large (because **R** is a 61×61 matrix (assuming the standard genetic code). Only a few models of evolution facilitate this approach [76, 127, 128]. However, the approach does address the issue of correlations among neighboring sites within codons.

3. The third approach (Fig. 10c) is a compromise between the previous two approaches. Each codon position is treated independently and assigned its own substitution model. The advantage of this approach is that codon-site-specific characteristics may be partly accommodated (e.g., nucleotide content and the rates of evolution are often found to vary across codon positions), but the issue of correlation among neighboring sites within each codon is not adequately addressed.

An alternative to these approaches would be to translate the codons to amino acids before further analysis. Like the first approach, the proteins may be regarded as sequences of independently evolving sites and analyzed as such using one of the available amino acid substitution models [129–140]. The approach is attractive because it accommodates correlations among neighboring codon sites, but it is slower than the first approach (because $\mathbf{R}$ is a 20×20 matrix) and it does not account for correlations among neighboring amino acids.

Using protein-coding genes obtained from viral and eukaryote genomes, Shapiro et al. [141] compared the first three approaches and found that the second approach is superior to the other two—however, the second approach came at a high computational cost. This study also found that the third approach is an attractive alternative to the second approach.

The first three approaches can be extended to account for the structure and function of gene products and the fact that a nucleotide sequence may encode several gene products. One way to account for the structure and function of a gene product is to incorporate extra categories (Fig. 10d). For example, Hyman et al. [142] used six categories to account for differences between the first and second codon positions (third codon position was ignored) as well as differences among the codons (the discriminating factor being whether the corresponding amino acid would end up in the: (a) lumen between the mitochondrial membranes, (b) inner mitochondrial membrane, or (c) mitochondrial matrix). The same approach can be used in conjunction with codon-based substitution models. In some cases, a nucleotide has more than one function (e.g., it may encode more than one product), in which case complex models may be required to approximate the evolutionary process. In the case of mitochondrial and viral genomes, some protein-coding genes overlap (Fig. 10e), in which case complex models are available [72, 79].

Sometimes, there are reasons to suspect that some sites may have been temporarily invariable. In such situations, it may be beneficial to use statistical tests developed by Lockhart et al. [82] and a phylogenetic method developed by Galtier [83]. But these tests rely on prior knowledge about the sequences, allowing the

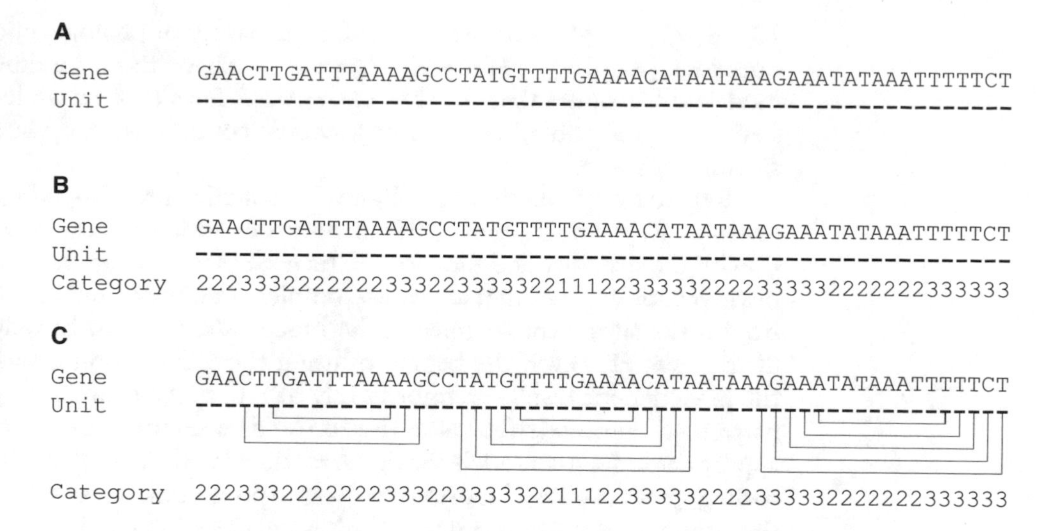

Fig. 11 Models used to describe the relationship among sites in RNA-coding genes. An RNA-coding gene may be regarded as a sequence of independently evolving units, where each unit is (**a**) a site, or (**b**) a site assigned its own model of evolution, given the role it serves in the gene product (here, category 1 corresponds to a model assigned to sites encoding the anticodon, category 2 corresponds to a model assigned to sites that encode loops in the gene product, and category 3 corresponds to a model assigned to sites encoding the stems in the gene product). A more advanced approach uses structural information to link nucleotides that match each other in the gene product (**c**) (*thin lines* connect these pairs of nucleotides)

investigators to partition their sequences into evolutionarily sound groups, and such information may not always be available.

In the context of RNA-coding genes, the sequences could be viewed as independently evolving sites (Fig. 11a), but that approach would ignore the structure and function of the gene product. A more appropriate approach would be to (a) divide the sites according to the features they encode in the gene product and (b) assign a Markov model to the sites in each partition. For example, for alignments of transfer RNA-coding genes it would be relevant to partition the sites into three categories, one for sites encoding the anticodon, another for sites encoding loops, and a third for sites encoding stems (Fig. 11b). This strategy may be further extended by incorporating knowledge of sites that encode the stems of RNA molecules (Fig. 11c)—although such sites may be at a distance from one another on the gene, their evolution is still likely to be correlated because of their role in forming the stems of RNA molecules.

The sites that form the stems in RNA molecules have most likely evolved in a correlated manner, and for this reason the Markov models assigned to these sites should consider the substitutions between pairs of nucleotides rather than single nucleotides (i.e., the Markov models should consider changes between 16 possible pairs of nucleotides: AA, AC, …, TT). Several Markov models have been developed for pairs of nucleotides [70, 71,

73–75, 77, 78, 80] and used to address a variety of phylogenetic questions [81, 126, 142–144]. However, each of these Markov models still assumes that the sites in the DNA have evolved under stationary, reversible, and homogeneous conditions, an issue already discussed.

Regardless of whether the alignment contains protein-coding genes or RNA-coding genes, the inclusion of additional information on the structure and function of the gene products may lead to phylogenetic analyses that are more complex than they would have been if the sites were assumed to be independent and identically distributed. However, the benefit of using this information is that the phylogenetic results are more likely to reflect the evolutionary pattern and processes that gave rise to the data. Finding the most appropriate Markov models for different sites in RNA- or protein-coding genes can be a laborious and error-prone task, especially if the sites have not evolved under stationary, reversible, and homogeneous conditions. Moreover, there is always the possibility that the extra parameters used to approximate the evolutionary processes simply fit random noise in the data rather than the underlying trends [80]. To address these problems, it is often useful to compare the alternative models by using statistical methods, like the parametric bootstrap [122, 125] or permutation tests [126], both of which are tree-dependent methods—below we will return to this issue.

3.5 Choosing a Time-reversible Substitution Model

If a set of sites in an alignment were found to have evolved independently under stationary, reversible, and homogeneous conditions, then there is a big family of time-reversible Markov models available for analysis of these data. Finding the most appropriate Markov model from this family of models is easy due to the fact that many of the models are nested, implying that the likelihood-ratio test [122] may be used to determine whether the alternative hypothesis, H_1, provides a significantly better fit to the data than the null hypothesis, H_0. This model-selection method became practically possible in 1998 for nucleotide sequences [145] and in 2005 for amino acid sequences [146]. In both cases, the method allows for RHAS, thus catering for some of the differences found among sites in phylogenetic data.

Although the abovementioned model-selection method appears attractive, there are reasons for concern. For example, it assumes that at least one of the models compared is correct, an assumption that would be violated in most cases. Other problems include those arising when: (a) multiple tests are done on the same data and the tests are non-independent; (b) sample size (i.e., number of sites) is small; and (c) non-nested models are compared (for an informative discussion of the problems, *see* Refs. [147, 148]). Finally, it is assumed that the tree used during the comparisons of models is the most likely tree for every model compared, which might not be the case.

Some of these problems are easily dealt with by other model-selection methods. Within the context of likelihood, alternative models of evolution may be compared using the Akaike Information Criterion (AIC) [149], where AIC for a given model, **R**, is a function of the maximized log-likelihood on **R** and the number of estimable parameters, K (e.g., nucleotide frequency, conditional rates of change, proportion of invariant sites, rate variation among sites, and number of edges in the tree):

$$\text{AIC} = -2max(logL(\text{data}|\mathbf{R})) + 2K. \quad (12)$$

If the sample size, l, is small (the exact meaning of sample size is currently unclear but it is occasionally thought to be approximately equal to the number of sites in the aligned data) compared to the number of estimable parameters (e.g., $l/K < 40$), then the corrected AIC (AICc) is recommended [150]:

$$\text{AICc} = \text{AIC} + \frac{2K(K+1)}{l-K-1}. \quad (13)$$

The AIC may be regarded as the amount of information lost by using **R** to approximate the evolutionary processes and $2K$ may be regarded as a penalty for allowing $2K$ parameters; hence, the best-fitting Markov model corresponds to the smallest value of AIC (or AICc).

Within the Bayesian context, alternative models of evolution may be compared using the Bayesian Information Criterion (BIC) [151], the Bayes factor (BF) [152–154], posterior probabilities (PP) [155, 156] and decision theory (DT) [157], where, for example,

$$\text{BF}_{ij} = \frac{Pr(\text{data}|\mathbf{R}_i)}{Pr(\text{data}|\mathbf{R}_j)} \quad (14)$$

and

$$\text{BIC} = -2max(logL(\text{data}|\mathbf{R})) + Klogl. \quad (15)$$

A common feature of the model-selection methods that calculate BF and PP is that the likelihood of a given model is calculated by integrating over parameter space; hence, the methods often rely on computationally intensive techniques to obtain the likelihood of the model. The model-selection method that calculates PP is more attractive than that which calculates BF because the former facilitates concurrent comparisons of multiple models, including non-nested models. However, the drawback of the model-selection method that calculates PP is that, although some Markov models may appear more realistic than other such models, it is difficult to quantify the prior probability of alternative models [148].

The method for calculating BIC is more tractable than those for calculating BF and PP, and provided the prior probability is uniform for the models under consideration, the BIC statistic is also easier to interpret than the BF statistic [148]. The prior probability is unlikely to be uniform, however, implying that interpretation of the BIC and DT statistics (the latter is an extension of the former [157]) may be more difficult.

There is evidence that the use of different model-selection methods may result in different Markov models being selected for phylogenetic analyses of the same data [158], so there is a need for more information on how to select model-selection methods. Based on practical and theoretical considerations, Posada and Buckley [148] suggested that model-selection methods should: (a) be able to compare non-nested models; (b) allow for simultaneous comparison of multiple models; (c) not depend on significance levels; (d) incorporate topological uncertainty; (e) be tractable; (f) allow for model averaging; (g) provide the possibility of specifying priors for models and model parameters; and (h) be designed to approximate, rather than to identify, truth. Based on these criteria and with reference to a large body of literature on philosophical and applied aspects of model selection, Posada and Buckley [148] concluded that the hierarchical likelihood-ratio test is not the optimal approach for model selection in phylogenetics and that the AIC and Bayesian approaches provide important advantages, including the ability to simultaneously compare multiple nested and non-nested models.

Given these considerations and the ease with which estimates of AIC, AICc and BIC can be computed from alignments of nucleotides or amino acids, a number of useful programs for identifying optimal models of evolution for such data have been developed [11, 12, 104, 159]. The different implementations of the standard model-selection method, which rely on the estimates of AIC, AICc or BIC, are easy to use and are now increasingly part of the standard phylogenetic protocol. The standard model-selection method has been extended to consider partitioned data [13, 14], which is particularly relevant for phylogenetic analysis of multi-gene data sets. Despite the ease with which these programs can produce what appears to be reliable results, there are good reasons to interpret the results with care. For example, when AIC, AICc and BIC are calculated standard model-selection procedures, the calculations are preceded by estimation of a phylogenetic tree, and it is this tree that is used during the estimation of AIC, AICc and BIC for the different models of evolution. In other words, the tree is not free to change during the model selection. Because this approach may lead to biased estimates, it is better to use procedures that allow the tree to change during the model selection. Another concern pertains to the manners in which RHAS is modeled. Usually, three models are considered: (a) the I model (a proportion of sites are assumed to be invariable and the remaining sites are assumed to have evolved under the same conditions); (b) the Γ_4 model (RHAS is modeled using a discrete Γ distribution with 4 rate categories); and (c) the I+Γ_4 model (a hybrid of the first two models). The first of these models is useful because it is likely to be correct that some sites are invariable but the assumption that the remaining sites have evolved under the same conditions may not be correct in many cases. The second of these models assumes that the RHAS is best approximated by a Γ distribution, which might not be correct. The third of these models combines different models that are designed to address the same problem and should not be used [67, 93, 160]. Other ways to model RHAS have been proposed [42, 161–163] but many users of phylogenetic programs have not yet embraced these models.

3.6 General Approaches to Model Selection

Having inferred the best tree for a given data set using a model of evolution, it is always a good idea to evaluate the fit between tree, model(s), and data. This cannot be accomplished using the non-parametric bootstrap (because it measures variability of the estimate obtained using a phylogenetic method) but can be done using the parametric bootstrap. The use of the parametric bootstrap to test the appropriateness of a given Markov model was proposed by Goldman [125], and is a modification of Cox's [164] test, which considers non-nested models. The test can be performed as follows:

1. For a given model, **R**, use the original alignment to obtain the log-likelihood, log L, and the maximum-likelihood estimates of the free parameters;
2. Calculate the unconstrained log-likelihood, log L^*, for the original alignment using the following equation:

$$logL^* = \sum_{i=1}^{N} log\left(\frac{N_i}{N}\right), \tag{16}$$

where N is the number of sites in the alignment and N_i is the number of times that the pattern at column i occurs in the alignment;

3. Calculate $\delta_{obs} = \log L^* - \log L$;
4. Use the inferred tree and the optimized parameter values (obtained during **step 1**) to generate 1000 pseudo-data sets;
5. For each pseudo-data set, $j = 1, \ldots, 1000$, obtain log L_j (the log-likelihood under **R**), log L_j^*, and $\delta_j = \log L_j^* - \log L_j$;
6. Estimate p, the proportion of times where $\delta_j > \delta_{\text{obs}}$. A large p-value supports the hypothesis that **R** is sufficient to explain the evolutionary process underpinning the data while a small p-value provides evidence against this hypothesis.

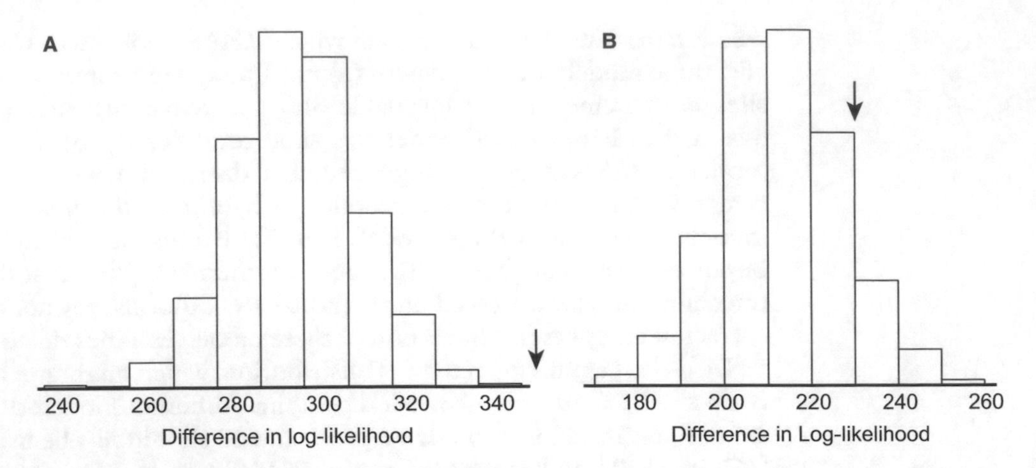

Fig. 12 Examples of the results from two parametric bootstrap analyses. (**a**) Parametric bootstrap results under the GTR+Γ model based on 1000 simulations. For each bootstrap replicate, the difference in log-likelihoods was obtained by subtracting the log-likelihood under the GTR+Γ model from the unconstrained log-likelihood. The *arrow* indicates the difference in log-likelihood for the actual data under the GTR+Γ model. (**b**) Parametric bootstrap results under the BH+I model based on 1000 simulations. For each bootstrap replicate, the difference in log-likelihoods was obtained by subtracting the log-likelihood under the BH+I model from the unconstrained log-likelihood. The *arrow* indicates the difference in log-likelihood for the actual data under the BH+I model

For time-reversible models of evolution, pseudo-data may be obtained using Seq-Gen [165] or INDELible [166]; for more general models of evolution, the data may be obtained using other purpose-developed programs [30, 34, 42, 50, 167]. Foster [30] and Jayaswal et al. [34] used the parametric bootstrap to show that the models of evolution used provide a good fit even though the sequences appear to have evolved under conditions that are not stationary, reversible, and homogeneous. Figure 12 shows the results of a parametric bootstrap analysis for the bacterial sequences analyzed by Jayaswal et al. [34]. If these data had been analyzed assuming time-reversible models, ModelTest [145] would have selected the GTR+Γ model as the most appropriate. However, the graph in Fig. 12a shows that the difference between the unconstrained log-likelihood and log-likelihood under the GTR+Γ model is significant, implying that the GTR+Γ model fails to adequately describe the complex conditions under which these data evolved. On the other hand, a similar analysis using the BH+I model [34] led to a result showing that the BH+I model is an adequate approximation to the evolutionary processes that gave rise to these bacterial sequences (Fig. 12b).

Sometimes, it may be useful to compare the hypothesis that the sites of an alignment have evolved under identical and independent conditions to the hypothesis that the nucleotides encode a molecule that necessitates the alignment be partitioned and then analyzed using partition-specific Markov models. Given this challenge a permutation test [168] may be useful. A suitable strategy for comparing two such hypotheses is described as follows:

1. For a given set of models, $\mathbf{R}_1, \mathbf{R}_2, \ldots$, each of which is applied to its own partition of the alignment, calculate the log-likelihood, $\log L_{obs}$, of the original alignment;
2. Generate 1000 pseudo-data sets by randomizing the order of columns in the data;
3. For each pseudo-data set, $j = 1, \cdots, 1000$, calculate $\log L_j$ (i.e., the log-likelihood of the pseudo-data under $\mathbf{R}_1, \mathbf{R}_2, \ldots$);
4. Estimate p, the proportion of times where $\log L_j > \log L_{\text{obs}}$. A small p-value supports the notion that the sites should be partitioned and analyzed using separate models for each partition.

Telford et al. [126] used a modification of this approach to show that the evolution of the small subunit ribosomal RNA genes of Bilateria can be explained best by using two Markov models, one for the loop-coding sites and another for the stem-coding sites. They also found that paired-sites models [70, 73, 74] were significantly better at describing the evolution of stem-coding sites than a model that assumes independence of the stem-forming sites.

The permutation test presented above is applicable in the general case. It may be regarded as an alternative to the model-selection methods designed for partitioned data [13, 14] that have evolved under stationary, reversible, and homogeneous conditions. However, for partitioned data that has evolved under more complex conditions, the permutation test is the better one. For nucleotide sequences, the HAL-HAS model [42] may be a suitable alternative.

4 Discussion

It is clear that model selection plays a central role in molecular phylogenetics, especially in the context of distance-based, likelihood-based and Bayesian phylogenetic methods. Given that different model-selection methods may select different models and that application of poorly fitting models affects many aspects of phylogenetic studies (including estimates of phylogeny, substitution rates, posterior probabilities, and bootstrap values), it is important to know the advantages and disadvantages of the available model-selection methods.

It is equally important to know that any model of evolution that we can construct to analyze a set of sequences is unlikely to be the true model. Rather, the models that we infer, using prior knowledge of the data and the model-selection methods available, are at best reasonable approximations of the underlying evolutionary processes. The process of identifying optimal models of evolution for phylogenetic studies, therefore, should be considered as a method of approximating, rather than identifying, the evolutionary processes [147, 148]. When identifying an appropriate approximation to the evolutionary processes, it is important to strike a balance between bias and variance—the problem of bias may arise when too few parameters are used to approximate the evolutionary processes and the problem of variance may arise if too many parameters are used [147, 148]. Using model-selection methods that include a penalty for including more than the necessary number of parameters, therefore, appears very appealing.

Having identified a suitable model of evolution and subsequently inferred the phylogeny, it is wise to use the parametric bootstrap to assess whether the estimates of evolutionary patterns and evolutionary processes are consistent with the data. This is not done as often as it should be, although this trend might be changing (*see* for example Ref. [169]). With reference to the parametric bootstrap, it is important to remember that a good fit does not guarantee that the tree and model are correct. For example, Jayaswal et al. [34] obtained a good fit between tree, model and data for the bacterial 16S ribosomal RNA even though the model used did not consider structural and functional information about the gene product.

Finally, it is clear that there is a need for more efficient and reliable methods to identify appropriate Markov models for phylogenetic studies, in particular for data that have not evolved under stationary, reversible, and homogeneous conditions. There is also still a need for phylogenetic methods that (1) allow the partitions of alignments to be analyzed using a combination of nonreversible Markov models and (2) allow the parameters of each model to be optimized independently (except for the length of edges in each tree).

References

1. Zakharov EV, Caterino MS, Sperling FAH (2004) Molecular phylogeny, historical biogeography, and divergence time estimates for swallowtail butterflies of the genus Papilio (Lepidoptera: Papilionidae). Syst Biol 53:193–215
2. Brochier C, Forterre P, Gribaldo S (2005) An emerging phylogenetic core of Archaea: phylogenies of transcription and translation machineries converge following addition of new genome sequences. BMC Evol Biol 5:36
3. Hardy MP, Owczarek CM, Jermiin LS et al (2004) Characterization of the type I interferon locus and identification of novel genes. Genomics 84:331–345
4. de Queiroz K, Gauthier J (1994) Toward a phylogenetic system of biological nomenclature. Trends Ecol Evol 9:27–31
5. Board PG, Coggan M, Chelnavayagam G et al (2000) Identification, characterization and crystal structure of the Omega class of glutathione transferases. J Biol Chem 275:24798–24806
6. Pagel M (1999) Inferring the historical patterns of biological evolution. Nature 401:877–884
7. Charleston MA, Robertson DL (2002) Preferential host switching by primate lentiviruses can account for phylogenetic similarity with the primate phylogeny. Syst Biol 51:528–535
8. Jermann TM, Opitz JG, Stackhouse J et a (1995) Reconstructing the evolutionary history of the artiodactyl ribonuclease superfamily. Nature 374:57–59
9. Eisen JA (1998) Phylogenomics: improving functional predictions for uncharacterized

genes by evolutionary analysis. Genome Res 8:163–167

10. Misof B, Liu SL, Meusemann K et al (2014) Phylogenomics resolves the timing and pattern of insect evolution. Science 346:763–767
11. Darriba D, Taboada GL, Doallo R et al (2011) ProtTest 3: fast selection of best-fit models of protein evolution. Bioinformatics 27:1164–1165
12. Darriba D, Taboada GL, Doallo R et al (2012) jModelTest 2: more models, new heuristics and parallel computing. Nat Methods 9:772
13. Lanfear R, Calcott B, Ho SYW et al (2012) Partitionfinder: combined selection of partitioning schemes and substitution models for phylogenetic analyses. Mol Biol Evol 29:1695–1701
14. Lanfear R, Calcott B, Kainer D et al (2014) Selecting optimal partitioning schemes for phylogenomic datasets. BMC Evol Biol 14:82
15. Jermiin LS, Ho JWK, Lau KW et al (2009) SeqVis: a tool for detecting compositional heterogeneity among aligned nucleotide sequences. In: Posada D (ed) Bioinformatics for DNA sequence analysis. Humana Press, Totowa, NJ, pp 65–91
16. Barry D, Hartigan JA (1987) Statistical analysis of hominoid molecular evolution. Stat Sci 2:191–210
17. Reeves J (1992) Heterogeneity in the substitution process of amino acid sites of proteins coded for by the mitochondrial DNA. J Mol Evol 35:17–31
18. Steel MA, Lockhart PJ, Penny D (1993) Confidence in evolutionary trees from biological sequence data. Nature 364:440–442
19. Lake JA (1994) Reconstructing evolutionary trees from DNA and protein sequences: paralinear distances. Proc Natl Acad Sci U S A 91:1455–1459
20. Lockhart PJ, Steel MA, Hendy MD et al (1994) Recovering evolutionary trees under a more realistic model of sequence evolution. Mol Biol Evol 11:605–612
21. Steel MA (1994) Recovering a tree from the leaf colourations it generates under a Markov model. Appl Math Lett 7:19–23
22. Galtier N, Gouy M (1995) Inferring phylogenies from DNA sequences of unequal base compositions. Proc Natl Acad Sci U S A 92:11317–11321
23. Steel MA, Lockhart PJ, Penny D (1995) A frequency-dependent significance test for parsimony. Mol Phylogenet Evol 4:64–71
24. Yang Z, Roberts D (1995) On the use of nucleic acid sequences to infer early branches in the tree of life. Mol Biol Evol 12:451–458
25. Gu X, Li W-H (1996) Bias-corrected paralinear and logdet distances and tests of molecular clocks and phylogenies under nonstationary nucleotide frequencies. Mol Biol Evol 13:1375–1383
26. Gu X, Li W-H (1998) Estimation of evolutionary distances under stationary and nonstationary models of nucleotide substitution. Proc Natl Acad Sci U S A 95:5899–5905
27. Galtier N, Gouy M (1998) Inferring pattern and process: maximum-likelihood implementation of a nonhomogenous model of DNA sequence evolution for phylogenetic analysis. Mol Biol Evol 15:871–879
28. Galtier N, Tourasse N, Gouy M (1999) A nonhyperthermophilic common ancestor to extant life forms. Science 283:220–221
29. Tamura K, Kumar S (2002) Evolutionary distance estimation under heterogeneous substitution pattern among lineages. Mol Biol Evol 19:1727–1736
30. Foster PG (2004) Modelling compositional heterogeneity. Syst Biol 53:485–495
31. Thollesson M (2004) LDDist: a Perl module for calculating LogDet pair-wise distances for protein and nucleotide sequences. Bioinformatics 20:416–418
32. Jayaswal V, Jermiin LS, Robinson J (2005) Estimation of phylogeny using a general Markov model. Evol Bioinf Online 1:62–80
33. Blanquart S, Lartillot N (2006) A Bayesian compound stochastic process for modeling nonstationary and nonhomogeneous sequence evolution. Mol Biol Evol 23:2058–2071
34. Jayaswal V, Robinson J, Jermiin LS (2007) Estimation of phylogeny and invariant sites under the General Markov model of nucleotide sequence evolution. Syst Biol 56:155–162
35. Blanquart S, Lartillot N (2008) A site- and time-heterogeneous model of amino acid replacement. Mol Biol Evol 25:842–858
36. Dutheil J, Boussau B (2008) Non-homogeneous models of sequence evolution in the Bio++ suite of libraries and programs. BMC Evol Biol 8:255
37. Jayaswal V, Jermiin LS, Poladian L et al (2011) Two stationary, non-homogeneous Markov models of nucleotide sequence evolution. Syst Biol 60:74–86
38. Jayaswal V, Ababneh F, Jermiin LS et al (2011) Reducing model complexity when

the evolutionary process over an edge is modeled as a homogeneous Markov process. Mol Biol Evol 28:3045–3059

39. Dutheil JY, Galtier N, Romiguier J et al (2012) Efficient selection of branch-specific models of sequence evolution. Mol Biol Evol 29:1861–1874
40. Zou LW, Susko E, Field C et al (2012) Fitting nonstationary general-time-reversible models to obtain edge-lengths and frequencies for the Barry-Hartigan model. Syst Biol 61:927–940
41. Groussin M, Boussau B, Gouy M (2013) A branch-heterogeneous model of protein evolution for efficient inference of ancestral sequences. Syst Biol 62:523–538
42. Jayaswal V, Wong TKF, Robinson J et al (2014) Mixture models of nucleotide sequence evolution that account for heterogeneity in the substitution process across sites and across lineages. Syst Biol 63:726–742
43. Woodhams MD, Fernandez-Sanchez J, Sumner JG (2015) A new hierarchy of phylogenetic models consistent with heterogeneous substitution rates. Syst Biol 64:638–650
44. Jermiin LS, Jayaswal V, Ababneh F et al (2008) Phylogenetic model evaluation. In: Keith J (ed) Bioinformatics: data, sequence analysis, and evolution. Humana Press, Totowa, NJ, pp 331–364
45. Sullivan J, Arellano EA, Rogers DS (2000) Comparative phylogeography of Mesoamerican highland rodents: concerted versus independent responses to past climatic fluctuations. Am Nat 155:755–768
46. Demboski JR, Sullivan J (2003) Extensive mtDNA variation within the yellow-pine chipmunk, Tamias amoenus (Rodentia: Sciuridae), and phylogeographic inferences for northwestern North America. Mol Phylogenet Evol 26:389–408
47. Carstens BC, Stevenson AL, Degenhardt JD et al (2004) Testing nested phylogenetic and phylogeographic hypotheses in the Plethodon vandykei species group. Syst Biol 53:781–792
48. Penny D, Hendy MD, Steel MA (1992) Progress with methods for constructing evolutionary trees. Trends Ecol Evol 7:73–79
49. Tavaré S (1986) Some probabilistic and statistical problems on the analysis of DNA sequences. Lect Math Life Sci 17:57–86
50. Ababneh F, Jermiin LS, Robinson J (2006) Generation of the exact distribution and simulation of matched nucleotide sequences on a phylogenetic tree. J Math Model Algor 5:291–308
51. Bryant D, Galtier N, Poursat M-A (2005) Likelihood calculation in molecular phylogenetics. In: Gascuel O (ed) Mathematics of evolution and phylogeny. Oxford University Press, Oxford, pp 33–62
52. Ullah I, Sjöstrand J, Andersson P et al (2015) Integrating sequence evolution into probabilistic orthology analysis. Syst Biol 64:969–982
53. Drouin G, Prat F, Ell M et al (1999) Detecting and characterizing gene conversion between multigene family members. Mol Biol Evol 16:1369–1390
54. Posada D, Crandall KA (2001) Evaluation of methods for detecting recombination from DNA sequences: computer simulations. Proc Natl Acad Sci U S A 98:13757–13762
55. Posada D (2002) Evaluation of methods for detecting recombination from DNA sequences: empirical data. Mol Biol Evol 19:708–717
56. Martin DP, Williamson C, Posada D (2005) RDP2: recombination detection and analysis from sequence alignments. Bioinformatics 21:260–262
57. Bruen TC, Philippe H, Bryant D (2006) A simple and robust statistical test for detecting the presence of recombination. Genetics 172:2665–2681
58. Ragan MA (2001) On surrogate methods for detecting lateral gene transfer. FEMS Microbiol Lett 201:187–191
59. Dufraigne C, Fertil B, Lespinats S et al (2005) Detection and characterization of horizontal transfers in prokaryotes using genomic signature. Nucleic Acids Res 33:e6
60. Azad RK, Lawrence JG (2005) Use of artificial genomes in assessing methods for atypical gene detection. PLoS Comp Biol 1:461–473
61. Tsirigos A, Rigoutsos I (2005) A new computational method for the detection of horizontal gene transfer events. Nucleic Acids Res 33:922–933
62. Ragan MA, Harlow TJ, Beiko RG (2006) Do different surrogate methods detect lateral genetic transfer events of different relative ages? Trends Microbiol 14:4–8
63. Beiko RG, Hamilton N (2006) Phylogenetic identification of lateral genetic transfer events. BMC Evol Biol 6:15
64. Sjöstrand J, Tofigh A, Daubin V et al (2014) A Bayesian method for analyzing lateral gene transfer. Syst Biol 63:409–420
65. Fitch WM (1986) An estimation of the number of invariable sites is necessary for the accurate estimation of the number of nucleotide substitutions since a common ancestor. Prog Clin Biol Res 218:149–159
66. Lockhart PJ, Larkum AWD, Steel MA et al (1996) Evolution of chlorophyll and bacteriochlorophyll: the problem of invariant sites in

sequence analysis. Proc Natl Acad Sci U S A 93:1930–1934

67. Yang Z (1996) Among-site rate variation and its impact on phylogenetic analysis. Trends Ecol Evol 11:367–372
68. Waddell PJ, Steel MA (1997) General time reversible distances with unequal rates across sites: mixing G and inverse Gaussian distributions with invariant sites. Mol Phylogenet Evol 8:398–414
69. Gowri-Shankar V, Rattray M (2006) Compositional heterogeneity across sites: effects on phylogenetic inference and modelling the correlations between base frequencies and substitution rate. Mol Biol Evol 23:352–364
70. Schöniger M, von Haeseler A (1994) A stochastic model for the evolution of autocorrelated DNA sequences. Mol Phylogenet Evol 3:240–247
71. Tillier ERM (1994) Maximum likelihood with multiparameter models of substitution. J Mol Evol 39:409–417
72. Hein J, Støvlbœk J (1995) A maximum-likelihood approach to analyzing nonoverlapping and overlapping reading frames. J Mol Evol 40:181–190
73. Muse SV (1995) Evolutionary analyses of DNA sequences subject to constraints on secondary structure. Genetics 139:1429–1439
74. Rzhetsky A (1995) Estimating substitution rates in ribosomal RNA genes. Genetics 141:771–783
75. Tillier ERM, Collins RA (1995) Neighbor joining and maximum likelihood with RNA sequences: addressing the interdependence of sites. Mol Biol Evol 12:7–15
76. Pedersen A-MK, Wiuf C, Christiansen FB (1998) A codon-based model designed to describe lentiviral evolution. Mol Biol Evol 15:1069–1081
77. Tillier ERM, Collins RA (1998) High apparent rate of simultaneous compensatory base-pair substitutions in ribosomal RNA. Genetics 148:1993–2002
78. Higgs PG (2000) RNA secondary structure: physical and computational aspects. Q Rev Biophys 30:199–253
79. Pedersen A-MK, Jensen JL (2001) A dependent-rates model and an MCMC-based methodology for the maximum-likelihood analysis of sequences with overlapping frames. Mol Biol Evol 18:763–776
80. Savill NJ, Hoyle DC, Higgs PG (2001) RNA sequence evolution with secondary structure constraints: comparison of substitution rate models using maximum-likelihood methods. Genetics 157:399–411
81. Jow H, Hudelot C, Rattray M et al (2002) Bayesian phylogenerics using an RNA substitution model applied to early mammalian evolution. Mol Biol Evol 19:1591–1601
82. Lockhart PJ, Steel MA, Barbrook AC et al (1998) A covariotide model explains apparent phylogenetic structure of oxygenic photosynthetic lineages. Mol Biol Evol 15:1183–1188
83. Galtier N (2001) Maximum-likelihood phylogenetic analysis under a covarion-like model. Mol Biol Evol 18:866–873
84. Pupko T, Galtier N (2002) A covarion-based method for detecting molecular adaptation: application to the evolution of primate mitochondrial genomes. Proc R Soc B 269:1313–1316
85. Susko E, Inagaki Y, Field C et al (2002) Testing for differences in rates-across-sites distributions in phylogenetic subtrees. Mol Biol Evol 19:1514–1523
86. Wang HC, Spencer M, Susko E et al (2007) Testing for covarion-like evolution in protein sequences. Mol Biol Evol 24:294–305
87. Wang HC, Susko E, Spencer M et al (2008) Topological estimation biases with covarion evolution. J Mol Evol 66:50–60
88. Wu JH, Susko E (2009) General heterotachy and distance method adjustments. Mol Biol Evol 26:2689–2697
89. Wang HC, Susko E, Roger AJ (2009) PROCOV: maximum likelihood estimation of protein phylogeny under covarion models and site-specific covarion pattern analysis. BMC Evol Biol 9:225
90. Wang HC, Susko E, Roger AJ (2011) Fast statistical tests for detecting heterotachy in protein evolution. Mol Biol Evol 28:2305–2315
91. Wu JH, Susko E (2011) A test for heterotachy using multiple pairs of sequences. Mol Biol Evol 28:1661–1673
92. Kolmogoroff A (1936) Zur theorie der Markoffschen ketten. Math Annal 112:155–160
93. Yang Z (2014) Molecular evolution: a statistical approach. Oxford University Press, Oxford
94. Jukes TH, Cantor CR (1969) Evolution of protein molecules. In: Munro HN (ed) Mammalian protein metabolism. Academic, New York, pp 21–132
95. Lanave C, Preparata G, Saccone C et al (1984) A new method for calculating evolutionary substitution rates. J Mol Evol 20:86–93
96. Naylor GPJ, Brown WM (1998) Amphioxus mitochondrial DNA, chordate phylogeny,

and the limits of inference based on comparisons of sequences. Syst Biol 47:61–76
97. Grundy WN, Naylor GJP (1999) Phylogenetic inference from conserved sites alignments. J Exp Zool 285:128–139
98. Li CH, Matthes-Rosana KA, Garcia M et al (2012) Phylogenetics of Chondrichthyes and the problem of rooting phylogenies with distant outgroups. Mol Phylogenet Evol 63:365–373
99. Campbell MA, Chen WJ, Lopez JA (2013) Are flatfishes (Pleuronectiformes) monophyletic? Mol Phylogenet Evol 69:664–673
100. Ho SYW, Jermiin LS (2004) Tracing the decay of the historical signal in biological sequence data. Syst Biol 53:623–637
101. Lartillot N, Philippe H (2004) A Bayesian mixture model for across-site heterogeneities in the amino-acid replacement process. Mol Biol Evol 21:1095–1109
102. Le SQ, Dang CC, Gascuel O (2012) Modeling protein evolution with several amino acid replacement matrices depending on site rates. Mol Biol Evol 29:2921–2936
103. Lartillot N, Rodrigue N, Stubbs D et al (2013) PhyloBayes MPI: phylogenetic reconstruction with infinite mixtures of profiles in a parallel environment. Syst Biol 62:611–615
104. Nguyen L-T, Schmidt HA, Von Haeseler A et al (2015) IQ-TREE: a fast and effective stochastic algorithm for estimating maximum-likelihood phylogenies. Mol Biol Evol 32:268–274
105. Jermiin LS, Ho SYW, Ababneh F et al (2004) The biasing effect of compositional heterogeneity on phylogenetic estimates may be underestimated. Syst Biol 53:638–643
106. Ababneh F, Jermiin LS, Ma C et al (2006) Matched-pairs tests of homogeneity with applications to homologous nucleotide sequences. Bioinformatics 22:1225–1231
107. Ho JWK, Adams CE, Lew JB et al (2006) SeqVis: visualization of compositional heterogeneity in large alignments of nucleotides. Bioinformatics 22:2162–2163
108. Lanave C, Pesole G (1993) Stationary MARKOV processes in the evolution of biological macromolecules. Binary 5:191–195
109. Rzhetsky A, Nei M (1995) Tests of applicability of several substitution models for DNA sequence data. Mol Biol Evol 12:131–151
110. Waddell PJ, Cao Y, Hauf J et al (1999) Using novel phylogenetic methods to evaluate mammalian mtDNA, including amino acid-invariant sites-LogDet plus site stripping, to detect internal conflicts in the data, with special reference to the positions of hedgehog, armadillo, and elephant. Syst Biol 48:31–53
111. Bowker AH (1948) A test for symmetry in contingency tables. J Am Stat Assoc 43:572–574
112. Stuart A (1955) A test for homogeneity of the marginal distributions in a two-way classification. Biometrika 42:412–416
113. Holm S (1979) A simple sequentially rejective multiple test procedure. Scand J Stat 6:65–70
114. Cannings C, Edwards AWF (1968) Natural selection and the de Finetti diagram. Ann Hum Genet 31:421–428
115. Bourlat SJ, Juliusdottir T, Lowe CJ et al (2006) Deuterostome phylogeny reveals monophyletic chordates and the new phylum Xenoturbellida. Nature 444:85–88
116. Fitch WM, Margoliash E (1967) Construction of phylogenetic trees. Science 155:279–284
117. Cavalli-Sforza LL, Edwards AWF (1967) Phylogenetic analysis: models and estimation procedures. Am J Hum Genet 19:233–257
118. Saitou N, Nei M (1987) The neighbor-joining method: a new method for reconstructing phylogenetic trees. Mol Biol Evol 4:406–425
119. Gascuel O (1997) BIONJ: an improved version of the NJ algorithm based on a simple model of sequence data. Mol Biol Evol 14:685–695
120. Zou L, Susko E, Field C et al (2011) The parameters of the Barry-Hartigan model are statistically non identifiable. Syst Biol 60:872–875
121. Minin VN, Suchard MA (2008) Fast, accurate and simulation-free stochastic mapping. Philos Trans R Soc Lond B 363:3985–3995
122. Huelsenbeck JP, Rannala B (1997) Phylogenetic methods come of age: testing hypotheses in an evolutionary context. Science 276:227–232
123. Whelan S, Goldman N (1999) Distributions of statistics used for the comparison of models of sequence evolution in phylogenetics. Mol Biol Evol 16:11292–11299
124. Goldman N, Whelan S (2000) Statistical tests of gamma-distributed rate heterogeneity in models of sequence evolution in phylogenetics. Mol Biol Evol 17:975–978
125. Goldman N (1993) Statistical tests of models of DNA substitution. J Mol Evol 36:182–198
126. Telford MJ, Wise MJ, Gowri-Shankar V (2005) Consideration of RNA secondary structure significantly improves likelihood-based estimates

of phylogeny: examples from the bilateria. Mol Biol Evol 22:1129–1136
127. Goldman N, Yang Z (1994) A codon-based model of nucleotide substitution for protein-coding DNA sequences. Mol Biol Evol 11:725–736
128. Muse SV, Gaut BS (1994) A likelihood approach for comparing synonymous and nonsynonymous nucleotide substitution rates, with application to the chloroplast genome. Mol Biol Evol 11:715–724
129. Dayhoff MO, Schwartz RM, Orcutt BC (eds) (1978) A model of evolutionary change in proteins. National Biomedical Research Foundation, National Biomedical Research Foundation, Washington, DC
130. Jones DT, Taylor WR, Thornton JM (1992) The rapid generation of mutation data matrices from protein sequences. CABIOS 8:275–282
131. Henikoff S, Henikoff JG (1992) Amino acid substitution matrices from protein blocks. Proc Natl Acad Sci U S A 89:10915–10919
132. Adachi J, Hasegawa M (1996) Model of amino acid substitution in proteins encoded by mitochondrial DNA. J Mol Evol 42:459–468
133. Cao Y, Janke A, Waddell PJ et al (1998) Conflict among individual mitochondrial proteins in resolving the phylogeny of eutherian orders. J Mol Evol 47:307–322
134. Yang Z, Nielsen R, Hasegawa M (1998) Models of amino acid substitution and applications to mitochondrial protein evolution. Mol Biol Evol 15:1600–1611
135. Müller T, Vingron M (2000) Modeling amino acid replacement. J Comp Biol 7:761–776
136. Adachi J, Waddell PJ, Martin W et al (2000) Plastid genome phylogeny and a model of amino acid substitution for proteins encoded by chloroplast DNA. J Mol Evol 50:348–358
137. Whelan S, Goldman N (2001) A general empirical model of protein evolution derived from multiple protein families using a maximum likelihood approach. Mol Biol Evol 18:691–699
138. Dimmic MW, Rest JS, Mindell DP et al (2002) RtREV: an amino acid substitution matrix for inference of retrovirus and reverse transcriptase phylogeny. J Mol Evol 55:65–73
139. Abascal F, Posada D, Zardoya R (2007) MtArt: a new model of amino acid replacement for Arthropoda. Mol Biol Evol 24:1–5
140. Le SQ, Gascuel O (2008) An improved general amino acid replacement matrix. Mol Biol Evol 25:1307–1320
141. Shapiro B, Rambaut A, Drummond AJ (2005) Choosing appropriate substitution models for the phylogenetic analysis of protein-coding sequences. Mol Biol Evol 23:7–9
142. Hyman IT, Ho SYW, Jermiin LS (2007) Molecular phylogeny of Australian Helicarionidae, Microcystidae and related groups (Gastropoda: Pulmonata: Stylommatophora) based on mitochondrial DNA. Mol Phylogenet Evol 45:792–812
143. Hudelot C, Gowri-Shankar V, Jow H et al (2003) RNA-based phylogenetic methods: application to mammalian mitochondrial RNA sequences. Mol Phylogenet Evol 28:241–252
144. Murray S, Flø Jørgensen M, Ho SYW et al (2005) Improving the analysis of dinoflagelate phylogeny based on rDNA. Protist 156:269–286
145. Posada D, Crandall KA (1998) MODELTEST: testing the model of DNA substitution. Bioinformatics 14:817–818
146. Abascal F, Zardoya R, Posada D (2005) ProtTest: selection of best-fit models of protein evolution. Bioinformatics 21:2104–2105
147. Burnham KP, Anderson DR (2002) Model selection and multimodel inference: a practical information-theoretic approach. Springer, New York
148. Posada D, Buckley TR (2004) Model selection and model averaging in phylogenetics: advantages of akaike information criterion and bayesian approaches over likelihood ratio tests. Syst Biol 53:793–808
149. Akaike H (1974) A new look at the statistical model identification. IEEE Trans Auto Cont 19:716–723
150. Sugiura N (1978) Further analysis of the data by Akaike's information criterion and the finite corrections. Comm Stat A Theor Meth 7:13–26
151. Schwarz G (1978) Estimating the dimension of a model. Ann Stat 6:461–464
152. Suchard MA, Weiss RE, Sinsheimer JS (2001) Bayesian selection of continuous-time Markov chain evolutionary models. Mol Biol Evol 18:1001–1013
153. Aris-Brosou S, Yang Z (2002) Effects of models of rate evolution on estimation of divergence dates with special reference to the metazoan 18S ribosomal RNA phylogeny. Syst Biol 51:703–714
154. Nylander JA, Ronquist F, Huelsenbeck JP et al (2004) Bayesian phylogenetic analysis of combined data. Syst Biol 53:47–67

155. Kass RE, Raftery AE (1995) Bayes factors. J Am Stat Assoc 90:773–795
156. Raftery AE (1996) Hypothesis testing and model selection. In: Gilks WR, Richardson S, Spiegelhalter DJ (eds) Markov chain Monte Carlo in practice. Chapman & Hall, London, pp 163–167
157. Minin V, Abdo Z, Joyce P et al (2003) Performance-based selection of likelihood models for phylogenetic estimation. Syst Biol 52:674–683
158. Posada D, Crandall KA (2001) Selecting methods of nucleotide substitution: An application to human immunedeficiency virus 1 (HIV-1). Mol Biol Evol 18:897–906
159. Posada D (2008) jModelTest: phylogenetic model averaging. Mol Biol Evol 25:1253–1256
160. Yang Z (2006) Computational molecular evolution. Oxford University Press, Oxford
161. Yang Z, Kumar S, Nei M (1995) A new method of inference of ancestral nucleotide and amino acid sequences. Genetics 141:1641–1650
162. Susko E, Field C, Blouin C et al (2003) Estimation of rates-across-sites distributions in phylogenetic substitution models. Syst Biol 52:594–603
163. Soubrier J, Steel M, Lee MSY et al (2012) The influence of rate heterogeneity among sites on the time dependence of molecular rates. Mol Biol Evol 29:3345–3358
164. Cox DR (1962) Further results on tests of separate families of hypotheses. J R Stat Soc B 24:406–424
165. Rambaut A, Grassly NC (1997) Seq-Gen: an application for the Monte Carlo simulation of DNA sequence evolution along phylogenetic trees. CABIOS 13:235–238
166. Fletcher W, Yang ZH (2009) INDELible: a flexible simulator of biological sequence evolution. Mol Biol Evol 26:1879–1888
167. Jermiin LS, Ho SYW, Ababneh F et al (2003) *Hetero*: a program to simulate the evolution of DNA on a four-taxon tree. Appl Bioinformatics 2:159–163
168. Felsenstein J (2004) Inferring phylogenies. Sinauer Associates, Sunderland, MA
169. Rokas A, Krüger D, Carroll SB (2005) Animal evolution and the molecular signature of radiations compressed in time. Science 310:1933–1938

Chapter 16

Scaling Up the Phylogenetic Detection of Lateral Gene Transfer Events

Cheong Xin Chan, Robert G. Beiko, and Mark A. Ragan

Abstract

Lateral genetic transfer (LGT) is the process by which genetic material moves between organisms (and viruses) in the biosphere. Among the many approaches developed for the inference of LGT events from DNA sequence data, methods based on the comparison of phylogenetic trees remain the gold standard for many types of problem. Identifying LGT events from sequenced genomes typically involves a series of steps in which homologous sequences are identified and aligned, phylogenetic trees are inferred, and their topologies are compared to identify unexpected or conflicting relationships. These types of approach have been used to elucidate the nature and extent of LGT and its physiological and ecological consequences throughout the Tree of Life. Advances in DNA sequencing technology have led to enormous increases in the number of sequenced genomes, including ultra-deep sampling of specific taxonomic groups and single cell-based sequencing of unculturable "microbial dark matter." Environmental shotgun sequencing enables the study of LGT among organisms that share the same habitat.

This abundance of genomic data offers new opportunities for scientific discovery, but poses two key problems. As ever more genomes are generated, the assembly and annotation of each individual genome receives less scrutiny; and with so many genomes available it is tempting to include them all in a single analysis, but thousands of genomes and millions of genes can overwhelm key algorithms in the analysis pipeline. Identifying LGT events of interest therefore depends on choosing the right dataset, and on algorithms that appropriately balance speed and accuracy given the size and composition of the chosen set of genomes.

Key words Lateral genetic transfer, Horizontal genetic transfer, Phylogenetic analysis, Phylogenomics, Multiple sequence alignment, Orthology

1 Introduction

LGT is now recognized as a major contributor to the evolution and functional diversification of bacteria, archaea, and many eukaryotic microbes. Since the publication of the *Haemophilus influenzae* genome in 1995 [1], the number of sequenced genomes has risen rapidly. Sequenced genomes have been used for the inference of LGT since the first few became available. From these first few genomes it was already clear that gene content variation was

Jonathan M. Keith (ed.), *Bioinformatics: Volume I: Data, Sequence Analysis, and Evolution*, Methods in Molecular Biology, vol. 1525, DOI 10.1007/978-1-4939-6622-6_16, © Springer Science+Business Media New York 2017

large, even among strains of the same species [2], and that LGT was a significant force in the remodeling of microbial genomes [3–5]. The thousands of genomes now available allow for targeted analyses of specific groups including pathogens [6, 7], as well as studies that aim to infer LGT broadly across the Tree of Life, using either all available genomes [8–10] or representative subsets [11]. Initial studies of the human microbiome suggest that rates of LGT may be higher in host-associated settings than in any other environment [12]. Many microbes can regulate their uptake of DNA from the environment, and processes such as biofilm formation [13] and inflammation [14] may induce a massive increase in the rates of LGT.

Detection of LGT events relies on statistical methods to identify DNA sequences either with different properties than the majority of the genome, or with unusual patterns of distribution or relatedness across a set of genomes. Although LGT events need not be delimited by gene [15] or domain [16] boundaries, in many studies the gene (or protein) is the unit of analysis owing to interest in which gene-encoded functions have been acquired via LGT. Initially, DNA of lateral origin bears features of the donor genome (e.g., G+C content, codon usage, or higher order compositional features) which may be anomalous in the new genomic content. Composition-based approaches have been used to identify genes of foreign origin and carry the advantage that inference can be made from a single genome. Phylogenetic approaches have their own distinct advantages: the best build on many decades of theory and practice, employ explicit models of the evolutionary process, are reasonably precise in delineating the lateral region, and can detect events of different ages from very recent to more ancient. Since compositional anomalies eventually "ameliorate" [17] to those of the host genome, phylogenetic methods are needed to detect more-ancient transfer events [8, 18]. However, phylogenetic approaches as typically applied require the inference of orthologous sets of genes or proteins to proceed. All approaches run up against the challenges of overlapping and/or superimposed events, and loss of signal at deeper divergences.

In this chapter, we outline a computational workflow of methods for assembling and aligning sets of putatively orthologous sequences, inferring phylogenetic trees and, by comparisons with a reference topology, identifying *prima facie* instances of LGT. Recent work [10] has highlighted the need for algorithms that scale to tens of thousands of genomes while making use of current best-practice algorithms. All the methods we present here are meant to scale well with increasing dataset size, and many have been applied to thousands and even millions of sequences. The notes not only present further details, but where appropriate also call attention to limitations of these methods and to alternative approaches.

2 Systems, Software, and Databases

2.1 Sources of Data

The inference methods we focus on depend on the availability of sequenced genomes (completed or draft), with gene predictions. We will not review algorithms for sequence assembly or gene prediction here, and refer the reader instead to reviews of these important steps in this book (Chapters 2 and 11) and elsewhere [19–21].

The principal source of sequenced genomes remains the GenBank resource [22]. As of 12 September 2016 the NCBI Genome database lists over 73,600 genomes of bacteria and archaea, from *Abiotrophia defectiva* strain ATCC 49176 to *Zymomonas mobilis* ATCC 29192 in various states of assembly, including ~5800 completed genomes. Also available are over 13,000 viral and plasmid sequences, and about 3500 genomes of microbial eukaryotes. Other microbial eukaryote genomes are also available from the Joint Genome Institute's Genome Portal (http://genome.jgi.doe.gov/). These genomes with associated gene predictions can be acquired in a number of ways, including in bulk using UNIX commands such as 'rsync' and 'wget.' Other resources include the Pathosystems Resource Integration Center [23], which offers function and pathway annotation alongside the sequenced genomes. The Genomes OnLine Database [24] includes standards-compliant metadata, including the type of environment and geographic coordinates of isolation.

2.2 Software Programs

All programs described in this chapter are freely available:

BLAST [25]: http://blast.ncbi.nlm.nih.gov/; alternatives include UBLAST [26] and RAPSearch2 [27]

MCL [28]: http://micans.org/mcl/

MUSCLE [29]: http://www.drive5.com/muscle/

MAFFT [30]: http://mafft.cbrc.jp/alignment/software/

BMGE [31]: https://wiki.gacrc.uga.edu/wiki/BMGE

Gblocks [32]: http://molevol.cmima.csic.es/castresana/Gblocks.html

MrBayes [33]: http://mrbayes.sourceforge.net/

RAxML [34]: http://sco.h-its.org/exelixis/web/software/raxml/index.html

CLANN [35]: http://chriscreevey.github.io/clann/

SPR supertree [36]: http://kiwi.cs.dal.ca/Software/index.php/SPRSupertrees

phytools [37]: http://cran.r-project.org/web/packages/phytools/

3 Methods

The overall approach described in this Chapter is illustrated in Fig. 1.

3.1 Clustering of Sequences into Sets of Putative Homologs and Orthologs

1. All-versus-all BLAST (or the more efficient UBLAST or RAP-Search2) is carried out on the set of predicted proteins from all genomes in the analysis. Each pairwise BLAST with expectation score $e \leq 10^{-3}$ is kept and is normalized by dividing by its self-score, yielding the set of significant edges (by this criterion). If the genomes are closely related (e.g., conspecific), this and subsequent steps should instead be carried out on gene (not protein) sets, with corresponding changes in parameter values where required.
2. This edge set is used as input for Markov clustering using MCL. Validation of MCL on the Protein Data Bank suggested that an inflation parameter (I) of 1.1 is appropriate (*see* **Note 1**). As MCL is memory intensive, it may be necessary to carry out

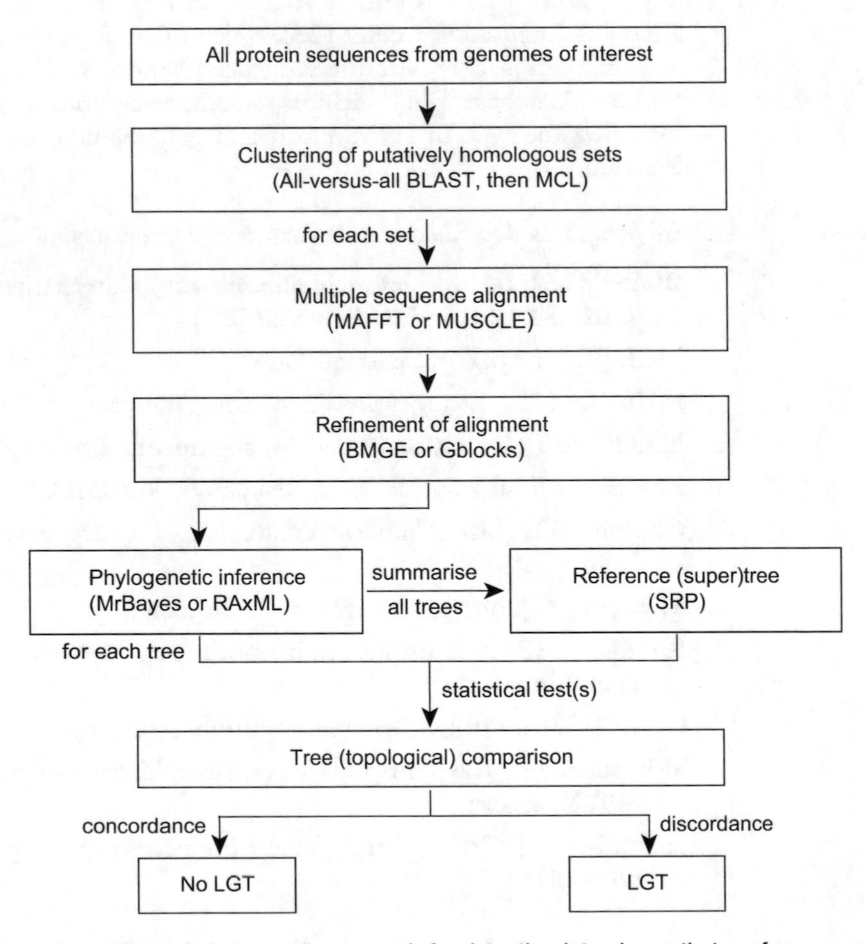

Fig. 1 An overall workflow of phylogenetic approach for detecting lateral genetic transfer

this step on specialized hardware for larger datasets (more than a few hundred thousand sequences). The resulting Markov clusters are interpreted as sets of putatively homologous proteins, from which orthologous relationships can be extracted.

3. Each Markov cluster is then subjected to single-linkage clustering to delineate putatively orthologous subsets. For this purpose, we define a maximally representative cluster (MRC) as the largest subcluster that contains no more than one protein from any given genome [8, 38]. An MRC is formed by lowering the normalized BLAST threshold until the next protein would duplicate a genome already represented in the subcluster, or until all proteins in the Markov cluster have been included. MRCs are interpreted as putative sets of orthologs (orthogroups); those with $\geq$4 proteins are useful for inference of phylogenetic trees, and are taken forward to the next step. Alternatively, for sets of closely related genomes it may be possible to use positional conservation to help delineate putative orthogroups [39].

3.2 Multiple Sequence Alignment

1. Each MRC of protein sequences is kept in FASTA format, the most commonly accepted file format across computational tools for sequence analysis.
2. Next, multiple sequence alignment is performed on each MRC. Many tools are available for this purpose; MUSCLE [29] and MAFFT [30] are the most popular programs to date. Running MUSCLE or MAFFT (mafft-linsi) at default settings should yield good results in most cases. In MUSCLE, the maximum number of iterations during the refinement process (-maxiters option) is 16 by default, and can be increased for sets of very dissimilar (highly divergent) sequences. The aligned sequences are in FASTA format by default. Other tools, e.g., FSA and Clustal Omega, are tailored for very large sequence sets (e.g., >100 sequences).
3. For each alignment, ambiguously aligned regions and ragged ends can be removed using BMGE [31]. The default parameters (including use of the BLOSUM62 model) are likely to be sufficient for most protein sequence alignments. The default output format is PHYLIP sequential, but other formats can be easily specified using the -o option. Another popular program for this purpose is GBlocks [32], but its default settings are too strict for normal use, so it is usually necessary to adjust its parameter settings depending on the number of sequences within a set [40].

3.3 Inference of Trees

1. The number of sequences to be analyzed (overall, and in the largest orthogroups) is an important consideration in determining the algorithmic framework and software selected for

phylogenetic inference. If computational resources are available, for datasets up to 100 sequences we recommend MrBayes [33] (**step 2** later). However, in general Bayesian approaches become computationally infeasible as data size exceeds 100 sequences (*see* **Note 2**). For datasets >100 we recommend RAxML [34], described in **step 3** later.

2. For inference using MrBayes, each trimmed alignment is converted into NEXUS format using Readseq, to which a MrBayes command block is appended. Based on earlier calibration and sensitivity testing of MRCs of gene sequences from prokaryotes [40], we recommend the following settings:
   (a) MCMC run: the minimum values set for key parameters are nchains = 4 ngen = 1000000 samplefreq = 100 burnin = 2500. The default heating parameter at 0.5 is used. In this instance, each MCMC run involves four Markov chains at one million generations each, with sampling frequency of every 100th generation (i.e., a total of 10,000 samples) and burn-in threshold at 2500 samples. For sequence sets of size N, we recommend that the number of replicates (nruns) approximate $N/10$ (i.e., nruns = 3 when $N = 32$; nruns = 2 by default). A burn-in threshold (burnin) of at least 25 % of the total samples (in this example, 2500 of 10,000 total samples) is recommended, to allow convergence of tree likelihoods. The average standard deviation of split frequencies <0.01 in the MrBayes output is a good indication of convergence. A lack of convergence across replicate runs usually suggests that longer runs are needed. This burn-in threshold can be further refined by convergence diagnostics based on stabilization of likelihood values from a given Markov chain [40].
   (b) Priors of tree topology, branch lengths, and proportion of invariant sites: default settings in MrBayes are recommended.
   (c) Prior of substitution model: the choice of substitution model is estimated from observed data (prset aamodelpr = mixed). Among-site rate variation is modeled by a four-category discrete approximation of the gamma distribution [41] (lset rates = gamma ngammacat = 4).
   (d) Sump, post-MCMC analysis: use sump burnin = 2500 to summarize the state of the Markov chain (e.g., average likelihood of trees) among the post-burn-in samples. With this parameter setting, the first 2500 samples are not considered.
   (e) Sumt, post-MCMC analysis: use sumt burnin-2500 to summarize the post-burn-in trees (as earlier, with this

parameter setting the first 2500 sampled trees are not considered). This summary includes posterior probabilities on trees and bipartitions, and average branch lengths. The default consensus type (contype = halfcompat) yields a 50 % majority-rule tree; this tree may not be perfectly bifurcating (i.e., binary). If binary trees are required, contype = allcompat can be used. In MrBayes v3.2 [33], conformat = simple needs to be specified to generate a consensus tree file that can be read in common tree-viewing packages or for downstream scripting purposes.

(f) To ensure that the results have indeed arisen from the data and not from the choice of prior, the analysis could be rerun without data (mcmc data = no) or on a sample of the datasets, after which the resulting MCMC parameters can be compared.

3. For inference using RAxML [34], each trimmed alignment is converted into PHYLIP 3.2 (sequential) format using Readseq (*see* **Note 3**). We recommend the following settings in a single command line for each RAxML run:

   (a) Choice of model: -m PROTGAMMAWAG specifies empirical amino acid substitution model based on Whelan and Goldman [42] and among-site rate variation modeled under a gamma distribution [41].

   (b) Run parameters: input file in PHYLIP format needs to be specified using the -s option, and multithreaded runs can be specified to speed up the run time, e.g., -T 4 will run using four threads.

   (c) Support on bipartitions: In RAxML, this is done using nonparametric bootstrap analysis. The specification "-b 16897 -N 100" turns on nonparametric bootstrapping via a random seed number (here 16897), with 100 bootstrap samples. For a quicker alternative, a rapid bootstrap algorithm can be invoked by specifying "-f a -x 16897 -N 100."

   (d) Output specification: an output suffix needs to be specified using -n. All output files (including the list of trees from bootstrap sampling and the best tree with bootstrap support values) generated will bear this same suffix.

4. Phylogenetic approaches that do not require multiple sequence alignment have recently been proposed. In one major variant, each sequence in the set is represented as a distribution of its component subsequences of a predefined length (known variously as words, *k*-mers or *n*-grams), and a measure of similarity is calculated between each pair of sequences based on the number or value of words that they share in common [43, 44]. In another variant of the approach, pairwise similarity is

based on conserved match length of common or unique substrings [45, 46]. In either case, the pairwise measures of similarity can be treated as a distance and (with or without normalization) used as input into neighbor-joining, e.g., in PHYLIP [47]. *See* [48, 49] for reviews. Alignment-free approaches tend to be highly scalable, and their potential in accurate inference of phylogenetics has recently been demonstrated in a number of biological scenarios [47, 50]. While the "standard" phylogenetic approach is described here, the scalable alternative approach could potentially become a new standard for dealing with large datasets [51].

3.4 Reference Topology

A reference hypothesis of organism or genome relationships is required, for which a supertree brings the advantage of being based on the same dataset. From the set of well-supported bipartitions (Subheading 3.3, **step 3** or **4**), a supertree can be constructed using software such as CLANN [35] or the "phytool" package [37]. Several supertree methods are available including MRP [52, 53], the most popular method to date. We recommend the recently published SPR supertree method [36], which scales well to very large datasets and supports multifurcating nodes in trees; the published SPR supertree tool also allows direct inference of LGT (later). Alternatively, the reference hypothesis might be a tree inferred from well-characterized phylogenetic markers, e.g., 16S ribosomal RNA sequences, from the weighted or unweighted proportions of genes held in common, or from shared physiological or morphological characteristics [40].

3.5 Inference of Lateral Transfer Events via Topological Comparisons of Trees

1. Topological comparisons between the rooted reference tree (Subheading 3.4) and inferred gene- or protein-family trees (Subheading 3.3) are used to infer *prima facie* instances of LGT. Tree files should again be in Newick format, optionally with branch lengths (which are ignored) and statistical support values. Gene trees may or may not be rooted.
2. As with supertree inference, SPR operations are a natural representation for LGT events. EEEP [54] was an early method that attempted to permute the reference tree with successive SPR operations to yield a topology consistent with a given gene tree; however, its permutation approach was memory and time consuming. A better approach is to construct a maximum agreement forest (MAF) by bisecting edges in the reference and gene trees to generate subtrees, until there is no topological disagreement between the forests resulting from the two trees. This approach leads to very efficient algorithms that can operate on even very large trees [55]. The SPRSupertrees software [36] can perform an LGT analysis given only the full set of gene-family trees, in which case it will infer the reference

topology as in Subheading 3.4; or it can additionally be provided with a reference tree as input, in which case it will proceed directly to the inference of LGT events.

3. To analyze LGT affecting only certain taxa or genomes, use the "LGT Analysis" option in SPRSupertrees to summarize inferred transfers for which the recipient taxa are members of the group of interest. This allows the inference of highways of gene sharing among lineages. It is important to recognize that the SPR approach to LGT inference (earlier) will return one most parsimonious solution, i.e., a path comprising a minimal number of LGT events. In many cases, more than one set of paths may be equally good, but SPRsupertrees will identify only one of the candidates.
4. Whereas the lineages of vertical (classical parent-to-offspring) descent that have given rise to genomes can be represented as a tree (e.g., the reference tree), the lateral edges connecting these lineages constitute a network. Densely connected regions of this LGT network (cliques and near-cliques) thus demarcate *genetic exchange communities* [56]. By labeling the nodes (organisms, genomes, environments) and/or edges (inferred number, size or antiquity of transfers; type of vector) we can gain insight into (lateral) biology.

4 Notes

1. The value of the inflation parameter I in MCL [28] is a key contributing factor to the granularity of the resulting clusters. A high inflation value results in less-granular clusters (i.e., many small clusters), as opposed to a low inflation value, which results in less-granular clusters (i.e., fewer but large clusters). While there is no upper limit, the value commonly ranges from 1.1 to 5.0. Generally an inflation value smaller than the default value 2 is suitable for highly divergent dataset (e.g., sequences from taxa across all bacterial groupings), whereas a higher value ($\geq$2) might be more suitable for highly similar dataset (e.g., sequences from different strains of a single species). Please refer to the MCL manual (http://micans.org/mcl/) for more information about cluster validation and finding the optimal inflation value.
2. Bayesian phylogenetic approaches are highly computationally intensive (thus not feasible for large datasets), but Bayesian inference provides trees that are readily interpretable, as it yields the posterior probability of a tree (and a substitution model) given an alignment. We recommend the use of parallelized implementation of MrBayes (the MPI-enabled version) where possible. Other tools for Bayesian phylogenetic inference are

available. ExaBayes [57] is highly parallelizable, but it runs slower than MrBayes when dealing with protein data. Those who prefer a graphics user interface to command line can consider using MrBayes implemented within BEAST [58].

3. In the phylogenetic context, Bayesian approaches give the posterior probability of a distribution of trees, given the data (the alignment) and a model of sequence change. By contrast, ML computes the likelihood of the data given a tree and a substitution model. Nonparametric bootstrapping can be used to indicate support for bipartitions, but the relationship between bootstrap proportions and probability remains elusive. RAxML, the most popular ML tool in phylogenetics, scales to datasets of thousands of sequences by implementing algorithmic shortcuts for topology and subtree support. PhyML [59] and FastTree [60] are faster and even more scalable, but less accurate; we recommend them only if computation power is otherwise limiting.

Acknowledgements

The authors acknowledge the collaboration of Robert Charlebois, Aaron Darling, Tim Harlow, Elizabeth Skippington, Chris Whidden, Simon Wong, and Norbert Zeh. CXC is supported by a University of Queensland Early Career Researcher Grant. RGB acknowledges the support of the Canada Research Chairs program. The SPR supertree work was supported by the Canadian Natural Sciences and Engineering Research Council, the Dalhousie Killam Trusts, and the Canadian Institutes for Health Research. MAR acknowledges support from the Australian Research Council, the James S. McDonnell Foundation, and The University of Queensland.

References

1. Fleischmann RD, Adams MD, White O et al (1995) Whole-genome random sequencing and assembly of *Haemophilus influenzae* Rd. Science 269:496–512
2. Welch RA, Burland V, Plunkett G et al (2002) Extensive mosaic structure revealed by the complete genome sequence of uropathogenic *Escherichia coli*. Proc Natl Acad Sci U S A 99:17020–17024
3. Gogarten JP, Townsend JP (2005) Horizontal gene transfer, genome innovation and evolution. Nat Rev Microbiol 3:679–687
4. Keeling PJ, Palmer JD (2008) Horizontal gene transfer in eukaryotic evolution. Nat Rev Genet 9:605–618
5. Ochman H, Lawrence JG, Groisman EA (2000) Lateral gene transfer and the nature of bacterial innovation. Nature 405:299–304
6. Chan CX, Beiko RG, Ragan MA (2011) Lateral transfer of genes and gene fragments in *Staphylococcus* extends beyond mobile elements. J Bacteriol 193:3964–3977
7. Young BC, Golubchik T, Batty EM et al (2012) Evolutionary dynamics of *Staphylococcus aureus* during progression from carriage to disease. Proc Natl Acad Sci U S A 109:4550–4555
8. Beiko RG, Harlow TJ, Ragan MA (2005) Highways of gene sharing in prokaryotes. Proc Natl Acad Sci U S A 102:14332–14337

9. Puigbò P, Wolf YI, Koonin EV (2010) The tree and net components of prokaryote evolution. Genome Biol Evol 2:745–756
10. Beiko RG (2011) Telling the whole story in a 10,000-genome world. Biol Direct 6:34
11. Yutin N, Puigbò P, Koonin EV et al (2012) Phylogenomics of prokaryotic ribosomal proteins. PLoS One 7:e36972
12. Smillie CS, Smith MB, Friedman J et al (2011) Ecology drives a global network of gene exchange connecting the human microbiome. Nature 480:241–244
13. Ehrlich GD, Ahmed A, Earl J et al (2010) The distributed genome hypothesis as a rubric for understanding evolution in situ during chronic bacterial biofilm infectious processes. FEMS Immunol Med Microbiol 59:269–279
14. Stecher B, Denzler R, Maier L et al (2012) Gut inflammation can boost horizontal gene transfer between pathogenic and commensal Enterobacteriaceae. Proc Natl Acad Sci U S A 109:1269–1274
15. Chan CX, Beiko RG, Darling AE et al (2009) Lateral transfer of genes and gene fragments in prokaryotes. Genome Biol Evol 1:429–438
16. Chan CX, Darling AE, Beiko RG et al (2009) Are protein domains modules of lateral genetic transfer? PLoS One 4:e4524
17. Lawrence JG, Ochman H (1997) Amelioration of bacterial genomes: rates of change and exchange. J Mol Evol 44:383–397
18. Ragan MA, Harlow TJ, Beiko RG (2006) Do different surrogate methods detect lateral genetic transfer events of different relative ages? Trends Microbiol 14:4–8
19. Stein L (2001) Genome annotation: from sequence to biology. Nat Rev Genet 2:493–503
20. El-Metwally S, Hamza T, Zakaria M et al (2013) Next-generation sequence assembly: four stages of data processing and computational challenges. PLoS Comput Biol 9: e1003345
21. Richardson EJ, Watson M (2013) The automatic annotation of bacterial genomes. Brief Bioinform 14:1–12
22. Benson DA, Cavanaugh M, Clark K et al (2013) GenBank. Nucleic Acids Res 41: D36–D42
23. Wattam AR, Abraham D, Dalay O et al (2014) PATRIC, the bacterial bioinformatics database and analysis resource. Nucleic Acids Res 42: D581–D591
24. Pagani I, Liolios K, Jansson J et al (2012) The Genomes OnLine Database (GOLD) v. 4: status of genomic and metagenomic projects and their associated metadata. Nucleic Acids Res 40:D571–D579
25. Camacho C, Coulouris G, Avagyan V et al (2009) BLAST+: architecture and applications. BMC Bioinformatics 10:421
26. Edgar RC (2010) Search and clustering orders of magnitude faster than BLAST. Bioinformatics 26:2460–2461
27. Zhao Y, Tang H, Ye Y (2012) RAPSearch2: a fast and memory-efficient protein similarity search tool for next-generation sequencing data. Bioinformatics 28:125–126
28. Enright AJ, Van Dongen S, Ouzounis CA (2002) An efficient algorithm for large-scale detection of protein families. Nucleic Acids Res 30:1575–1584
29. Edgar RC (2004) MUSCLE: multiple sequence alignment with high accuracy and high throughput. Nucleic Acids Res 32:1792–1797
30. Katoh K, Standley DM (2013) MAFFT multiple sequence alignment software version 7: improvements in performance and usability. Mol Biol Evol 30:772–780
31. Criscuolo A, Gribaldo S (2010) BMGE (Block Mapping and Gathering with Entropy): a new software for selection of phylogenetic informative regions from multiple sequence alignments. BMC Evol Biol 10:210
32. Talavera G, Castresana J (2007) Improvement of phylogenies after removing divergent and ambiguously aligned blocks from protein sequence alignments. Syst Biol 56:564–577
33. Ronquist F, Teslenko M, van der Mark P et al (2012) MrBayes 3.2: efficient Bayesian phylogenetic inference and model choice across a large model space. Syst Biol 61:539–542
34. Stamatakis A (2014) RAxML version 8: a tool for phylogenetic analysis and post-analysis of large phylogenies. Bioinformatics 30:1312–1313
35. Creevey CJ, McInerney JO (2005) CLANN: investigating phylogenetic information through supertree analyses. Bioinformatics 21:390–392
36. Whidden C, Zeh N, Beiko RG (2014) Supertrees based on the subtree prune-and-regraft distance. Syst Biol 63:566–581
37. Revell LJ (2012) phytools: an R package for phylogenetic comparative biology (and other things). Methods Ecol Evol 3:217–223
38. Harlow TJ, Gogarten JP, Ragan MA (2004) A hybrid clustering approach to recognition of protein families in 114 microbial genomes. BMC Bioinformatics 5:45

39. Skippington E, Ragan MA (2011) Within-species lateral genetic transfer and the evolution of transcriptional regulation in *Escherichia coli* and *Shigella*. BMC Genomics 12:532
40. Beiko RG, Ragan MA (2008) Detecting lateral genetic transfer: a phylogenetic approach. Methods Mol Biol 452:457–469
41. Yang Z (1994) Estimating the pattern of nucleotide substitution. J Mol Evol 39:105–111
42. Whelan S, Goldman N (2001) A general empirical model of protein evolution derived from multiple protein families using a maximum-likelihood approach. Mol Biol Evol 18:691–699
43. Reinert G, Chew D, Sun F et al (2009) Alignment-free sequence comparison (I): statistics and power. J Comput Biol 16:1615–1634
44. Wan L, Reinert G, Sun F et al (2010) Alignment-free sequence comparison (II): theoretical power of comparison statistics. J Comput Biol 17:1467–1490
45. Ulitsky I, Burstein D, Tuller T et al (2006) The average common substring approach to phylogenomic reconstruction. J Comput Biol 13:336–350
46. Domazet-Lošo M, Haubold B (2009) Efficient estimation of pairwise distances between genomes. Bioinformatics 25:3221–3227
47. Chan CX, Bernard G, Poirion O et al (2014) Inferring phylogenies of evolving sequences without multiple sequence alignment. Sci Rep 4:6504
48. Bonham-Carter O, Steele J, Bastola D (2013) Alignment-free genetic sequence comparisons: a review of recent approaches by word analysis. Brief Bioinform 15:890–905
49. Haubold B (2014) Alignment-free phylogenetics and population genetics. Brief Bioinform 15:407–418
50. Ragan MA, Bernard G, Chan CX (2014) Molecular phylogenetics before sequences: oligonucleotide catalogs as *k*-mer spectra. RNA Biol 11:176–185
51. Chan CX, Ragan MA (2013) Next-generation phylogenomics. Biol Direct 8:3
52. Baum BR (1992) Combining trees as a way of combining data sets for phylogenetic inference, and the desirability of combining gene trees. Taxon 41:3–10
53. Ragan MA (1992) Phylogenetic inference based on matrix representation of trees. Mol Phylogenet Evol 1:53–58
54. Beiko RG, Hamilton N (2006) Phylogenetic identification of lateral genetic transfer events. BMC Evol Biol 6:15
55. Whidden C, Beiko R, Zeh N (2013) Fixed-parameter algorithms for maximum agreement forests. SIAM J Comput 42:1431–1466
56. Skippington E, Ragan MA (2011) Lateral genetic transfer and the construction of genetic exchange communities. FEMS Microbiol Rev 35:707–735
57. Aberer AJ, Kobert K, Stamatakis A (2014) ExaBayes: massively parallel Bayesian tree inference for the whole-genome era. Mol Biol Evol 31 (10):2553–2556
58. Drummond AJ, Suchard MA, Xie D et al (2012) Bayesian phylogenetics with BEAUti and the BEAST 1.7. Mol Biol Evol 29:1969–1973
59. Guindon S, Dufayard JF, Lefort V et al (2010) New algorithms and methods to estimate maximum-likelihood phylogenies: assessing the performance of PhyML 3.0. Syst Biol 59:307–321
60. Price MN, Dehal PS, Arkin AP (2010) FastTree 2—approximately maximum-likelihood trees for large alignments. PLoS One 5:e9490

Chapter 17

Detecting and Analyzing Genetic Recombination Using RDP4

Darren P. Martin, Ben Murrell, Arjun Khoosal, and Brejnev Muhire

Abstract

Recombination between nucleotide sequences is a major process influencing the evolution of most species on Earth. The evolutionary value of recombination has been widely debated and so too has its influence on evolutionary analysis methods that assume nucleotide sequences replicate without recombining. When nucleic acids recombine, the evolution of the daughter or recombinant molecule cannot be accurately described by a single phylogeny. This simple fact can seriously undermine the accuracy of any phylogenetics-based analytical approach which assumes that the evolutionary history of a set of recombining sequences can be adequately described by a single phylogenetic tree. There are presently a large number of available methods and associated computer programs for analyzing and characterizing recombination in various classes of nucleotide sequence datasets. Here we examine the use of some of these methods to derive and test recombination hypotheses using multiple sequence alignments.

Key words Recombination, Gene conversion, Breakpoints, Phylogenetic trees, RDP4

1 Introduction

Many methods have been developed for the detection and analysis of recombination in nucleotide sequence data (for a reasonably comprehensive list see http://www.bioinf.manchester.ac.uk/recombination/programs.shtml). While most of these methods provide some statistical indication of how much evidence of recombination is present within a group of nucleotide sequences, some will also infer likely recombination breakpoint positions, identify recombinants and their possible parental sequences, or estimate recombination rates [1].

Detecting recombination is often very straightforward. If, for example, three nucleotide sequences are considered, two of them will generally be more closely related to one another than either is to the third. In the absence of recombination one would expect these relationships to be maintained continuously across the lengths of the three sequences. To detect recombination all that is

Jonathan M. Keith (ed.), *Bioinformatics: Volume I: Data, Sequence Analysis, and Evolution*, Methods in Molecular Biology, vol. 1525, DOI 10.1007/978-1-4939-6622-6_17, © Springer Science+Business Media New York 2017

required is a means of identifying situations where this expectation is provably wrong.

A multitude of approaches may be adopted to analyze and characterize recombination. These range from the very simple but fast "heuristic" pattern recognition methods, to the immensely sophisticated but more computationally expensive fully "parametric" methods.

Given a sample of aligned nucleotide sequences, one would ideally like to use a computer program that will accurately indicate whether there is any evidence of recombination in these sequences and, if such evidence exists, provide the exact recombination history of every nucleotide in every sequence. Unfortunately none of the available recombination analysis programs are capable of accomplishing this. It is also very improbable that any exact recombination history could ever be reliably inferred from any but the simplest datasets. Given a group of recombining nucleotide sequences the most comprehensive of attempts at determining their recombination history would yield a set of recombination hypotheses that parsimoniously describe the series of events needed to account for all recombination signals detectable in the sequences.

The aim of this chapter is to convey how the computer program RDP4 (available from http://web.cbio.uct.ac.za/~darren/rdp.html) [2] can be productively used to:

1. Identify within multiple sequence alignments the most obvious recombinant sequences and the approximate locations of their recombination breakpoints.
2. Produce recombination-free datasets for downstream analyses.
3. Identify whether factors such as genome organization or selection against misfolded chimeric proteins have detectably influenced recombination breakpoint patterns.

2 Program Usage

2.1 Data Files

RDP4 will accept nucleotide sequence alignments in many common formats including FASTA, CLUSTAL, MEGA, NEXUS, GDE, PHYLIP, and DNAMAN. Although the program performs optimally on alignments containing between 4 and 1000 sequences of up to 50 kb in length, it can also be used to analyze between 4 and 100 sequences of up to 5 Mb in length.

A wide variety of datasets can be productively analyzed for recombination providing care is taken during their assembly. The optimal size of a dataset depends on the degree of sequence diversity present therein. As recombination can only be detected if it occurs in sequences that are not identical to one another, it is important that datasets intended for recombination analysis contain enough diversity. For datasets containing sequences with very low degrees of diversity, increasing the lengths of the sequences can increase the number of variable nucleotide sites that inform recombination detection.

It is inadvisable to simply construct the largest dataset possible. It is particularly unwise to analyze datasets containing sequences that cannot be accurately aligned, since misaligned tracts of sequence will very frequently yield spurious evidence of recombination. Certain datasets may be too divergent to accurately align (i.e., some of the sequences might share pairwise identities <70 %). In such cases, a computer program such as SDT (available from http://web.cbio.uct.ac.za/~brejnev/ and distributed with RDP4) [3] can be used to split these into sequence subsets where each subset contains only sequences that share a specified degree of similarity. Since most available sequence alignment programs will at least partially misalign sequences that are less than ~80 % similar, it is advisable to manually edit the produced sequence alignments to prepare these for recombination analysis. IMPALE (available from http://web.cbio.uct.ac.za/~arjun and distributed with RDP4) is an alignment editor which has been engineered to streamline this process.

Another reason for not simply choosing the largest possible dataset for a recombination analysis is that exploratory searches for recombination signals require repetitive statistical testing, with the number of tests performed increasing both cubically with the number, and linearly with the lengths of sequences being examined. A multiple testing correction is therefore absolutely required to prevent false inference of recombination. Unfortunately, guarding against false positives often means discarding some evidence of genuine recombination. RDP4 applies a Bonferroni multiple testing correction which gets cubically more severe with the numbers of sequences being analyzed. What this means is that at a point dependent on the diversity of the sequences under analysis, the extra recombination signals that will potentially be detectable by increasing the number of sequences in a dataset will be outweighed by the increasing severity of multiple testing correction needed to guard against false positives.

2.2 Program Settings

Before screening a nucleotide sequence alignment for recombination it may be advisable to adjust various program settings that can influence how RDP4 will search for and analyze recombination signals. All of the program's settings can be changed using the "Options" button at the top of the main program window (Fig. 1).

Under the "General" tab in the "Analyze Sequences Using:" section (Fig. 2), it is possible to select the methods that will be used to detect recombination. In most cases, however, it is advisable to use the default selections (RDP, GENECONV, and MAXCHI)

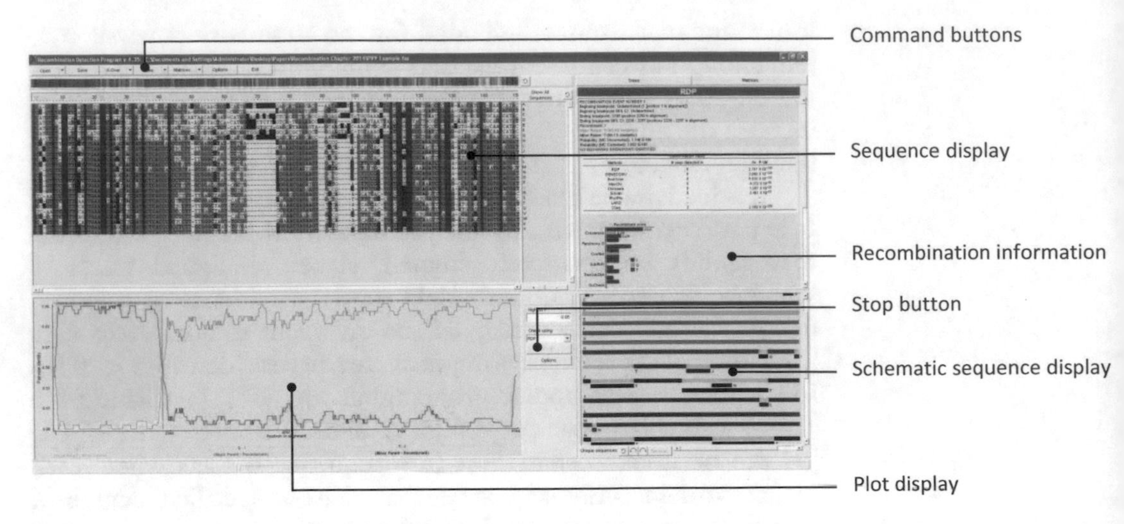

Fig. 1 Important components of the RDP4 interface

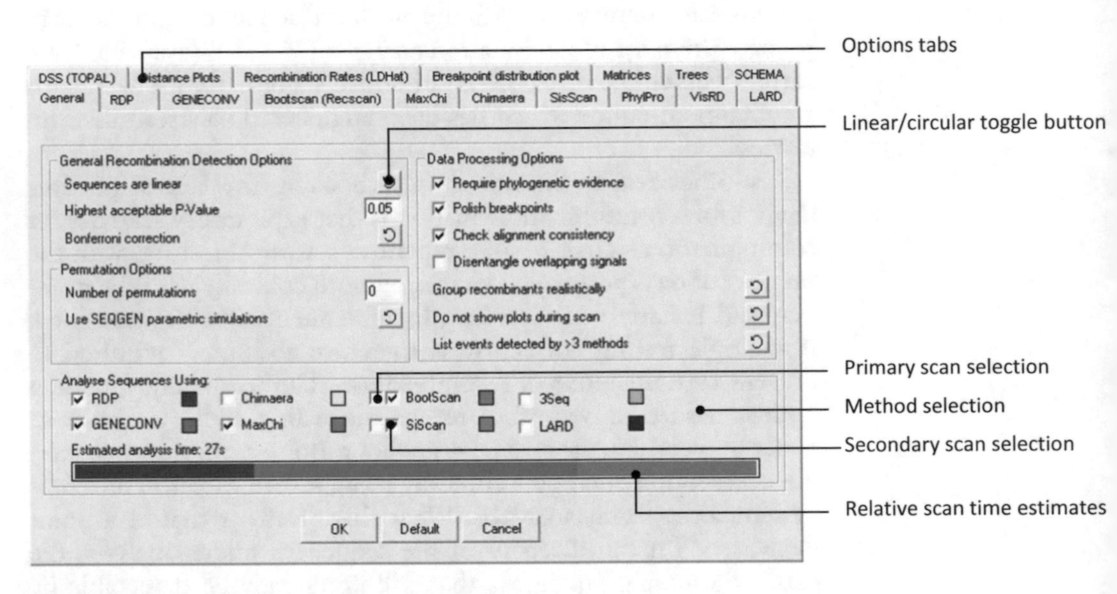

Fig. 2 The General settings section of the Options window

[4–6] since these provide the best balance between detection power, analysis speed, accurate breakpoint inference, and accurate recombinant sequence identification (Fig. 3). If, however, small datasets (<50 sequences) are being analyzed and/or analysis time is not an issue, then additional methods (such as CHIMAERA, RECSCAN, 3SEQ, and SISCAN) [9–12] could be added. Note that although the names "RECSCAN" and "BOOTSCAN" are used interchangeably in the RDP4 interface, we will exclusively refer here to the RDP4 version of the "BOOTSCAN" method as "RECSCAN" to differentiate it from the various other versions of this method that are used in other computer programs. Note also

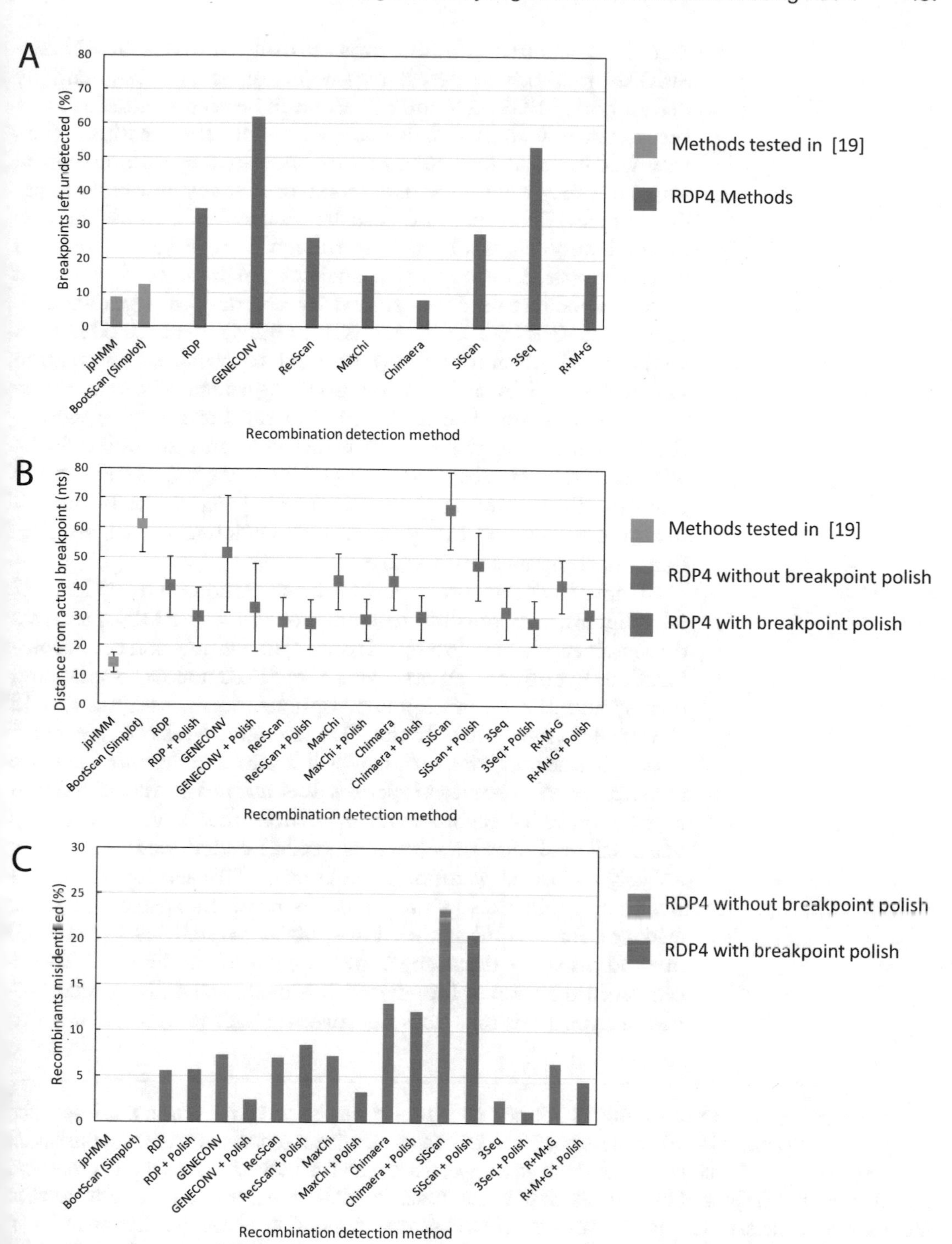

Fig. 3 The inference power and accuracy of various recombination detection methods. Seven of the recombination detection methods implemented in RDP4 were compared individually (RDP, GENECONV, RECSCAN, MAXCHI, CHIMAERA, SISCAN, and 3SEQ) and in combination (R + G + M referring to the RDP, GENECOV, and MAXCHI methods used in primary screening mode and all of the methods used in secondary screening mode as per the RDP4 default settings) with that of both the jpHMM [7] method and the Simplot

that there are two possible ways of using the RECSCAN and SISCAN methods to detect recombination in an alignment. By default both RECSCAN and SISCAN will be used to automatically check recombination signals detected by all other methods (i.e., they will be used for secondary or confirmatory recombination scans) but they will not be used to explore for any new recombination signals. These methods can be selected to explore for new recombination signals by ticking the left box beside the method name (be warned though that analyses can become very slow if these methods are used for exploratory screening of large datasets). The RDP, GENECONV, MAXCHI, 3SEQ, and CHIMAERA methods will all automatically be used to check recombination signals detected by all other methods regardless of whether they are selected or not. The LARD [13] method can only be used to check signals detected by other methods and should only be selected if datasets are very small (<20 sequences). A very rough estimate of the anticipated analysis time is given at the bottom of the options menu so that you can judge whether particular selections are computationally viable.

Under the "Data processing options" section on the "General" tab (Fig. 2), it is possible to configure the way RDP4 processes detectable recombination signals during its formulation of a recombination hypothesis. Apart from the "Disentangle overlapping events" and "Polish breakpoints" options, default settings should almost always be used. If the "Disentangle overlapping events" option is selected, the program will attempt to ensure that the recombination hypothesis it derives does not invoke recombination between pairs of recombinant sequences that have breakpoints which fall at similar sites (such as would be identified as relatively unlikely reciprocal recombination events). This setting works well when recombination in the dataset is relatively sparse and some evidence for recombination hot-spots is present. However, the method used to disentangle overlapping recombination events can get into a circular loop where it is unable to derive a recombination hypothesis that does not involve reciprocal recombination.

Fig. 3 (Continued) version of the BOOTSCAN method [8] using simulated HIV recombinant datasets and analysis results published in [7]. (**a**) Recombination detection power (lower scores are better). (**b**) Breakpoint inference accuracy without (in *blue*) and with (in *orange*) the "Polish breakpoints" setting (lower scores are better). Note that the jpHMM method has close to the maximum accuracy achievable for a recombination breakpoint site inference test. (**c**) The accuracy of recombinant identification without (in *blue*) and with (in *orange*) the "Polish breakpoints" setting (lower is better). Note that the jpHMM method and the Simplot version of the BOOTSCAN method were used to screen known simulated recombinants against a set of known non-recombinant reference sequences and therefore could not be directly compared to the RDP4 methods with respect to recombinant identification accuracy. For almost all of the methods in RDP4 the "Polish breakpoints" setting has a large positive impact on the accuracy with which both breakpoint sites are inferred, and recombinant sequences are identified

It is therefore recommended that the "Disentangle overlapping events" setting should be tried first. If the analysis appears to get stuck (i.e., it is taking much longer than the predicted analysis time), it should be restarted without this setting.

The "Polish breakpoints" setting can increase the run time threefold. In order to have RDP4 produce results in a reasonable time, it might be necessary to disable "Polish breakpoints" when analyzing some large datasets (i.e., those with either >800 sequences or sequences that are >100 kb in length). It is, however, advisable that this setting be left on whenever possible since it has a substantial impact on the accuracy with which RDP4 identifies both recombination breakpoints and recombinant sequences (Fig. 3).

For the method-specific options, the only settings that should ever be changed from their default values are window and step sizes. The optimal window size will vary slightly from method to method and from dataset to dataset. It is important to note that whereas the RDP, GENECONV, MAXCHI, and CHIMAERA methods only examine variable nucleotide positions in triplets of sequences that are sampled from the alignment, the RECSCAN, SISCAN, and LARD methods examine all variable and conserved positions. Be aware (a) that window sizes used for the CHIMAERA and MAXCHI methods should be approximately twice as large as those used for the RDP method, and (b) that window sizes used for the SISCAN and RECSCAN methods should be approximately the same. Ideally window sizes should be small enough to ensure that recombination events involving exchanges of small tracts of sequence (<200 bp) will be detectable in the most divergent of the sequences being examined. The optimal window size to detect a recombination event involving a 200 bp sequence transfer will be 200 bp for RECSCAN and SISCAN and will vary for the other methods depending on the number of nucleotide differences between the parental and the recombinant sequences. The MAXCHI and CHIMAERA methods can be set to run with a variable window size that will get bigger and smaller with lower and higher degrees of parental sequence divergence, respectively. Although window size settings can have an impact on the preliminary recombination hypothesis formulated during the automated recombination signal screening stage of an analysis, the subsequent (and usually necessary) phase of manually testing and refining analysis results should largely counteract any "settings biases" that occur during the automated screening stage.

2.3 Producing a Preliminary Recombination Hypothesis

Once analysis settings have been selected and a multiple sequence alignment has been loaded in RDP4, an automated exploratory search for recombination signals can be carried out by pressing the "X-Over" button (Fig. 1). Note that the automated exploratory search consists of two main phases: the first involving the detection of recombination signals in the alignment and the second involving inference of the numbers and characteristics of the unique recombination events that generated these signals. If an automated search for recombination is taking far longer than anticipated, the "STOP" button can be pressed to terminate the search (Fig. 1). When the STOP button is pressed, analysis results will be given up till the point where the search was stopped. From this point it is possible to continue the analysis, either by restarting the automated recombination scan with different settings, or by proceeding to the manual examination phase of the analysis that is outlined in Subheading 2.5.

2.4 Making a Recombination-Free Dataset

When an automated analysis has been either terminated or run to completion, a set of colored blocks will be presented in the "Schematic sequence display" on the bottom right panel of the program (Fig. 1). These blocks graphically represent the recombination events that RDP4 has detected and characterized. For each sequence in the dataset, the name of the sequence and a colored strip are displayed. Beneath some of these strips (and corresponding to lightened sections of the colored strips) are a series of colored blocks. Each of these blocks represents a proposed recombination event. If the mouse pointer is moved over any of these blocks, information relating to the represented recombination event will be displayed in the "recombination information display" on the top right panel of the screen (Fig. 1). This information includes:

1. Possible recombination breakpoints.
2. Names of sequences in the dataset that are most closely related to the parents of the recombinant sequence.
3. The approximate probability that the apparent recombination signal arose due to convergent mutation rather than recombination.
4. The number of sequences in the dataset carrying similar recombination signals (in the "Confirmation table").
5. A bar graph showing evidence used by the program to identify the recombinant.

The most important bit of information displayed here is, however, the series of warnings that the program gives in capitalized red letters. These will indicate when RDP4 is reasonably unsure about some of the conclusions it has reached. The program will issue a warning if (a) one or both of the inferred breakpoint positions are probably inaccurate, (b) the wrong sequence may have been identified as the recombinant, (c) it is possible that an alignment error has generated a false positive recombination signal, (d) there is only one sequence in the dataset resembling one of the recombinant's parental sequences, and (e) only trace evidence of a recombination event is present within the currently specified sequence (*see* **Note 1**).

At this point various types of recombination-free nucleotide sequence alignments can be saved by simply right clicking on the sequence display (Fig. 1) and selecting the appropriate option. These options include alignments with recombinant sequences completely removed, alignments with only the tracts of sequence between recombination breakpoints removed (and replaced by gap characters), distributed alignments where the recombinant sequences are split up into their component parts, and alignments split into multiple subalignments at the detected breakpoint positions.

If all that is needed is a reasonably recombination-free dataset that is suitable for some other downstream purpose, then saving a recombination-free dataset will be the end point of the analysis and there will be no need to go any further. If, however, inference of the actual patterns of recombination underlying the sequences in a dataset is desired, then it is recommended that further manual checking of the analysis results be carried out (*see* Subheadings 2.5–2.10).

2.5 Navigating Through the Analysis Results

The automated output given by RDP4 is nothing more than a preliminary hypothesis describing a small fraction of the recombination events that have occurred during the evolutionary histories of the sequences being analyzed. It is very important to understand that the program is fallible. The program's failures will be of three major types:

1. inaccurate identification of recombination breakpoint positions;
2. incorrect identification of parental sequences as recombinants; and
3. incorrect identification of groups of recombinants that have descended from the same recombinant ancestor.

Unfortunately there are no automated tools in RDP4 that will reliably indicate whether the preliminary results that it yields contain these errors. It is very likely that, unless the initial automated RDP4 results indicate that there only a few recombinant sequences (<20 % of the sequences in the dataset are recombinants), the program will have made some mistakes interpreting the patterns of recombination it has detected. The size and importance of the mistakes will scale with the number of unique recombination events the program detects. It is especially important to understand that mistakes made early on in an analysis (such as in the first 10 % of unique recombination events the program characterizes) will be more impactful than those made in the end stages of an analysis. This arises because RDP4 identifies and characterizes the easiest to detect recombination events first and leaves the interpretation of the least obvious recombination signals until last. Once, for example, a mistake has been made identifying which of the sequences is recombinant for a specific recombination event, the probability that the program will make additional mistakes of this type during the characterization of all subsequent recombination events will be increased.

If the accurate characterization of historical recombination events is desired, it is strongly recommended that manual refinements are made to the recombination hypotheses produced by RDP4. It is specifically advised that these manual refinements are implemented in a specifically ordered pattern, one recombination event at a time. As mistakes early on in the analysis are likely to be more consequential than mistakes toward the end, refinements should be made to recombination events in precisely the same order as that of their identification by RDP4 during the formulation of its preliminary recombination hypotheses. To aid in this process, all of the unique recombination events that are detected by RDP4 within a particular dataset will be numbered in order from the first to the last that the program characterized.

There are a number of ways in which to navigate in RDP4 from one recombination event to the next. Left clicking on a background region of the schematic sequence display will enable ordered navigation through the detected recombination events using the "Pg Up" and "Pg Dn" keys. Alternatively, clicking on the left and right arrows at the bottom of the schematic sequence display will move backwards and forwards through the list of detected recombination events. It is also possible to go straight to any particular recombination event simply by right clicking on the background of the schematic sequence display and selecting the "Go to event > Select number" option.

For each successive recombination event that is examined, a graph of the clearest recombination signal used to detect that recombination event will be presented in the "plot display" on the bottom left panel, while other information relevant to this same event will be presented in the "recombination information display" in the top right panel (Fig. 1).

2.6 Checking the Accuracy of Breakpoint Identification

Graphs in the plot display (Fig. 1) are useful for checking the accuracy of recombination breakpoint estimation. Light and dark-gray shaded areas of the graphs, respectively, indicate the 99 % and 95 % confidence intervals of breakpoint locations as determined by the BURT method (Breakpoint Uncertainty in Recombined Triplets). While the positions of breakpoints are indicated by intersecting lines in SISCAN, RDP, and RECSCAN plots, they are instead indicated by peaks in MAXCHI (see plot display in Fig. 4) and CHIMAERA plots, and conversely are represented as alternating peaks and troughs in 3SEQ plots. If the breakpoint positions that are indicated by different methods do not match, this will imply that there is a fair degree of uncertainty regarding the position of the breakpoints. The various recombination detection

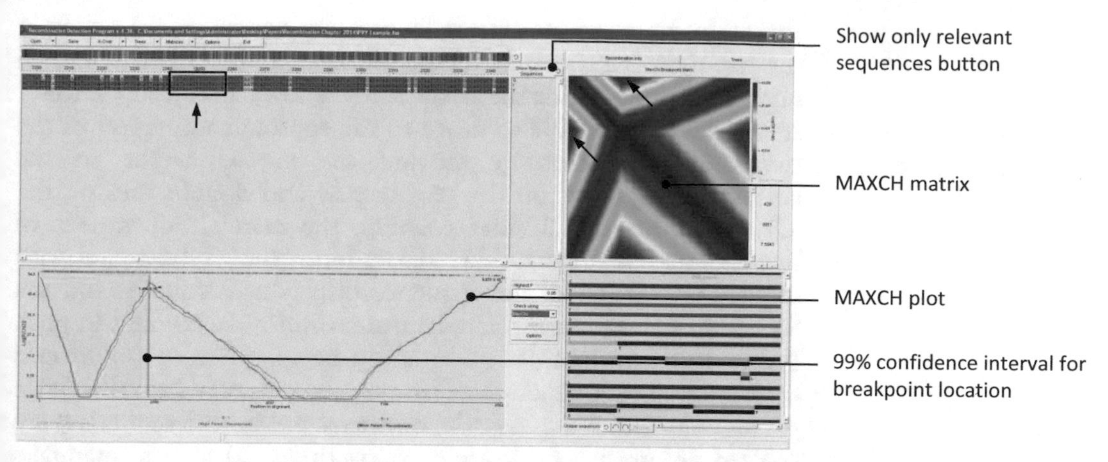

Fig. 4 Various tools for checking the accuracy of breakpoint prediction. The peaks of MAXCHI plots (in the *bottom left panel*) should coincide with breakpoint positions (indicated by the *left* and *right bounds* of the *pink area*). Breakpoint sites should also fall within the *gray area* (indicating the 99 % confidence interval of the breakpoint locations as determined by the BURT method). A MAXCHI matrix (*top right panel*) is similar to a MAXCHI plot but it expresses maximum Chi square *p*-values associated with every possible pair of breakpoints. The "peaks" in the matrix are the *dark red regions. Arrows* on the matrix indicate peaks corresponding with the pair of breakpoints identified by RDP4. The patterns of polymorphic sites in a sequence triplet (*top left panel*) can help indicate the range of invariant nucleotide sites where a breakpoint position likely occurs (grayed-out sites that are indicated by the *black box*). The site identified as the breakpoint position is indicated by a *vertical arrow*

methods implemented in RDP4 do not all have the same inferential accuracy and power in the identification of the precise recombination breakpoint sites. Although the MAXCHI and CHIMAERA methods have the highest overall power for detecting the presence of recombination (i.e., whether recombination breakpoints are present or not), the best methods for pinpointing the precise locations of recombination breakpoints are the RECSCAN, 3SEQ, MAXCHI, and CHIMAERA methods (Fig. 3).

Breakpoints very often occur in pairs and when the sequences being analyzed are circular they are always paired. MAXCHI and LARD matrices may be used to identify the locations of such paired breakpoints and display their relative statistical supports graphically (Fig. 4). To view a MAXCHI matrix, press the matrices button on the top right-hand panel (Fig. 1). Move the mouse pointer into the matrix window and press the left mouse button. Select the MAXCHI matrix option from the menu that appears (Fig. 4). Interpretation of MAXCHI and LARD matrices is relatively simple in that the most probable (although not necessarily correct) breakpoint pairs correspond with matrix cells that have the best associated *p*-values.

If the positions of one or both of the breakpoints need to be adjusted, this can be done via the sequence display panel (Fig. 1). First though, the "show relevant sequences" version of this display should be activated by repeatedly pressing on the curled arrow in the top right-hand corner of the sequence display panel (Fig. 4) until the caption beside the arrow reads "show relevant sequences." It will then be possible to move to the approximate region of the suspected breakpoint by moving the mouse cursor to the corresponding point on the plot display and double clicking the left mouse button. Besides enabling the manual adjustment of breakpoints at nucleotide resolution, the "show relevant sequences" version of the sequence display uses color to aid the accuracy of this process. Invariant nucleotides are colored in gray. Polymorphic nucleotides in common between recombinants and their inferred major and minor parental sequences (where the major parent contributes the majority of the recombinant's sequence) are colored in green and purple, respectively. Ideally, a manually adjusted breakpoint position should be placed between two variable nucleotides, one of which indicates a closer relationship to the first parent and the other indicating a closer relationship to the second. A breakpoint can be placed at a particular nucleotide position by pointing the mouse cursor at the nucleotide and pressing the right mouse button. On the menu that appears, two of the options will involve placing either a "beginning" (i.e., left or 5′) or an "ending" (i.e., right or 3′) breakpoint at this position. Selecting one of these options will adjust the position of the relevant recombination breakpoint in all the sequences carrying evidence of it.

2.7 Checking the Accuracy of Recombinant Sequence Identification

The next thing to consider for a particular detected recombination event is whether RDP4 has correctly identified the recombinant sequence. This can be very difficult to assess. The program uses a range of phylogenetic and genetic distance-based tests to infer which of the three sequences used to detect a recombination signal is the recombinant. Very often different tests will indicate that different sequences are most likely recombinant. RDP4 therefore uses a weighted consensus of these tests when it automatically identifies recombinant sequences (*see* **Note 2**). The results of these tests are displayed, together with the weighted consensus, as a series of bar graphs in the "recombination information display" (Fig. 1) on the top left panel. RDP4 will display a warning if the tests do not clearly indicate which sequence is recombinant. This warning should not be disregarded; time and care should be afforded to determine whether the recombinant has been misidentified as one of the suggested parental sequences.

The best way to assess whether a recombinant sequence has been correctly identified is to compare phylogenetic trees constructed from the portion of the alignment between the inferred breakpoints with those constructed from the remainder of the alignment. By default RDP4 will automatically construct UPGMA trees for each of the two sections of the alignment whenever a particular recombination event is selected for more detailed

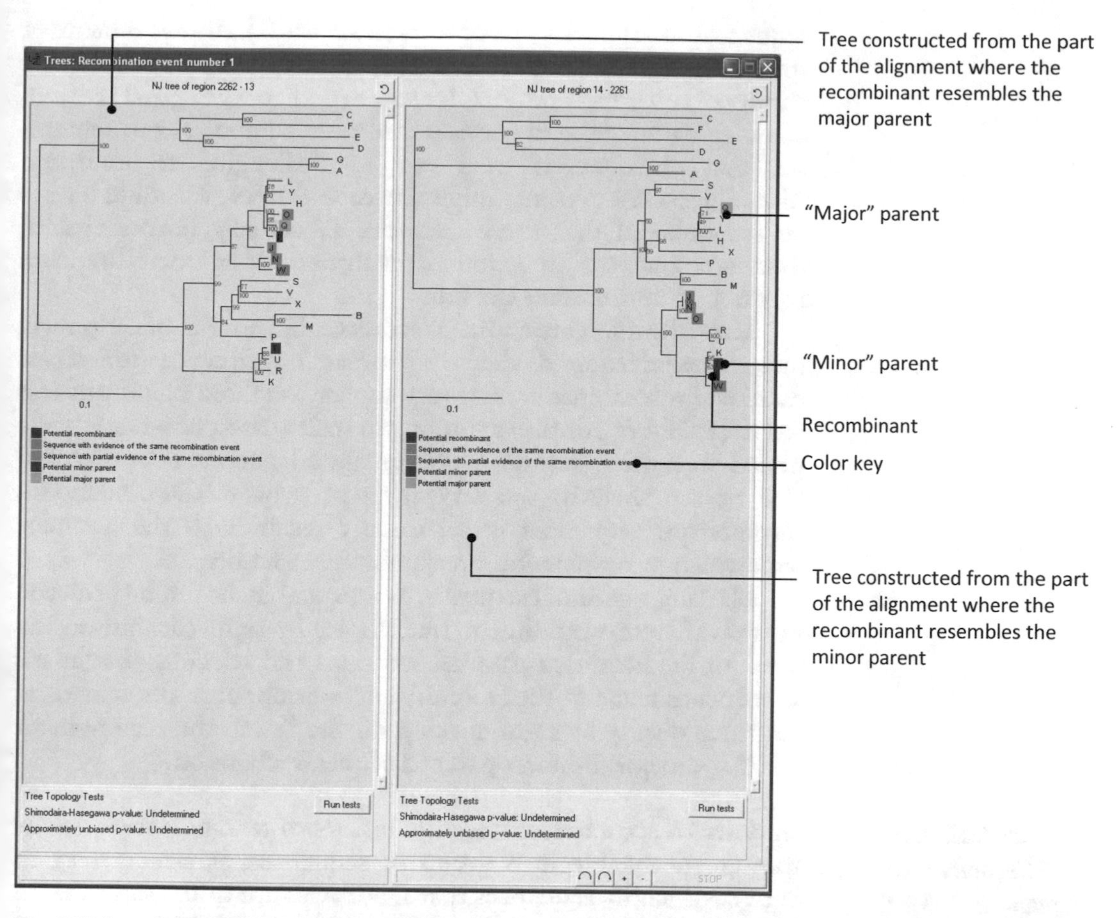

Fig. 5 Using phylogenetic trees to determine which sequence(s) is (are) recombinant. In this example RDP4 has inferred that five sequences ("O," "I," "J," "N," and "W") are descended from a common recombinant ancestor with parental sequences resembling the "Q" and "T" sequences. Note that these five sequences do not form a monophyletic group in either tree. This indicates either that RDP4 has "over-grouped" the sequences or that other recombination events elsewhere in these sequences may be obscuring the phylogenetic relationships between them (as turns out to be the case in this example)

analysis. These trees can be viewed side by side by pressing the "Trees" button at the top of the screen (Figs. 4 and 5). It is possible to either change the default tree that RDP4 will construct to a FastNJ tree (which will give an indication of branch supports) [14], or change individual trees to neighbor joining (constructed using PHYLIP) [15], least squares (constructed using FastTree2) [14], maximum likelihood (constructed using PHYML, RaxML, or FastTree2) [16, 17], or Bayesian trees (constructed with MrBayes) [18] by right clicking on a tree and selecting the appropriate option on the menu that appears.

If it is determined that RDP4 might have incorrectly identified the recombinant, it is advisable that care be taken when modifying the choice that the program has made. There are many factors that might seriously complicate the identification of recombinant sequences using phylogenetic trees (*see* **Note 3**). Always remember that the program has used a phylogenetic approach along with a battery of other tests and has, for some (perhaps very good) reason, come up with a different decision. Also, it is important to remember that phylogenetic trees can be challenging to interpret. Although RDP4 presents midpoint rooted trees, it should always be remembered that these trees are all actually unrooted and, therefore, that the direction of evolution along their branches may not be immediately obvious.

Manually resolving an incorrect recombinant identification will most often be required when the program has either identified two candidate recombinants with very similar weighted consensus test scores (RDP4 will display a warning in such a case), or when a badly placed breakpoint position was manually adjusted in the preceding step of the analysis (the accuracy with which RDP4 identifies recombinant sequences is particularly sensitive to the accuracy with which recombination breakpoints are identified).

Marking a parental sequence as a recombinant can be achieved in two different ways: in the tree display by right clicking on the name of the identified parental sequence and selecting the "make < sequence name > the recombinant" option; or in the schematic sequence display by right clicking on the "swap the recombinant and the <major/minor> parental sequences" option.

2.8 Evaluating How Well Recombination Signals Have Been Grouped into Recombination Events

In order to accurately retrace the history of recombination events that are detectable in a group of sequences, it is necessary to correctly identify sequences that share evidence of the same ancestral recombination event(s). RDP4 will often mistakenly group sequences that are the descendants of different ancestral recombinants. Conversely, RDP will seldom mistakenly identify two sequences carrying evidence of the same recombination event as carrying evidence of two different recombination events. Although the descendants of an ancestral recombinant might be expected to all have nearly identical recombination breakpoint patterns and cluster together within phylogenetic trees, this is not always the case. For example, some sequences may contain only partial evidence of a particular recombination event because a second, newer recombination event overprinted part of the evidence from the older recombination event (*see* **Note 4**).

Besides searching for the synchronized movement of sequence clusters within phylogenetic trees constructed from different parts of sequence alignments, another way that sequences with evidence of the same ancestral recombination events can be identified is by comparing RECSCAN or RDP plots that are generated with the same parental sequences but with different potential recombinant sequences. In the phylogenetic trees that RDP4 displays, sequences are highlighted in red, pink, and purple whenever they are inferred to be carrying evidence of a particular recombination event (Fig. 5)

Right clicking on one of the sequence names that are highlighted in red or purple and selecting the "Recheck plot with <sequence name> as recombinant" option will construct a graph that can be used to visually assess whether the selected sequence shares evidence of the same recombination signal as that found in the sequence that is highlighted in red. A colored bar will be displayed above this plot which will indicate how closely the plot constructed with the selected sequence resembles that constructed using the sequence highlighted in red. Whereas blue/green colors along this bar represent a good match, orange/red colors represent a bad match. If the bar has a red or orange color on either side of both of the inferred recombination breakpoint locations, there is a good chance that the two signals being compared represent different unique recombination events. If, however, the bar is blue on both sides of either one or both of the breakpoint locations, then the two signals likely represent the same ancestral recombination event.

If it is found that RDP4 has either missed evidence that a group of sequences have all descended from a common ancestral recombinant, or incorrectly identified that a group of recombinant sequences have all descended from a common recombinant ancestor, this can be manually rectified. Recombination signals evident within a collection of sequences can be either grouped into a single recombination event or ungrouped to form separate recombination events by right clicking on the name of the sequence in the tree display and selecting the "mark <sequence name> as having/not having evidence of this recombination event" option. Two different recombination events can also be merged by right clicking on a colored block representing the event in the bottom right schematic sequence display and selecting the "Merge events" option.

2.9 Accepting a Verified Recombination Event

Having either refined or confirmed the characteristics of a particular recombination event (i.e., its breakpoint positions, and the identity of the recombinants carrying evidence of the recombination event) it is necessary to specify that the recombination event has been suitably verified. This is done by right clicking either on the name of a recombinant sequence in the tree display or on the flashing colored block representing the recombination event in the bottom right schematic sequence display and selecting one of the "accept" options. If more than one sequence carries evidence of a particular recombination event, the appropriate option will be to accept the event in all of the sequences in which the recombination event was detected. This option should only be chosen if due care was taken in the previous step to ensure that all of the sequences highlighted in pink/purple in the tree display carry reasonable evidence of the current recombination event. Once a recombination event is accepted, a red border will appear around the colored block (and all others) representing that recombination event in the bottom right schematic sequence display panel. This will inform RDP4 during subsequent recombination scanning cycles that all subsequent searches of the dataset for evidence of recombination should be started with the current event preaccepted (i.e., RDP4 will not have to rediscover these events).

If there is significant doubt about the accuracy of an inferred breakpoint position, if it is unclear whether the correct sequence has been identified as a recombinant, or if it is evident that the detected recombination signal may be a false positive (i.e., for example, likely caused by sequence misalignment), the recombination event can be either rejected or left unaccepted. To reject a recombination event, right click either on the name of the sequence in the tree display or on a colored block that represents the event in the bottom right schematic sequence display and select one of the "reject" options. When events are either accepted or rejected, RDP4 will by default skip these during its navigation through the remaining events that need to be manually verified.

If the automatically identified breakpoint positions and recombinant sequence(s) were left unmodified and the recombination event is accepted, then it is time to move on to the next recombination event in the list. This can be accomplished by pressing the "Pg Dn" button or by pressing the right arrow at the bottom of either the schematic sequence display or the side-by-side tree display panel.

If, however, the automatically identified breakpoint positions or recombinant sequence(s) were modified and these modifications were accepted, then it is important to rescan the data to account for these modifications. The reason for this is that the breakpoint and recombinant identification hypotheses proposed by RDP4 for all subsequent recombination events may have been negatively influenced by the (presumably) incorrectly identified breakpoint positions and/or recombinant sequences. The manual changes that were made, no matter how minor, could influence (hopefully positively) the way that RDP4 interprets the remainder of the recombination events that are detectable within a dataset.

A rescan can be initiated either by pressing the "Rescan" button at the bottom of the schematic sequence display (it should be red and flashing if a rescan is warranted), or by right clicking on a gray area of the schematic sequence display and selecting the "Rescan and reidentify recombinant sequences for all unaccepted events" option from the menu that appears.

Once RDP4 has rescanned the sequences, the process described between Subheadings 2.6 and 2.9 should be repeated until every detected recombination event has been verified.

2.10 Saving Analysis Results

This manual verification procedure can become extremely tedious for large datasets and it is advisable that analysis results be regularly saved in .rdp format. This can be performed by pressing the "Save" button on the menu bar at the top of the screen (Fig. 1) and selecting ".rdp" (RDP project file) as the format in which the data should be saved. RDP project files can be reloaded in RDP so that manual verification part of a recombination analysis can be carried out over multiple sessions.

When an analysis is completed, the final results can also be saved in a table format by pressing the "Save" button and selecting the ". csv" format option. This format will produce a tabulated results file that can be opened in any spreadsheet program (such as Microsoft Excel). Note, however, that RDP4 will not be able to reload analysis results from a ".csv" file.

It is also possible to save or copy sequence alignments, trees, plots, and matrices in various formats by right clicking on these in their respective program windows and selecting either the "Save as..." or "Copy" options presented.

3 Examples

3.1 Producing a Preliminary Recombination Hypothesis

Load the example alignment file "PVY Example.fas" (this and all other example files referred to here can be found in the directory where you have installed RDP4) and press the "Options" button (Fig. 1). The example sequences we will be analyzing are linear virus genomes, so in the "General Recombination Detection Options" section under the "General settings" tab, change the "Sequences are circular" setting to "Sequences are linear" (Fig. 2). Besides this change, we will use the default RDP4 settings for this example. Press the "OK" button at the bottom of the options form. Press the "X-Over" button (Fig. 1) and wait for the automated analysis to complete (it should take approximately 8 min).

3.2 Navigating Through the Results

Press the left mouse button when the mouse pointer is on a background grayed area of the schematic sequence display (Fig. 1). This focuses the program on the display. Pressing either the "Pg Up," "Pg Dn," or "space bar" keys on your computer keyboard will allow you to navigate through the detected recombination events in an ordered fashion (alternatively you can use the arrow buttons beneath the schematic sequence display to do the same thing). Immediately after finishing the automated analysis, pressing the "Pg Dn" button will take you to the first recombination event identified by RDP4. Pressing it again will take you to the second event, and so forth. Pressing the "Pg Up" button will take you to the previous event. Pressing the space bar will take you to the recombination event with the best associated *p*-value.

Starting with the first event (press the "Pg Up" or "Pg Dn" button until information on "recombination event 1" is displayed in the recombination information panel) you will see a graph drawn on the "plot display" (Fig. 1). The exact type of graph that is plotted will depend on the recombination method that was used to detect the recombination event at hand. In this case the plot should be one that is produced using the RDP method. The yellow, purple, and green plotted lines each represent comparisons between different pairs of sequences in the sequence triplet (in this case "Q" vs. "T," "I" vs. "T," and "I" vs. "Q," respectively) that was used to detect the recombination event. The part of the graph between alignment positions 1 and 2250 indicates that the sequence identified as the recombinant, "I," is most closely related to sequence "T" in this region. Over the remainder of the graph, between alignment positions 2251 and 9594, sequence "I" is much more closely related to sequence "Q" than it is to sequence "T." The probability that this pattern of relatedness between "I," "Q," and "T" could occur without recombination under neutral selection is approximately 1 in 10^{198}. This is very strong evidence of recombination. Turn your attention to the confirmation table in the recombination information display (Fig. 1). Note that all methods other than LARD and PHYLPRO [19] have also indicated that the shifting relationship between "I" and these other two sequences is good evidence of recombination. All associated *p*-values are smaller than 10^{-28}. The LARD and PHYLPRO methods were not used during the automated exploratory scan for recombination signals and this is the reason that they have no associated *p*-value. The large differences between the reported *p*-values for the other methods are largely due to the fact that the different methods consider slightly different signals in the alignment and have different approaches to approximating the probability that apparent recombination signals are caused by accidental convergent mutation rather than recombination.

Note that there is a warning (in red capitalized letters) in the recombination information display. Because you are analyzing linear sequences, RDP4 is telling you that only the "Ending" (or 3′) breakpoint is an actual recombination breakpoint (the other "breakpoint" listed is the beginning of the sequence).

3.3 Checking the Accuracy of Breakpoint Identification

It is important that you check the accuracy with which RDP4 has identified the recombination breakpoint positions. The RDP method used to detect this recombination signal has a lower degree of breakpoint inference accuracy than the RECSCAN, MAXCHI, CHIMAERA, and 3SEQ methods (Fig. 3). To see a MAXCHI graph for event 1, press the "check using" listbox on the right-hand side of the plot display (Fig. 1). One of the options listed is to construct a MAXCHI plot. Select this option and see whether the peaks on any of the three lines plotted correspond with the left and right borders of the pink area. Look at graphs for some of the other methods. The left and right boundaries of the pink area should match positions in the RDP, RECSCAN, SISCAN, and DISTANCE plots where two of the three plotted lines intersect

As with the MAXCHI plot, the left and right boundaries of the pink area should match peaks in at least one of the lines in CHIMAERA, TOPAL [20], PHYLPRO, and LARD plots. For this recombination event all of the methods seem to indicate the recombination breakpoint at position 2250 has been correctly identified.

The actual breakpoint position in this example might, however, not be as obvious as you think. Note that in the recombination information display, five different sequences have been identified as descendants of the same recombinant (you can tell this by looking at the confirmation table in the recombination information display). Press the "Trees" button at the top of the screen (Fig. 1). Five of the sequences in the trees displayed (Fig. 5) are highlighted in red, pink, or purple. These are all sequences that potentially also carry evidence of recombination event 1. The sequence in red, "I," is currently selected. Move the mouse pointer over "W" and press the right mouse button. Select the "Go to W" option. This will center the schematic sequence display on "W." Move the mouse pointer over the left most colored block representing the recombination event 1 signal in "W." Look at the recombination information display. Note that the "Ending" breakpoint position is identified here as 2261 and not 2250. This is a small but important difference. Press the "show relevant sequences" button (Fig. 4) and use the scroll bar at the bottom of the sequence display to move to position 2261. The color coding of the nucleotides now corresponds with the colors of the lines in the plot later. You will see that at position 2258, "W" and "T" share an A nucleotide, and also that the breakpoint is inferred to lie three nucleotides to the right of this point in "W" (instead of eight nucleotides to the left of this point as in "I"). Now go back to the corresponding representation of recombination event 1 in "I" and left click on it. Look at the sequence display and you will see that at position 2258 sequence "I" has a G residue that is shared with sequence "Q."

Clearly the breakpoint position should be somewhere in the region between 2250 and 2261, but its precise location is unknown. Let us, just for the sake of this example, adjust the breakpoint position to nucleotide 2261. To do this move the mouse pointer to nucleotide 2261 of the middle sequence in the sequence display and right click on it. One of the options offered will be to "Place ending breakpoint here." Select this option, so that when you look at the representations of this event in the schematic sequence display you will see that they all report the breakpoint position as 2261.

3.4 Checking the Accuracy of Recombinant Sequence Identification

Look at the bar graphs in the recombination information display (Fig. 1). The first set of three bars indicate the consensus "Recombinant scores" of sequences "I" (0.667), "Q" (0.077), and "T" (0.256). These scores are the weighted consensus of a series of tests (each indicated by a set of three bars in the graph) to determine which sequence out of the three is the recombinant. In this case it is apparent that "I" is probably the recombinant.

It is useful to consider the phylogenetic trees in order to validate the status of "I" as the recombinant. Bring up the side-by-side tree display by pressing the "Trees" button in the command button panel (Fig. 1). Right click on an empty area in one of the two windows. The menu that appears has an option to "Change tree type"—select this and then select the neighbor joining tree type. Now change the tree in the other window to neighbor joining too. You should see the same trees as in Fig. 5. For now focus on the sequences in the tree that are highlighted in red ("I"), green ("Q"), and blue ("T") and ignore those highlighted in pink ("W") and purple ("J," "N," and "O"). Comparing the locations of these three sequences in the trees should indicate that the sequence highlighted in red has "moved" from the "Q" clade of the tree to the "T" clade (*see* **Note 5**). It seems that the program was correct in its identification of this sequence as the recombinant. You therefore do not need to change anything.

For the sake of this example, however, right click on the flashing block in the schematic sequence display and select the "Make the minor parent (T) the recombinant" option. Observe how the flashing block is "moved" to "T." Look at the tree and see how the interpretation of this recombination event has changed. Now make "I" the recombinant again.

3.5 Evaluating RDP4's Grouping of Recombination Events

There is apparently some evidence that recombination event 1 may have occurred in the common ancestor of five sequences in the dataset—those sequences currently highlighted in purple/pink/red in the trees ("I," "W," "J," "N," and "O"). It is, however, also apparent that the five identified recombinant sequences neither all cluster within the phylogenetic trees, nor all move together between the phylogenetic trees. This fact suggests that RDP4 may have "over-grouped" these sequences and that they may in fact be carrying evidence of multiple different independent recombination events. If you look at these five sequences in the schematic sequence display you will, however, immediately see the probable reason that these sequences do not move together between the two phylogenetic trees: RDP4 has identified additional recombination events in all of these sequences other than "I." In such cases it is not expected that a group of sequences carrying evidence of the same ancestral recombination event will all cluster together in both of the trees.

RDP4 provides another tool with which you can check whether two sequences carry evidence of the same ancestral recombination event. In either one of the trees right click on sequence "W" and select the "Recheck the plot with W as the recombinant" option. This will compare the plots produced using the currently selected sequence (in this case sequence "I"—the one in red) with that of sequence "W." The result of this comparison is displayed

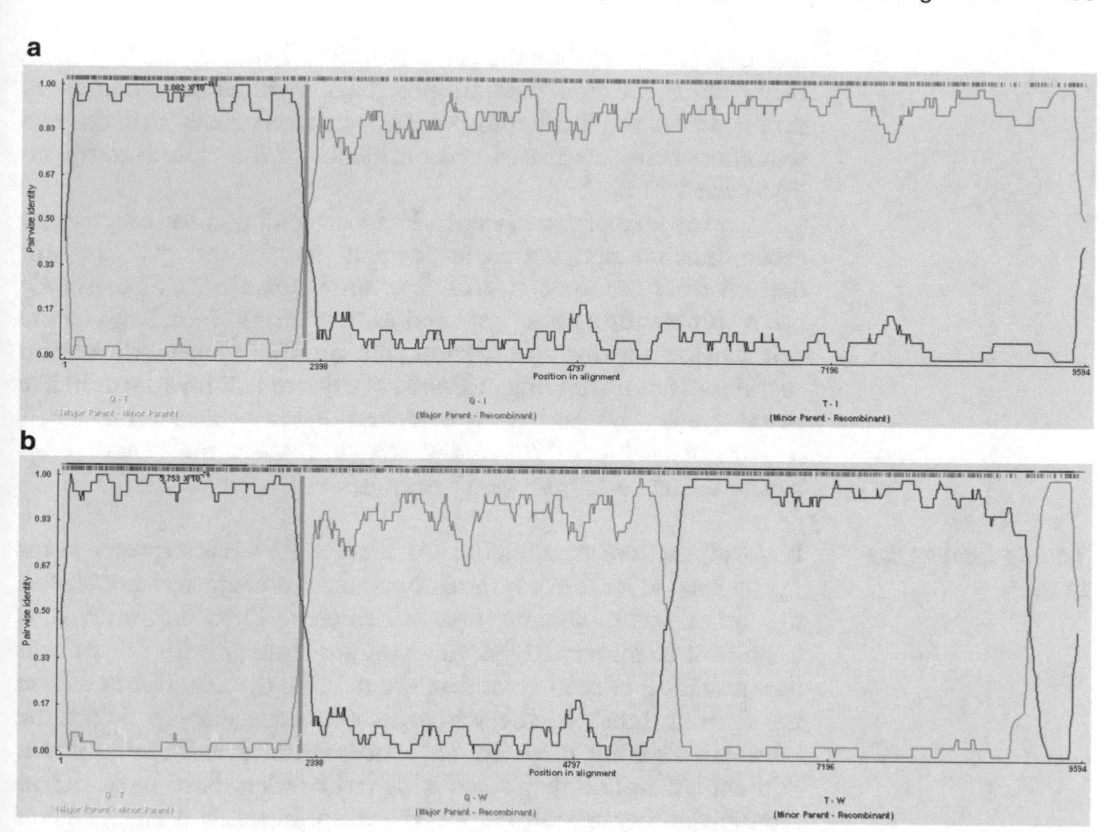

Fig. 6 Comparing recombination signals to determine whether two recombinants are descended from a common recombinant ancestor. (**a**) An RDP method plot for recombination event number 1 in the example dataset. (**b**) A similar RDP method plot to that in A but with sequence "W" replacing sequence "I" in the scanned sequence triplet. The *colored line* above the plot is a graphical representation of how closely the plot in (**b**) resembles that in (**a**). Note that across the recombination breakpoints the two plots are nearly identical (the *blue color* in the bar expresses this similarity) implying that "I" and "W" probably both descended from the same recombinant ancestor. Note also that in (**b**) the *deep red color* in the part of the *colored line* corresponding to sequence coordinates ~5000 to ~9000 clearly indicates that "W" likely carries evidence of a second large recombination event that is not shared with "I"

graphically in the form of a multicolored line above the plot in the graph display (Fig. 6). Whereas blue along this line indicates regions of sequence where recombination signals are very similar, dark red indicates regions of sequence where recombination signals are very dissimilar. You will notice when comparing "W" with "I" that the line becomes very red between positions 5680 and 9099. If you look at sequence "W" in the schematic sequence display on the bottom left (click on "W" in the trees and select the "go to W" option) you observe that RDP4 has detected a second recombination event in "W" with breakpoints corresponding to these alignment coordinates. If the portion of sequence spanning one or both recombination breakpoints (i.e., the left and right bounds of the pink area in the plot display) corresponds with a blue/green line (as it does in this case) then the pattern of sites shared by the sequences being compared and their supposed parental sequences are very similar across the breakpoint(s). Thus it is very likely that the two sequences being compared carry evidence of the same recombination event.

For the sake of this example, let us pretend that the very similar recombination signals that are evident in "I" and "W" are not derived from the same ancestral recombination event. To exclude the recombination event detected in "W" from event 1, go to the side-by-side tree display and right click on "W." Choose the option to "Mark W as not having evidence of this event." If you would like to reinclude "W" as having evidence of recombination event 1, then right click on "W" in the tree and select the "Mark W as having evidence of this event" option.

3.6 Completing the Analysis

Because you have changed the way that RDP4 has interpreted event 1, you need to let the program reformulate its characterization of all the other recombination events detected. First, however, it is important to inform RDP4 that you are content with the current interpretation of recombination event 1. To do this, right click on the flashing block in the schematic sequence display. Select the "Accept this event in all five sequences where it is found" option. You should notice that a series of red borders have been drawn around the blocks representing the recombination event 1 signals in sequences "I," "W," "J," "N," and "O." Now either click on the flashing "Re-scan" button beneath the schematic sequence display or right click anywhere in the schematic sequence display and select the "Re-Scan and re-identify recombinant sequences for all unaccepted events" option.

When RDP4 has finished reanalyzing the remaining recombination signals press the "Pg Dn" button on your keyboard and you can start evaluating recombination event 2. Continue until you reach the end of the analysis. You may notice that the program skips event 2. The recombination signal corresponding to recombination event 2 has been identified by RDP4 as being attributable to sequence misalignment. To see recombination event 2 you will need to click on the "options" button at the top of the screen, move to the "General" tab, in the "Data Processing Options" section, press the button besides the "list events detected by >1 method" label until the label reads "list all events." If you now look at sequence "B" in the schematic sequence display, you should notice a gray block labeled "unknown" under the line representing this sequence: this block is representative of "recombination event 2."

Recombination event 3 is detected in sequences "J," "N," and "W" but it is clear that the sizes of the recombinationally derived fragments in these three sequences differ substantially. Whereas all three of the sequences have similar recombination signals across the beginning (or 5′) breakpoint (approximately at position 5680), they all have completely different recombination signals across the ending (or 3′) breakpoint. This likely indicates that following an initial recombination event subsequent recombination events have "overprinted" the 3′ breakpoint (identified here as recombination events 5 and 8 in "J" and "W," respectively; *see* **Note 4**).

Recombination events 9 and on become progressively more difficult to interpret. This is primarily because they involve either recombination between very closely related sequences (such as recombination event 11 which has a 5′ breakpoint with a high degree of uncertainty), or recombination between parental viruses that are only distantly related to sequences within the analyzed dataset (such as recombination event 9 for which there is no sequence in the dataset that closely resembles the major parent).

3.7 Further Analyses

If large numbers of recombination breakpoints have been detected, you may want to test whether the distributions of these breakpoints indicate the presence of recombination hot- or cold-spots within the sequences that have been analyzed. To demonstrate this, open the file "HIV Example.rdp" (it can be found in the directory where you have installed RDP4). Press the arrow beside the "X-Over" button and select the "breakpoint distribution plot" menu option. After a minute or two the program will display the plot indicated in (Fig. 7). The black line in this plot represents the numbers of recombination breakpoints (individually indicated by vertical lines above the plot) that fall within 200 nucleotides of the genome coordinates indicated on the *x*-axis (in this case corresponding to nucleotide sites within the first sequence in the analyzed dataset, A1.KE.94). The gray and white areas, respectively, represent the 95 % and 99 % confidence intervals of the expected degrees of breakpoint clustering under random recombination. Whereas genome coordinates at which the black line spikes up above the white/gray area are statistically supported recombination hot-spots, those where the black line dips below the white/gray area are statistically supported recombination cold-spots.

If you have a GenBank file on hand that corresponds with one of the sequences in a dataset that has been analyzed for recombination, and this file contains information on the locations of gene boundaries, then it is also possible to test for associations between recombination breakpoint distributions and genome organization. Once again, open the file "HIV Example.rdp." When it is loaded press the "Open" button again and select the file "HXB2 Genbank File.txt" (it can be found in the folder where you installed RDP4). This file simply contains a plain text version of the GenBank file for sequence "B.FR.83.HXB2" that is accessible using the following URL: http://www.ncbi.nlm.nih.gov/nuccore/K03455.1. When this file is loaded into RDP4 you should notice that a series of arrows are added to the colored similarity map above the sequence display. Press the arrow beside the "X-Over" button and again select the

a

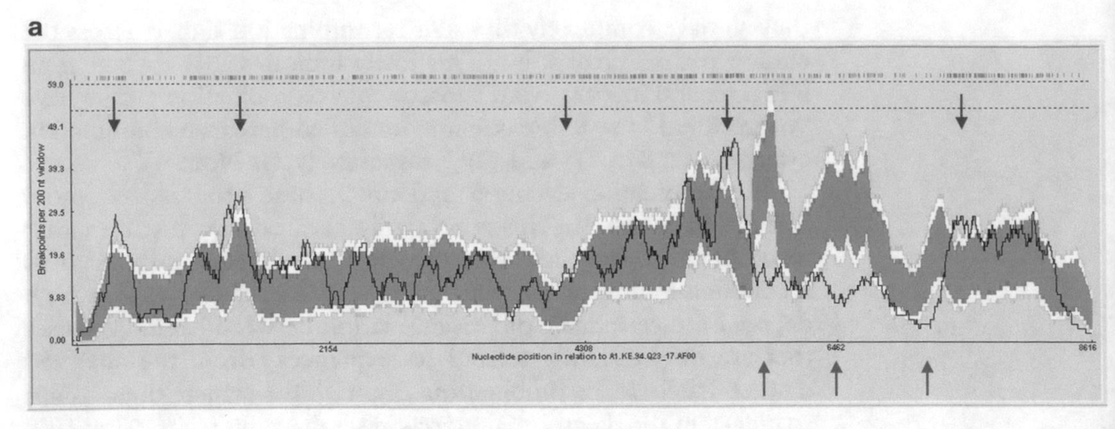

b

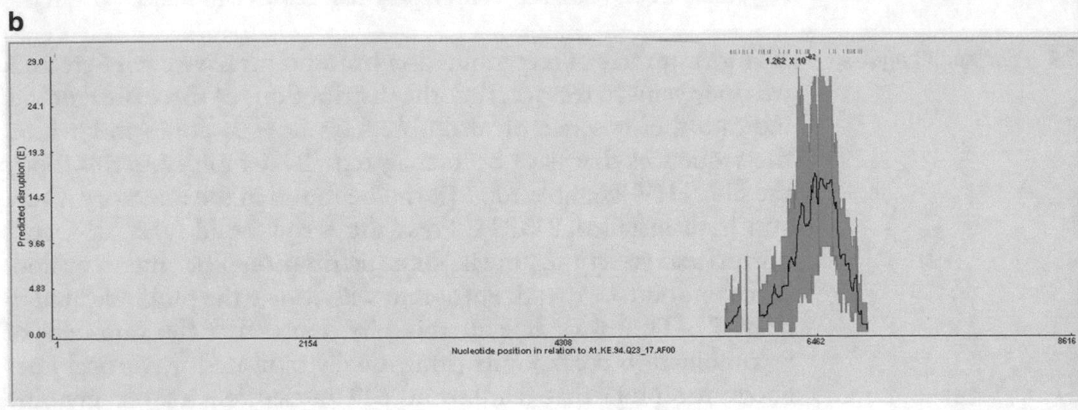

Fig. 7 Recombination breakpoint distribution and recombination-induced protein folding disruption plots. (**a**) Recombination breakpoint hot- (*red arrows*) and cold-spots (*blue arrows*) detectable within HIV-1M genomes represented in the file, HIV Example.rdp (after [21]). The *black plot* represents the inferred clustering of recombination breakpoints identified within the HIV-1M sequences analyzed here. *Dark* and *gray areas*, respectively, represent the 95 % and 99 % bounds of the expected degrees of breakpoint clustering under random recombination. *Vertical lines* above the plot indicate the positions of recombination breakpoints. (**b**) Expected degrees of folding disruption within chimeric envelope proteins of simulated HIV-1M recombinants. Whereas the *black line* represents the mean folding disruptions inferred for the simulated chimeric envelope proteins, the *gray area* represents the range of folding disruptions observed in these simulated proteins. The *vertical black lines* above the plots represent the locations of recombination breakpoint sites that were detected in the envelope genes of HIV genomes represented in file "HIV Example.rdp"

"breakpoint distribution plot" menu option. This time, in addition to displaying the breakpoint distribution plot, extra information is tabulated in the recombination information panel in the top right. This extra information relates to the relative clustering of breakpoints between (1) coding and noncoding regions, (2) between different genes, and (3) between different parts of genes (Fig. 8). In this table the specific genome regions that are being compared in each row are graphically depicted in orange and blue. Numbers in the table are colored according to the genome regions that they relate to. By examining the table, it is evident that: (a) detectable recombination

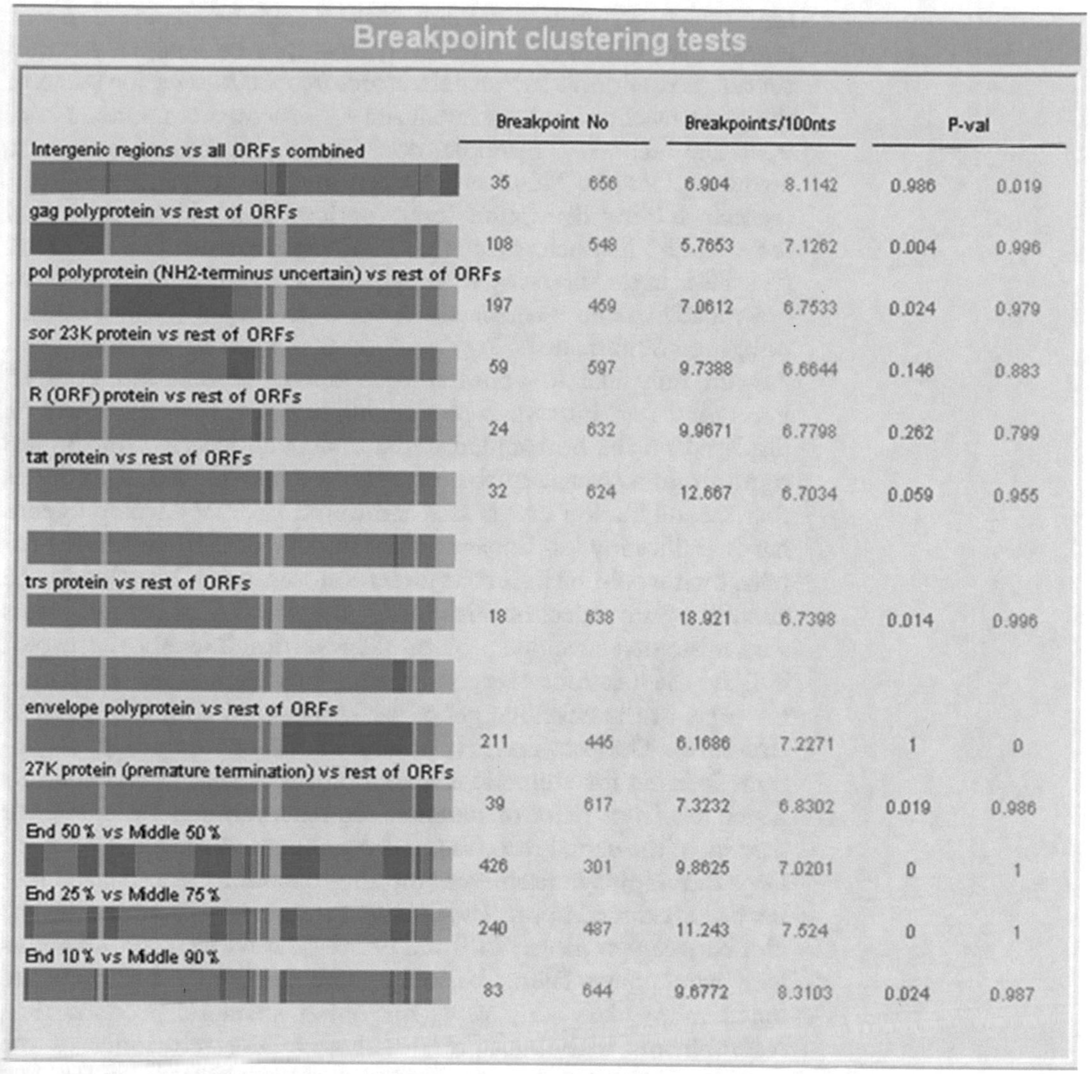

Breakpoint clustering tests

	Breakpoint No		Breakpoints/100nts		P-val	
Intergenic regions vs all ORFs combined	35	656	6.904	8.1142	0.986	0.019
gag polyprotein vs rest of ORFs	108	548	5.7653	7.1262	0.004	0.996
pol polyprotein (NH2-terminus uncertain) vs rest of ORFs	197	459	7.0612	6.7533	0.024	0.979
sor 23K protein vs rest of ORFs	59	597	9.7388	6.6644	0.146	0.883
R (ORF) protein vs rest of ORFs	24	632	9.9671	6.7798	0.262	0.799
tat protein vs rest of ORFs	32	624	12.667	6.7034	0.059	0.955
trs protein vs rest of ORFs	18	638	18.921	6.7398	0.014	0.996
envelope polyprotein vs rest of ORFs	211	445	6.1686	7.2271	1	0
27K protein (premature termination) vs rest of ORFs	39	617	7.3232	6.8302	0.019	0.986
End 50% vs Middle 50%	426	301	9.8625	7.0201	0	1
End 25% vs Middle 75%	240	487	11.243	7.524	0	1
End 10% vs Middle 90%	83	644	9.6772	8.3103	0.024	0.987

Fig. 8 Testing for associations between genome arrangement and breakpoint distributions. Breakpoint clustering is compared between the genome regions represented in *blue* and *orange*. In this example (found in the file "HIV Example.rdp") it is clear from the last three rows of the table that that HIV-1M breakpoints tend to cluster far more around the edges of genes (in *blue* with low associated *p*-values) than they do within the central parts of genes (in *orange*)

breakpoints are significantly more clustered within coding regions than they are within non-coding regions ($p = 0.019$ in row one), (b) the gag and pol polyproteins contain significantly higher densities of detectable recombination breakpoints than other HIV genes (see rows 2 and 3), and (c) detectable recombination breakpoints tend to occur significantly more frequently on the edges of genes than they do in the middle parts of genes (see the last three rows).

It is also possible to test whether the observed recombination breakpoints have tended to fall at locations within genes that have less impact on protein folding than would be expected under random recombination. However, this type of analysis requires that high resolution 3D protein structures are known for some of the proteins encoded by an analyzed set of sequences. Once again, open the file "HIV Example.rdp." When it is loaded, press the arrow besides the "X-Over" button and select the "SCHEMA (protein folding disruption test)" option. You will be prompted for a ".pdb" file. Select the file "HIV env structure (2B4C).pdb" (this file is in the directory where you installed RDP but can also be downloaded from Protein Data Bank at http://www.rcsb.org/pdb/home/home.do). You will then be asked whether you "would only like to consider accepted recombination events." Press the "Yes" button. A plot resembling that in Fig. 7b will be displayed on the bottom left and a table will be given on the top right program panel. In this case the *p*-value of ~0.013 indicates that recombination events that are found in HIV envelope genes have significantly less impact on the folding of HIV envelope proteins than would be expected under random recombination in the absence of any selection disfavoring the survival of recombinants with misfolded envelope proteins. The vertical lines above the plot indicate the locations of recombination breakpoints that fall within the parts of the envelope gene that correspond with the analyzed structures. The plot itself represents the range of folding disruptions inferred for chimeric envelope proteins that were randomly generated from pairs of parental sequences which resemble the parents of the actual recombinants considered during this analysis. Gaps in the plot indicate genome sites encoding amino acids that were not included in the 2B4C envelope structure. The peak in the plot between positions 6300 and 6650 indicates that recombinants with breakpoints falling between these genome coordinates are much more likely to express misfolded envelope proteins than recombinants with breakpoints falling in the remainder of the envelope gene. The low *p*-value indicated by this test implies that there is a significant tendency for breakpoint sites in the envelope genes of actual recombinant HIV genomes to fall outside the region where they will maximally disrupt protein folding.

4 Notes

1. When RDP4 attempts to infer the number of unique detectable recombination events in an alignment, it focuses on an individual recombination signal and tries to determine which other sequences in the alignment carry a similar recombination signal. Often multiple sequences will display evidence of carrying a similar recombination signal (i.e., they may all have descended from the same ancestral recombinant sequence). Sometimes this signal might be strong enough to yield a statistically significant

p-value in some of the sequences, but not in others. These weaker signals are referred to as "trace" signals and are listed as such in both the "recombination information" part of the RDP4 display and the phylogenetic trees. The program constructs these trees to help you decide which sequences are recombinant and which are relatives of parental sequences.

2. There is also no guarantee that the relative weighting of the tests is accurate. Tests using recombinant HIV sequences (for which parental viruses have until recently been epidemiologically separated) have been used to weight the different tests. However, this weighting might not be appropriate for all datasets. Also, even in tests with HIV, the program only has an approximately 90 % success rate when it comes to correctly identifying recombinant sequences (Fig. 4).
3. RDP4 scans every sequence triplet in an alignment to identify recombination signals. Whenever more than a single recombination event is detectable in a dataset, it is possible that some triplets could contain two or three different recombinant sequences. This can be a particularly serious problem if the breakpoints are close together since the program will possibly infer a recombination event with two breakpoints, each of which comes from a different recombinant sequence in the triplet. This will seriously influence how effectively trees can be used to judge which of the sequences is "most" recombinant (they could all be recombinant but with none of them having the inferred breakpoint pair). Even when the correct breakpoint pairs are identified, when two of the sequences in a triplet are recombinant, the nonrecombinant parent will have an increased probability of being misidentified as the recombinant.
4. If recombination has occurred frequently within the history of a sample of sequences, evidence could exist of older recombination events being "overprinted" by more recent ones. When recombination events are partially overprinted in some of the sequences that are descended from an ancestral recombinant but not in others, there is a good chance that RDP4 will not group the recombination signals properly and it may infer that two recombination events have occurred instead of one. It can be particularly difficult to interpret the phylogenetic signals caused by partially or completely overprinted recombination events and although RDP4 might correctly infer the occurrence of two recombination events, it could still misidentify the exact sequence exchanges that took place.
5. RDP4 allows you to mark particular sequences in the trees that it displays. This can be very useful for tracking the "movement" of particular sequences between the clades of different trees. You can mark a sequence by moving the mouse pointer over the

name of the sequence in the tree and pressing the left mouse button. The same sequence is then marked both in the current tree, and in all of the other trees. It is also possible to clear markings or automatically color sequence names (so that they are the same colors as those displayed in the schematic sequence display on the bottom right panel). This can be accomplished by selecting the appropriate option on the menu that appears whenever you press the right mouse button while the mouse pointer is over one of the tree displays.

References

1. Martin DP, Lemey P, Posada D (2011) Analysing recombination in nucleotide sequences. Mol Ecol Resour 11:943–955
2. Martin DP, Williamson C, Posada D (2005) RDP2: recombination detection and analysis from sequence alignments. Bioinformatics 21:260–262
3. Muhire B, Martin DP, Brown JK et al (2013) A genome-wide pairwise-identity-based proposal for the classification of viruses in the genus Mastrevirus (family Geminiviridae). Arch Virol 158:1411–1424
4. Martin D, Rybicki E (2000) RDP: detection of recombination amongst aligned sequences. Bioinformatics 16:562–563
5. Padidam M, Sawyer S, Fauquet CM (1999) Possible emergence of new geminiviruses by frequent recombination. Virology 265:218–225
6. Smith JM (1992) Analyzing the mosaic structure of genes. J Mol Evol 34:126–129
7. Schultz A-K, Zhang M, Leitner T et al (2006) A jumping profile Hidden Markov Model and applications to recombination sites in HIV and HCV genomes. BMC Bioinformatics 7:265
8. Lole KS, Bollinger RC, Paranjape RS et al (1999) Full-length human immunodeficiency virus type 1 genomes from subtype C-infected seroconverters in India, with evidence of intersubtype recombination. J Virol 73:152–160
9. Posada D, Crandall KA (2001) Evaluation of methods for detecting recombination from DNA sequences: computer simulations. Proc Natl Acad Sci U S A 98:13757–13762
10. Martin DP, Posada D, Crandall KA, Williamson C (2005) A modified bootscan algorithm for automated identification of recombinant sequences and recombination breakpoints. AIDS Res Hum Retroviruses 21:98–102
11. Boni MF, Posada D, Feldman MW (2007) An exact nonparametric method for inferring mosaic structure in sequence triplets. Genetics 176:1035–1047
12. Gibbs MJ, Armstrong JS, Gibbs AJ (2000) Sister-scanning: a Monte Carlo procedure for assessing signals in recombinant sequences. Bioinformatics 16:573–582
13. Holmes EC, Worobey M, Rambaut A (1999) Phylogenetic evidence for recombination in dengue virus. Mol Biol Evol 16:405–409
14. Price MN, Dehal PS, Arkin AP (2010) FastTree 2–approximately maximum-likelihood trees for large alignments. PLoS One 5:e9490
15. Felsenstein J (1989) PHYLIP—Phylogeny Inference Package (version 3.2). Cladistics 5:163–166
16. Guindon S, Gascuel O (2003) A simple, fast, and accurate algorithm to estimate large phylogenies by maximum likelihood. Syst Biol 52:696–704
17. Stamatakis A (2006) RAxML-VI-HPC: maximum likelihood-based phylogenetic analyses with thousands of taxa and mixed models. Bioinformatics 22:2688–2690
18. Ronquist F, Huelsenbeck JP (2003) MrBayes 3: Bayesian phylogenetic inference under mixed models. Bioinformatics 19:1572–1574
19. Weiller GF (1998) Phylogenetic profiles: a graphical method for detecting genetic recombinations in homologous sequences. Mol Biol Evol 15:326–335
20. McGuire G, Wright F (2000) TOPAL 2.0: improved detection of mosaic sequences within multiple alignments. Bioinformatics 16:130–134
21. Simon-Loriere E, Galetto R, Hamoudi M et al (2009) Molecular mechanisms of recombination restriction in the envelope gene of the human immunodeficiency virus. PLoS Pathog 5:e1000418

Chapter 18

Species Tree Estimation from Genome-Wide Data with guenomu

Leonardo de Oliveira Martins and David Posada

Abstract

The history of particular genes and that of the species that carry them can be different for a variety of reasons. In particular, gene trees and species trees can differ due to well-known evolutionary processes such as gene duplication and loss, lateral gene transfer, or incomplete lineage sorting. Species tree reconstruction methods have been developed to take this incongruence into account; these can be divided grossly into supertree and supermatrix approaches. Here we introduce a new Bayesian hierarchical model that we have recently developed and implemented in the program guenomu. The new model considers multiple sources of gene tree/species tree disagreement. Guenomu takes as input posterior distributions of unrooted gene tree topologies for multiple gene families, in order to estimate the posterior distribution of rooted species tree topologies.

Key words Supertree, Supermatrix, Tree reconciliation, Bayesian phylogenetics, Tree distance, Species tree, Duplications and losses, Incomplete lineage sorting

1 Estimation of Species Trees

If we define a species as a group of interbreeding individuals, then a species tree will describe the series of reproductive isolation events leading to the extant composition of species. However, this process will seldom be in perfect agreement with the tree of a randomly chosen gene, since evolutionary models for individual genes typically only take into account point mutations and insertions/deletions. At the species level, on the other hand, we also observe the duplication of whole genes or chromosomes, acquisition of new material by lateral transfer from unrelated species, as well as the loss of genomic regions. Furthermore, within a given extant or ancestral population, the segregation pattern of a particular allele does not always match exactly the speciation events, due to ancestral polymorphisms or deep coalescences [1], a phenomena called incomplete lineage sorting. All of these biological phenomena can lead to

Jonathan M. Keith (ed.), *Bioinformatics: Volume I: Data, Sequence Analysis, and Evolution*, Methods in Molecular Biology, vol. 1525, DOI 10.1007/978-1-4939-6622-6_18, © Springer Science+Business Media New York 2017

gene trees in disagreement with the history of the species, even assuming we can reconstruct the gene phylogenies without error.

Many methods have been developed to infer the evolutionary history of the species as a whole. One strategy is to judiciously select a "core" gene set that can be assumed to be a good representation of the phylogeny of the species, and then proceed with standard phylogenetic reconstruction approaches [2, 3]. Other attempts include analyzing the phylogenetic profiles (presence or absence of gene families) [4, 5], protein domain organization [6], conserved gene pairs [7, 8], and gene order information [9, 10].

In general, the most successful species tree method should make most use of the available genetic information, and therefore should incorporate gene tree reconstruction while accounting for the disagreement between the gene and species phylogenetic histories. Such species methods can be broadly divided between the supertree [11, 12] and supermatrix [13, 14] approaches, which are represented visually in Fig. 1. What we will call gene families are actually the homologous sets—the collections of sequences that can be aligned and safely assumed to have a common ancestor. Supermatrix methods depend on an exact correspondence between sequences from different genes, such that they can be concatenated. Supertree methods also have been historically limited to one individual from each species per gene family, although recent improvements have made it possible to analyze more general data sets, as the distance matrix-based inference under the coalescent [15] or the Gene Tree Parsimony reconciliation [16, 17].

2 Bayesian Inference of Species Trees

The Bayesian paradigm offers an elegant solution to the problem of reconciling gene trees and species trees, since it makes the hierarchical relation between them explicit. In the same way as the dependence between the gene alignment D and the gene tree G can be described probabilistically by the phylogenetic likelihood function $P(D|G)$ [18], the gene tree can also be treated as a stochastic parameter depending on a species tree S, as in $P(G|S)$—where the particular form of $P(G|S)$ varies depending on our assumptions about the G–S disagreement. This will lead to a hierarchical Bayesian model of the form $P(S|D) \propto P(D|G)\ P(G|S)\ P(S)$, where multigene data sets can be naturally partitioned as $P(S|D) \propto P(S) \prod P(D_i\ |G_i)\ P(G_i\ |S)$ (where we exclude other parameters for clarity).

This idea has been implemented, with more or less success under a few models—for a review, see [19]. A full Bayesian inference model for incomplete lineage sorting has been developed under the multispecies coalescent model [20, 21], which however has a heavy computational burden. Using a birth–death model for the evolution of genes inside the species tree, a Bayesian model for

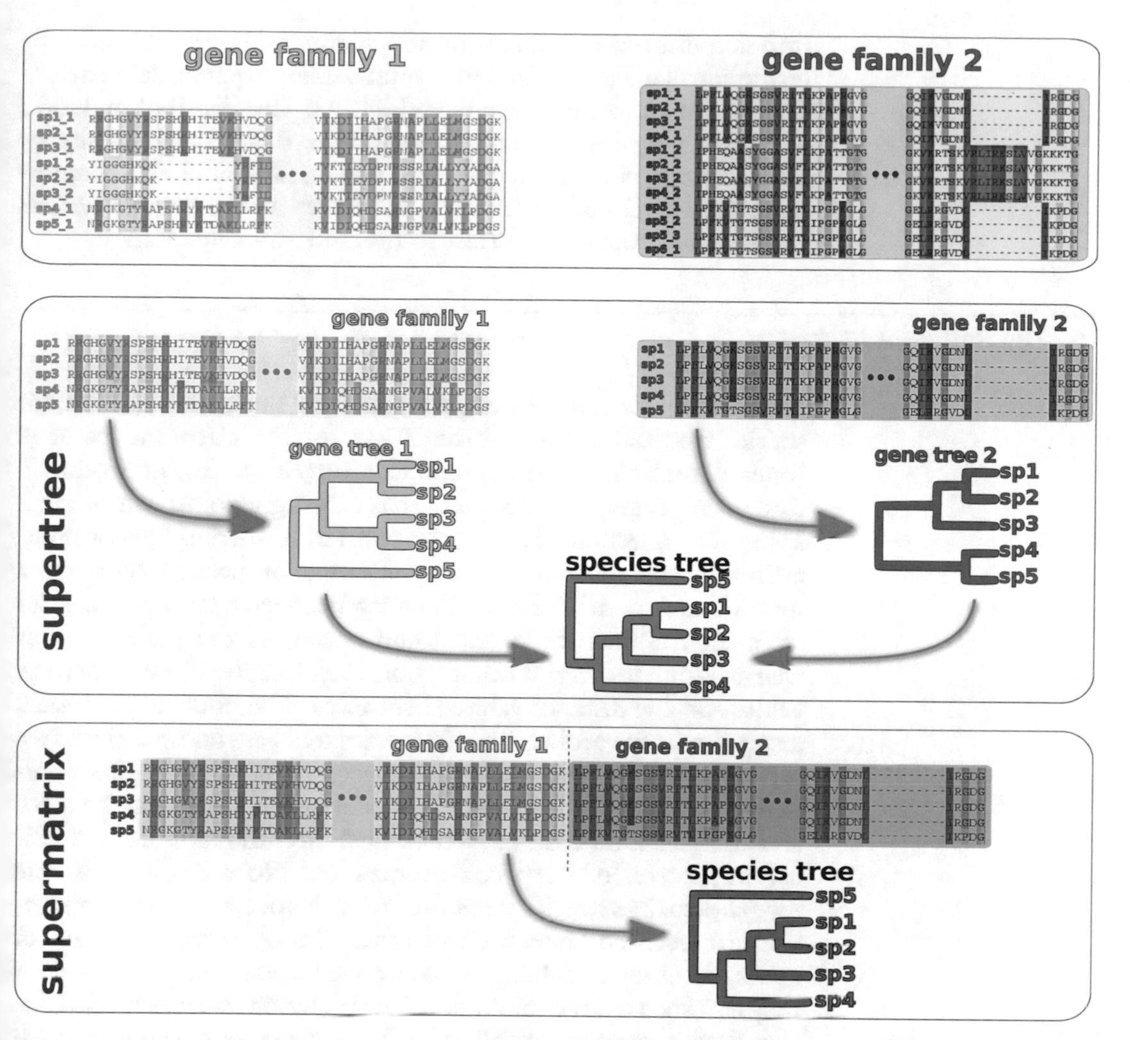

Fig. 1 Comparison between the supertree and supermatrix approaches. In the *top panel* we have an example data set composed of two gene families (homologous sets), already aligned. In this example we have several sequences from the same species in each gene family, which might represent paralogs within an individual sample or more than one sampled individual from the species. A supertree approach (*middle panel*) would then estimate the gene tree for each alignment independently and then summarize the information from all resulting gene trees into a single tree. A supermatrix approach, on the other hand, would first concatenate all gene family alignments into one single alignment, which would then be used in the phylogenetic reconstruction. Notice how for the supermatrix approach it is essential to find the correspondent sequences for all gene families, a limitation also imposed by several supertree methods. Therefore, the user must decide which representative from each species to use, for all gene families. That is, does the sequence *sp1_1* from *gene family 1* correspond to *sp1_1* or *sp1_2* from *gene family 2*?

gene tree estimation has been devised [22], being restricted however to a fixed species tree. A maximum-likelihood approach based on a simplified version of this birth–death model has recently been shown to allow for species tree estimation under a hierarchical model [23]. Recently, we developed a Bayesian model that takes into account several sources of disagreement between G and S by assuming that P(G|S) follows a multivariate exponential distribution over distances between G and S [24]. This distribution works as a penalty against species and gene trees that are very dissimilar under several biologically plausible scenarios. Under this assumption, the model can work with data sets where the orthology relations are unknown, in contrast to the other implementations.

3 The Software *guenomu*

Here we will focus on our novel hierarchical Bayesian procedure for species tree estimation, which is part of the guenomu package (openly available at http://bitbucket.org/leomrtns/guenomu/). The main program in this package is called guenomu, which takes as input a collection of gene trees (posterior distributions or point estimates) and generates a posterior sample of species trees under a multivariate model for the similarity between genes and species trees. At the same time, guenomu resamples the gene trees, as well as estimates several other parameters like the distance penalty values and the distance values themselves. The multivariate model assumes that the probability of a species tree generating a given tree is proportional to the similarity between them according to a predefined set of distance measures. Several such distances can be used, and reconciliation costs are a natural choice: the number of duplications, losses, or deep coalescences (*see* **Note 1**). That is, the model penalizes species trees that must invoke too many duplications or deep coalescences to explain the observed evolution of genes. Another possibility is to use well-known distances even if they lack such a direct biological interpretation, such as the Robinson–Foulds distance (*see* **Note 2**). In fact guenomu can work with all these distances at once, with the only caveat that it neglects branch length information—the multivariate model looks only at the topological disagreement. Another limitation is that it is best suited for distances (or dissimilarity measures, in general) that can work with trees of different sizes, since the gene trees can have several sequences from the same species. To emphasize the generality of these gene data sets, we will call them "gene families" in what follows.

3.1 Guenomu Programs

The guenomu package is composed of several programs, of which the main one is the Bayesian sampler, called guenomu. Besides the main MCMC sampler and analyzer, the package also includes a few auxiliary programs that can be executed independently from the main program. We will discuss mainly the Bayesian sampler, but following is a description of other auxiliary functionality offered by the package.

3.1.1 Bayesian Sampler

Given a collection of gene families, guenomu will estimate the posterior distribution of species trees together with other parameters pertaining to the model. By using a resampling technique, the species tree inference is done through two Bayesian procedures, where first the gene family alignments are used one at a time to sample the distributions of gene trees, and then these individual gene distributions are incorporated into a second, multigene Bayesian model. guenomu is responsible for the later, multigene Bayesian sampling, that will sample species trees respecting the distance constraints imposed by the multivariate model. The user is then free to use her favorite single gene phylogenetic inference algorithm to estimate the gene tree distributions (e.g., MrBayes [25], PhyloBayes [26], BEAST [27]), which will then be given as input to guenomu.

Even gene trees that did not come from a Bayesian analysis can be used in guenomu, and a set of compatible species trees will still be inferred. However, besides the probabilistic interpretation being lost in such an attempt, care must be taken to preserve the uncertainty in the gene tree estimation. In other words, even bootstrap replicates should be preferred over point estimates, since guenomu relies on the gene tree uncertainty to explore the space of species trees. The primary output of guenomu is a file or a set of files in a compact binary format that must be interpreted by the same program guenomu in order to output human-readable files.

3.1.2 Maximum-Likelihood (ML) Estimation of Species Trees

Usually an MCMC Bayesian sampler can also be used as a ML estimator by simply controlling the temperature of the Monte Carlo chains, as in simulated annealing [28]. This can be easily achieved by changing a few options in the same program guenomu, as we will see shortly.

3.1.3 Summary Statistics Coalescent Species Trees

One important auxiliary functionality of the guenomu package is offered through the program bmc2_maxtree, which estimates the species tree under the multispecies coalescent using several distance-based algorithms, namely, GLASS [29], SD [30], STEAC [31], and MAC [15]. It only needs a file with the species names (as for guenomu, as we will see later) and the files with the gene trees, and it will output four species tree estimates, one under each modification of the main algorithm. These algorithms rely strongly on the branch lengths of the gene trees (*see* **Note 3**), but if these are absent then the program will assume that they are all equal to one.

3.1.4 Pairwise Distances Between Trees

Another useful auxiliary program is bmc2_tree, which calculates all pairwise distances between a set of gene trees and a set of species trees. Given two tree files, it will create a file named "pairwise.txt" with a table consisting of the gene tree and species tree indices, followed by the list of distances between them.

As usual, the program assumes that the names of each species can be found within the gene leaves, and therefore it is sensitive to the order: the gene tree file must be given before the species tree file, in the command line. Furthermore, it neglects the root location of the gene trees, since the distances are defined between unrooted gene trees and rooted species trees—hence it does consider the root of the species trees, when defined.

3.1.5 Simulation of Tree Inference Error

The guenomu program can benefit from the uncertainty in the estimation of individual gene trees, since it allows the model to consider alternative trees that, however unlikely in isolation, can provide a better fit when analyzed in conjunction with other gene families. For each given input gene tree with branch lengths, the auxiliary program bmc2_addTreeNoise tries to generate a collection of trees similar to it, based on the premise that shorter branch lengths are more likely to be wrong. In other words, it will transform each tree from an input file into a distribution of trees, increasing the space of possibilities. It also works on trees without branch lengths and can be used to preprocess the gene tree files estimated by the user before the Bayesian analysis conducted by guenomu.

4 Input Files

The guenomu sampler estimates the set of species trees compatible with the phylogenetic information from a set of gene families, which are represented by the so-called input gene tree distributions. These distributions are ideally a good representation of their posterior distributions as estimated by a gene-wise independent Bayesian phylogenetic inference, using state-of-the-art software such as MrBayes [25], PhyloBayes [26], or BEAST [27]. Therefore, previous to the species tree estimation with guenomu, the user should already have tree estimates for each gene family.

The idea behind guenomu is that the researcher should not have to decide beforehand which genes, partitions, or sequences are appropriate, that is, which sequences comprise a locus and which to discard as paralogs, for instance. Therefore, as long as the sequences can be assumed to come from a common ancestor, be it from individuals from a population or from paralogy or speciation, they should be included in the alignment that we call a "gene family."

Guenomu accepts as input gene tree distributions tree files of one of two forms:

1. Standard nexus format, where each tree represents a sample from the posterior distribution of a gene-wise independent phylogenetic inference program (such as MrBayes, PhyloBayes, or BEAST).

2. A compact nexus format, where only topologically distinct trees are represented, together with their representativity (frequency). This file is a valid nexus tree file, but where frequencies (or weights) are annotated as comments alongside each tree. This is, e.g., the .trprobs file output by MrBayes' sumt command. We will therefore call such files "trprobs" files.

Files from the compact form represent the exact same information as standard nexus files, where the weight values describe how many times each tree was observed in the posterior distribution output by phylogenetic inference programs. Both forms can have the "TRANSLATE" table (from the nexus specification) or not, and branch lengths are allowed despite the fact that they will be ignored by guenomu's sampler. The trees must not contain multifurcations (except for a deepest trifurcation, characteristic of unrooted trees), and it treats all gene trees as unrooted, even if explicitly defined as rooted.

In practice the input gene tree distributions don't need to be generated by a Bayesian analysis and can actually be bootstrap replicates or even a point estimate as the maximum likelihood tree—but we warn that in such cases the probabilistic interpretation is lost.

5 Running guenomu

You can run guenomu by including all settings into a control file (and run it like "guenomu run.ctrl"), or by setting the parameters as command line arguments. You can also include the default parameters in the control file, and then overwrite some of these parameters in the command line (which take precedence). The control file can contain a series of arguments and accepts commentaries inside square brackets, which are then ignored by the program. Absent parameters are replaced by default values.

The only exception is the collection of gene tree files, and a list with the names of all species, that must be given by the user. Many programs require a gene-wise mapping between each gene leaf and the species it represents, but guenomu only assumes that the species name can be found within the gene leaf name and does the mapping automatically. For instance, if one has for a given gene family, three sequences "a1," "b1," and "x" all from *Escherichia coli* (they might be paralogous sequences, or individuals from the same locus, or a combination of both), then it is enough to rename the sequences to something like "ecoli_a1," "ecoli_b1," and "ecoli_x" while including "ecoli" into the list of species and remembering that in this case all gene families must use the name "ecoli" to refer to this species. It is allowed for a given species not to be found in some gene family,

```
begin_list_of_tree_files        [ nexus files with distribution of gene family trees]
  gene001.tre
  gene002.tre
  gene003.tre
end_of_list [ of tree files ]   [ <<this command must be in separate line>> ]

begin_list_of_species_names          [ names of species]
  ecoli homo pongo mus gallus danio
end_of_list [ of species names ]     [ <<this command must be in separate line>> ]
```

Fig. 2 Excerpt of an example control file for guenomu showing the list of gene tree files and the list of species names that should be found within the leaves of the gene trees

```
param_file_with_tree_files = list_of_trees.txt          [ <filename> ]
param_file_with_species_names = list_of_species.txt [ <filename> ]
```

Fig. 3 Control file options for guenomu that describe the file names containing lists of gene trees and of species names

but a gene family member that cannot be mapped to any species in the list will result in error.

Both lists (of gene tree files and species names) can be given as a text file or added directly into the control file. If included directly in the control file, then they must be between the commands "begin_list_of_" and "end_of_list," as in the example of Fig. 2.

Notice how the comments in brackets are ignored, but still may contain useful tips. Alternatively, if the lists of gene tree files and species names are written in files—let us say, "list_of_trees.txt" and "list_of_species.txt," respectively, then these file names can be included in the control file, as shown in Fig. 3.

Or, equivalently, these file names can be given at run time to guenomu as arguments "–genetrees=list_of_genes.txt –species-=list_of_species.txt." With the exception of the "begin_list_of_" parameters, all other control file options expect a specific number of values after the equals sign. Other relevant options in the control file are (with examples of values):

- **param_reconciliation_prior = 0.0001**: should be set to a reasonable mean value for the distance penalty parameters. Must be a very small value, to account for our expectation of few disagreements, and furthermore this value is shared across distances (since they are all rescaled to the interval [0, 1]). Corresponds exactly to the parameter controlling the hyperhyperprior P (lambda_zero) (*see* Ref. 24).
- **param_n_generations = 10000 500000**: number of iterations for the burn-in stage followed by the number of iterations of the main sampling stage of the Bayesian posterior sampling. In this example, the program will run for 10k iterations without sampling at all, and then it will run for another 500k iterations

where now it will sample at regular intervals to compose the posterior distribution.

- **param_n_samples = 1000**: number of posterior samples to save. That is, if we ask for 500k iterations after the burn-in (as in the example earlier), then it will save the state of the chain at every 500 iterations, to form the requested sample size of 1000.
- **param_anneal = 10 1000 0.5 20**: values for the simulated annealing stage that is run even before the burn-in stage. They are, respectively, the number of annealing cycles we want, followed by the number of iterations for each cycle, followed by the initial and final (inverse) temperatures of the chain, within each cycle. During this stage no information is saved, unless the program is in "optimization mode," as we will see later.
- **param_execute_action = importance**: which action should be executed by guenomu, to run the importance sampler (regular MCMC run) or to analyze its results, generating human-readable output files. The options are then "importance" or "analyse" ("analyze" is also accepted).
- **param_use_distances = 1111000**: which group of distances should be included in the penalty model, in bit-string representation. The bit-string representation is composed of a one in the *n*th position if the *n*th distance is used, and zero otherwise. The distances are, from left to right: (1) number of duplications, (2) number of losses, (3) number of deep coalescences, (4) the mulRF distance [32], (5) Hdist_1, (6) Hdist_2, and (7) the approximate SPR distance. The Hdist distances are very experimental and are based on a matching between branches [33]. The approximate SPR distance was developed for detection of recombination [34] and may help resolve horizontal gene transfer [35], but this is also experimental since, as with the Hdist, it cannot handle the so-called multrees—that is, gene trees with more than one leaf mapping to the same species. That is why in our example these bits are set to zero.

All of the earlier options have command line equivalents, which can be consulted by running the program with the help argument "guenomu –help." For instance, we find that in practice it is easier to set the execution mode for the program at the command line than at the control file. In this case, instead of setting the param_execute_action option, we run the program with the option "-z 0" for running the MCMC sampler and then run it once more, this time with the option "-z 1" to analyze its output.

5.1 Optimization by Simulated Annealing

If the number of iterations for the main sampling stage of the Bayesian sampling is set to zero, then the program will behave differently: it will assume that we are interested in the simulated annealing results. That is, we are using guenomu as an optimization tool, and it will store the chain status at the final iteration of each cycle. The simulated annealing will sample from a 'modified' distribution, which is the Bayesian posterior distribution exponentiated to a value (the inverse of its temperature, on a thermodynamic interpretation). In the simulated annealing step, several cycles are performed serially where each cycle is composed of many iterations with an initial and a final temperature. That is, within each cycle the temperature changes across iterations, and the chain state at the end of one cycle is used as the initial state for the next one.

The simulated annealing step was devised to control the initial state of the chain, for the posterior sampling. That is, even within the Bayesian sampling one can benefit from the simulated annealing: if one wants to start the chain at a really random initial value, it is enough to set both the initial and final temperatures to a value below one—which will allow the chain to explore freely even regions of very low probability. This is a safe choice for convergence checks. On the other hand, if one is interested in starting the sample at a good initial estimate of trees, then it might be worth setting the final temperature to a high value, such that only moves increasing the posterior probability are accepted. Increasing the number of cycles allows for the chain to escape local optima.

If one is interested in using guenomu as an optimization tool, then we suggest that several cycles of optimization are employed to avoid local optima, and that the initial and final (inverse) temperatures should span a large interval. The output given by guenomu (after running it with the option "-z 1," for instance) will have the optimal estimated species tree at the end of each cycle. We might then look at the most frequent or best species tree overall, remembering that the optimal values are the set of genes and species trees that minimize their overall distances between each other. In this scenario, guenomu's results are therefore a generalization of the SPR supertree approach [35], the mulRF supertree [32], or the GTP methods [17], and can be directly compared by choosing only the appropriate distances.

5.2 Output

The main sampler will store its output with all trees and parameters in a binary, compact format that cannot be used by other programs. This file is named "job0.checkpoint.bin," and if you are running the parallel version then each job will generate its own file, for example, "job1.checkpoint.bin," "job2.checkpoint.bin," etc. Such files must be interpreted by another run of guenomu (with option "-z 1"), which will generate a single output file with all numeric parameters as well as the set of posterior species and gene family trees.

5.2.1 Posterior Sample of Numeric Parameters

The output file with the discrete and continuous numeric parameters from all posterior samples is called "params.txt." One example file is shown in Fig. 4. It is a tab-formatted file with a header row

iter	sptree	dup_hyperprior	los_hyperprior	lnl	cup	los	lnl_0	dup_0	los_0	dup_lambda_0	los_lambda_0	lnl_1	dup_1	los_1	dup_lambda_1	los_lambda_1
0	0	0.0010	0.0010	-164.3669	32	18	-89.2878	14	14	0.0224	0.0275	-75.0791	18	4	0.0242	0.0152
100	1	0.0010	0.0010	-170.8870	32	17	-98.9733	15	15	0.0246	0.0213	-71.9138	17	2	0.0231	0.0087
200	2	0.0010	0.0010	-157.7817	31	15	-81.6237	14	13	0.0238	0.0210	-76.1579	17	2	0.0215	0.0055
300	0	0.0010	0.0010	-182.9566	31	16	-105.1715	14	14	0.0182	0.0197	-77.7851	17	2	0.0204	0.0065
400	2	0.0010	0.0010	-151.9971	31	15	-76.6495	14	13	0.0365	0.0249	-75.3475	17	2	0.0210	0.0082
500	3	0.0010	0.0010	-170.8747	30	16	-102.2037	13	14	0.0178	0.0187	-68.6710	17	2	0.0259	0.0081
600	4	0.0010	0.0010	-173.4719	31	16	-92.0407	14	14	0.0181	0.0257	-81.4312	17	2	0.0169	0.0064
700	3	0.0010	0.0010	-180.3990	30	16	-97.3111	13	14	0.0250	0.0182	-83.0878	17	2	0.0169	0.0051
800	4	0.0010	0.0010	-169.7152	31	16	-89.5524	14	14	0.0223	0.0228	-80.1628	17	2	0.0196	0.0086
900	5	0.0010	0.0010	-166.6404	31	17	-100.8945	14	15	0.0198	0.0242	-65.7459	17	2	0.0189	0.0090
1,000	6	0.0010	0.0010	-170.3713	31	16	-102.8706	14	14	0.0218	0.0192	-67.5008	17	2	0.0224	0.0067
1,100	6	0.0010	0.0010	-176.8397	31	16	-97.5535	14	14	0.0192	0.0200	-79.2862	17	2	0.0206	0.0068
1,200	7	0.0010	0.0010	-162.7474	3[illegible]	16	-85.9524	14	14	0.0202	0.0281	-76.7950	17	2	0.0221	0.0075
1,300	8	0.0010	0.0010	-195.0731	3[illegible]	17	-108.2947	14	16	0.0207	0.0216	-86.7785	16	1	0.0202	0.0057
1,400	9	0.0010	0.0010	-163.1009	30	17	-97.6203	14	16	0.0204	0.0257	-65.4807	16	1	0.0190	0.0037
1,500	9	0.0010	0.0010	-167.0248	30	17	-93.8525	14	16	0.0245	0.0221	-73.1724	16	1	0.0232	0.0030
1,600	8	0.0010	0.0010	-183.2758	30	17	-108.3877	14	16	0.0147	0.0217	-74.8881	16	1	0.0227	0.0059
1,700	9	0.0010	0.0010	-162.0510	30	17	-94.5557	14	16	0.0253	0.0209	-67.4952	16	1	0.0213	0.0047
1,800	8	0.0010	0.0010	-170.0601	30	17	-91.0428	14	16	0.0226	0.0210	-79.0173	16	1	0.0236	0.0057
1,900	9	0.0010	0.0010	-178.3039	30	17	-108.5818	14	16	0.0203	0.0191	-69.7220	16	1	0.0209	0.0042
2,000	8	0.0010	0.0010	-166.8673	30	17	-96.4994	14	16	0.0201	0.0237	-70.3678	16	1	0.0190	0.0062
2,100	8	0.0010	0.0010	-154.5166	30	17	-87.4300	14	16	0.0232	0.0292	-67.0866	16	1	0.0258	0.0065
2,200	9	0.0010	0.0010	-169.2316	30	17	-91.5442	14	16	0.0309	0.0209	-77.6874	16	1	0.0200	0.0029

Fig. 4 Example of file params.txt with posterior samples of parameters from the MCMC chain. The first seven columns are the global parameters, while the remaining columns represent the parameters per gene family. For this example only duplications and losses were used for two gene families

followed by samples from the posterior distribution, one iteration per row. The header row contains the parameter names, where the order for gene-wise parameters follows the same order as input in the list of gene tree files. Each column represents one variable or statistic (function of the variables) from the posterior sample.

Notice that only iterations after the burn-in period are sampled, and that furthermore they are already "thinned" to avoid correlation between successive samples—this is accomplished by selecting a number of iterations much larger than the number of posterior samples to save in the parameter input options. The first columns are variables relating to the overall model, followed by gene-wise variables. Therefore, the total number of columns depends on the number of gene families being analyzed. The overall parameters are, respectively:

1. **Iter**—the iteration number of the MCMC run.
2. **sptree**—the ID of the species trees, as they appear on output file species.tre.
3. **dup_hyperprior, los_hyperprior, dco_hyperprior, rfd_hyperprior**—the hyperhyperparameters from the multivariate exponential model, that is, the genome-wise parameters controlling the gene-wise penalties for each distance. The prefixes are the possible distances, namely, duplications ("dup"), losses ("los"), deep coalescences ("dco"), and the mulRF distance ("rfd").
4. **lnl**—the unscaled posterior probability.
5. **dup, los, dco, rfd**—the sum of distances over all gene families.

While the set of parameters for gene X are as follows:

1. **lnl_X**—the contribution of this gene family to the unscaled posterior probability.
2. **dup_X, los_X, dco_X, rfd_X**—the distances between the gene and species trees, for this iteration.
3. **dup_lambda_X, los_lambda_X, dco_lambda_X, rfd_lambda _X**—the estimated lambda of the exponential distribution, that is, the penalty for this gene family.

This file can be used directly by convergence diagnostic programs such as Tracer [36] or Coda for R [37], where the behavior of the global parameters is especially relevant to evaluate if the MCMC has been run for long enough. Within a single run the parameters should appear to be stationary when plotted against the iterations, and their distributions should be similar across independent runs. These are necessary but not sufficient conditions for good convergence, and programs such as Tracer and Coda can better assess how the posterior distribution has been explored by our sampler.

5.2.2 Output Trees

Besides the output file with numeric parameters, guenomu also outputs the resampled gene tree files as well as the posterior distribution of species trees in several formats. The resampled gene trees will contain the same trees as in the input tree files, but with their posterior frequencies, that is, taking into account information from other gene families. These files will have the same name as the input, but with the prefix "post" and in the trprobs format, such that, for example, an input gene tree file named "gene001.tre" will originate a posterior file called "post.gene001.trprobs."

The posterior species tree distribution is output to three files: "species.tre," "species.trprobs," and "unrooted.trprobs." The model in guenomu works with rooted species trees, and therefore the files "species.tre" and "species.trprobs" are comprised of rooted species trees. The difference between these two files is the format, where "species.tre" is in standard nexus format and contains all sampled species trees in the order in which they appeared in the MCMC chain. The file "species.trprobs," however, is in the compact nexus format where only distinct trees are represented, together with their posterior frequencies. The trees inside this file will be named "tree_0", "tree_1," etc., where the number is the same as in the "sptree" column of the numeric parameters file, and is used to identify them. This number also appears as comments inside the "species.tre" file, to allow for the mapping of trees between both files. Please be careful since most phylogenetic analysis software cannot handle the trprobs format properly, and therefore the standard nexus file should be used to estimate the consensus tree or in other analyses.

If one is interested in obtaining a consensus tree from the "species.tre" file, then we suggest a few options: (1) the "sumt" option of MrBayes [25], (2) the "consensus()" function of the ape library for R [38], (3) the "consensus" method of the dendropy module for python [39]. We do, however, suggest to always check also the posterior frequencies directly from the trprobs files to have an idea about the sharpness of the distributions. In Fig. 5 we show an example of the file "species.tre" that can be compared to the file "species.trprobs" from Fig. 6.

Notice how in this file the trees are ordered by frequency, where for instance the tree number 3 is the so-called maximum a posteriori (MAP) tree, with frequency of 17.5 %. Notice, also, that some of the trees (e.g., tree_2, tree_3, tree_7 and tree_12) differ only in the root location.

The file "unrooted.trprobs" also contains the posterior distribution of species trees in the compact trprobs format, but this time neglecting the information about the root location. That is, it

```
#NEXUS

Begin trees;
 Translate
  1 schizosaccharomyces_pombe,
  2 saccharomyces_cerevisiae,
  3 drosophila_yakuba,
  4 monosiga_brevicollis,
  5 caenorhabditis_elegans,
  6 schistosoma_mansoni,
  7 drosophila_melanogaster,
  8 gallus_gallus,
  9 homo_sapiens
;
tree tree_0 [idx = 0] =  (((((6,5),(2,1)),(7,3)),4),8),9);
tree tree_1 [idx = 1] =  ((((((2,1),5),6),(7,3)),4),9),8);
tree tree_2 [idx = 2] =  ((((7,3),(2,1)),(6,5)),4),(9,8));
tree tree_3 [idx = 3] =  (((((7,3),(2,1)),(6,5)),4),9),8);
tree tree_4 [idx = 4] =  ((((((2,1),5),6),(7,3)),4),8),9);
tree tree_5 [idx = 5] =  ((((6,5),(2,1)),(7,3)),4),(9,8));
tree tree_6 [idx = 2] =  ((((7,3),(2,1)),(6,5)),4),(9,8));
tree tree_7 [idx = 2] =  ((((7,3),(2,1)),(6,5)),4),(9,8));
tree tree_8 [idx = 0] =  (((((6,5),(2,1)),(7,3)),4),8),9);
tree tree_9 [idx = 6] =  (((((6,5),(2,1)),(7,3)),4),9),8);

(...)

tree tree_1000 [idx = 6] =  (((((6,5),(2,1)),(7,3)),4),9),8

End;
```

Fig. 5 Example of "species.tre" file output by guenomu. In this file we have on tree topology per MCMC iteration, where identical trees can be identified by th descriptor "idx" (which is a comment according to the nexus format and therefor does not interfere with other programs)

summarizes the posterior distribution assuming that the sampled species trees are unrooted. It is helpful to compare it with "species. trprobs" in order to verify if the samples differ only in their rooting, in which case we might conclude when the data does not allow the model to distinguish between different root locations. In Fig. 7 we show the "unrooted.trprobs" file equivalent to the rooted file described in Fig. 6.

The trees are still rooted according to the newick format, but its root node can be safely ignored. In the earlier example we can see how the information from trees 2, 3, 7, and 12 from file "species. trprobs" are condensed into the same unrooted tree (tree_0001), with an accumulated frequency of 48.6 %.

Note that guenomu is very fast, such that a single run on a data set with 447 gene family tree distributions and 37 species [40] would take less than 6 h using a single processor.

```
#NEXUS

Begin trees;
 Translate
  1  schizosaccharomyces_pombe,
  2  saccharomyces_cerevisiae,
  3  drosophila_yakuba,
  4  monosiga_brevicollis,
  5  caenorhabditis_elegans,
  6  schistosoma_mansoni,
  7  drosophila_melanogaster,
  8  gallus_gallus,
  9  homo_sapiens
;
tree tree_3  [p = 0.1748, P = 0.1748] = [&W 0.1748251] ((((((7,3),(2,1)),(6,5)),4),9),8);
tree tree_0  [p = 0.1658, P = 0.3407] = [&W 0.1658341] ((((((6,5),(2,1)),(7,3)),4),8),9);
tree tree_2  [p = 0.1618, P = 0.5025] = [&W 0.1618381] (((((7,3),(2,1)),(6,5)),4),(9,8));
tree tree_6  [p = 0.1499, P = 0.6523] = [&W 0.1498501] ((((((6,5),(2,1)),(7,3)),4),9),8);
tree tree_7  [p = 0.1479, P = 0.8002] = [&W 0.1478521] ((((((7,3),(2,1)),(6,5)),4),8),9);
tree tree_5  [p = 0.1469, P = 0.9471] = [&W 0.1468531] (((((6,5),(2,1)),(7,3)),4),(9,8));
tree tree_4  [p = 0.0110, P = 0.9580] = [&W 0.0109890] (((((((2,1),5),6),(7,3)),4),8),9);
tree tree_9  [p = 0.0100, P = 0.9680] = [&W 0.0099900] (((((((2,1),6),5),(7,3)),4),8),9);
tree tree_1  [p = 0.0090, P = 0.9770] = [&W 0.0089910] (((((((2,1),5),6),(7,3)),4),9),8);
tree tree_10 [p = 0.0080, P = 0.9850] = [&W 0.0079920] ((((((2,1),5),6),(7,3)),4),(9,8));
tree tree_11 [p = 0.0070, P = 0.9920] = [&W 0.0069930] ((((((2,1),6),5),(7,3)),4),(9,8));
tree tree_8  [p = 0.0060, P = 0.9980] = [&W 0.0059940] (((((((2,1),6),5),(7,3)),4),9),8);
tree tree_12 [p = 0.0020, P = 1.0000] = [&W 0.0019980] (((((7,3),(2,1)),(6,5)),(9,8)),4);

End;
```

Fig. 6 Example of guenomu's "species.trprobs" output file. Unlike the "species.tre" file (*see* Fig. 5), distinct topologies appear only once, and furthermore are ordered according to their posterior frequencies (described within comments)

```
#NEXUS

Begin trees;
 Translate
    1  schizosaccharomyces_pombe,
    2  saccharomyces_cerevisiae,
    3  drosophila_yakuba,
    4  monosiga_brevicollis,
    5  caenorhabditis_elegans,
    6  schistosoma_mansoni,
    7  drosophila_melanogaster,
    8  gallus_gallus,
    9  homo_sapiens
;
tree tree_0001 [p = 0.4865, P = 0.4865]= [&W 0.4865134](((((7,3),(2,1)),(6,5)),4),(9,8));
tree tree_0002 [p = 0.4625, P = 0.9491]= [&W 0.4625374]((((((6,5),(2,1)),(7,3)),4),8),9);
tree tree_0003 [p = 0.0280, P = 0.9770]= [&W 0.0279720](((((((2,1),5),6),(7,3)),4),9),8);
tree tree_0004 [p = 0.0230, P = 1.0000]= [&W 0.0229770](((((((2,1),6),5),(7,3)),4),9),8);

End;
```

Fig. 7 Example of "unrooted.trprobs" output file from the guenomu software. This file has the same information as "species.trprobs" (Fig. 6) but where the root location of each topology was neglected—therefore joining several trees that differ only in the root location. Notice how the trees are still represented as rooted (deepest node is a dichotomy), just to ease reading and visualization

6 Conclusion

The evolutionary history of the species cannot be postulated to follow the phylogeny of a few genes without making too many arbitrary assumptions, while the accumulation of data poses new challenges to extracting as much information as possible from genomes. Recent Bayesian methods promise to bridge this gap between single-gene and genomic tree inference, but at a high computational cost. On the other hand, novel supertree methods can tackle large data sets, offering however a limited view of the possible evolutionary pathways. Here we show a model, implemented in the program guenomu, that tries to integrate strengths from both approaches, allowing at the same time for full utilization of complex data sets under minimal assumptions.

7 Notes

1. Parsimony reconciliation algorithms try to find the minimal cost under an evolutionary scenario with duplications and/or deep coalescences (incomplete lineage sortings). Both scenarios are calculated independently but make use of the same algorithm, namely, the least common ancestor (LCA) mapping. The LCA mapping associates to each node in a rooted gene tree its equivalent node (representing a branch) on the rooted species tree. Under a model of duplications, parent-descendant gene nodes mapped to the same species node represent a duplication, while under the coalescent model the number of "extra" lineages (gene nodes within the same species branch) represents deep coalescences. Gene losses can be deduced from the most parsimonious duplication scenario. Here, different leaves from the gene tree can be mapped to the same leaf in the species tree.
2. The Robinson–Foulds distance, or *symmetric* distance, is a measure of the number of branches from one tree that do not have a correspondence on the other tree (as described by the bipartitions they generate). This distance assumes a perfect correspondence between the leaves of one tree and the other. However, a generalization exists for the case where one of the trees (the gene tree) has several leaves corresponding to the same leaf in the other tree (the species tree). This generalized RF distance relies in extending the species tree by adding the duplicated leaves as an external polytomy.
3. The matrix distance-based methods GLASS, SD, STEAC, and MAC are very similar in that they start by generating a distance matrix between species for each gene family (using patristic distances between leaves), and then summarize these matrice

into one species-level matrix, from which a dendrogram is estimated. They all assume a coalescent model in which the gene/species tree disagreement comes from incomplete lineage sorting. Conveniently, they can handle arbitrarily large gene families by defining the gene-level distance between species as the minimum (GLASS and SD) or the average (MAC and STEAC) of the patristic distances between leaves from the same pair of species. The overall species-level distances can also be estimated as the minimum (GLASS and MAC) or the average (STEAC and SD) across gene-level values.

References

1. Rannala B, Yang Z (2008) Phylogenetic inference using whole genomes. Annu Rev Genomics Hum Genet 9:217–231
2. Woese CR (1987) Bacterial evolution. Microbiol Rev 51:221–271
3. Brown JR, Doolittle WF (1997) Archaea and the prokaryote-to-eukaryote transition. Microbiol Mol Biol Rev 61:456–502
4. Fitz-Gibbon ST, House CH (1999) Whole genome-based phylogenetic analysis of free-living microorganisms. Nucleic Acids Res 27:4218–4222
5. Snel B, Bork P, Huynen MA (1999) Genome phylogeny based on gene content. Nat Genet 21:108–110
6. Fukami-Kobayashi K, Minezaki Y, Tateno Y, Nishikawa K (2007) A tree of life based on protein domain organizations. Mol Biol Evol 24:1181–1189
7. Grishin NV, Wolf YI, Koonin EV (2000) From complete genomes to measures of substitution rate variability within and between proteins. Genome Res 10:991–1000
8. Clarke GDP, Beiko RG, Ragan MA, Charlebois RL (2002) Inferring genome trees by using a filter to eliminate phylogenetically discordant sequences and a distance matrix based on mean normalized BLASTP scores. J Bacteriol 184:2072–2080
9. Housworth EA, Postlethwait J (2002) Measures of synteny conservation between species pairs. Genetics 162:441–448
10. Lin Y, Moret BME (2008) Estimating true evolutionary distances under the DCJ model. Bioinformatics 24:i114–i122
11. Gordon A (1986) Consensus supertrees: the synthesis of rooted trees containing overlapping sets of labeled leaves. J Classif 348:335–348
12. Ragan MA (1992) Phylogenetic inference based on matrix representation of trees. Mol Phylogenet Evol 1:53–58
13. Kluge AG (1989) A concern for evidence and a phylogenetic hypothesis of relationships among Epicrates (Boidae, Serpentes). Syst Zool 38:7–25
14. de Queiroz A, Gatesy J (2007) The supermatrix approach to systematics. Trends Ecol Evol 22:34–41
15. Helmkamp LJ, Jewett EM, Rosenberg NA (2012) Improvements to a class of distance matrix methods for inferring species trees from gene trees. J Comput Biol 19:632–649
16. Slowinksi J, Page RDM (1999) How should species trees be inferred from molecular sequence data? Syst Biol 48:814–825
17. Chaudhary R, Bansal MS, Wehe A, Fernández-Baca D, Eulenstein O (2010) iGTP: a software package for large-scale gene tree parsimony analysis. BMC Bioinformatics 11:574
18. Felsenstein J (1981) Evolutionary trees from DNA sequences: a maximum likelihood approach. J Mol Evol 17:368–376
19. Chaudhary R, Boussau B, Burleigh JG, Fernandez-Baca D (2014) Assessing approaches for inferring species trees from multi-copy genes. Syst Biol 64:325–339
20. Heled J, Drummond AJ (2010) Bayesian inference of species trees from multilocus data. Mol Biol Evol 27:570–580
21. Liu L, Pearl DK (2007) Species trees from gene trees: reconstructing Bayesian posterior distributions of a species phylogeny using estimated gene tree distributions. Syst Biol 56:504–514
22. Akerborg O, Sennblad B, Arvestad L, Lagergren J (2009) Simultaneous Bayesian gene tree reconstruction and reconciliation analysis. Proc Natl Acad Sci U S A 106:5714–5719
23. Boussau B, Szöll GJ, Duret L, Gouy M, Tannier E, Daubin V (2013) Genome-scale coestimation of species and gene trees. Genome Res 23:323–330
24. De Oliveira Martins L, Mallo D, Posada D (2014) A Bayesian supertree model for

genome-wide species tree reconstruction. Syst Biol 65(3):397–416. doi:10.1093/sysbio/syu082

25. Ronquist F, Teslenko M, van der Mark P, Ayres DL, Darling A, Höhna S, Larget B, Liu L, Suchard MA, Huelsenbeck JP (2012) MrBayes 3.2: efficient Bayesian phylogenetic inference and model choice across a large model space. Syst Biol 61:539–542
26. Lartillot N, Lepage T, Blanquart S (2009) PhyloBayes 3: a Bayesian software package for phylogenetic reconstruction and molecular dating. Bioinformatics 25:2286–2288
27. Drummond AJ, Suchard MA, Xie D, Rambaut A (2012) Bayesian phylogenetics with BEAUti and the BEAST 17. Mol Biol Evol 29:1969–1973
28. Rubenthaler S, Rydén T, Wiktorsson M (2009) Fast simulated annealing in rd with an application to maximum likelihood estimation in state-space models. Stoch Proc Appl 119:1912–1931
29. Mossel E, Roch S (2008) Incomplete lineage sorting: consistent phylogeny estimation from multiple loci. IEEE/ACM Trans Comput Biol Bioinf 7:166–171
30. Maddison WP, Knowles LL (2006) Inferring phylogeny despite incomplete lineage sorting. Syst Biol 55:21–30
31. Liu L, Yu L, Pearl DK, Edwards SV (2009) Estimating species phylogenies using coalescence times among sequences. Syst Biol 58:468–477
32. Chaudhary R, Fernández-Baca D, Burleigh JG (2015) MulRF: a software package for phylogenetic analysis using multi-copy gene trees. Bioinformatics 31:432–433
33. Nye TMW, Liò P, Gilks WR (2006) A novel algorithm and web-based tool for comparing two alternative phylogenetic trees. Bioinformatics 22:117–119
34. de Oliveira Martins L, Leal É, Kishino H (2008) Phylogenetic detection of recombination with a Bayesian prior on the distance between trees. PLoS One 3:e2651
35. Whidden C, Zeh N, Beiko RG (2014) Supertrees based on the subtree prune-and-regraft distance. Syst Biol 63:566–581
36. Rambaut A, Suchard MA, Xie D, Drummond A (2013) Tracer v1.5. Available at http://beast.bio.ed.ac.uk/tracer
37. Plummer M, Best N, Cowles K, Vines K (2006) Coda: convergence diagnosis and output analysis for mcmc. R News 6:7–11
38. Paradis E, Claude J, Strimmer K (2004) APE: analyses of phylogenetics and evolution in R language. Bioinformatics 20:289–290
39. Sukumaran J, Holder MT (2010) DendroPy: a python library for phylogenetic computing. Bioinformatics 26:1569–1571
40. Song S, Liu L, Edwards SV, Wu S (2012) Resolving conflict in eutherian mammal phylogeny using phylogenomics and the multispecies coalescent model. Proc Natl Acad Sci U S A 109:14942–14947

Erratum to: Sequence Segmentation with `changeptGUI`

Edward Tasker and Jonathan M. Keith

Jonathan M. Keith (ed.), *Bioinformatics: Volume I: Data, Sequence Analysis, and Evolution*, Methods in Molecular Biology, vol. 1525, DOI 10.1007/978-1-4939-6622-6_12, © Springer Science+Business Media New York 2017

DOI 10.1007/978-1-4939-6622-6_12

The ESM files in the original version of Chapter 12 have been updated and can be found here
https://static-content.springer.com/esm/chp%3A10.1007%2F978-1-4939-6622-6_12/MediaObjects/316167_2_En_12_MOESM1_ESM.zip

The updated online version of the chapter can be found under
http://dx.doi.org/10.1007/978-1-4939-6622-6_12

Jonathan M. Keith (ed.), *Bioinformatics: Volume I: Data, Sequence Analysis, and Evolution*, Methods in Molecular Biology, vol. 1525, DOI 10.1007/978-1-4939-6622-6_19, © Springer Science+Business Media New York 2017

Index

Jonathan M. Keith (ed.), *Bioinformatics: Volume I: Data, Sequence Analysis, and Evolution*, Methods in Molecular Biology, vol. 1525, DOI 10.1007/978-1-4939-6622-6, © Springer Science+Business Media New York 2017

C

J

K

L

M

N

Q

R

S

T

U

V

W

X

Y

Z

Printed in the United States
By Bookmasters